Encyclopedia of
Genetics
Revised Edition

Encyclopedia of **Genetics**

Revised Edition

Volume 1

Aggression – Hybridization and Introgression

Editor, Revised Edition
Bryan D. Ness, Ph.D.
Pacific Union College
Department of Biology

Editor, First Edition
Jeffrey A. Knight, Ph.D.
Mount Holyoke College
Department of Biological Sciences

Salem Press, Inc.
Pasadena, California
Hackensack, New Jersey

Editor in Chief: Dawn P. Dawson

Managing Editor: Christina J. Moose   *Acquisitions Editor:* Mark Rehn
Developmental Editor: Tracy Irons-Georges   *Photograph Editor:* Philip Bader
Research Supervisor: Jeffry Jensen   *Page Design:* James Hutson
Assistant Editors: Andrea E. Miller,   *Layout:* William Zimmerman
Desirée Dreeuws

Library of Congress Cataloging-in-Publication Data

Encyclopedia of genetics / editor, revised edition, Bryan D. Ness ;
editor, first edition, Jeffrey A. Knight. — Rev. ed.
    p.   ; cm.
Includes bibliographical references and index.
    ISBN 1-58765-149-1 (set : alk. paper) — ISBN 1-58765-150-5 (v.1 : alk. paper) —
ISBN 1-58765-151-3 (v.2 : alk. paper) —
    1. Genetics—Encyclopedias.
    [DNLM: 1. Genetics—Encyclopedias—English. QH 427 E56 2004] I. Ness, Bryan D.
QH427.E53 2004
576.5′03—dc22

                                                                    2003026056

First Printing

Contents

Publisher's Note

The award-winning *Encyclopedia of Genetics* was originally created in 1999 to provide the general reader with a thorough yet accessible overview of one of modern science's most vital and intriguing fields. This 2004 *Revised Edition* adds 64 new overview essays on current topics to the original entries, reflecting the rapid developments in an exciting and often controversial branch of science that is increasingly shaping our world.

Of the 172 original entries (168 overviews plus 4 appendices), 7 overviews ("Biotechnology," "Genetic Medicine," "Immune Deficiency Disorders," "Lethal Alleles," "Meiosis," "Methane-Producing Bacteria," and "Sheep Cloning") have been dropped because they have been superseded by other, updated or new, essays; 26 have been replaced because they were so out of date as to demand completely new coverage (two of these, "Aging" and "Mitosis and Meiosis," are from other Salem publications, deemed more up to date than the original entries); and 131 were moderately to heavily revised and updated by the editor as deemed appropriate.

All essays' bibliographies are new or are fully updated. The four appendices have been updated and two new ones, "Web Sites" and "Nobel Prizes for Discoveries in Genetics," have been added. Featured in this new edition are 25 new "sidebars," or mini-essays (500 words each). These sidebars, appearing in shaded boxes, offer coverage of particularly significant and current subtopics appended to the overview essays. A list displaying the status of the essays, designed to assist librarians in comparing the two editions, appears at the end of this note.

The result, in two volumes, is 223 overviews, 25 sidebars, and 6 appendices: a 30 percent increase based on number of overview essays, and an increase of more than 40 percent in overall word length. The set surveys this continually evolving discipline from a variety of perspectives, offering historical and technical background along with a balanced discussion of recent discoveries and developments. Basics of biology—from the molecular and cellular levels through the organismal level, from Mendelian principles to the latest on DNA sequencing technology—constitute the core coverage. Medical topics comprise a significant number of essays, as the genetic predisposition for many illnesses and syndromes has increasingly come to light. Genetic technologies that promise a world without hunger, disease, and disability—and promise to rewrite human values—are addressed as well. The encyclopedia's scope embraces the key social and ethical questions raised by these new genetic frontiers: from cloning to stem cells to genetically modified foods and organisms.

Each essay follows a standard format, including ready-reference top matter and the following standard features:

- **Fields of study** lists one or more of a dozen subdisciplines of genetics or biology under which the topic falls.

- **Significance** provides a definition and summary of the topic's importance.

- **Key terms**, concepts central to the topic, are next identified and defined.

- **Subheads** break the main body of each essay into clearly marked subtopics.

- The **contributor's byline** lists the biologist or other area expert who wrote the essay.

- The **See also** section lists cross-references to other essays of interest.

- **Further Reading** lists sources for further study with annotations; all of these biographical sections have been fully updated and reformatted to include the latest relevant works and full citation data for easy library access.

- **Web Sites of Interest**, finally, includes annotated entries for the most authoritative free sites on the Internet, including the sponsoring organization and URL. This section, which appears in more than half the essays, was included for topics (such as diseases and syndromes) for which authoritative government agencies, professional or academic societies, or support organizations were available, with only the occasional nod to a particularly useful personal URL. All sites were accessed by the editors in August, 2003.

A series of appendices supplements the overview essays: An updated **Time Line of Major Developments in Genetics** offers a chronological overview of the field's development. **Nobel Prizes for Discoveries in Genetics** lists Nobel Prize winners (all prizes) whose contributions altered the history of genetics. An updated **Biographical Dictionary of Important Geneticists** has more than doubled in size, now including all Nobel laureates whose discoveries in genetics garnered them the award, as well as numerous others. The updated **Glossary** provides definitions of more than 500 commonly used terms and important concepts. The updated general **Bibliography** references important works in each field of study, joining with individual articles' "Further Reading" sections to offer plentiful citations to recently published sources for additional reasearch. The impor-

tance of the Internet to bioinformatics and to general education in genetics is reflected in the new **Web Sites** appendix.

The articles in the *Encyclopedia of Genetics, Revised Edition* are arranged alphabetically by title; an alphabetical list of contents appears at the beginning of each volume. To help readers locate topics of interest by area of study, a Category Index, a Personages Index, and a comprehensive Subject Index are included at the end of volume 2. Nearly 100 diagrams, charts, graphs, drawings, and tables elucidate complex concepts, and more than one hundred photographs illustrate the text.

We wish to thank the many biologists and other scholars who contributed both to the original edition and this revised edition; their names and academic affiliations appear in the Contributor List that follows. Special credit is due the editor of the *Revised Edition*, Dr. Bryan D. Ness of the Department of Biology, Pacific Union College. Professor Ness paid close attention to the contents of every essay, carefully updating all of the original text, elucidating complex concepts for the general reader, and making valuable contributions to the project on all levels.

Preface to the Revised Edition

In the five years since publication of the first edition of *Encyclopedia of Genetics*, the field of genetics has continued to expand, if possible, exponentially. The volume of data flowing from genetics research is so great that new methods of organizing and analyzing it are still being devised. As with any expanding field, practical applications have lagged behind predictions. The Human Genome Project, now completed, which was predicted to usher in a new era in medical genetics, has so far had only minor effects, and the many genetically modified (GM) crops that were supposed to revolutionize agriculture have caused more controversy than success. Yet, like most technologies of the past that took time to gain acceptance, the knowledge and technologies flowing from the Human Genome Project will almost certainly gain influence and acceptance over the next decades. The slow and steady application of these technologies will eventually have a world-changing impact on all aspects of life.

The current revised edition is an attempt to increase the coverage begun in the first edition and to cover as many of the new developments in genetics as possible, a daunting task considering that new discoveries seem to happen weekly. By way of perspective, consider what has occurred in the field of genetics in the past five years. The Human Genome Project, probably the most widely publicized genetic project of the past century, saw the completion of the human genome sequence two years ahead of its original schedule—appropriately, fifty years after the discovery by James Watson and Francis Crick of the double-helical structure of DNA. The field of genomics—the study of the sequence and structure of the genomes of various organisms—has now solidly entered the public consciousness and has spawned the related specialty proteomics, the study of the proteins expressed by genomes, which focuses on differences among cell types as well as differences between gene expression in health and disease. Although these studies have not yet transformed medicine, as more data accumu-

late and are analyzed, medicine will certainly become a more exact science, enabling therapy to be tailored to a person's genotype.

As spinoffs of the Human Genome Project, the genomes of many "model" organisms have also been sequenced, a process that has been accelerated by the development of technologies such as the polymerase chain reaction and automated sequencers. Only a handful of small genomes had been sequenced at the time the first edition of *Encyclopedia of Genetics* was published five years ago; today, hundreds of genomes have been sequenced, including some larger genomes such as those of *Arabidopsis thaliana* (the model mustard plant), *Drosophila melanogaster* (the fruit fly), and *Caenorhabditis elegans* (the model round worm). Many other genome sequencing projects are under way, and as more genomes are sequenced, geneticists will obtain ever clearer insights into how our genes make us who we are—not to mention how our genomes relate to those of other organisms (the focus of comparative genomics) and hence how such organisms can be manipulated genetically to our benefit.

Since the cloning of Dolly the sheep in 1996, a variety of other mammals have been cloned, including mice and, most recently, a horse. This brave new world of cloning has spawned the imaginations of filmmakers and writers as surely as space exploration did in earlier generations—fueling wild speculations about the possibility of reviving ancient life-forms, such as the dinosaurs of Michael Crichton and Stephen Spielberg's *Jurassic Park* (1993), and, more recently, claims of human cloning that are plausible if not probable, though fraught with both technical and ethical obstacles. Shortly after Dolly was cloned, for example, it was discovered that her telomeres (the ends of her chromosomes) were shorter than normal for a juvenile sheep. Telomeres are known to shorten throughout the life spans of many organisms and have been implicated as part of the cause, or at least one of the effects, of aging. With her shortened telomeres, the big question was, how would it affect her longevity? In February of 2003 the answer came when Dolly had to be euthanized because she was suffering significantly from a form of arthritis usually seen only in older sheep, as well as advanced lung disease. Her early decline, at six years of age, cast doubt on the hoped-for success of cloning mammals.

Of course, the cloning of Dolly fueled increased speculation about the feasibility of human cloning. Considering Dolly's premature death and other health issues, most geneticists do not consider the technology ready for human cloning. Ethically, human cloning, indeed organismal cloning of all types, is extremely controversial—with myriad implications economically, socially, politically, and ecologically (as human manipulation supplants traditional methods of natural selection as a force in evolution)—but it is even more so when the high probability of producing a debilitated human clone is considered.

Along with the budding debate over cloning is another, related issue that arose simultaneously at the beginning of the new millennium: the use of fetal stem cells in research. Some geneticists believe that stem cells— "totipotent" cells, capable of differentiating into essentially any other kind of cell—may have potential for treating a variety of neurological diseases such as Alzheimer's and Parkinson's, as well as conditions requiring new organs that might be generated by implanting stem cells with the genetic instructions to develop into ears or kidneys that could be grown on animals especially designated for this purpose. Unfortunately, the best stem cells come from developing fetuses, and the ethics of harvesting fetal stem cells from aborted fetuses is hotly debated. The much-touted potential of using adult stem cells instead is clouded by both political and social agendas and the fact that they have not yet proven to be nearly as versatile or easy to culture as fetal stem cells. In the United States, research on fetal stem cells has been limited to a small number of cell lines existing at the time that the federal government addressed the issue in 2001; it was determined to withhold funding from any research group that harvests fetal stem cells as a part of their work. Research continues on the limited stem cell lines and on alternatives such as the less controversial adult source of stem cells.

Regardless of the eventual decisions regarding human cloning and research using fetal stem cells, the ethical questions raised by both endeavors clearly run parallel. With the endless potential uses of such technologies, where might they lead? Where does the "necessity" to solve human suffering end and the "brave new world" of self-proliferation, designer traits, worker and military subclasses, perfect progeny, and potential ecological disaster begin? What obligations do we have to fulfill the new promise of genetic science to alleviate human suffering, and what obligations do we have to limit that promise out of concern for greater detriment?

This last consideration is now a very real concern. GM crop plants were once considered the Holy Grail of agricultural genetics, seen as a solution to everything from more effectively battling pests and weeds to correcting Third World nutritional deficiencies and utilizing marginal habitats such as saline soil. Much of the early headiness surrounding the potential of high-yield and transgenic crops has dissipated and opposition has increased against the use of such plants. Objections range from fears over human health and safety to ecological and economic concerns. Most Europeans have rejected GM foods completely, and even consumers in the United States are uncomfortable using them—or at least feel that it is their right to be informed, through labeling, before making purchase decisions. If the many concerns expressed by consumers cannot be addressed, GM crops may not be embraced widely for some time, if ever.

With these advances, and many others not mentioned, the general public has become overwhelmed with the implications. Although the terms are casually thrown around in the media and strident statements are made, the general public has become increasingly uncomfortable with a technology they barely understand. As a result, opposition from many directions continues to build, epitomized by both the federal ban on stem cell research mentioned above and the Human Cloning Prohibition Act that is now making its way through Congress.

Especially with regard to GM foods, many geneticists consider the root problem to be lack of knowledge about genetics on the part of nonscientists. Although terms such as "DNA," "cloning," "GM food," and "gene therapy" have entered the public vocabulary, many people gain their understanding of these technologies from the science fiction of books, film, and television rather than from the science on which they are founded. Those who are opposed to genetic technology have taken advantage of this "safety-first" attitude and have spread fear about genetic technology, coining inflammatory terms like "Franken foods" and citing the law of "unintended consequences." Consequently, many nonscientists misunderstand the issues and mistrust the scientists doing the work, picturing them as ambitious and amoral rather than as responsible researchers. Part of the problem lies with geneticists themselves, many of whom find it difficult to communicate with nonscientists. Geneticists need to learn to communicate in a way that the general public will understand. Without a broad understanding by the general public, genetics will continue to be viewed by many in a negative light.

It is in this spirit that the current volumes were updated and expanded. The articles contained herein are written specifically with the nonscientist in mind—and specifically to explain, as simply as possible, some of the science behind the technologies and issues described above. All original topics were considered for updating, many were heavily revised, 26 were entirely replaced, and 64 are entirely new. In other cases, recent advances put a spotlight on topics—such as Bioinformatics, Biological Weapons, Smallpox, and Xenotransplants—that were too minor or obscure to have been included in the previous edition.

Other improvements in this edition include the addition of 25 "sidebars," elucidating particularly important and timely subtopics; a new time line of Nobel laureates whose work in genetics garnered them the award; a "Biographical Dictionary of Important Geneticists" that is twice its original size; more than 100 new definitions added to the Glossary; an updated "Time Line of Major Developments in Genetics"; a heavily expanded and re-categorized

general Bibliography; and a select list of genetics Web sites aimed at students and nonspecialists. In addition, every essay carries a "Further Reading" section that has been thoroughly reviewed, updated, and annotated—to which we have added more than 400 new books and articles published since the first edition. In recognition of the influence of the Internet on student research, we for the first time include, in more than half the essays, a section headed "Web Sites of Interest" targeted at the essay's topic. Finally, every effort has been made to make the essays user-friendly, easy to read, and clear, with the goal of improving the presentation and understandability of all the old, as well as the new, essays.

Even as this edition goes to press, it is already beginning to go out of date. Genetics is a dynamic field that deserves close attention as we begin the new millennium. Nevertheless, the basic scientific principles presented here will provide science students with insight into the topics on which they can build, and that fundamental understanding will repay students and general readers regardless of their ultimate occupations or career interests. Such an understanding behooves all of us: The potential of genetic principles to completely alter the way we live and interact with the environment is profound. We will see changes in the way doctors diagnose and treat disease, new GM crops and animals, powerful new forensics techniques, and unique ways to solve environmental and societal problems. There are potential dangers as well, and these must be carefully analyzed and vetted in the public forum. As we attempt to navigate our genetic future, knowledge will be essential if we are to take full advantage of the positive benefits, and prevent the negative consequences, of genetic technology. It is hoped that the information contained here will open a new world of understanding to the nonspecialist and encourage further exploration of the amazing world of genetics, which offers frontiers for exploration just as real as those of space or oceanic exploration—and in many ways more personal and tangible.

—*Bryan D. Ness*
August, 2003

Preface to the First Edition

The science of genetics, once the purview only of serious students and professionals, has in recent decades come of age and entered the mainstream of modern life. An unparalleled explosion of new discoveries, powerful new molecular techniques, and practical applications of theories and research findings has brought genetics and its related disciplines to the forefront of public consciousness. The successful cloning of "Dolly" the sheep has sparked widespread public interest and debate and raised new questions about the ethics of this and other genetic technologies. Gene therapy has made the transition from science fiction to reality and is used to treat serious diseases, and there is increasing demand for the newest health professionals, genetic counselors, at hospitals and medical centers around the world. As we celebrate the new millennium, it is perhaps worth noting that the young science of genetics celebrates its one hundredth birthday.

Among many other events of historical importance, the year 1900 marked the rediscovery of the Austrian monk Gregor Mendel's experimental work on the inheritance of traits in the garden pea. Mendel had published his results thirty-four years earlier, but his work attracted little attention and soon faded into obscurity. By the close of the nineteenth century, however, much had happened on the scientific front. Chromosomes had been discovered, and the cellular processes of mitosis and meiosis had been observed under the microscope. The physical bases for understanding Mendel's principles of inheritance had been established, and the great significance of his pioneering work could finally be appreciated. The so-called chromosome theory of heredity was born, and the age of transmission genetics had arrived.

The first great geneticist to emerge (and some would still call him the greatest of the twentieth century) was Thomas Hunt Morgan, who established his "fly laboratory" at Columbia University and began studying the principles of transmission genetics, using the fruit fly as a model organism. All the major principles of transmission genetics, including single and multifactorial inheritance, chromosome mapping, linkage and recombination, sex linkage, mutagenesis, and chromosomal aberrations, were first investigated by Morgan and his students.

The subdisciplines of bacterial and molecular genetics had their beginnings in the 1940's, when bacteria and their viruses became favored genetic systems for research because of their relative simplicity and the ease with which they could be grown and manipulated in the laboratory. In particular, the common intestinal bacterium *Escherichia coli* was studied intensely, and today far more is known about the biology of this single-celled organism than about any other living system. In 1952, James Watson and Francis Crick provided the molecular model for the chemical structure of DNA, the genetic material, and the next twenty years saw great progress in the understanding of the molecular nature of essential cellular processes such as DNA replication, protein synthesis, and the control of bacterial gene expression.

The 1970's witnessed the discovery of a unique class of enzymes known as restriction endonucleases, which set the stage for the development of the exciting new technology known by various names as cloning, genetic engineering, or recombinant DNA technology. Since that time, research has progressed rapidly on several fronts, with the development of genetic solutions to many practical problems in the fields of medicine, agriculture, plant and animal breeding, and environmental biology. With the help of the new technology, many of the essential questions in cell and molecular biology that were first addressed in bacteria and viruses in the 1950's and 1960's can now be effectively studied in practically any organism.

And what are the major problems remaining to be solved? No doubt there are many, some of which cannot even be articulated given the present state of scientific understanding. Two important questions, however, are drawing disproportionate shares of attention in the current sphere of basic research. One of these is the problem variously referred to as "the second genetic code" or "protein folding." Scientists know how a particular molecule of DNA, with a known sequence of nucleotide subunits, can cause the production of a particular unique protein composed of a known sequence of amino acid subunits. What is not understood, however, is the process by which that protein will spontaneously fold into a characteristic three-dimensional shape in which each amino acid interacts with other amino acids to produce a functional protein that has the proper pockets, ridges, holes, protuberances, and other features that it needs in order to be biologically active. If all the rules for protein folding were known, it would be possible to program a computer to create an instant three-dimensional picture of the protein resulting from any given sequence of amino acids. Such knowledge would have great applications, both for understanding the mechanisms of action of known proteins and for designing new drugs for therapeutic or industrial use.

The second "big question" at the forefront of experimental genetic inquiry relates to the control of gene expression in humans and other higher organisms. In other words, what factors come into play in turning on or turning off genes at the proper times, either during an individual cell cycle or during the developmental cycle of an organism? How is gene expression controlled differentially—that is, how are different sets of genes turned on or off in different tissues in the same organism at the same time? Many human genetic diseases are now known or suspected to be caused by errors in gene expression—that is, too much or too little of a particular protein is made in the critical tissues at the critical developmental times—so the answers to these and related questions are sure to suggest new possibilities for gene therapy or other treatments.

The purpose of these reference volumes is twofold. First, the editors seek to highlight some of the most exciting new advances and applications of genetic research, particularly in the fields of human medical genetics and agriculture. Second, we hope to provide a solid basis for understanding the fundamental principles of genetics as they have been developed over this first one hundred years, along with an appreciation of the historical context in which the most important discoveries were made. It is our hope that such an understanding and appreciation might help to inspire a new generation of geneticists who will continue to expand the boundaries of scientific knowledge well into the next millennium.

—*Jeffrey A. Knight*

Contributor List

Barbara J. Abraham
Hampton University

Linda R. Adkison
Mercer University School of Medicine

Richard Adler
University of Michigan, Dearborn

Oluwatoyin O. Akinwunmi
Muskingum College

Michele Arduengo
Independent Scholar

J. Craig Bailey
University of North Carolina, Wilmington

Carl L. Bankston III
Tulane University

Kenneth D. Belanger
Colgate University

D. B. Benner
East Tennessee State University

Alvin K. Benson
Utah Valley State College

Gerald K. Bergtrom
University of Wisconsin, Milwaukee

Massimo D. Bezoari
Huntingdon College

Barbara Brennessel
Wheaton College

Douglas H. Brown
Wellesley College

Stuart M. Brown
New York University School of Medicine

Thomas L. Brown
Wright State University

Fred Buchstein
John Carroll University

Paul R. Cabe
Washington and Lee University

James J. Campanella
Montclair State University

Rebecca Cann
University of Hawaii, Manoa

Laurie F. Caslake
Lafayette College

J. Aaron Cassill
University of Texas, San Antonio

Stephen Cessna
Eastern Mennonite University

Robert Chandler
Union College

Kerry L. Cheesman
Capital University

Richard W. Cheney, Jr.
Christopher Newport University

Stacie R. Chismark
Heartland Community College

Jaime S. Colomé
California Polytechnic State University, San Luis Obispo

Joyce A. Corban
Wright State University

Stephen S. Daggett
Avila University

Jennifer Spies Davis
Shorter College

Patrick J. DeLuca
Mount Saint Mary College

David K. Elliott
Northern Arizona University

Daniel J. Fairbanks
Brigham Young University

Phillip A. Farber
Bloomsburg University of Pennsylvania

James L. Farmer
Brigham Young University

Linda E. Fisher
University of Michigan, Dearborn

Chet S. Fornari
DePauw University

Kimberly Y. Z. Forrest
Slippery Rock University of Pennsylvania

Daniel R. Gallie
University of California, Riverside

W. W. Gearheart
Piedmont Technical College

John R. Geiser
Western Michigan University

Soraya Ghayourmanesh
Independent Scholar

Sibdas Ghosh
University of Wisconsin, Whitewater

Sander Gliboff
Indiana University

James S. Godde
Monmouth College

D. R. Gossett
Louisiana State University, Shreveport

Daniel G. Graetzer
University of Washington Medical Center

Dennis W. Grogan
University of Cincinnati

Anne Grove
*Louisiana State University,
Baton Rouge*

Patrick G. Guilfoile
Bemidji State University

Randall K. Harris
William Carey College

H. Bradford Hawley
Wright State University

Robert Haynes
Albany State University

Jane F. Hill
Independent Scholar

Carl W. Hoagstrom
Ohio Northern University

Vicki J. Isola
Hope College

Domingo M. Jariel
*Louisiana State University,
Eunice*

Karen E. Kalumuck
Exploratorium

Manjit S. Kang
Louisiana State University

Susan J. Karcher
Purdue University

Armand M. Karow
Xytex Corporation

Roger H. Kennett
Wheaton College

Stephen T. Kilpatrick
*University of Pittsburgh,
Johnstown*

Samuel V. A. Kisseadoo
Hampton University

Jeffrey Knight
Mount Holyoke College

Audrey Krumbach
Huntingdon College

Steven A. Kuhl
Indiana Purdue University

William R. Lamberson
University of Missouri, Columbia

Kate Lapczynski
Motlow State Community College

Craig S. Laufer
Hood College

David M. Lawrence
*J. Sargeant Reynolds Community
College*

Michael R. Lentz
University of North Florida

Doug McElroy
Western Kentucky University

Sarah Lea McGuire
Millsaps College

Michael J. Mclachlan
University of South Carolina

Nancy Farm Männikkö
Independent Scholar

Sarah Crawford Martinelli
*Southern Connecticut State
University*

Lee Anne Martínez
*Colorado State University,
Pueblo*

Grace D. Matzen
Molloy College

Ulrich Melcher
Oklahoma State University

Ralph R. Meyer
University of Cincinnati

Randall L. Milstein
Oregon State University

Eli C. Minkoff
Bates College

Paul Moglia
*South Nassau Communities
Hospital*

Thomas J. Montagno
Simmons College

Beth A. Montelone
Kansas State University

Robin Kamienny Montvilo
Rhode Island College

Randy Moore
Wright State University

Nancy Morvillo
Florida Southern College

Donald J. Nash
Colorado State University

Mary A. Nastuk
Wellesley College

Leah C. Nesbitt
Huntingdon College

Bryan D. Ness
Pacific Union College

Henry R. Owen
Eastern Illinois University

Robert J. Paradowski
Rochester Institute of Technology

Massimo Pigliucci
*University of Tennessee,
Knoxville*

Nancy A. Piotrowski
University of California, Berkeley

Frank E. Price
Hamilton College

Diane C. Rein
Purdue University

Mary Beth Ridenhour
*State University of New York,
Potsdam*

Connie Rizzo
Pace University

James L. Robinson
University of Illinois at Urbana-Champaign

James N. Robinson
Huntingdon College

Charles W. Rogers
Southwestern Oklahoma State University

David Wijss Rudge
Western Michigan University

Paul C. St. Amand
Kansas State University

Virginia L. Salmon
Northeast State Technical Community College

Mary K. Sandford
University of North Carolina, Greensboro

Lisa M. Sardinia
Pacific University

Cathy Schaeff
American University

Elizabeth D. Schafer
Independent Scholar

Matthew M. Schmidt
State University of New York, Empire State College

Tom E. Scola
University of Wisconsin, Whitewater

Rebecca Lovell Scott
Massachusetts College of Pharmacy & Health Sciences

Rose Secrest
Independent Scholar

Bonnie L. Seidel-Rogol
Plattsburgh State University

Nancy N. Shontz
Grand Valley State University

R. Baird Shuman
University of Illinois at Urbana-Champaign

Sanford S. Singer
University of Dayton

Robert A. Sinnott
Larreacorp, Ltd.

David A. Smith
Lock Haven University

Dwight G. Smith
Southern Connecticut State University

Roger Smith
Independent Scholar

Lisa Levin Sobczak
Independent Scholar

F. Christopher Sowers
Wilkes Community College

Sharon Wallace Stark
Monmouth University

Joan C. Stevenson
Western Washington University

Jamalynne Stuck
Western Kentucky University

James N. Thompson, Jr.
University of Oklahoma

Leslie V. Tischauser
Prairie State College

Charles L. Vigue
University of New Haven

Peter J. Waddell
University of South Carolina

Matthew J. F. Waterman
Eastern Nazarene College

Marcia J. Weiss
Point Park College

Patricia G. Wheeler
Indiana University School of Medicine

Kayla Williams
Huntingdon College

Steven D. Wilt
Kentucky Wesleyan College

Michael Windelspecht
Appalachian State University

James A. Wise
Hampton University

R. C. Woodruff
Bowling Green University

Ming Y. Zheng
Gordon College

Alphabetical List of Contents

Volume 1

Volume 2

Encyclopedia of
Genetics
Revised Edition

Aggression

Field of study: Human genetics and social issues

Significance: *Aggression refers to behavior directed toward causing harm to others. Aggressive antisocial behavior is highly heritable, and antisocial behavior (ASB) during childhood is a good predictor of ASB in adulthood and crime. Physical acts of aggression are sometimes distinguished from the more context-sensitive "covert" ASBs, including theft, truancy, and negative peer interactions.*

Key terms

ANTISOCIAL BEHAVIOR (ASB): behavior that violates rules or conventions of society and/or personal rights

IMPULSIVITY: a tendency to act quickly without planning or a clear goal in mind

IRRITABILITY: a tendency to overreact to minor stimuli; short-temperedness or volatility

LIABILITY: the risk of exhibiting a behavior; the higher one's score for a measure of liability, the greater is one's the risk of exhibiting the behavior

SEROTONIN: a neurotransmitter, 5-hydroxytryptamine (5-HT), present in blood platelets, the gastrointestinal tract, and certain regions of the brain, which plays role in initiating sleep, blood clotting, and stimulating the heartbeat, and levels of which have been correlated with aggressive behavior as well as depression and panic disorder

Aggression and Related Behaviors

Aggression or agonistic behavior in animals is usually an adaptive response to specific environmental situations during competition for resources, as in establishing dominance and a territory or in sexual competition. Rat and mice studies indicate it is partly genetic, because selective breeding produces strains that differ in levels of aggression. Human aggression can also represent a variety of natural responses to challenging situations. Measures of aggression vary, but of greatest concern are antisocial behaviors (ASBs) such as crime and delinquency and whether some individuals are more likely to engage in these behaviors than others.

The earliest evidence for a genetic contribution to these complex behaviors comes from twin and adoptee studies. Genes also increase the liability for many clinical conditions that include aggressive behaviors, such as conduct disorder (physically aggressive acts such as bullying or forced sexual activity) and antisocial personality disorder (persistent violation of social norms, including criminal behavior) and for personality traits that often accompany aggression, such as impulsivity and irritability. Differences in measuring ASBs partly account for the variability in heritability estimates, which range from 7 to 81 percent, but many studies indicate a heritability for genetic influences of 0.40-0.50, a minor influence of shared environment, and a much more significant influence of nonshared environment (environment unique to the individual).

Aggression and Human Development

Aggressive behavior develops in children through a complex interaction of many environmental and biological factors. Also increasing liability for aggression and perhaps criminality are such factors as low socioeconomic status and parental psychopathology. A consistent finding is that the measure of the activity of the central nervous system's serotonin correlates inversely with levels of lifetime aggression, tendency to physically assault, irritability, and impulsivity. Some of the implicated genes regulate serotonin synthesis, release, and reuptake as well as metabolism and receptor activation, and vary from individual to individual. Serotonergic dysfunction is also noted in alcoholism with aggression and in suicide attempters and completers. Brain injuries can also exacerbate tendencies to exhibit ASBs.

Some aggression, however, is a normal part of development. Thus, Terrie Moffitt and colleagues distinguish between "adolescent-limited aggression"—times when most adolescents are rebelling against adult authority—and "life-course persistent" ASB, which likely reflects neuropsychological deficits and specific temperaments that are often exacerbated in unsupportive family settings. Genetic factors play a smaller role in adolescent delinquency and are consistent with aggression at this age as a developmental response to social context.

Sex Differences

A significant feature of ASB is a marked difference between the sexes. Males exhibit higher levels of physical aggression and violence at every age in all situations except in the context of partner violence (where females exceed males). More males than females are diagnosed with conduct disorder at every age. More males than females begin acts of theft and violence at every age. Males also exhibit higher rates of risk factors, such as impaired neurocognitive status, increased hyperactivity, and difficulties with peers. Females are rarely identified with the life-course persistent form of ASB; the male:female sex ratio is 10:1. Antisocial male and female adolescents tend to associate and often marry and reproduce at younger ages. The role that hormones, particularly testosterone, may play in these differences is not clear.

Social Significance

There is much controversy surrounding the efforts to identify genes associated with aggression or crime, especially now that genome sequencing is easier than ever. Many demand that the privacy of individuals be protected because the presence of specific genes does not dictate behavioral outcomes: Genes do not determine socially defined behaviors but only act on physiological systems. In addition, what constitutes acceptable or unacceptable behavior for individuals is culturally defined. Biological and environmental risk factors may increase an individual's liability to commit an act of aggression or crime, but the behavior must be interpreted within its specific context. Criminal law presumes that behavior is a function of free will, and most attempts to use genes as a mitigating factor in the courtroom have been unsuccessful. Efforts to prevent crime and violence must include consideration of all factors. Family milieu and parental competence are just as important as impaired cognitive mechanisms such as reduced serotonin activity. An imbalance in brain chemistry leading to impulsivity or aggression may be ameliorated by a supportive home setting, by medication, or by adequate nutrition.

—*Joan C. Stevenson*

See also: Aging; Behavior; Biological Determinism; Criminality; DNA Fingerprinting; Forensic Genetics; Sociobiology; Steroid Hormones; XYY Syndrome.

Further Reading

Bock, Gregory R., and Jamie A. Goode. *Genetics of Criminal and Antisocial Behaviour.* New York: John Wiley & Sons, 1996. This symposium was held at the Ciba Foundation in London in 1995 and includes a representative sample of the research foci in this arena, followed by discussions.

Fishbein, Diana H., ed. *The Science, Treatment, and Prevention of Antisocial Behaviors: Application to the Criminal Justice System.* Kingston, N.J.: Civic Research Institute, 2000. An excellent set of reviews on aggression and the many associated behaviors and mental disorders.

Lesch, Klaus Peter, and Ursula Merschdorf. "Impulsivity, Aggression, and Serotonin: A Molecular Psychobiological Perspective." *Behavioral Sciences and the Law* 18, no. 5 (2000): 581-604. A wonderful review of all the interacting factors, including all the elements of the serotonin system.

Moffitt, Terrie E., Avshalom Caspi, Michael Rutter, and Phil A. Silva. *Sex Differences in Antisocial Behaviour: Conduct Disorder, Delinquency, and Violence in the Dunedin Longitudinal Study.* New York: Cambridge University Press, 2001. Sex differences are documented as children grow up.

Roush, Wade. "Conflict Marks Crime Conference." *Science* 269, no. 5232 (1995): 1808-1809. An excellent description of the pros and cons of genetic research on ASB.

Web Site of Interest

National Institutes of Health, National Institute of Mental Health. http://www.nimh.nih .gov/publicat/violenceresfact.cfm. Provides information on child and adolescent violence and antisocial behavior, including research into the possible genetic factors of aggression.

Aging

Field of study: Human genetics and social issues

Significance: *In the light of modern science and medicine, it has become apparent that the roots of aging lie in genes; therefore, the genetic changes that take place during aging are the source of the major theories of aging currently being proposed.*

Key terms

ANTIOXIDANT: a molecule that preferentially reacts with free radicals, thus keeping them from reacting with other molecules that might cause cellular damage

FREE RADICAL: a highly reactive form of oxygen in which a single oxygen atom has a free, unpaired electron; free radicals are common by-products of chemical reactions

MITOCHONDRIAL DNA (mtDNA): the genome of the mitochondria, which contain many of the genes required for mitochondrial function

PLEIOTROPY: a form of genetic expression in which a gene has multiple effects; for example, the mutant gene responsible for cystic fibrosis causes clogging of the lungs, sterility, and excessive salt in perspiration, among other symptoms

Why Study Aging?

Biologists have long suspected that the mechanisms of aging would never be understood fully until a better understanding of genetics was obtained. As genetic information has exploded, a number of theories of aging have emerged. Each of these theories has focused on a different aspect of the genetic changes observed in aging cells and organisms. Animal models, from simple organisms such as *Tetrahymena* (a single-celled, ciliated protozoan) and *Caenorhabditis* (a nematode worm) to more complex organisms like *Drosophila* (fruit fly) and mice, have been used extensively in efforts to understand the genetics of aging. The study of mammalian cells in culture and the genetic analysis of human progeroid syndromes (that is, premature aging syndromes) such as Werner's syndrome and diseases of old age such as Alzheimer's disease have also improved the understanding of aging. From these data, several theories of aging have been proposed.

Genetic Changes Observed in Aging Cells

Most of the changes thus far observed represent some kind of degeneration or loss of function. Many comparisons between cells from younger and older individuals have shown that more mutations are consistently present in older cells. In fact, older cells seem to show greater genetic instability in general, leading to chromosome deletions, inversions, and other defects. As these errors accumulate, the cell cycle slows down, decreasing the ability of cells to proliferate rapidly. These genetic problems are partly a result of a gradual accumulation of mutations, but the appearance of new mutations seems to accelerate with age due to an apparent reduced effectiveness of DNA repair mechanisms.

Cells that are artificially cultured have been shown to undergo a predictable number of cell divisions before finally becoming senescent, a state where the cells simply persist and cease dividing. This phenomenon was first established by Leonard Hayflick in the early 1960's when he found that human fibroblast cells would divide up to about fifty times and no more. This phenomenon is now called the Hayflick limit. The number of divisions possible varies depending on the type of cell, the original age of the cell, and the species of organism from which the original cell was derived. It is particularly relevant that a fibroblast cell from a fetus will easily approach the fifty-division limit, whereas a fibroblast cell from an adult over age fifty may be capable of only a few divisions before reaching senescence.

The underlying genetic explanation for the Hayflick limit appears to involve regions near the ends of chromosomes called telomeres. Telomeres are composed of thousands of copies of a repetitive DNA sequence and are a required part of the ends of chromosomes due to certain limitations in the process of DNA replication. Each time a cell divides, it must replicate all of the chromosomes. The process of replication inevitably leads to loss of a portion of each telomere, so that with each new cell division the telomeres get shorter. When the telomeres get to a certain critical length, DNA replication seems to no longer be possible, and the cell enters senescence. Although the process discussed above is fairly consistent with most studies, the mechanism whereby a cell knows it has reached the limit is unknown.

A result of these genetic changes in aging humans is that illnesses of all kinds are more common, partly because the immune system seems to function more slowly and less efficiently with age. Other diseases, like cancer, are a direct result of the relentless accumulation of mutations. Cancers generally develop after a series of mutations or chromosomal rearrangements have occurred that cause the mutation of or inappropriate expression of proto-oncogenes. Proto-oncogenes are normal genes that are involved in regulating the cell cycle and often are responsible for moving the cell forward toward mitosis (cell division). Mutations in proto-oncogenes transform them into oncogenes (cancer genes), which results in uncontrolled cell division, along with the other traits displayed by cancer cells.

Progeroid Syndromes as Models of Aging

Several progeroid syndromes have been studied closely in hopes of finding clues to the underlying genetic mechanisms of aging. Although such studies are useful, they are limited in the sense that they display only some of the characteristics of aging. Also, because they are typically due to a single mutant gene, they represent a gross simplification of the aging process. Recent genetic analyses have identified the specific genetic defects for some of the progeroid syndromes, but often this has only led to more questions.

Down syndrome is the most common progeroid syndrome and is usually caused by possession of an extra copy of chromosome 21 (also called trisomy 21). Affected individuals display rapid aging for a number of traits such as atherosclerosis and cataracts, although the severity of the effects varies greatly. The most notable progeroid symptom is the development of Alzheimer's disease-like changes in the brain such as senile plaques and neurofibrillary tangles. One of the genes sometimes involved in

Alzheimer's disease is located on chromosome 21, possibly accounting for the common symptoms.

Werner's syndrome is a very rare autosomal recessive disease. The primary symptoms are severe atherosclerosis and a high incidence of cancer, including some unusual sarcomas and connective tissue cancers. Other degenerative changes include premature graying, muscle atrophy, osteoporosis, cataracts, and calcification of heart valves and soft tissues. Death, usually by atherosclerosis, often occurs by fifty or sixty years of age. The gene responsible for Werner's syndrome has been isolated and encodes a DNA helicase (called WRN DNA helicase), an enzyme that is involved in helping DNA strands to separate during the process of replication. The faulty enzyme is believed to cause the process of replication to stall at the replication fork, the place where DNA replication is actively taking place, which leads to a higher-than-normal mutation rate in the DNA, although more work is needed to be sure of its mechanism.

Hutchinson-Gilford progeria shows even more rapid and pronounced premature aging. Effects begin even in early childhood with balding, loss of subcutaneous fat, and skin wrinkling, especially noticeable in the facial features. Later, bone loss and atherosclerosis appear, and most affected individuals die before the age of twenty-five. The genetic inheritance pattern for Hutchinson-Gilford progeria is still debated, but evidence suggests it may be due to a very rare autosomal dominant gene, which may represent a defect in a DNA repair system.

Cockayne syndrome, another very rare autosomal recessive defect, displays loss of subcutaneous fat, skin photosensitivity (especially to ultraviolet, or UV, light), and neurodegeneration. Age of death can vary but seems to center around forty years of age. The specific genetic defect is known and involves the action of a few different proteins. At the molecular level, the major problems all relate to some aspect of transcription, the making of messenger RNA (mRNA) from the DNA template, which can also affect some aspects of DNA repair.

Another, somewhat less rare, autosomal recessive defect is ataxia telangiectasia. It displays

In April, 2003, fifteen-year-old John Tacket announced the discovery of a gene that causes the disease he suffers from, progeria, a syndrome that accelerates aging. (AP/Wide World Photos)

a whole suite of premature aging symptoms, including neurodegeneration, immunodeficiency, graying, skin wrinkling, and cancers, especially leukemias and lymphomas. Death usually occurs between forty and fifty years of age. The specific defect is known to be loss of a protein kinase, an enzyme that normally adds phosphate groups to other proteins. In this case, the kinase appears to be involved in regulating the cell cycle, and its loss causes shortening of telomeres and defects in the repair of double-stranded breaks in DNA. One of the proteins it appears to normally phosphorylate is *p53*, a tumor-suppressor gene whose loss is often associated with various forms of cancer.

Although the genes involved in the various progeroid syndromes are varied, they do seem to fall into some common functional types. Most have something to do with DNA replication, transcription, or repair. Other genes are involved in control of some part of the cell cycle. Although many other genes remain to be

discovered, they will likely also be involved with DNA or the cell cycle in some way. Based on many of the common symptoms of aging, these findings are not too surprising.

Genetic Models of Aging

The increasing understanding of molecular genetics has prompted biologists to propose a number of models of aging. Each of the models is consistent with some aspect of cellular genetics, but none of the models, as yet, is consistent with all evidence. Some biologists have suggested that a combination of several models may be required to adequately explain the process of aging. In many ways, understanding of the genetic causes of aging is in its infancy, and geneticists are still unable to agree on even the probable number of genes involved in aging. Even the extent to which genes control aging at all has been debated. Early studies based on correlations between time of death of parents and offspring or on the age of death of twins suggested that genes accounted for 40 to 70 percent of the heritability of longevity. More recent research on twins has suggested that genes may only account for 35 percent or less of the observed variability in longevity, and for twins reared apart the genetic effects appear to be even less.

Genetic theories of aging can be classified as either genome-based or mutation-based. Genome-based theories include the classic idea that longevity is programmed, as well as some evolution-based theories such as antagonistic pleiotropy, first proposed by George C. Williams, and the disposable soma theory. Mutation-based theories are based on the simple concept that genetic systems gradually fall apart from "wear and tear." The differences among mutation-based theories generally involve the causes of the mutations and the particular genetic systems involved. Even though genome-based and mutation-based theories seem to be distinct, there is actually some overlap. For example, the antagonistic pleiotropy theory (a genome-based theory) predicts that selection will "weed out" lethal mutations whose effects are felt during the reproductive years, but that later in life lethal mutations will accumulate (a mutation-based theory) because selection has no effect after the reproductive years.

Genome-Based Theories of Aging

The oldest genome-based theory of aging, sometimes called programmed senescence, suggested that life span is genetically determined. In other words, cells (and by extrapolation, the entire organism) live for a genetically predetermined length of time. The passing of time is measured by some kind of cellular clock and when the predetermined time is reached, cells go into a self-destruct sequence that eventually causes the death of the organism. Evidence for this model comes from the discovery that animal cells, when grown in culture, are only able to divide a limited number of times, the so-called Hayflick limit discussed above, and then they senesce and eventually die. Further evidence comes from developmental studies where it has been discovered that some cells die spontaneously in a process called apoptosis. A process similar to apoptosis could be responsible

for cell death at old age. The existence of a cellular clock is consistent with the discovery that telomeres shorten as cells age.

In spite of the consistency of the experimental evidence, this model fails on theoretical grounds. Programmed senescence, like any complex biological process, would be required to have evolved by natural selection, but natural selection can only act on traits that are expressed during the reproductive years. Because senescence happens after the reproductive years, it cannot have developed by natural selection. In addition, even if natural selection could have been involved, what advantage would programmed senescence have for a species?

Because of the hurdles presented by natural selection, the preferred alternative genome-based theory is called antagonistic pleiotropy. Genes that increase the chances of survival before and during the reproductive years are detrimental in the postreproductive years. Because natural selection has no effect on genes after reproduction, these detrimental effects are not "weeded" out of the population. There is some physiological support for this in that sex hormones, which are required for reproduction earlier in life, cause negative effects later in life, such as osteoporosis in women and increased cancer risks in both sexes.

The disposable soma theory is similar but is based on a broader physiological base. It has been noted that there is a strong negative correlation among a broad range of species between metabolic rate and longevity. In general, the higher the average metabolic rate, the shorter lived the species. In addition, the need to reproduce usually results in a higher metabolic rate during the reproductive years than in later years. The price for this high early metabolic rate is that systems burn out sooner. This theory is not entirely genome-based, but also has a mutation-based component. Data on mutation rates seem to show a high correlation between high metabolic rate and high mutation rates.

One of the by-products of metabolism is the production of free oxygen radicals, single oxygen atoms with an unpaired electron. These free radicals are highly reactive and not only cause destruction of proteins and other molecules, but also cause mutations in DNA. So the high metabolic rate during the reproductive years causes a high incidence of damaging DNA mutations which lead to many of the diseases of old age. After reproduction, natural selection no longer has use for the body, so it gradually falls apart as the mutations build up. Unfortunately, all attempts so far to assay the extent of the mutations produced have led to the conclusion that not enough mutations exist to be the sole cause of the changes observed in aging.

Mutation-Based Theories of Aging

The basic premise of all the mutation-based theories of aging is that the buildup of mutations eventually leads to senescence and death, the ultimate cause being cancer or the breakdown of a critical system. The major support for these kinds of theories comes from a number of recent studies that have found a larger number of genetic mutations in elderly individuals than in younger individuals, the same pattern being observed even when the same individual is assayed at different ages. The differences among the various mutation-based theories have to do with what causes the mutations and what kinds of DNA are primarily affected. As mentioned above, the disposable soma theory also relies, in part, on mutation-based theories.

The most general mutation-based theory is the somatic mutation/DNA damage theory, which relies on background radiation and other mutagens in the environment as the cause of mutations. Over time, the buildup of these mutations begins to cause failure of critical biochemical pathways and eventually causes death. This theory is consistent with experimental evidence from the irradiation of laboratory animals. Irradiation causes DNA damage, which, if not repaired, leads to mutations. The higher the dose of radiation, the more mutations result. It has also been noted that there is some correlation between the efficiency of DNA repair and life span. Further support comes from observations of individuals with more serious DNA repair deficiencies, such as those affected by xeroderma pigmentosum. Individuals with xeroderma pigmentosum have almost no ability to repair the type of DNA damage caused by exposure to UV light, and as a result they de-

velop skin cancer very easily, which typically leads to death.

The major flaw in this theory is that it predicts that senescence should be a random process, which it is not. A related theory called error catastrophe also predicts that mutations will build up over time, eventually leading to death, but it suffers from the same flaw. Elderly individuals do seem to possess greater amounts of abnormal proteins, but that does not mean that these must be the ultimate cause of death.

The free radical theory of aging is more promising and is probably one of the most familiar theories to the general public. This theory has also received much more attention from researchers. The primary culprit in this theory is free oxygen radicals, which are highly reactive and cause damage to proteins, DNA, and RNA. Free radicals are a natural by-product of many cellular reactions and most specifically of the reactions involved in respiration. In fact, the higher the metabolic rate, the more free radicals will likely be produced. Although this theory also involves a random process, it is a more consistent and predictable process, and through time it can potentially build on itself, causing accelerated DNA damage with greater age.

Significant attention has focused on mitochondrial DNA (mtDNA). Because free radicals are produced in greater abundance in respiration, which takes place primarily in the mitochondria, mtDNA should show more mutations than nuclear DNA. In addition, as DNA damage occurs, the biochemical pathways involved in respiration should become less efficient, which would theoretically lead to even greater numbers of free radicals being produced, which would, in turn, cause more damage. This kind of positive feedback cycle would eventually reach a point where the cells could not produce enough energy to meet their needs and they would senesce. Assays of mtDNA have shown a greater number of mutations in the elderly, and it is a well-known phenomenon that mitochondria are less efficient in the elderly. Muscle weakness is one of the symptoms of these changes.

The free radical theory has some appeal, in the sense that ingestion of increased amounts of antioxidants in the diet would be expected to reduce the number of free radicals and thus potentially delay aging. Although antioxidants have been used in this way for some time, no significant increase in life span has been observed, although it does appear that cancer incidence may be reduced.

From Theory to Practice

Many of the genetic theories of aging are intriguing and even seem to be consistent with experimental evidence from many sources, but none of them adequately addresses longevity at the organismal level. Although telomeres shorten with age in individual cells, cells continue to divide into old age, and humans do not seem to die because all, or most, of their cells are no longer able to divide. Cells from older individuals do have more mutations than cells from younger individuals, but the number of mutations observed does not seem adequate to account for the large suite of problems present in old age. Mitochondria, on average, do function more poorly in older individuals and their mtDNA does display a larger number of mutations, but many mitochondria remain high functioning and appear to be adequate to sustain life.

Essentially, geneticists have opened a crack in the door to a better understanding of the causes of aging, and the theories presented here are probably correct in part, but much more research is needed to sharpen the understanding of this process. The hope of geneticists, and of society in general, is to learn how to increase longevity. Presently, it seems all that is possible is to help a larger number of people approach the practical limit of 120 years through lifestyle modification and medical intervention. Going significantly beyond 120 years is probably a genetic problem that will not be solved for some time.

—*Bryan Ness*

See also: Alzheimer's Disease; Autoimmune Disorders; Biochemical Mutations; Biological Clocks; Biological Determinism; Cancer; Chemical Mutagens; Developmental Genetics; Diabetes; DNA Repair; Genetic Engineering: Medical Applications; Heart Disease; Human Genetics; Human Growth Hormone; Immunogenetics; Insurance; Mitochondrial Genes; Mutation and Mutagenesis; Oncogenes; Stem Cells; Telomeres; Tumor-Suppressor Genes.

Further Reading

Arking, Robert, ed. *Biology of Aging: Observations and Principles*. 2d ed. Sunderland, Mass.: Sinauer, 2001. A revised edition of a 1990 text that examines such topics as defining and measuring aging, changes in populations, genetic determinants of longevity, and aging as an intracellular process.

Austad, Steven N. *Why We Age: What Science Is Discovering About the Body's Journey Throughout Life*. New York: John Wiley & Sons, 1997. A review of the latest biological research and theories of aging, including an assessment of the oldest attainable age for humans.

Hekimi, Siegfried, ed. *The Molecular Genetics of Aging*. New York: Springer, 2000. Part of the Results and Problems in Cell Differentiation series. Illustrated.

Macieira-Coelho, Alvaro. *Biology of Aging*. New York: Springer, 2002. A solid text that includes many figures, tables, charts, and illustrations.

Manuck, Stephen B., et al., eds. *Behavior, Health, and Aging*. Mahwah, N.J.: Lawrence Erlbaum, 2000. Examines a host of health care dilemmas associated with the elderly. One section considers the basic tenets of genetic and molecular biology, including some of the methods of looking at heritable differences in health and well-being. Illustrated.

Medina, John J. *The Clock of Ages: Why We Age, How We Age—Winding Back the Clock*. New York: Cambridge University Press, 1996. A book written especially for the general reader. Covers aging on a system-by-system basis and includes a large section on the genetics of aging.

Ricklefs, Robert E., and Caleb E. Finch. *Aging: A Natural History*. New York: W. H. Freeman, 1995. A good general introduction to the biology of aging by two biologists who specialize in aging research.

Rusting, Ricki L. "Why Do We Age?" *Scientific American* 267 (December, 1992). Summarizes the changes that occur with aging and the roles of oxidants and free radicals.

Timiras, Paola S. *Physiological Basis of Aging and Geriatrics.* 3d ed. Boca Raton, Fla.: CRC Press, 2003. Divided into three main sections, this text addresses the basic processes of biogerontology, surveys the aging of body systems, and provides a synopsis of pharmacologic, nutritional, and physical exercise guidelines for preserving physical and mental health into senescence. Illustrated with numerous tables and graphs.

Toussaint, Olivier, et al., eds. *Molecular and Cellular Gerontology.* New York: New York Academy of Sciences, 2000. Elucidates the molecular mechanisms of aging.

Yu, Byung Pal, ed. *Free Radicals in Aging.* Boca Raton, Fla.: CRC Press, 1993. An in-depth discussion of the importance of free radicals in aging.

Web Sites of Interest

Alliance for Aging Research. http://www.aging research.org. Provides information on genetics and the aging process, including how the Human Genome Project will affect the future of health and health care.

American Geriatrics Society. http://www .americangeriatrics.org. The national society for health care providers for older persons, posting information on genetic screening for such disorders as Alzheimer's disease.

Centagenetix. http://www.centagenetix.com. This group's mission is to support better understanding of the aging process and associated diseases; the site offers a scientific overview, media center, and information on related careers.

National Institue on Aging. http://www.nia.nih .gov. Supports research programs on the biology and genetics of aging, as well as information on aging for the public.

Albinism

Field of study: Diseases and syndromes

Significance: *Albinism—the absence of pigment such as melanin in eyes, skin, hair, scales, or feathers—is a direct result of decreased or nonexistent pigmentation of the skin, hair, and eyes. Albino humans are susceptible to sunburns and skin cancer, while albino animals lack the ability to adjust to environments in which nonalbino animals thrive.*

Key terms

MELANISM: the opposite of albinism, a condition that leads to the overproduction of melanin

PHOTOPHOBIA: a condition, often observed in albinos, in which sunlight is painful to the eyes

PIEBALDISM: a condition involving the patchy absence of skin pigment seen in partial albinos

Occurrence and Symptoms

Tyrosine, an amino acid, is normally converted by the body to a variety of pigments called melanins, which give an organism its characteristic colors in areas such as the skin, hair, and eyes. Albinism results when the body is unable to produce melanin because of defects in the metabolism of tyrosine. Those with albinism can be divided into two subgroups: tyrosinase-negative (those who lack the enzyme tyrosinase) and tyrosinase-positive (those in whom tyrosinase is present but inactive). The most serious case is that of complete albinism or tyrosinase-negative oculocutaneous albinism, in which there is a total absence of pigment. People with this condition have white hair, colorless skin, red irises, and serious vision defects. The red irises are caused by the lack of pigmentation in the retina and subsequent light reflection from the blood present in the retina. These people also display rapid eye movements (nystagmus) and suffer from photophobia, decreased visual acuity, and, in the long run, functional blindness. People with this disorder sunburn easily, since their skin does not tan. Partial albinos have a condition known as piebaldism, characterized by the patchy absence of skin pigment in places such as the hair, the forehead, the elbows, and the knees.

Several complex diseases are associated with albinism. Waardenberg syndrome is identified by the presence of a white forelock (a lock of hair that grows on the forehead) or the absence of pigment in one or both irises, Chediak-

Higashi syndrome is characterized by a partial lack of pigmentation of the skin, and tuberous sclerosis patients have only small, localized depigmented areas. A more serious case is the Hermansky-Pudlak syndrome, a disorder that includes bleeding.

Ocular albinism is inherited and involves the lack of melanin only in the eye while the rest of the body shows normal or near-normal coloration. The condition reduces visual acuity from 20/60 to 20/400, with African Americans occasionally showing acuity as good as 20/25. Other problems include strabismus (crossed eyes or "lazy eye"), sensitivity to brightness, and nystagmus. The color of the iris may be any of the normal colors, but an optician can easily detect the condition by shining a light from the side of the eye. In ocular albinos, the light shines through the iris because of the absence of the light-absorbing pigment. Children with this condition have difficulty reading what is on a blackboard unless they are very close to it. Surgery and the application of optical aids appear to have had positive results in correcting such problems.

Albinism has long been studied in humans and captive animals. It has also been detected in wild animals, but such animals often have little chance of survival because they cannot develop normal camouflage colors, important for protection from predators. Animals in which albinism has been recorded include deer, giraffes, squirrels, frogs, parrots, robins, turtles, trout, and lobsters. Partial albinism has also been reported in wildlife. In other cases, such as the black panther of Asia, too much melanin is formed and the disorder is called melanism.

Albinism has also been observed in plants, but their life span rarely goes beyond seedline state, because without the green pigment chlorophyll, they cannot obtain energy using photosynthesis. A few species of plants, such as Indian pipes (*Monotropa*), are normally albino and obtain their energy and nutrition from decaying material in the soil.

Impact and Applications

Albinism appears in various forms and may be passed to offspring through autosomal recessive, autosomal dominant, or X-linked modes of inheritance. In the autosomal recessive case, both parents of a child with autosomal recessive albinism are carriers; that is, they each have one copy of the recessive form of the gene and are therefore not albino themselves. When both parents are carriers, there is a one-in-four chance that the child will inherit the condition. On the other hand, X-linked albinism occurs almost exclusively in males, and mothers who carry the gene will pass it on 50 percent of the time.

Albinism has not been found to affect expected life span among humans but can affect lifestyle. Treatment of the disease involves reduction of the discomfort the sun creates. Thus photophobia may be relieved by sunglasses that filter ultraviolet light, while sunburn may be re-

At the Santa Lucia school in Guatemala City, an albino girl, Maria del Carmen Quel, eats a snack as she plays on a swing. Albinism is frequently associated with blindness. (AP/Wide World Photos)

duced by the use of sun protection factor (SPF) sunscreens and by covering the skin with clothing. Since albinism is basically an inherited condition, genetic counseling is of great value to individuals with a family history of albinism.

—*Soraya Ghayourmanesh*

See also: Biochemical Mutations; Complete Dominance; Dihybrid Inheritance; Inborn Errors of Metabolism; Monohybrid Inheritance.

Further Reading

Gahl, William A., et al. "Genetic Defects and Clinical Characteristics of Patients with a Form of Oculocutaneous Albinism (Hermansky-Pudlak Syndrome)." *New England Journal of Medicine* 338, no. 18 (April 30, 1998): 125. Discusses several aspects of Hermansky-Pudlak syndrome. Details the diagnosis of this syndrome in forty-nine patients of Puerto Rican descent and patients from the mainland United States. Two charts.

Gershoni-Baruch, R., et al. *Journal of the American Academy of Dermatology* 24 (1991). Describes the hair bulb tyrosinase test.

King, R. A., V. J. Hearing, D. J. Creel, and W. S. Oetting. "Albinism." In *The Metabolic and Molecular Basis of Inherited Disease*, edited by C. R. Scriver et al. New York: McGraw-Hill, 1995. A solid introduction and overview of albinism.

Pollier, Pascale. *Journal of Audiovisual Media in Medicine* 24, no. 3 (September, 2001): 127. Examines the medical, biological, and genetic causes of albinism and provides notes from the author's attendance at a conference of the Albinism Fellowship.

Scriver, Charles, et al., eds. *The Metabolic and Molecular Bases of Inherited Disease*. 8th ed. 4 vols. New York: McGraw-Hill, 2001. These authoritative volumes on genetic inheritance, by some of the biggest names in the field, survey all aspects of genetic disease. The eighth edition has been thoroughly updated; more than half of the contents are new.

Tomita, Yasuchi. "Molecular Bases of Congenital Hypopigmentary Disorders in Humans and Oculocutaneous Albinism 1 in Japan." *Pigment Cell Research* 13, no. 5 (October, 2000): 130. Presents a study that identified the molecular bases of congenital hypopigmentary disorders in humans and oculocutaneous albinism (OCA)-1 in Japan. Piebaldism, Waardenburg syndrome, Hermansky-Pudlak syndrome, and tyrosinase gene-related OCA-1 are closely examined.

Witkop, C. J., Jr. *Clinical Dermatology* 7 (1989). Includes an overview of albinism.

Web Sites of Interest

International Albinism Center, University of Minnesota. http://www.cbc.umn.edu/iac. Run by a team of research professionals with a variety of specialities in human albinism—clinical genetics, molecular biology, ophthalmology, dermatology, and biochemistry—who are attempting to understand the cause and effect of albinism and other forms of pigment loss. Papers, fact sheets, glossary, other resources. Links to the Albinism Database, which lists mutations associated with albinism.

National Organization for Albinism and Hypopigmentation. http://www.albinism.org. A volunteer organization for albinos and for those who care for people with albinism, providing resources for self-help and promoting research and education.

Alcoholism

Field of study: Diseases and syndromes

Significance: *Alcohol is one of the most widely consumed substances of abuse worldwide. Because alcohol dependence can be life-threatening, its potential genetic basis is of great interest, and a variety of genes have been found linked to its physiologic markers and to the diagnosis of alcohol dependence.*

Key terms

ALCOHOL DEPENDENCE: a medical diagnosis given when there is repeated use of alcohol over the course of at least a year, despite the presence of negative consequences, such as tolerance, withdrawal, uncontrolled use, unsuccessful efforts to quit, considerable time spent getting or using the drug, and a decrease in other important activities

CIRRHOSIS: a disease of the liver, marked by the development of scar tissue that interferes with organ functioning, that can result from chronic alcohol consumption

FETAL ALCOHOL SYNDROME: a medical condition resulting from alcohol use by a mother while pregnant, usually evidenced by facial abnormalities and mental impairments in the child and sometimes resulting in fetal death

Defining Alcoholism

"Alcoholism" is a word that is used to convey that a person is experiencing serious problems related to the use of alcohol. The technical diagnosis of alcohol dependence with physiological dependence is the diagnosis that corresponds to the notion of alcoholism. That diagnosis, formally explained in the *Diagnostic and Statistical Manual of Mental Disorders: DSM-IV-TR,* or *DSM* (rev. 4th ed., 2000), issued by the American Psychiatric Association, refers to physical and psychological reliance on alcohol, despite the presence of problems associated with its use.

Alcohol-related problems are typically studied and tracked by epidemiologists, physicians, psychologists, public health professionals, and basic scientists. Basic scientists tend to observe the heritability of problem acquisition and expression in lab animals, while the other professionals tend to track the problem in humans via clinical observations and research involving reports of family histories or the review of medical records.

Due to changes in how the DSM has developed its problem definitions over time, researchers must use care to examine how definitions of alcoholism have changed over time. Also, the social stigma associated with alcoholism may have caused over- and underrecognition of the problem in some groups. For instance, more men are recognized to have alcoholism than women. Alcoholism is also known to have its strongest genetic findings for men of alcoholic fathers. However, only about 50 percent of male cases are explained by genetics; thus, there is also a strong environmental component to this problem.

Regarding women, however, recent evidence suggests that a family history of mood disorders, such as depression, may also be linked to how this disorder may be inherited. It may be that the heritability of depression and alcohol have something in common, that there are differences in heritability by gender, or perhaps that symptom expression by clients or recognition by professionals varies by gender as a result of other factors. Stigma, for instance, may be relevant; historically, women with alcohol problems have often been misdiagnosed with depression.

Symptom expression of alcohol problems may also differ by culture and ethnicity, because people of different cultures vary in terms of how they express physical and mental ailments. Different ethnic and racial groups may have different biological responses to therapeutic drugs and drugs such as alcohol. Some groups may even enjoy greater protection against alcoholism as a result of their genetics. Asians, for example, tend to be unable to tolerate alcohol because they generally lack an enzyme to process it out of the body. In contrast, there may be differential vulnerability to alcoholism itself, as well as differential vulnerabilities to certain types of organ damage related to alcoholism. For instance, vulnerability to cirrhosis, cardiomyopathy, pancreatitis, and Wernicke-Korsakoff's syndrome also might be heritable and may vary by ethnicity. Latino men, for example, tend to show greater susceptibility to alcohol-related liver damage than do white men.

There is also the issue of early alcohol exposure and how such early exposure can interact with genetics to cause problems in development. Fetal alcohol syndrome, for example, can result in a child's having mild to severe facial and dental abnormalities, mental impairments, or problems related to the skeletal and the cardiovascular systems. Problems with vision, hearing, and attention are also common. Children of alcoholic fathers also can have difficulties in learning, language, and temperament. Causes of such problems are multiple, including the contributions made by the individual's genes as well as the environmental effects of growing up in a home that may be unstable as a result of problems in the father. In sum, parents who drink may increase the likelihood that their children will develop alcoholism both through genes and through nongenetic environmental circumstances.

Alcohol Research

To date, some important physiological markers linked to alcoholism have included event-related potentials (ERPs) in electroencephalographic performance (EEGs), frontal lobe functioning, enzymes responsible for hepatic alcohol metabolism (such as alcohol dehydrogenase and aldehyde dehydrogenase), and inhibitory receptors such as gamma-aminiobutyric acid (GABA) receptors. There are a variety of genes linked to such physiologic markers and to the diagnosis of alcohol dependence, including *ADH2*2, ADH3*1, ALDH2*2, CYP 2EI, GABRA6, GABRA1, COMT, DRD4, DRD2,* and *D2.*

Future Directions

The presence of alcohol in modern life may have genetic roots. Historically, it helped those who could tolerate its taste and effects to survive and be selected for when others who could not do so perished as a result of consuming contaminated water. Alcohol has a complex relationship to human life, and alcoholism will be studied for some time. Continued study of the genes associated with different patterns of alcohol problems, protective genetic effects in populations with exceptionally low rates of alcoholism, and genetically based interventions (such as matching pharmacotherapies to different populations of individuals to forestall the development of the problem) are assured. The study of genetics and alcoholism is also likely to encourage growth in the field of ethnopharmacology, the study of how different therapeutic drugs differentially affect members of specific ethnic groups.

—*Nancy A. Piotrowski*

See also: Aggression; Behavior; Congenital Defects; Criminality; Eugenics; Genetic Testing: Ethical and Economic Issues; Heriditary Diseases; Thalidomide and Other Teratogens.

Further Reading

American Psychiatric Association. *Diagnostic and Statistical Manual of Mental Disorders: DSM-IV-TR.* Rev. 4th ed. Washington, D.C.: Author, 2000. This American professional manual describes all major psychiatric disorders. There is a chapter devoted to substance use disorders.

Plomin, Robert, and Gerald E. McClearn, eds. *Nature, Nurture, and Psychology.* Washington, D.C.: American Psychological Association, 1993. The topic of alcoholism is discussed, among other topics, with an emphasis on comparing the roles of genetics versus social processes and the environment.

Web Site of Interest

National Institute on Alcohol Abuse and Alcoholism, ETOH. http://etoh.niaaa.nih.gov. ETOH is the chemical abbreviation for ethyl alcohol. This site includes reports related to alcohol dependence, including epidemiology, etiology, prevention, policy, and treatment.

Allergies

Field of study: Immunogenetics
Significance: *In economically developed countries, allergies are responsible for a large portion of illnesses and medical expenses. Many allergies have genetic components and thus tend to "run" in families; the identification of such hereditary factors can help in diagnosis and in family planning. Moreover, research into the causes of allergies may lead to a more precise understanding of how the immune system functions. This may lead ultimately to the development of better drugs to treat allergies.*

Key terms

ANTIBODY: a protein made by the body in response to an antigen; antibodies or immunoglobulins are specific for each antigen
ANTIGEN: any substance that, when injected into the body, causes antibody formation that reacts specifically to that substance; also known as an allergen or an immunogen
HYPERSENSITIVITY: an exaggerated response of the immune system to an antigen beyond what is considered "normal"; a synonym for allergy

IMMUNE SYSTEM: the defense mechanism of the body against foreign matter (bacteria, viruses, and parasites); it is composed of different types of cells and chemical substances

The Basic Information About Allergies

Sneezing, sniffling, and wheezing are the symptoms most often associated with allergies. Allergies, or hypersensitivities, are the human body's exaggerated response to a foreign substance such as pollen. Hypersensitivity reactions can be immediate (hay fever) or delayed (contact dermatitis—for example, a reaction to latex or poison ivy) depending upon the body's immune reaction to the antigen.

Essentially, there are three stages of an allergic reaction. The first stage causes no symptoms. It is the immune system's initial contact with the antigen. The cells of the immune system react to the antigen by producing IgE antibodies that attach to mast cells and eosinophils (two cell types of the immune system) that are circulating in the blood. When the same antigen is encountered a second time and attaches to two adjacent IgE antibodies on a mast cell, the mast cell is said to be "activated." During this second stage, the mast cell releases chemi-

cal substances (such as histamines, prostaglandins, and leukotrienes) that are responsible for many of the common allergic symptoms. The third and final stage of an allergic reaction is the prolonged immune activity caused by the chemical substances released by cells of the immune system. This prolonged or late-phase reaction can cause the immune system to continue to react and cause tissue damage.

Based on varying responses to antigens, researchers Peter Gell and Robert Coombs have classified allergies into four types: I (anaphylaxis), II (cytotoxic), III (immune complex), and IV (cell-mediated). Type I hypersensitivity—anaphylaxis, from the Greek *ana* (against) and *phylaxis* (protection), or "the opposite of protected"—can be further divided into either systemic or local response. Systemic anaphylaxis is the whole body's response to an antigen such as a bee sting. Because of the amount of chemical substances released by the cells of the immune system, the body reacts immediately by a drop in blood pressure (leading to shock), difficulty in breathing, and swelling of the airways. If not treated immediately, anaphylactic shock can be fatal. Localized anaphylactic reactions (atopy) are the most familiar of the hypersensi-

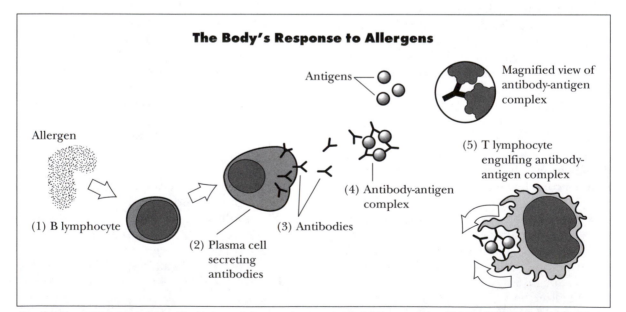

The Body's Response to Allergens

Antigens

Magnified view of antibody-antigen complex

Allergen

(5) T lymphocyte engulfing antibody-antigen complex

(4) Antibody-antigen complex

(1) B lymphocyte

(3) Antibodies

(2) Plasma cell secreting antibodies

An allergic reaction is caused when foreign materials, or antigens, enter the immune system, which produces B lymphocytes (1) that cause blood plasma cells to secrete antibodies (2). The antibodies (3) link with antigens to form antibody-antigen complexes (4), which then are engulfed and destroyed by a T lymphocyte (5). (Hans & Cassidy, Inc.)

tivities. The symptoms are dependent upon the route the antigen uses to enter the body. For airborne antigens such as house dust, pollens, and animal dander, symptoms may include hay fever (itchy eyes, runny nose, sneezing, and coughing) or bronchial asthma (wheezing, coughing, and difficulty breathing). Other atopic symptoms may include hives, itchy skin, and diarrhea. Food allergies are also examples of an atopic reaction.

Type II (cytotoxic) hypersensitivity reaction involves the binding of an antigen and antibody complex to a cell that destroys the target cell. Examples of this type of hypersensitivity are incompatible blood groups (giving type B blood to a person who has type A blood), hemolytic anemia (destruction of red blood cells), and hemolytic disease of a newborn (the mother produces antibodies against the fetus based on a protein found in the blood).

Type III (immune complex) hypersensitivity reaction involves the depositing of immune complexes (an antigen bound to an antibody) on the walls of blood vessels, causing inflammation and tissue damage. Glomerulonephritis, inflammation of the blood vessels in the kidneys, is a type III hypersensitivity reaction. This disease is believed to be a reaction to a particular bacterial infection.

Contact dermatitis (for example, a reaction to poison ivy, latex, cosmetics, or jewelry) is a common example of the last category, type IV (cell-mediated) hypersensitivity. T cells (a cell type of the immune system) initially react with the antigen; upon a second exposure to the antigen, clones of the same T cell release chemical factors that cause a reaction to the antigen. This release of chemicals results in a red, itchy rash or hives. The reaction to *Mycobacterium tuberculosis* (the bacteria that causes tuberculosis) is a type IV hypersensitivity. The immune system has also been known to attack itself and cause disease. These disorders (autoimmune disorders), such as multiple sclerosis, juvenile diabetes, and systemic lupus erythematosus, are only beginning to be understood.

Impact and Applications

Treatment of allergies may include avoidance of the antigen, use of antihistamines (drugs that block the release of histamine from mast cells) and anti-inflammatories (steroids), and desensitization (allergy shots). Efforts by scientists to learn how the immune system functions and why it overreacts to antigens will lead to the development of better, less toxic drugs to combat allergies and their symptoms.

—Mary Beth Ridenhour

See also: Autoimmune Disorders; Genetic Engineering: Risks; Immunogenetics; Synthetic Antibodies.

Further Reading

Cutler, Ellen W. *Winning the War Against Asthma and Allergies.* Albany, N.Y.: Delmar, 1998. This clearly written book provides practical information on all aspects of allergies—what they are, their causes, testing, diagnosis, and treatment, including nontraditional therapies. Preventive measures are covered, as are scenarios for various allergy elimination therapies.

Joneja, Janice M. V., and Leonard Bielory. *Understanding Allergy, Sensitivity, and Immunity.* New Brunswick, N.J.: Rutgers University Press, 1990. The authors provide extensive discussion of allergies and the roles played by the immune system. They describe the means by which one can learn to cope with allergies and discuss various testing methods for the identification of allergens.

Kuby, Janis. *Immunology.* 4th ed. New York: W. H. Freeman, 2000. The section on hypersensitivity in this immunology textbook is well written and includes a mixture of detail and overview of the subject. Particularly useful are discussions of the various types of hypersensitivity reactions. Some knowledge of biology is useful.

Lichtenstein, Lawrence. "Allergy and the Immune System." *Scientific American* 269 (September, 1993). A basic explanation of the way human allergies work.

Life, Death, and the Immune System. New York: W. H. Freeman, 1994. This comprehensive collection of articles from *Scientific American* provides basic information and research directions on autoimmune disorders and allergies as well as an excellent discussion of the immune system in general.

Mak, Tak W., and John J. L. Simard. *Handbook of Immune Response Genes*. New York: Plenum Press, 1998. Contains chapters on such topics as genes of the major histocompatibility complex and of the antigen-processing pathways, autoimmune disorders, cytokines of the immune system, and leukocyte cluster of differentiation antigens. Illustrated.

Steinman, Lawrence. "Autoimmune Disease." *Scientific American* 269 (September, 1993). Describes what happens when the immune system develops an autoimmune disorder that causes it to see the body as foreign and attack itself.

Theofilopoulos, A. N., ed. *Genes and Genetics of Autoimmunity*. New York: Karger, 1999. Examines a broad spectrum of topics related to autoimmunity, including the chapters "Immunoglobulin Transgenes in B Lymphocyte Development, Tolerance, and Autoimmunity," "The Role of Cytokines in Autoimmunity," "Genetics of Human Lupus," and "Genetics of Multiple Sclerosis." Illustrated.

Tortora, Gerard, et al. *Microbiology: An Introduction*. San Francisco: Benjamin Cummings, 2000. Provides a general overview of the human immune system and allergies via such topics as microbial metabolism; microbial genetics; viruses, viroids, and prions; disorders associated with the immune system; and metabolic pathways.

Walsh, William. *The Food Allergy Book*. New York: J. Wiley, 2000. In this excellent guide to one prevalent form of allergy, the author presents useful background information on food allergies and a pragmatic guide to identifying and eliminating food allergens from your diet.

Web Sites of Interest

Allergic Diseases Resource Center. http://www.worldallergy.org/allergicdiseasecenter.shtml. Part of the World Allergy Organization, this site provides scientific and medical information on allergic diseases, research updates, and more.

National Institue of Allergy and Infectious Diseases. http://www.niaid.nih.gov/default.htm. An arm of the National Institutes of Health, linking to reports and information on immunology, allergies, gene therapy, HIV/AIDS, and more.

Altruism

Field of study: Population genetics

Significance: *In a strictly Darwinian system, actions that reduce the success of individual reproduction should be selected against; however, altruism, which occurs at a cost to the altruist, is observed regularly in natural populations. This paradox may be resolved if the cost of altruism is offset by the reproductive success of relatives with which altruists share genes. Kin selection results in selection for altruistic behaviors which, if directed at relatives, preserve inclusive reproductive success, and thus Darwinian fitness.*

Key terms

ALTRUISM: behavior that benefits others at the evolutionary (reproductive) cost of the altruist

EVOLUTION: a change in the frequency of alleles resulting from the differential reproduction of individuals

HAPLODIPLOIDY: a system of sex determination in which males are haploid (developing from unfertilized eggs) and females are diploid

INCLUSIVE FITNESS: an individual's total genetic contribution to future generations, comprising both direct fitness, which results from individual reproduction, and indirect fitness, which results from the reproduction of close relatives

KIN SELECTION: an evolutionary mechanism manifest in selection for behaviors that increase the inclusive fitness of altruists

MATERNAL ALTRUISM: altruism on the part of mothers toward offspring as well as between and among members of groups comprising closely related females

NATURAL SELECTION: a process whereby environmental factors influence the survival and reproductive success of individuals; natural selection leads to genetic changes in populations over time

RECIPROCAL ALTRUISM: mutual exchange of altruistic acts typically associated with highly cohesive social groups

Reproductive Success = Survival

If evolutionary outcomes in a Darwinian world are described as natural economies, then individual reproduction is the currency of these economies and of natural selection. Given both naturally occurring genetic variation among individuals and a certain environmental dynamic, it follows that some individuals will be better adapted to locally changing environments than others. Such differential adaptation is expressed as a difference in the frequency with which individual genes pass into future generations. This simple scenario fulfills the genetic definition of evolution—change in allele frequencies in natural populations—by explaining environmental influences on these changes. Note that this argument emphasizes, as its central postulate, the importance of individual reproduction rather than simple survival. Survival of the fittest is therefore more properly viewed as the differential propagation of genes.

A challenge to such a scenario is the paradox of altruism. Altruism is defined as any behavior that benefits another at a cost to the altruist. Charles Darwin himself suggested that this problem was a "special difficulty . . . which at first appeared . . . insuperable, and actually fatal to [the] whole theory" of natural selection. The individual who pushes siblings from the track as he himself is killed by the rushing locomotive is an altruist; the colony sentinel that issues an alarm call to her cohort to take cover, despite the risk of drawing the attention of an approaching predator, is also acting altruistically. These behaviors make no sense in Darwin's economy, since they appear to decrease the likelihood of individual reproduction—unless, as W. D. Hamilton suggested, Darwinian success is not limited to the success of individual bodies harboring particular genes but may be extended to include the reproductive success of relatives who share genes with the altruist. Hamilton defined inclusive fitness as the sum of an individual's own fitness plus the influence that individual has on the fitness of relatives. Kin selection is the evolutionary mechanism that selects for behaviors that increase the inclusive fitness of altruists. Even though there are potential costs to altruistic behavior, the evolutionary economy of an altruist operates in the black because actors profit (beyond associated costs) by helping others who share their genes. The bottom line is that altruists increase their inclusive fitness through the reproduction of others.

Evidence of Kin Selection

One of the best evidences for kin selection is the social structure of certain groups of insects, including the *Hymenoptera* (ants, bees, and wasps). A unique system of sex determination (haplodiploidy) in which females are diploid and males are haploid predisposes some group members to behave altruistically. In certain bees, for example, the queen is diploid and fertile. Worker bees are female, diploid, and sterile. Drones are male, developed from unfertilized eggs, and haploid. Such a situation makes for unusual patterns of genetic relationship among hive members. In diploid systems the genetic relation between parents and offspring and among offspring is symmetrical. Offspring receive half of their genetic complement from their mother and half from the father; sons and daughters are related to each parent by $\frac{1}{2}$ and sibs (siblings) are related to each other by $\frac{1}{2}$. In the haplodiploid system such genetic relationships are asymmetric. Drones are haploid and receive half of the queen's genome. Workers are diploid and share 100 percent of their paternal genes and, on average, half of their maternal genes with their sisters. Sisters are therefore related to each other by $\frac{3}{4}$. Because sisters and their brothers share no paternal genes, and on average half of their maternal genes, they are related to drones by only $\frac{1}{4}$. In this economy it makes sense that workers should act altruistically to assist the queen in the production of sisters. What would appear to be purely altruistic acts, on the part of workers, result in greater inclusive success than if the workers had reproduced themselves. In contrast, drones contribute little to community welfare and serve only to fertilize the queen. Note that in this system there is no conscious

The altruistic behaviors of honeybees and some other animal species may be a result of selection for behaviors that place the group, rather than the individual, at a reproductive advantage. (AP/Wide World Photos)

decision on the part of workers not to reproduce; their sterility is an inherent part of this unusual system of sex determination.

A Test of Predictions

One prediction made by the kind of kin selection described above is that, assuming the queen produces male and female offspring in equal proportion, female workers should invest three times the energy in caring for sisters that they do for brothers. Because queens are related to both male and female offspring equally, one would predict that eggs are equally divided between the sexes. Because workers are related to their sisters by $\frac{3}{4}$ and to their brothers by $\frac{1}{4}$, one would predict that they should invest three times the energy in care of eggs eventually yielding sisters that they do in the care of eggs eventually yielding brothers. Remarkably, it has been shown that certain worker ants are able to identify and then selectively care for eggs containing sisters. Kin recognition has also been studied in the house mouse, *Mus*

musculus domesticus, and in some cases individuals can distinguish full sibs from half sibs on the basis of their major histocompatibility complexes (glycoproteins important in immune system function). The specific MHC type is fairly unique for each mouse, but related individuals will have similar patterns and share some specific MHC glycoproteins. MHC glycoproteins are found in mouse urine, and individuals can distinguish these molecules by smell. Consistent with the foregoing hypothesis, the degree of female altruism toward the offspring of close relatives was predicted by the degree of relation based on MHC type and type recognition.

Maternal Altruism

Altruism may be observed in a variety of natural systems in which groups comprise individuals who share a high degree of genetic relatedness. A classic example of this sort occurs with Belding's ground squirrels. Males tend to disperse from colonies, while females remain to create highly related maternal groups. Members of such maternal groups demonstrate altruistic behaviors such as alarm calling to warn relatives of danger. Although truly altruistic in the sense that alarm callers may incur risk of personal injury or death, they can be reasonably assured of breaking even in this economy as long as their genes live on in the bodies of those they have saved by their actions.

Reciprocal Altruism

It would seem that altruism based on Hamilton's argument of inclusive fitness would be precluded by human social organization. Scientists have predicted, however, that reciprocal altruism should exist in systems characterized by a high frequency of interaction among member individuals and life spans long enough to allow the recipients of altruistic acts to repay altruists. Note that the theoretical basis for the existence of reciprocal altruism differs from that for kin selection, and that any system in which evidence for reciprocity is found must necessarily include the development of a complex web of sophisticated social interaction. Such systems would be expected to foster traits expressing the panoply of human emotion and the development of certain moral architectures and group cohesion.

—David A. Smith

See also: Behavior; Evolutionary Biology; Homosexuality; Natural Selection; Population Genetics; Sociobiology.

Further Reading

Freeman, Scott, and Jon C. Herron. "Kin Selection and Social Behavior." In *Evolutionary Analysis*. Upper Saddle River, N.J.: Prentice Hall, 2001. A well-written and logical analysis of altruistic behavior. Arguments are supported with data and analysis from the primary literature.

Gould, Stephen Jay. "So Cleverly Kind an Animal." In *Ever Since Darwin*. New York: W. W. Norton, 1973. An elegantly expressed description of altruism and haplodiploidy in social insects.

Volpe, E. Peter, and Peter A. Rosenbaum. "Natural Selection and Social Behavior." In *Understanding Evolution*. 6th ed. Boston: McGraw-Hill, 2000. Nice analysis of the theoretical basis for kin selection including consideration of genetic asymmetries associated with haplodiploidy.

Alzheimer's Disease

Field of study: Diseases and syndromes
Significance: *Alzheimer's disease (AD) is the most common cause of irreversible dementia and accounts for approximately two-thirds of all dementia cases in the United States.*

Key terms

AMYLOID PLAQUES: plaques formed by protein fragments from amyloid precursor proteins

BETA-AMYLOID PEPTIDE: the main constituent of the neuritic plaques in the brains of Alzheimer's patients

DETERMINISTIC MUTATIONS: gene mutations associated with high risk for developing Alzheimer's

FAMILIAL ALZHEIMER'S DISEASE (FAD): inherited Alzheimer's disease

HIPPOCAMPUS: the area in the brain that encodes memory

NEUROFIBRILLARY TANGLES: abnormally twisted tau protein threads that lead to the death of brain cells

TAU PROTEIN: threads of protein in the cells of the brain that stabilize the brain's support structure

The Extent of AD

Alzheimer's disease (AD) is a progressive neurodegenerative disorder that causes a gradual, irreversible, decline in memory, language, visual-spatial perceptions, and judgment which are all the result of amyloid plaques, neurofibrillary tangles, and neuronal loss. Approximately four million Americans suffer with AD, a number that is expected to increase to almost 6 million by 2020. According to the Alzheimer's Association, 14 million Americans will be diagnosed with AD by 2050 if a cure is not found. Annually, $33 billion is lost by American businesses as a result of AD. At an annual cost of $100 billion, AD is the third most costly disease in the United States. Worldwide, the World Health Organization (WHO) estimates that by 2050, more than 22 million individuals worldwide will have developed AD, and some estimates are larger. AD currently accounts for between 50 and 75 percent of all dementias. Its prevalence increases from 1 percent at the age of sixty-five years to 20-35 percent by the age of eighty-five years. The average life span for AD sufferers ranges from eight to twenty years following diagnosis. As AD progresses, individuals become less and less able to perform activities of daily living because of progressive cognitive and social declines. AD will reach epidemic proportions as the human life span continues to increase. Understanding the genetic risks for developing diagnostics to identify AD and early intervention to treat AD will positively impact the quality of life of individuals who suffer with the disease.

Historical Perspectives on AD

Greeks and Romans first described symptoms of AD in their writings about dementias in old age. In 1906, a German physician, Dr. Alois Alzheimer, described plaques and neurofibrillary tangles in the brain of a mentally disturbed woman and identified them as a component in a type of acceleration in aging. Today, these plaques and tangles in the brain are hallmarks of AD that are identified on autopsy and are the only means to definitively diagnose AD. Early-onset AD was originally referred to as presenile dementia because it occurred in individuals younger than sixty or sixty-five years of age. Late-onset AD was referred to as senile dementia because it occurred in individuals older than eighty or eight-five years of age. Until recently, AD was considered a normal consequence of aging.

Genes Associated with AD

AD is not a normal part of aging. Over the past several years, scientists have discovered genetic links to two main types of AD. In the late 1980's scientists discovered amyloid precursor protein (APP). Alpha-, beta-, and gamma-secretase enzymes hang onto APP. The beta and gamma enzymes produce a sticky protein called beta-amyloid (A-beta). A-beta builds up

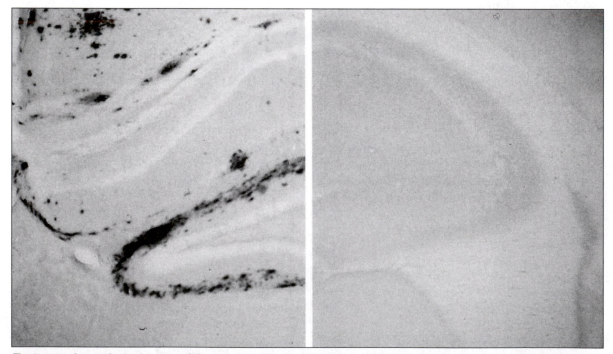

Two images of mouse-brain tissue, one (left) engineered to produce the dark protein deposits that characterize Alzheimer's disease, and the other normal. (AP/Wide World Photos)

in the fluid surrounding the neurons and is responsible for the formation of amyloid placques in AD.

Early-onset AD is caused by mutations in *APP* that cause abnormal proteins to form: presenilin 1 and presenilin 2. *APP* mutations cause amyloid plaque to develop in the hippocampus. *APP* mutations on chromosome 21 were the first gene mutations to be described in association with early-onset AD. *APP* mutations on chromosome 14 also produce presenilin 1. *APP* mutations on chromosome 1 produce presenilin 2. Individuals with these deterministic mutations will develop early-onset AD. Early-onset AD is rare and usually affects individuals thirty to sixty years of age. Most cases of early-onset AD are inherited and are called familial AD (FAD).

Another type, late-onset AD, is the most common form of AD, occurring in individuals who are sixty-five years of age and older. Late-onset AD is caused by mutations in apolipoprotein E (ApoE). ApoE is the most common genetic risk factor for developing AD. Scientists believe that these mutations allow longer isoforms of the neurotoxic A-peptide, which lead to the death of neurons. poE binds to beta-amyloid. The gene that produces ApoE is in the region of chromosome 19. There are at least three forms: *e2* allele, *e3* allele, and *e4* allele. ApoE *e2* allele is rare and develops later in life. It may also protect individuals against AD. ApoE *e3* allelle is the most common allele. According to researchers it appears to be neutral in AD. ApoE *e4* allele occurs in approximately 40 percent of individuals who develop late-onset AD. Individuals with ApoE *e4* allele may develop AD even if there is no family history of AD. Scientists believe that protein products from ApoE *e4* allele bind to APP and form plaques.

Chromosome 10 has also been identified as possibly containing genes that increase the risk of developing AD. Though it is still not certain whether beta-amyloid plaques cause AD or are a by-product of AD, the formation of beta-amyloid from APP is a key process in AD.

Information has been released by the National Institutes of Health (NIH) that reported that an alteration in brain-derived neurotrophic factor (BDNF) affected memory, influenced the activation of the hippocampus, and decreased the interconnection in neurons and neuron health in humans.

Risk Factors for AD

Advancing age and heredity are the most important risk factors for the development of AD. While it is known that mutations in *APP* and presenilin 1 and presenilin 2 genes cause early-onset AD, causes of late-onset AD are not as clear-cut. Some scientists hypothesize that late-onset AD may be initiated by inefficient processing of APP or by enhanced degradation of the tau protein. Individuals with ApoE *e4* allele have a two to four times greater risk for developing AD than those without it. Still, only 30-40 percent of ApoE *e4* allele carriers develop AD. Infectious agents, environmental toxins, and metabolic errors that have not yet been identified may also be possible causes of late-onset AD.

Aside from genetic risks involved in the development of AD, other factors that increase the likelihood for developing AD include traumatic brain injury and lower socioeconomic status, being overweight, lower educational level, sedentary lifestyle, depression, elevated blood cholesterol levels, and vascular diseases such as hypertension, coronary artery disease, atrial fibrillation, and myocardial infarction. Females are also at greater risk for developing AD.

Implications and Interventions

Early diagnosis of AD is essential to ensure that proper treatment and early detection of other underlying diseases such as depression, drug interactions, vitamin deficiencies, or endocrinologic problems are ruled out. The Risk Evaluation and Education for Alzheimer's Disease (REVEAL) study investigated the impact of identifying individuals with the ApoE genotype. The ApoE genotype is the most powerful genetic risk factor for AD and may be instrumental in predicting the chance of developing AD. This study offers guidance for using genetic risk information to screen, evaluate, and educate families with relatives suffering from AD.

The National Institutes for Health Alzheimer's Disease Prevention Initiative (NIHADPI) was organized to identify factors that will assist

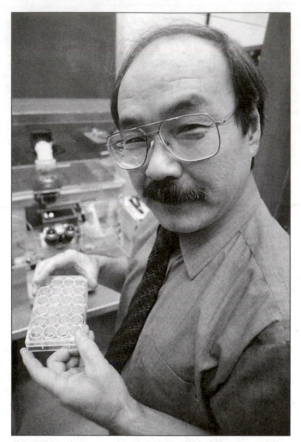

Etsuro Uemura, a professor of veterinary science who has been working on Alzheimer's disease since the early 1970's, before the disease was named, displays tissue cultures from rat brains that he has been using in his research. (AP/Wide World Photos)

in the early diagnosis of AD and in investigating pharmacological interventions that may assist in delaying or abating the development of AD. NIHADPI investigations include biological and epidemiological research, instrument development to identify high-risk individuals, clinical trials, and alternate strategies to treat behavioral disturbances in AD patients.

Clinical trials have identified positive effects from anti-inflammatory medications, statin medications, vitamin supplements, and diet in decreasing the risk for developing AD. Pharmaceutical agents such as acetylcholinesterase inhibitors have helped to slow the progression of AD by blocking the breakdown of neurotransmitters in the brain and to lessen symptoms of mild to moderate AD. Other studies

have shown that estrogen decreases the risk for women to develop AD by 30 to 40 percent. Its antioxidant and anti-inflammatory effects enhance neuron growth and therefore memory function. However, results from controlled trials on the effects of hormone replacement therapy, conducted by the Women's Health Initiative and reported in 2003, did not yield consistent findings of any beneficial effect of estrogen on cognitive function of women with AD, and these findings combined with a clearly increased risk for some women taking hormone replacement therapy cast the benefits of estrogen for postmenopausal women in doubt. The antioxidant effects of vitamin E and selegiline hydrochloride have been promising in slowing the rate of progression of AD. Ginkgo biloba has provided moderate cognitive improvement with few ill effects. Nonsteroidal anti-inflammatory medications may reduce AD risk by 30 to 60 percent. Finally, a synthetic form of beta-amyloid protein (AN-1792) vaccine is being investigated in clinical trials.

Because the ethical, legal, and social dilemmas surrounding AD cannot be ignored, issues of confidentiality are in the forefront for all genetic testing for AD. Confidentiality in genetic testing may be compromised if such testing becomes part of a person's medical records. Once medical records mention a patient's diagnosis of AD, employers, insurance companies, and other health care organizations can theoretically obtain information that could result in discriminatory actions such as refusal to hire, insure, or provide care.

Genetic counseling allows families to be aware of their genetic predisposition for AD. Since experts still do not know what benefits are gained by knowing that one is at risk for AD, there is great controversy about whether the benefits outweigh the detrimental impact that such knowledge holds.

—Sharon Wallace Stark

See also: Aging; Alcoholism; Behavior; Biological Clocks; Cancer; Diabetes; Down Syndrome; Genetic Testing: Ethical and Economic Issues; Heart Disease; Heriditary Diseases; Hypercholesterolemia; Insurance; Prion Diseases: Kuru and Creutzfeldt-Jakob Syndrome; Proteomics; Stem Cells; Telomeres.

Further Reading

Farrer, L., L. Cupples, J. Haines, et al. "Effects of Age, Gender, and Ethnicity on the Association Between Apolipoprotein E Genotype and Alzheimer Disease: A Meta-analysis." *JAMA* 278 (1997): 1349-1356. Designed for medical professionals, this paper discusses the most likely cause of the most common form of the disease, late-onset AD.

Food and Drug Administration. "Head Injury Linked to Increased Risk of Alzheimer's Disease." *FDA Consumer* (January/February, 2001): 8. Discusses research that focuses on the link between head injuries and dementias, including AD.

Gauthier, S., ed. *Clinical Diagnosis and Management of Alzheimer's Disease.* 2d ed. London: Martin Dunitz, 2001. A collection of discussions concerning symptoms, genetics, diagnosis, and treatment of AD.

Green, R. "Implications of Amyloid Precursor Protein and Subsequent Beta-Amyloid Production to the Pharmacotherapy of Alzheimer's Disease." *Pharmacotherapy* 22 (2002): 1547-1563. Identifies causes, genetic risks, diagnosis and treatment of AD.

Hamdy, Ronald, James Turnball, and Joellyn Edwards. *Alzheimer's Disease: A Handbook for Caregivers.* New York: Mosby, 1998. Causes, symptoms, stages, and treatment options for AD are discussed.

Leon, J., C. Cheng, and P. Neumann. "Alzheimer's Disease Care: Costs and Potential Savings." *Health Affiliates* (November/December, 1998): 206-216. Identifies the economic impact for caring for and treating those with AD and reasons for identifying a cure.

Mace, M., and P. Rabins. *The Thirty-six Hour Day: A Family Guide to Caring for Persons with Alzheimer Disease, Related Dementing Illnesses, and Memory Loss in Later Life.* Baltimore: Johns Hopkins University Press, 1999. Discusses what dementia is, physical and psychological effects on caregivers, financial and legal issues, and long-range care planning for Alzheimer's patients.

Powell, L., and K. Courtice. *Alzheimer's Disease: A Guide for Families and Caregivers.* Cambridge, Mass.: Perseus, 2001. Provides information about early signs, tests, diagnosis, and treatment for Alzheimer's disease.

St. George-Hyslop, Peter H. "Piecing Together Alzheimer's." *Scientific American* (December, 2000): 76-83. Good description of AD, including symptoms, support, and ongoing research in the quest for a cure.

Terry, R., R. Katzman, K. Bick, and S. Sisodia. *Alzheimer Disease.* 2d ed. Philadelphia: Lippincott Williams & Wilkins, 1999. An in-depth review of hereditary links, signs and symptoms, diagnosis, and treatment for Alzheimer's disease.

Web Sites of Interest

Alzheimer's Association. http://www.alz.org. This site provides a two-page genetics fact sheet and information about the Alzheimer's Disease Genetics Initiative, a study conducted by the Alzheimer's Association and the National Institute of Aging.

Alzheimer's Disease Education and Referral Center, National Institutes of Health. http://www.alzheimers.org. A good general starting place for information and links to standard resources. Includes a detailed page on the genetics of AD.

Dolan DNA Learning Center, Your Genes Your Health. http://www.ygyh.org. Sponsored by the Cold Spring Harbor Laboratory, this site, a component of the DNA Interactive Web site, offers information on more than a dozen inherited diseases and syndromes, including Alzheimer's disease.

Amniocentesis and Chorionic Villus Sampling

Field of study: Techniques and methodologies

Significance: *Amniocentesis is a procedure for removing amniotic fluid and fetal cells from a pregnant women. Chorionic villus sampling is a procedure to obtain fetal cells from placental tissue. These procedures, which can detect genetic disorders in the fetus, have broadened the possibilities for effective genetic counseling and for improving the child's health.*

Key terms

AMNIOTIC FLUID: the fluid in which the fetus is immersed during pregnancy

CHORIONIC VILLI: the fingerlike projections of the placenta that function in oxygen, nutrient, and waste transportation between a fetus and its mother

PRENATAL TESTING: testing that is done during pregnancy to examine the chromosomes or genes of a fetus to detect the presence or absence of a genetic disorder

Goals of Testing

The goal of prenatal testing is to provide at-risk families with information about the chances of having a child with a specific genetic disorder or birth defect. Only a small minority of such disorders can now be detected, but the list continues to grow.

The primary techniques for prenatal testing include amniocentesis, chorionic villus sampling (CVS), ultrasonography (in which high-frequency sound waves are used to "view" the fetus and obtain information about its position and structure), fetoscopy (a procedure that uti-lizes a fiber-optic instrument to obtain a direct image of the fetus), fetal blood sampling (in which blood cells of the fetus are obtained by inserting a needle directly into the umbilical cord), and screening for alpha fetoprotein (a fetal protein found in amniotic fluid, high levels of which may indicate the presence of neural tube defects).

Amniocentesis and chorionic villus sampling are not recommended for every pregnancy. Although the two procedures are relatively safe, they do not carry a zero risk factor and are not likely to be employed unless the risk of the procedure is lower than the risk factor for a birth defect in a specific pregnancy. The general risk for having a child with a significant birth defect is about 2 to 3 percent. Among the factors that indicate an increased risk of having a child with a birth defect are maternal age (the incidence of chromosomal defects in children increases sharply in pregnant women over age thirty-five), a previous child with a known chromosomal or genetic disorder, previous problems with spontaneous abortions or miscarriages, a history of genetic defects in the siblings or other relatives in one or both of the parents, a previous child with a neural tube defect, and marriage between closely related individuals such as first or second cousins.

Amniocentesis

Amniocentesis has been used safely and widely since 1967 and is used more often than other methods of prenatal testing. The procedure is usually performed on an outpatient basis between the fourteenth and eighteenth week of gestation. By this stage in the pregnancy, the volume of amniotic fluid is large enough to get an adequate sample. Also, it allows sufficient time for testing to be done in the laboratory, minimizing complications if it becomes necessary to perform a therapeutic abortion.

The skin of the abdomen is scrubbed, and a topical anesthetic may be applied. The exact location of the placenta and fetus is determined by ultrasound. A long, thin needle is inserted through the abdominal wall into the amniotic sac that encloses the fetus. A small amount of amniotic fluid is withdrawn. This fluid contains cells that have been sloughed off by the fetus.

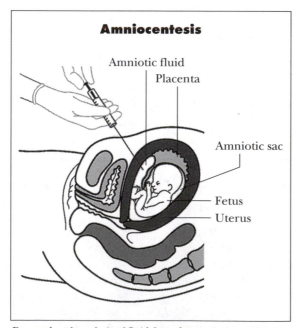

Amniocentesis

Amniotic fluid

Placenta

Amniotic sac

Fetus

Uterus

Removal and analysis of fluid from the amniotic sac that surrounds a fetus during gestation can be used to rule out or confirm the presence of serious birth defects or genetic diseases. (Hans & Cassidy, Inc.)

The cells must be cultured in the laboratory to produce a sufficient number for testing. A variety of chromosomal analyses and biochemical tests can then be carried out. Any sort of numerical chromosomal abnormality, such as Turner syndrome and Down syndrome, can be detected. Structural abnormalities of chromosomes, such as missing or extra pieces, also can be detected. Cri du chat syndrome is one such genetic disorder. Biochemical and DNA analyses can be carried out on the cells, and some specific genetic disorders can be detected in this manner. The biochemical assays are used to detect low levels of particular enzymes involved in specific biochemical defects. Although most genetic disorders cannot be diagnosed in this manner, the list is rapidly growing. Examples of some inborn errors are galactosemia, Hurler's syndrome, and Lesch-Nyhan syndrome.

It is also possible to analyze DNA directly in cells obtained from amniocentesis. Using specific genetic probes, it is possible to identify mutant genes associated with specific genetic diseases such as sickle-cell disease, cystic fibrosis, hemophilia A, and Duchenne muscular dystrophy. This approach is not yet possible for many genes since it is necessary to know the DNA sequence of the gene involved.

Chorionic Villus Sampling

Although amniocentesis has been a successful prenatal testing procedure, it does present some disadvantages. Perhaps the major disadvantage involves the need to perform it during the sixteenth week of pregnancy, which provides a fairly narrow window in which cells can be grown in culture, tests can be carried out, and procedures can be replicated, if necessary. If procedures run past the nineteenth or twentieth week of pregnancy, the physical and psychological complications associated with a late termination of pregnancy rise considerably. The technique of chorionic villus sampling addresses some of these problems.

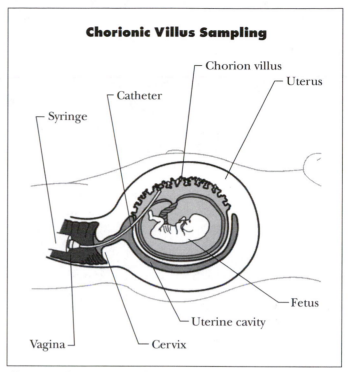

Chorionic villus sampling is one method of obtaining embryonic cells from a pregnant woman. Examination of these cells helps physicians determine fetal irregularities or defects, which allows time to assess the problem and make recommendations for treatment. (Hans & Cassidy, Inc.)

Like amniocentesis, chorionic villus sampling can be performed on an outpatient basis. After cleansing the vagina and cervix with an antiseptic, the physician uses ultrasound to guide the insertion of a catheter (a small, thin tube) through the cervix into the uterus. The catheter is placed in contact with the placenta, where the chorionic villi are located, and gentle suction is used to remove a small sample. Villus cells are produced by the fetus and comprise one of the outer layers of the placenta. Of major importance is the fact that this small sample of tissue contains millions of cells that can be used immediately for testing (recall that cells obtained during amniocentesis must be grown for a week or longer before testing can be done). This means that chromosomal analyses and some biochemical tests can be performed and results given to the patient before she leaves the physician's office. Chorionic villus sampling is usually performed between the ninth and twelfth weeks of pregnancy so that complete results are likely to be reported one month earlier than for amniocentesis. A termination of pregnancy after chorionic villus sampling is expected to have fewer complications than a termination performed at a later stage in pregnancy.

Impact and Applications

Both amniocentesis and chorionic villus sampling provide significant information to couples at risk for having a child with a genetic disorder or other type of birth defect. It is estimated that approximately one-half of the women over the age of thirty-five who are pregnant utilize amniocentesis or chorionic villus sampling. Thousands of women undergo some form of prenatal genetic testing each year. Amniocentesis increases the risk of spontaneous abortion only about 0.5 percent above the overall general risk. Chorionic villus sampling probably carries an increased risk of miscarriage of 1 to 2 percent above the overall general risk. It should be kept in mind, however, that these two techniques are usually not performed unless there is some additional risk already present in a specific pregnancy.

The techniques of amniocentesis and chorionic villus sampling may assure parents at risk of having a child with a genetic disorder that their child will be born without the disorder. Results may also be such that parents must be told that their child will definitely have a certain disorder. However, even if a couple elects to continue with a pregnancy, the tests provide useful information about the nature of the disorder and about treatments that might be used after birth to prepare for raising a child with a birth defect. It is to be expected that further developments in techniques will dramatically improve the convenience and safety of prenatal testing.

—Donald J. Nash

See also: Alcoholism; Bioethics; Congenital Defects; Cystic Fibrosis; Down Syndrome; Dwarfism; Eugenics; Fragile X Syndrome; Genetic Counseling; Genetic Engineering: Social and Ethical Issues; Genetic Screening; Genetic Testing; Genetic Testing: Ethical and Economic Issues; Hemophilia; Hereditary Diseases; Inborn Errors of Metabolism; Neural Tube De-

fects; Prader-Willi and Angelman Syndromes; Prenatal Diagnosis; RFLP Analysis; Sickle-Cell Disease; Tay-Sachs Disease; Thalidomide and Other Teratogens; Turner Syndrome.

Further Reading

Cummings, Michael R. *Human Heredity: Principles and Issues.* 5th ed. Pacific Grove, Calif.: Brooks/Cole, 2000. A good introduction to human genetics for nonspecialists.

Heyman, Bob, and Mette Henriksen. *Risk, Age, and Pregnancy: Case Study of Prenatal Genetic Screening and Testing.* New York: Palgrave, 2001. Offers a detailed case study of a prenatal genetic screening and testing system in one hospital, integrating observational, survey, and qualitative interview methods to provide insight into the risk management dilemmas faced by those who attempt to control genetic futures.

Parry, Vivienne. *Antenatal Testing Handbook: The Complete Guide to Testing in Pregnancy.* Collingdale, Pa.: DIANE, 1998. Provides a readable yet in-depth discussion of each antenatal test, including amniocentesis, examining what is available, what the tests are for, and the risks associated with each. Each test is explained in detail. An index of genetic diseases lists the most common handicapping conditions, together with their causes (where known) and their detection rates by method and stage of pregnancy.

Rapp, Rayna. *Testing Women, Testing the Fetus: The Social Impact of Amniocentesis in America.* New York: Routledge, 1999. Examines amniocentesis from an anthropological/sociological perspective, using dozens of anecdotes from women who have undergone the procedure.

Rothman, Barbara Katz. *The Tentative Pregnancy: How Amniocentesis Changes the Experience of Motherhood.* New York: Norton, 1993. Discusses how amniocentesis radically alters the way parents think about childbirth and parenthood, forcing them to confront an array of agonizing dilemmas. Surveys technological advances in the procedure and includes two appendices, "Using Earlier Tests" (on early amniocentesis and chorionic villus sampling) and "Maternal Serum Alpha Protein Testing."

Zallen, Doris Teichler. *Does It Run in the Family? A Consumer's Guide to DNA Testing for Genetic Disorders.* New Brunswick, N.J.: Rutgers University Press, 1997. A geneticist and science policy expert describes developments and applications, as well as the medical, psychological, and social implications, of genetic testing. Designed to help health care consumers ask the appropriate questions and determine what answers testing can, and cannot, provide. Glossary, lists of Web sites and support organizations.

Web Sites of Interest

American College of Obstetricians and Gynecologists. http://www.acog.org. Offers a wealth of information on procedures, conditions, and ethical considerations. Searchable by keyword

Association of Women's Health, Obstetric, and Neonatal Nurses. http://www.awhonn.org. Offers pages for education and practice resources as well as legal policy.

March of Dimes. http://www.marchofdimes.com. This site is searchable by keyword and includes information on the basics of amniocentesis and chorionic villus sampling and articles on how the two procedures relate to genetics.

Ancient DNA

Fields of study: Evolutionary biology; Molecular genetics

Significance: *Since the development of the polymerase chain reaction (PCR), it has become possible to amplify DNA sequences from extremely small samples and from samples that are greatly degraded, which has kindled interest in studying ancient DNA samples and comparing their gene sequences with those of related modern organisms. Attempts have been made to study DNA sequences from organisms trapped in ice, from mummified individuals, and even from fossils in which some of the original tissue remains. Results have so far been mixed, and gene sequences from some of the older samples have been discredited in many cases.*

Key terms

DNA POLYMERASE: the enzyme that produces a complementary strand using a DNA template

PRIMER: an oligonucleotide (short strand of nucleotides typically 18-30 bases long) used as a starting point for Taq polymerase to make a complementary copy of a DNA template strand; two primers are needed that flank the DNA sequence being amplified

TAQ POLYMERASE: DNA polymerase originally isolated from the hot spring bacteria *Thermus aquaticus*, which remains stable at temperatures close to boiling (100 degrees Celsius, or 212 degrees Fahrenheit); used as the DNA polymerase in PCR reactions

THERMAL CYCLER: a machine that can rapidly heat and cool reaction tubes; used for performing PCR reactions

Gene Amplification via Polymerase Chain Reaction

Anthropologists and taxonomists have long wished that they could go back in time, collect a few samples, and return to the present so they could compare past organisms with those alive today. Until recently, the best samples available were organisms frozen long ago, mummified, or fossilized. Such samples allowed for superficial comparisons of morphology (physical characteristics), but not much else. Comparisons among extant organisms, including humans, using DNA or protein sequences have now been done for half a century, but until the 1980's ancient samples typically yielded too little material to be useful.

In 1985, R. K. Saiki and Kary B. Mullis amplified the first gene using the polymerase chain reaction (PCR). This technique allows production of billions of copies of a specific DNA sequence, even with very small amounts of template DNA. PCR caused a revolution in the molecular genetics laboratory and was soon embraced by those studying ancient humans and other organisms. So much interest was spawned by the use of PCR to study ancient DNA samples that in 1993 filmmaker Stephen Spielberg directed and produced the movie *Jurassic Park*, based on the 1990 book of the same title by Michael Crichton. Although the science in *Ju-*

rassic Park goes well beyond what is possible, the film brought the idea of resurrecting ancient DNA samples before the general public.

Problems of Isolation and Analysis

The two greatest concerns when attempting to isolate ancient DNA are the degree to which it has degraded and potential contamination with more recent DNA. The amount of DNA degradation is a function of the age of the sample and the conditions under which it was preserved. Samples a few thousand years old will typically yield very usable DNA, whereas samples exceeding 100,000 years in age may be so degraded, regardless of preservation conditions, that no meaningful data can be obtained. Environmental conditions at the time the organism died also play a large part. A woolly mammoth frozen shortly after death will yield much more and better-preserved DNA than will the bones of a turtle weathered for many months or years before being buried. Insects trapped in amber would seem to have very well preserved DNA, as they were likely encased while still alive, but the age of such fossils may be so great that the DNA is still degraded beyond usefulness.

Of probably greater concern than DNA degradation is contamination with modern DNA. All fossils and ancient remains have the potential to be contaminated with all manner of more recent molecules of DNA, from pollen or bacteria to mold or the skin cells of the person extracting the DNA. Even minute quantities of DNA contamination can ruin the results of PCR analysis, which is capable of amplifying a DNA sequence with only a few template molecules from which to read. For these reasons, the conditions to which the sample was exposed prior to collection must be taken into account. If the sample was encased in ice or rock, the probability of contamination is minimal, whereas if the sample was exposed to the air, contamination can almost be assured. Thus, when collecting a sample, the researcher must immediately place it into an airtight, sterile container.

Once DNA extraction begins, the specimen must be completely protected from contamination by DNA from the researcher or anything else in the environment. The work area must be

sterile and the work surfaces are commonly exposed to damaging ultraviolet (UV) rays for sixty minutes or more in order to destroy any foreign DNA. Researchers wear sterile gloves and face masks, and all laboratory equipment and containers that come in contact with the samples or solutions must be sterile and are often exposed to UV light as well. A further precaution is to analyze ancient and modern DNA samples in completely different laboratories. Finally, after the DNA has been isolated, it must still be kept from subsequent opportunities for contamination.

The first criticism typically leveled against results of ancient DNA analyses is the potential for contamination. Guidelines have been proposed, but researchers have been slow in implementing them, as they are difficult to follow. There is still considerable debate about the veracity of many of the studies so far published, and some DNA initially believed to be ancient was later found to originate in contamination from contemporary sources. In spite of some of the problems with isolating and analyzing ancient DNA, it has produced some fruitful results.

Archaeological and Anthropological Discoveries

Probably the most fruitful area of ancient DNA analysis for archaeologists has been in the study of the origins of human diseases. Traditionally, determining whether an ancient human being suffered from a particular disease relied on circumstantial evidence: stunted growth, bone scars, and other circumstances subject to interpretation. Recovery of ancient bacterial DNA from Egyptian mummies has helped establish the presence of skeletal tuberculosis as long ago as about 3000 B.C.E.

Based on historical accounts, the plague, caused by the bacterium *Yersinia pestis*, has long been considered the cause of a repeated series of epidemics over the last two millennia. Unfortunately, without any adequate medical records to confirm plague epidemic accounts, confirmation is not possible. In 1998, researchers in France unearthed the skeletons of individuals from the sixteenth and eighteenth centuries that presumably died from the plague. Using

PCR to amplify a gene from *Y. pestis* from dental pulp, they obtained proof that the plague existed at the end of the sixteenth century in France. Other diseases that have been successfully identified in this way include leprosy, schistosomiasis, malaria, and Chagas' disease.

In 2003 archaeologists discovered a mummy, accompanied by two others, that might be the famous wife of Akhenaton and co-ruler of Egypt, Nefertiti. Extraction of DNA from all three mummies, to see if they are related, could help determine whether the mummy is Nefertiti. Although DNA evidence would not remove all doubt, added to circumstantial evidence presented by some anthropologists and historians, it would strengthen their case.

Archaeologists excavating an ancient Roman and early Byzantine site in Turkey found fish remains. When DNA was isolated and analyzed, it was discovered that the species of fish were not from the study area. Instead, they were fish that had to have come from Egypt or the Levant, hinting at potential ancient trade routes.

Dipping more deeply into the past, researchers have isolated DNA from the bones of ancient humans from Australia, Africa, Europe, and other parts of the world. The goal has been to analyze mitochondrial DNA (mtDNA) sequences and use them to help answer questions about human origins. This sort of DNA is preferred because it evolves at an appropriate rate and is more abundant in cells, so it is more likely to be present in ancient specimens. Although the results thus far have not settled some of the long-standing questions about human origins, they have shown that a remarkable similarity exists between ancient and modern DNA sequences.

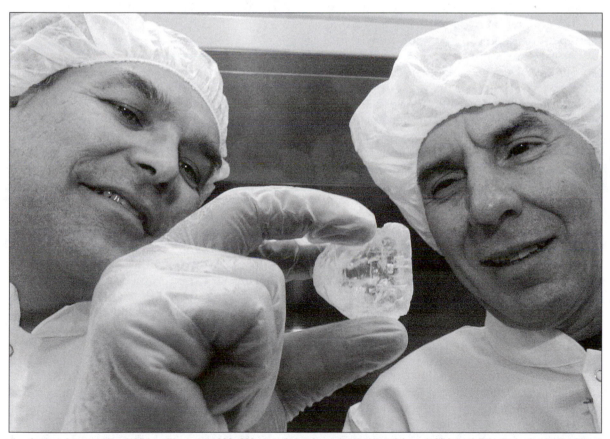

A salt crystal entrapping a 250-million-year-old bubble containing bacteria, excavated from 1,850 feet below ground near Carlsbad, New Mexico, offers scientists Russell Vreeland (left) and William Rosenzwieg the opportunity to study ancient DNA. (AP/Wide World Photos)

Ancient Dinosaur and Bacterial DNA

It had long been assumed that obtaining DNA from samples older than a few hundred thousand years would not be possible. DNA millions of years old was assumed to be so degraded that the fragments would be too small even for PCR to pull out sequences to analyze. In 1991, however, Edward Golenberg of Wayne State University isolated DNA from fossil magnolia leaves that were approximately 18 million years old. Just a year later, in 1992, researchers in California isolated DNA from a fossil bee trapped in amber and researchers at the American Museum of Natural History (New York City) isolated DNA from a termite trapped in amber. Although tantalizing, these studies have been questioned by those who suspect that the DNA that was isolated represents contamination rather than true ancient DNA. In some cases, reanalysis has shown that the supposed ancient DNA was an artifact or a contaminant. The likelihood that DNA could survive so long, even in amber, has also been questioned.

In 1994, Raúl Cano, at California Polytechnic State University, isolated 25-million-year-old DNA from bacteria in the gut of a bee trapped in amber. In 1999, Charles L. Greenblatt, at Hebrew University in Jerusalem, reported the isolation of DNA of several types of bacteria from 120-million-year-old amber. Comparisons of the DNA sequences of ribosomal RNA (rRNA) genes with modern bacteria revealed significant differences, lending support to the claim that Greenblatt and his colleagues had actually isolated ancient DNA and not contaminants.

A number of researchers have isolated DNA from dinosaur bones, ranging in age from 65 million to 80 million years old. The same doubts exist with these studies, particularly because there is no pure, undegraded DNA from ancient dinosaurs with which to compare the extracted DNA. Certainly, if the DNA is compared with bacterial or fungal DNA sequences and there is a match, then the DNA can be considered a contaminant. On the other hand, if the DNA resembles modern reptile or bird DNA, it may represent a contaminant as well but may also be dinosaur DNA, as claimed. Confidence in the results of ancient DNA analyses will come only as more samples are analyzed and DNA is shown to be capable of surviving so long, and even then there will still be room for skepticism. Working with such ancient samples and avoiding all possibility of contamination is extremely difficult.

Studying rates of evolution in a population over thousands of years has also been accomplished using ancient DNA. Adélie penguins have lived in colonies in the same areas of Antarctica for many thousands of years. Excavating various colony sites, researchers have collected partially fossilized bones covering a range of ages to almost seven thousand years old. By comparing the DNA sequences of a portion of mtDNA among the different aged samples and modern samples from live penguins, they have estimated evolutionary rates of change. Because these samples were only thousands of years old, the results are much more reliable than samples from fossils.

Future Prospects in Ancient DNA Research

Some limitation in the field of ancient DNA research will never be overcome, but as more research is done, methods of extraction and analysis of ancient DNA should improve. Also, as more samples are collected and analyzed, better methods for measuring the extent of DNA degradation and for determining whether contamination is present should be developed. However, doubt about the ability of DNA to survive longer than 100,000 years, in the best of conditions, has resulted in a drop in studies of very ancient DNA. It now appears that the most fruitful use of ancient DNA is from samples on the order of a few thousands to tens of thousands of years old, rather than DNA millions of years old. Archaeologists, anthropologists, and biologists studying relatively recent extinct animals will probably be the primary researchers analyzing ancient DNA.

—Bryan Ness

See also: DNA Structure and Function; Evolutionary Biology; Genetic Code; Molecular Clock Hypothesis; Mutation and Mutagenesis; Polymerase Chain Reaction; Punctuated Equilibrium; RNA Structure and Function; RNA World.

Further Reading

Arndt, Allan, et al. "Roman Trade Relationships at Sagalassos (Turkey) Elucidated by Ancient DNA of Fish Remains." *Journal of Archaeological Science* 30, no. 9 (2003): 1095-1105. Analysis of DNA isolated from ancient fish remains showed that the fish were from another location and must have represented foreign trade.

Desalle, Robert, and David Lindley. *The Science of "Jurassic Park" and "The Lost World."* New York: HarperCollins, 1998. A critical look at the science in the first two films based on Michael Crichton's novels, especially easy to read for nonscientists.

Herrmann, Bernd, and Susanne Hummel, eds. *Ancient DNA: Recovery and Analysis of Genetic Material from Paleographic, Archaeological, Museum, Medical, and Forensic Speciments.* New York: Springer-Verlag, 1994. Written when DNA fingerprinting was just coming to the fore and films such as *Jurrasic Park* were in theaters, this collection of papers by first-generation researchers reflects the broad applications of the technology, including paleontological investigations.

Hummel, Susanne. *Fingerprinting the Past: Research on Highly Degraded DNA and Its Applications.* New York: Springer-Verlag, 2002. Manual about typing ancient DNA.

Jones, Martin. *The Molecule Hunt: Archaeology and the Search for Ancient DNA.* New York: Arcade, 2002. A broad overview of the history of analyzing ancient DNA and other biological molecules. Includes a nice selection of the winners and losers in the hunt for ancient DNA.

Pennisi, Elizabeth. "A Shaggy Dog History." *Science* 298, no. 5598 (2003): 1540-1542. Reports on discoveries using DNA extracted from ancient dog remains that point to an Old World origin for New World dogs.

Wayne, Robert K., Jennifer A. Leonard, and Alan Cooper. "Full of Sound and Fury: The Recent History of Ancient DNA." *Annual Review of Ecology and Systematics* 30 (1999): 457-477. A comprehensive overview of what has happened in this highly controversial field. Includes some excellent examples of cases in which DNA originally considered ancient was found to be recent.

Animal Cloning

Field of study: Genetic engineering and biotechnology

Significance: *Animal cloning is the process of generating a genetic duplicate of an animal starting with one of its differentiated cells. Sheep, mice, cattle, goats, pigs, and a cat have been cloned. While currently an inefficient process that may pose risks to the clone, animal cloning offers the benefits of replicating valuable animals.*

Key terms

ASEXUAL REPRODUCTION: reproduction not requiring fusion of haploid gametes as a first step

CLONE: a genetic replica of a biological organism

DIFFERENTIATED CELL: a somatic cell with a specialized function

MITOCHONDRIAL GENOME: DNA found in mitochondria, coding for forty genes, involved in energy metabolism, and maternally inherited

NUCLEAR GENOME: DNA found in the nucleus, coding for 30,000 genes in higher organisms, half inherited from each parent

TELOMERE: a specialized structure at the chromosome end, which shortens in somatic cells with age

Clones and Cloning

Asexual reproduction occurs in numerous bacteria, fungi, and plants, as well as some animals, leading to genetically identical offspring or clones. In addition, humans can assist in such reproduction. For instance, cuttings from plants generate thousands of replicates. Dividing some animals, such as earthworms or flatworms, allows them to regenerate. However, most vertebrates, including all mammals, reproduce sexually, requiring fertilization of an ovum by sperm. In such species, clones occur, as in the case of identical twins, when an embryo splits into two early in development. This process can be instigated artificially using microsurgical techniques to divide a harvested early-stage embryo and reimplanting the halves into surrogate dams (mothers). While this can be considered animal cloning, the term should be reserved for cloning from nonembryonic cells.

Cloning Procedure

Animal cloning typically refers to mammals or other higher vertebrates and involves creating a duplicate animal starting from a differentiated cell. Although such a cell only has the ability to perform its specialized function, its nucleus retains all genetic information for the organism's development. Animal cloning requires that such information be reprogrammed into an undifferentiated cell that can re-initiate the developmental process from embryo to birth and beyond.

In theory, the process is straightforward. It consists of taking a differentiated cell from an adult animal, inserting its diploid nucleus into a donor ovum whose own haploid nucleus has been removed, initiating embryonic development of this ovum, inserting the resultant embryonic mass into a receptive surrogate dam and allowing it to proceed to term. In practice, the technique is difficult and was thought to be impossible until 1997. It also appears fraught with species specificity. Various differentiated cells have been used as the starting source; mammary cells were used in the first case, while skin fibroblasts and cumulus cells are now often used. The preparation of the anucleate ovum is an important step. A limitation to clon-

Dolly the Sheep

In 1997, the world was taken aback when a group of scientists headed by embryologist Ian Wilmut at the Roslin Institute in Scotland announced the successful cloning of a sheep named Dolly. Scientists had already cloned cows and sheep, but they had used embryo cells. Dolly was the first vertebrate cloned from the cell of an adult vertebrate.

The feat was accomplished by removing cells from the udder of a six-year-old ewe and placing them in a laboratory dish filled with nutrients, where they were left to grow for five days. Then the nutrients were reduced to 5 percent of what the cells needed to continue growing, which caused the cells to enter a state resembling suspended animation, making them more receptive to becoming dedifferentiated. When the nuclei of these cells were placed in the ova of host sheep, the cytoplasm of each ovum directed the nucleus it received to enter an undifferentiated state, thus causing the cell to develop into an embryo.

Of an initial 277 adult cells introduced into sheep ova, thirteen resulted in pregnancy, and only one, Dolly, was carried to full term. Dolly was a genetic replica of the sheep from whose udder the original cells were extracted. Environmental factors would make Dolly, like any other clone, individual, but genetically she would never have the individuality that an organism produced by usual reproductive means would possess. Over the next six years, she gave birth to several, apparently healthy, offspring. In 2002, at the age of six, Dolly became lame in her left hind leg, a victim of arthritis. Although sheep commonly suf-

fer arthritis, a veterinarian noted that both the location and the age of onset were uncommon. Then, in February, 2003, she was euthanized after the discovery of a progressive lung disease.

Dolly's health problems have led to speculations about premature aging in clones but are complicated by her unique experiences as well. As Wilmut noted, in the early years following the announcement of her cloning, she became something of a celebrity, which led to overfeeding by visitors and in turn a period of obesity, later corrected. More significant were the discovery of her arthritis and then her lung disease—conditions not uncommon in sheep but that tend to emerge later (sheep typically live to be eleven or twelve years old).

Theories of premature aging are supported by the fact that Dolly's telomeres were shorter than normal. These cell structures function as "caps" that prevent "fraying" at the ends of DNA cells. As a cell ages, its telomeres become progressively shorter, until finally they disappear altogether and are no longer able to protect the cell, which then dies.

Was Dolly older genetically than she was chronologically? The answer to the question of whether Dolly was completely "normal" or aged prematurely as a result of being a clone must await full investigation of her autopsy results, as well as tracking of her offspring's lives and monitoring of other vertebrate clones through their life spans.

—*R. Baird Shuman, updated by Christina J. Moose*

Dolly, the first animal cloned from an adult vertebrate cell, in 1997. She was euthanized in 2003 after complications from advanced arthritis and lung disease. (AP/Wide World Photos)

influence phenotypic expression. In addition, the prenatal environment can affect some traits. Coat color and color pattern are characteristics that can be developmentally influenced; the first cloned cat was not an exact duplicate of its progenitor in coloration. Some behavioral features are also impacted during intrauterine development.

Cloned Animals

The first cloned animal was a sheep named Dolly. While she was the only live offspring generated from 277 attempts, her birth showed that animal cloning was possible. Shortly thereafter, mice and cattle were cloned. Reproducible cloning of mice is more difficult than imagined, whereas more cattle were cloned in the first five years after Dolly's birth than any other species. Goats, pigs, and a cat were subsequently also cloned.

ing dogs appears to be the difficulty in obtaining ova suitable for nuclear transfer. The technique for inserting the nucleus is crucial, as is the conversion to the undifferentiated embryonic state. Transfer of the embryonic cells to a receptive surrogate dam is generally a well-developed technology, although more than four viable embryos are necessary to maintain pregnancy in pigs.

Furthermore, the genetic makeup of a putative clone must be verified, to ensure that it is indeed a replica of its progenitor and not an unintended offspring of either the donor of the ovum or the surrogate dam. DNA fingerprinting via microsatellite analysis at a number of polymorphic sites is an unambiguous way to establish its genetic identity.

Identicalness

Such a clone is not absolutely identical, because of mitochondrial differences and environmental effects. While the nuclear genome must be identical to its progenitor, the mitochondrial genome of the clone will invariably be different, because it comes from the ovum used. While mitochondria make a minor contribution to the total genetic makeup, they can

Problems and Potential Benefits

Prominent among the problems with animal cloning is its inefficiency. Although this may not be surprising as the technology is still under development, no more than 2 percent of embryos generated lead to viable offspring. Additionally, most cloned animals are larger than normal at birth, often requiring cesarian delivery, and some have increased morbidity and mortality. Some have had smaller telomeres and shorter lives. Dolly exhibited this trait and lived for only six years (although she was euthanized, she clearly would not have lived much longer)—half of the average life span. Conversely, some cloned mice do not exhibit shortened telomeres or premature aging, even through six consecutive cloned generations. Further research will establish whether these problems are inherent to cloning, are consequences of some aspect of the current procedure or are attributable to the small numbers of cloned animals studied.

At a news conference in July, 1998, Dr. Ryuzo Yanagimachi holds in his hands the first cloned mouse and its "parent." The cloning occurred nearly simultaneously with the cloning of Dolly the sheep. (AP/Wide World Photos)

The benefits of animal cloning would involve duplicating particularly valuable animals. Livestock with highly valued production characteristics could be targets for cloning. However, the technique is likely to be most beneficial in connection with transgenesis, to replicate animals that yield a therapeutic agent in high quantities or organs suitable for transplantation into humans. If animal cloning can be made efficient and trouble-free, its potential benefits could be fully developed.

—*James L. Robinson*

See also: Biopharmaceuticals; cDNA Libraries; Cloning; Cloning: Ethical Issues; Cloning Vectors; DNA Replication; DNA Sequencing Technology; Genetic Engineering; Genetic Engineering: Agricultural Applications; Genetic Engineering: Historical Development; Genetic Engineering: Industrial Applications; Genetic Engineering: Medical Applications; Genetic Engineering: Risks; Genetic Engineering: Social and Ethical Issues; Knockout Genetics and Knockout Mice; Mitochondrial Genes; Parthenogenesis; Polymerase Chain Reaction; Restriction Enzymes; Reverse Transcriptase; Shotgun Cloning; Telomeres; Transgenic Organisms; Xenotransplants.

Further Reading

Patterson, Lesley, William Richie, and Ian Wilmut. "Nuclear Transfer Technology in Cattle, Sheep and Swine." In *Transgenic Animal Technology, A Laboratory Handbook*, edited by Carl A. Pinkert. 2d ed. London: Academic Press, 2002. The detailed protocol needed to clone three livestock species, as well as the limitations to increased efficiency.

Pennisi, Elizabeth, and Gretchen Vogel. "Animal Cloning: Clones: A Hard Act to Follow." *Science* 288 (2000): 1722-1727. The status of animal cloning, three years after the announcement of Dolly. The problems, questions, and concerns are presented in a highly readable text.

Wilmut, Ian, Keith Campbell, and Colin Tudge. *The Second Creation: The Age of Biological Control by the Scientists That Cloned Dolly.* London: Headline, 2000. The story of the scientific collaboration between an agricultural scientist and a cell biologist, of their perseverance and the serendipity that led to the first cloned sheep, Dolly.

Web Sites of Interest

ActionBioScience.org. http://www.actionbio science.org/biotech/pecorino.html. Features the article "Animal Cloning: Old MacDonald's Farm Is Not What It Used To Be" and several useful links to the animal cloning debate.

Roslin Institute. http://www.roslin.ac.uk. The site of the oldest cloning group in the world, founded in 1919, which cloned the first animal, Dolly the sheep. Includes information on genomics and animal breeding.

Anthrax

Field of study: Bacterial genetics

Significance: *Anthrax has plagued humankind for thousands of years. Naturally occurring anthrax spores have caused disease in livestock and wildlife more often than in humans, but with the rise of genetic technologies anthrax has become amenable to manipulation as an agent of bioterrorism and biowarfare.*

Key terms

PLASMIDS: extrachromosomal DNA, found most commonly in bacteria, which can be transferred between bacterial cells

POLYMERASE CHAIN REACTION (PCR): a process in which a portion of DNA is selected and repeatedly replicated

SINGLE NUCLEOTIDE POLYMORPHISM (SNP): the difference in a single nucleotide between the DNA of individual organisms

VARIABLE NUMBER TANDEM REPEAT (VNTR): the difference in the number of tandem repeats (short sequences of DNA repeated over and over) between the DNA of individual organisms

History

A disease killing cattle in 1491 B.C.E., likely to have been anthrax, is recounted in the Book of Genesis. In Exodus 9, the Lord instructs Moses to take "handfuls of ashes of the furnace" and "sprinkle it toward the heaven in the sight of the Pharaoh." Moses performed the deed and "it became a boil breaking forth with blains upon man and upon beast." This may represent the first use of anthrax as a biological weapon. Greek peasants tending goats suffered from anthrax; the Greek word from which "anthrax" derives means coal, referring to the coal-black center of the skin lesion.

Anthrax became the first pathogenic bacillus to be seen microscopically when described in infected animal tissue by Aloys-Antoine Pollender in 1849. Studies by Robert Koch in 1876 resulted in the four postulates that form the basis for the study of infectious disease causation. In 1881, Louis Pasteur demonstrated the protective efficacy of a vaccine for sheep made with his attenuated vaccine strain.

The Disease

Anthrax is primarily a disease of herbivorous animals that can spread to humans through association with domesticated animals and their products. Herbivorous animals grazing in pastures with soil contaminated with anthrax endospores become infected when the spores gain entry through abrasions around the mouth and germinate in the surrounding tissues. Omnivores and carnivores can become infected by ingesting contaminated meat. Human infection is often a result of a close association with herbivores, particularly goats, sheep, or cattle (including their products of hair, wool, and hides).

The most common clinical illness in humans is skin infection (cutaneous anthrax), acquired when spores penetrate through cuts or abrasions. After an incubation period of three to five days, a papule develops, evolves into a vesicle, and ruptures, leaving an ulcer that dries to form the characteristic black scab. Inhaled spores reach the alveoli of the lung, where they are engulfed by macrophages and germinate into bacilli. Bacilli are carried to lymph nodes, where release and multiplication are followed by bloodstream invasion and the infection's spread to other parts of the body, including the brain, where it causes meningitis. The symptoms of the illness, which begin a few days after inhalation, resemble those of the flu and may be associated with substernal discomfort. Cough, fever, chills, and respiratory distress with raspy, labored breathing ensue. The least common type of infection is that of the gastrointestinal tract.

An effective vaccine is available for prevention, and antibiotics have been used when immediate protection is needed. Antibiotics can also successfully treat the infection. Inhalational anthrax is nearly always fatal if untreated, and even with treatment the mortality ranges from 40 to 80 percent. Mortality from treated cutaneous anthrax is less than 1 percent.

The Anthrax Bacterium

The *Bacillus anthracis* bacterium is large (1-1.2 × 3-10 microns), encapsulated, gram-positive, and rod-shaped. It produces spores and exotoxins (toxins that are released from the cells). Spores are ellipsoidal or oval (1-2 microns) and located within the bacilli. The endospores have no reproductive significance, as only one spore is formed by each bacillus and a germinated spore yields a single bacillus. Spores form in soil and dead tissue and with no measurable metabolism may remain dormant for years. They are resistant to drying, heat, and many disinfectants.

The genetic composition of *B. anthracis* differs little from the other *Bacillus* species, and studies have demonstrated remarkable similarity within *B. anthracis* strains. The resting stage of sporulation may have contributed to the extremely similar DNA of all strains of *B. anthracis*. The circular chromosomal DNA is com-

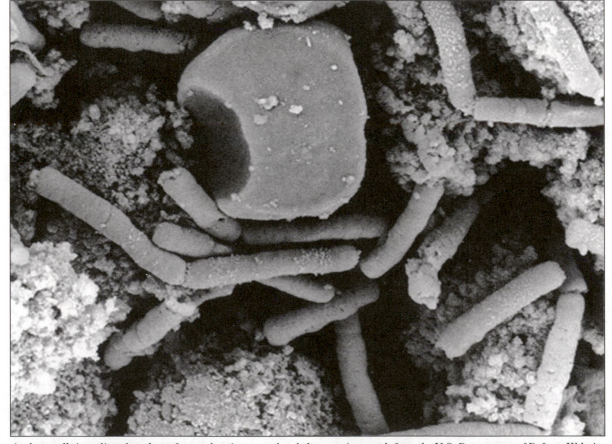

Anthrax cells invading the spleen of a monkey, in an undated electronmicrograph from the U.S. Department of Defense Web site. The United States and the Soviet Union developed biological weapons during the Cold War, and many of these lethal organisms remain housed in facilities around the world. (AP/Wide World Photos)

posed of 5.2 million base pairs and codes for metabolic function, cell repair, and the sequential process of sporulation. Comparative genome sequencing has uncovered only four differences between the single-copy chromosomal DNA of two strains. In addition to the single-copy DNA, comprising the majority of the genome, a remaining portion consists of repetitive DNA sequences that are either dispersed or clustered into satellites. The satellite repeats occur in tandem. The number of tandem repeats varies among different strains; six chromosomal marker loci have been identified by multiple-locus variable number tandem repeat (VNTR) analysis.

In addition to its chromosome, *B. anthracis* has two large plasmids that carry genes necessary for pathogenesis. The pXO1 plasmid has 181,654 base pairs and contains the structural genes for the anthrax toxins *cya* (edema factor), *lef* (lethal factor), and *pagA* (protective antigen). The pXO2 plasmid consists of 96,231 base pairs and carries three genes required for synthesis of the capsule. These plasmids contain a much greater number of single nucleotide polymorphisms (SNPs) and VNTRs among strains than the chromosomal genome. There are a variety of reference strains, such as Pasteur (which lacks the XO1 plasmid), Sterne (which lacks the XO2 plasmid), and Ames (which has both plasmids and is fully virulent).

Bioterrorism

Anthrax spores can be easily packaged to act as aerosoled (airborne) agents of war, and the genome may be bioengineered to alter the virulence or effectiveness of current vaccines. Knowledge of the genetic composition of *B. anthracis* has facilitated the investigation of anthrax attacks. In 1993, the Aum Shinrikyo cult aerosoled a suspension of anthrax near Tokyo, Japan. Molecular studies of the genome from this strain revealed it to be devoid of the pXO2 plasmid (Sterne strain), explaining why only a bad odor rather than illness was the fortunate consequence. In 2001, analysis of material from letter-based attacks with anthrax in the United States demonstrated the source to be the Ames strain. Furthermore, as a result of the extensive laboratory studies associated with these attacks, a sensitive and specific three-target (two-plasmid and one-chromosome) assay has been developed for rapid detection and identification of *B. anthracis*, including bioengineered strains, from both patients and the environment.

—H. Bradford Hawley

See also: Bacterial Genetics and Cell Structure; Bacterial Resistance and Super Bacteria; Biological Weapons; Plasmids; Smallpox.

Further Reading

Centers for Disease Control and Prevention. *Bioterrorism-Related Anthrax: Emerging Infectious Diseases* 8 (October, 2002): 1013-1183. Summarizes the investigation following the 2001 bioterrorism attacks in the United States.

Dixon, Terry C., et al. "Anthrax." *New England Journal of Medicine* 341 (September 9, 1999): 815-826. Details the disease and its pathogenesis.

Miller, Judith, Stephen Engelberg, and William Broad. *Germs: Biological Weapons and America's Secret War.* New York: Simon & Schuster, 2001. This book, written by three *New York Times* reporters, explores the ideas and actions of scientists and politicians involved in the past, present, and future of germ warfare. Forty-two pages of notes and a select bibliography.

Read, Timothy D., et al. "Comparative Genome Sequencing for Discovery of Novel Polymorphisms in *Bacillus anthracis*." *Science* 296 (June 14, 2002): 2028-2033. Describes the complete sequencing of the anthrax genome.

Web Sites of Interest

Centers for Disease Control, Public Health Emergency Preparedness and Response. http://www.bt.cdc.gov. This comprehensive site offers information on how to recognize illness caused by anthrax exposure and more. Available in Spanish.

Nature. http://www.nature.com. The online version of the premier science journal *Nature* includes links to research articles on the genetics of anthrax, including "Focus on Anthrax."

Antibodies

Field of study: Immunogenetics
Significance: *Antibodies provide the main line of defense (immunity) in all vertebrates against infections caused by bacteria, fungi, viruses, or other foreign agents. Antibodies are used as therapeutic agents to prevent specific diseases and to identify the presence of antigens in a wide range of diagnostic procedures. Large quantities of antibodies have also been produced in plants for use in human and plant immunotherapy. Because of their importance to human and animal health, antibodies are widely studied by geneticists seeking improved methods of antibody production.*

Key terms

B CELLS: a class of white blood cells (lymphocytes) derived from bone marrow responsible for antibody-directed immunity

B MEMORY CELLS: descendants of activated B cells that are long-lived and that synthesize large amounts of antibodies in response to a subsequent exposure to the antigen, thus playing an important role in secondary immunity

HELPER T CELLS: a class of white blood cells (lymphocytes) derived from bone marrow that prompts the production of antibodies by B cells in the presence of an antigen

LYMPHOCYTES: types of white blood cells (including B cells and T cells) that provide immunity

PLASMA CELLS: descendants of activated B cells that synthesize and secrete a single antibody type in large quantities and also play an important role in primary immunity

Antibody Structure

Antibodies are made up of a class of proteins called immunoglobulins (Ig's) produced by plasma cells (descendants of activated B cells) in response to a specific foreign molecule known as an antigen. Most antigens are also proteins or proteins combined with sugars. Antibodies recognize, bind to, and inactivate antigens that have been introduced into an organism by various pathogens such as bacteria, fungi, and viruses.

The simplest form of antibody molecule is a Y-shaped structure with two identical, long polypeptides (substances made up of many amino acids joined by chemical bonds) referred to as "heavy chains" and two identical, short polypeptides referred to as "light chains." These chains are held together by chemical bonds. The lower portion of each chain has a constant region made up of similar amino acids in all antibody molecules, even among different species. The remaining upper portion of each chain, known as the "variable region," differs in its amino acid sequence from other antibodies. The three-dimensional shape of the tips of the variable region (antigen-binding site) allows for the recognition and binding of target molecules (antigens). The high-affinity binding between antibody and antigen results from a combination of hydrophobic, ionic, and van der Waals forces. Antigen-binding sites have specific points of attachment on the antigen that are referred as "epitopes" or "antigenic determinants."

Antibody Diversity

There are five classes of antibodies (IgG, IgM, IgD, IgA, and IgE), each having a distinct structure, size, and function (see the table headed "Classes, Locations, and Functions of Antibodies"). IgG is the principal immunoglobulin and constitutes up to 80 percent of all antibodies in the serum.

The human body can manufacture a limitless number of antibodies, each of which can bind to a different antigen; however, human genomes have a limited number of genes that code for antibodies. It has been proposed that random recombination of DNA segments is responsible for antibody variability. For example, one class of genes (encoding light chain) contains three regions: the L-V (leader-variable) region (in which each variable region is separated by a leader sequence), the J (joining) region, and the C (constant) region. In the embryonic B cells, each gene consists of from one hundred to three hundred L-V regions, approximately six J regions, and one C region. These segments are widely separated on the chromosome. As the B cells mature, one of the L-V regions is randomly joined to one of the J regions and the adjacent C region by a recombination event. The remaining segments are cut from the chromosome and subsequently destroyed, resulting in a fusion gene encoding a specific light chain of an antibody. In mature B cells, this gene is then transcribed and translated into polypeptides that form a light chain of an antibody molecule. Genes for the other class of light chains as well as heavy chains are also made up of regions that undergo recombination during B-cell maturation. These random recombination events in each B cell during maturation lead to the production of billions of different antibody molecules. Each B cell has, however, been genetically programmed to produce only one of the many possible variants of the same antibody.

Production of Antibodies: Immune Response

Immunity is a state of bodily resistance brought about by the production of antibodies against an invasion by an antigen. The immune response is mediated by white blood cells known as lymphocytes that are made in the bone marrow. There are two types of lymphocytes: T cells, which are formed when lymphocytes migrate to the thymus gland, circulate in the blood, and become associated with lymph nodes and the spleen; and B cells, which are formed in bone marrow and move directly to the circulatory and the lymph systems. B cells are genetically programmed to produce antibodies. Each B cell synthesizes and secretes only one type of antibody, which has the ability to recognize with high affinity a discrete region (epitope or antigenic determinant) of an antigen. Generally, an antigen has several different epitopes, and each B cell produces a set of different antibodies corresponding to one of the many epitopes of the same antigen. All of the antibodies in this set, referred to as "polyclonal" antibodies, react with the same antigen.

The immune system is more effective at controlling infections than the nonspecific defense response (bodily defenses against infection—such as skin, fever, inflammation, phagocytes, natural killer cells, and some other antimicrobial substances—that are not part of the immune system proper). The immune system has three characteristic responses to antigens: diverse, which effectively neutralizes or destroys various foreign invaders, whether they are microbes, chemicals, dust, or pollen; specific, which effectively differentiates between harmful and harmless antigens; and anamnestic,

---

Classes, Locations, and Functions of Antibodies

Class	Location	Functions
IgG	Blood plasma, tissue fluid, fetuses	Produces primary and secondary immune responses; protects against bacteria, viruses, and toxins; passes through the placenta and enters fetal bloodstream, thus providing protection to fetuses.
IgM	Blood plasma	Acts as a B-cell surface receptor for antigens; fights bacteria in primary immune response; powerful agglutinating agent; includes anti-A and anti-B antibodies.
IgD	Surface of B cells	Prompts B cells to make antibodies (especially in infants).
IgA	Saliva, milk, urine, tears, respiratory and digestive systems	Protects surface linings of epithelial cells, digestive, respiratory, and urinary systems.
IgE	In secretion with IgA, skin, tonsils, respiratory and digestive systems	Acts as receptor for antigens causing mast cells (often found in connective tissues surrounding blood vessels), to secrete allergy mediators; excessive production causes allergic reactions (including hay fever and asthma).

which has a memory component that remembers and responds faster to a subsequent encounter with an antigen. The primary immune response involves the first combat with antigens, while the secondary immune response includes the memory component of a first assault. As a result, humans typically get some diseases (such as chicken pox) only once; other infections (such as cold and influenza) often recur because the causative viruses mutate, thus presenting a different antigenic face to the immune system each season.

An antibody-mediated immune response involves several stages: detection of antigens, activation of helper T cells, and antibody production by B cells. White blood cells known as macrophages continuously wander through the circulatory system and the interstitial spaces between cells searching for antigen molecules. Once an antigen is encountered, the invading molecule is engulfed and ingested by a macrophage. Helper T cells become activated by coming in contact with the antigen on the macrophage. In turn, an activated helper T cell identifies and activates a B cell. The activated T cells release cytokines (a class of biochemical signal molecules) that prompt the activated B cell to divide. Immediately, the activated B cell generates two types of daughter cells: plasma cells (each of which synthesizes and releases approximately two thousand to twenty thousand antibody molecules per second into the bloodstream during its life span of four to five days) and B memory cells (which have a life span of a few months to a year). The B memory cells are the component of the immune memory system that, in response to a second exposure to the same type of antigen, produces antibodies in larger quantities and at faster rates over a longer time frame than the primary immune response. A similar cascade of events occurs when a macrophage presents an antigen directly to a B cell.

Polyclonal and Monoclonal Antibodies

Plasma cells originating from different B cells manufacture distinct antibody molecules because each B cell was presented with a specific portion of the same antigen by a helper T cell or macrophage. Thus a set of polyclonal antibodies is released in response to an invasion by a foreign agent. Each member of this group of polyclonal antibodies will launch the assault against the foreign agent by recognizing different epitopes of the same antigen. The polyclonal nature of antibodies has been well recognized in the medical field.

In the case of multiple myeloma (a type of cancer), one B cell out of billions in the body proliferates in an uncontrolled manner. Eventually, this event compromises the total population of B cells of the body. The immune system will produce huge amounts of IgG originating from the same B cell, which recognizes only one specific epitope of an antigen; therefore, this person's immune system produces a set of antibodies referred to as "monoclonal" antibodies. Monoclonal antibodies form a population of identical antibodies that all recognize and are specific for one epitope on an antigen. Thus, someone with this condition may suffer frequent bacterial infections because of a lack of antibody diversity. Indeed, a bacterium whose antigens do not match the antibodies manufactured by the overabundant monoclonal B cells has a selective advantage.

The high-affinity binding capacity of antibodies with antigens has been employed in both therapeutic and diagnostic procedures. It is, however, unfortunate that the effectiveness of commercial preparations of polyclonal antibodies varies widely from batch to batch. In some instances of immunization, certain epitopes of a particular antigen are strong stimulators of antibody-producing cells, whereas at other times, the immune system responds more vigorously to different epitopes of the same antigen. Thus one batch of polyclonal antibodies may have a low level of antibody molecules directed against a major epitope and not be as effective as the previous batch. Consequently, it is desirable to produce a cell line that will produce monoclonal antibodies with a high affinity for a specific epitope on the antigen for commercial use. Such a cell line would provide a consistent and continual supply of identical (monoclonal) antibodies. Monoclonal antibodies can be produced by hybridoma cells, which are generated by the fusion of cancerous B cells and normal spleen cells obtained from mice immunized with a specific antigen. After initial selection of hybridoma clones, monoclonal antibody production is maintained in culture. In addition, the hybridoma cells can be injected into mice to induce tumors that, in turn, will release large quantities of fluid containing the antibody. This fluid containing monoclonal antibodies can be collected periodically and may be used immediately or stored for future use. Various systems used to produce monoclonal antibodies include cultured lymphoid cell lines, yeast cells, *Trichoderma reese* (ascomycetes), insect cells, *Escherichia coli*, and monkey and Chinese hamster ovary cells. Transgenic plants and plant cell cultures have been explored as potential systems for antibody expression.

Impact and Applications

The high-affinity binding capacity of antibodies may be used to inactivate antigens in vivo (within a living organism). The binding property of antibodies may also be employed in many therapeutic and diagnostic applications. In addition, it is a very effective tool in both immunological isolation and detection methods.

Monoclonal antibodies may outnumber all other products being explored by various biotechnology-oriented companies for the treatment and prevention of disease. For example, many strategies for the treatment of cancerous tumors as well as for the inhibition of human immunodeficiency virus (HIV) replication are based on the use of monoclonal antibodies. HIV is a retrovirus (a virus whose genetic material is ribonucleic acid, or RNA) that causes acquired immunodeficiency syndrome (AIDS). Advances in plant biotechnology have made it possible to use transgenic plants to produce monoclonal antibodies on a large scale for therapeutic or diagnostic use. Indeed, one of the most promising applications of plant-produced antibodies in immunotherapy is in passive immunization (for example, against *Streptococcus mutans*, the most common cause of tooth decay). Large doses of the antibody are required in multiple applications for passive immunotherapy to be effective. Transgenic antibody-producing plants may be one source that can supply huge quantities of antibodies in a safe

and cost-effective manner. It has been demonstrated that a hybrid IgA-IgG molecule produced by transgenic plants prevented colonization of *S. mutans* in culture, which appears to be how the antibody prevents colonization of this bacterium in vivo.

It has been estimated that antibodies expressed in soybeans at a level of 1 percent of total protein may cost approximately one hundred dollars per kilogram of antibody, which is relatively inexpensive in comparison with the cost of traditional antibiotics. Transgenic plants have also been used as bioreactors for the large-scale production of antibodies with no extensive purification schemes. In fact, antibodies have been expressed in transgenic tobacco roots and then accumulated in tobacco seeds. If this technology could be employed to obtain stable accumulation of antibodies in more edible plant organs such as potato tubers, it could potentially allow for long-term storage as well as a safe and easy delivery of specific antibodies for immunotherapeutic applications. In addition, plant-produced antibodies may be more desirable for human use than microbial-produced antibodies, because plant-produced antibodies undergo eukaryotic rather than the prokaryotic (bacterial) post-translational modifications. Human glycosylation (a biochemical process whereby sugars are attached onto the protein) is more closely related to that of plants than that of bacteria.

The potential use of antibody expression in plants for altering existing biochemical pathways has also been demonstrated. For example, germination mediated by phytochrome (a biochemical produced by plants) has been altered by utilizing plant-produced antibodies. In addition, antibodies expressed in plants have been successfully used to immunize host plants against pathogenic infection; for example, tobacco plants have already been immunized with antibodies against viral attack. This approach has great potential to replace the traditional methods (use of chemicals) in controlling pathogens.

—Sibdas Ghosh and Tom E. Scola

See also: Allergies; Autoimmune Disorders; Biopharmaceuticals; Blotting: Southern, Northern, and Western; Diabetes; Diphtheria;

Hybridomas and Monoclonal Antibodies; Immunogenetics; Molecular Genetics; Multiple Alleles; Oncogenes; Organ Transplants and HLA Genes; Synthetic Antibodies.

Further Reading

Browne, M. J., and P. L. Thurlby, eds. *Genomes, Molecular Biology, and Drug Discovery.* San Diego: Academic Press, 1996. Focuses on how the continuing advances in modern genome research and molecular biology, combined with new pharmacological and chemical strategies, help realize the achievement of practical therapeutic endpoint.

Glick, Bernard R., and Jack J. Pasternak, eds. *Molecular Biotechnology: Principles and Applications of Recombinant DNA.* Washington, D.C.: ASM Press, 1998. Discusses the structure and function of antibodies as well as the role of biotechnology in the use of antibodies. Covers both the underlying scientific principles and the wide-ranging industrial, agricultural, pharmaceutical, and biomedical applications of recombinant DNA technology. Numerous illustrations and figures in black and white and color.

Harlow, Ed, and David Lane, eds. *Using Antibodies: A Laboratory Manual.* Rev. ed. Cold Spring Harbor, N.Y.: Cold Spring Harbor Laboratory Press, 1999. Provides a detailed account of different methods involved in the production and application of antibodies. A standard manual.

Kontermann, Roland, and Stefan Dübel, eds. *Antibody Engineering.* New York: Springer, 2001. Serves as a lab manual for antibody engineers, demonstrating the state of the art and covering all essential technologies in the field. Designed both to lead beginners in this technology and to keep experienced engineers current with the most detailed protocols. Includes color and halftone illustrations.

Mayforth, Ruth D. *Designing Antibodies.* San Diego: Academic Press, 1993. Acts as a practical introduction to designing antibodies for use in medicine or science. Explains such aspects as making monoclonal antibodies, designing them for human therapy, targeting, idiotypes, and catalytic antibodies.

Raz, E. *Immunostimulatory DNA Sequences.* New York: Springer, 2001. Provides chapters on such topics as the introduction and discovery of immunostimulatory DNA sequences, mechanisms of immune stimulation by bacterial DNA, and multiple effects of immunostimulatory DNA on T cells and the role of type I interferons.

Smith, Mathew D. "Antibody Production in Plants." *Biotechnology Advances* 14 (1996). Summarizes production and applications of plant-produced antibodies.

Wang, Henry Y., and Tadayuki Imanaka, eds. *Antibody Expression and Engineering.* Washington, D.C.: American Chemical Society, 1995. Among other topics, examines antibody production and expression in insect cells, plants, myeloma and hybridoma cells, and proteins.

Antisense RNA

Field of study: Molecular genetics

Significance: *Antisense RNA and RNA interference are powerful modifiers of gene expression that act through RNA-RNA binding through complementary base pairing. This provides a flexible mechanism for specific gene regulation and has great potential for experimental studies and therapeutic action. RNA interference, a specialized form of antisense RNA, even mimics an immune system, for example, targeting RNA viruses within a cell. Processes involving antisense RNA appear in eukaryotes, eubacteria, and archaea.*

Key terms

ANTISENSE: a term referring to any strand of DNA or RNA that is complementary to a coding or regulatory sequence; for example, the strand opposite the coding strand (the sense strand) in DNA is called the antisense strand

DOWN-REGULATION: a process of gene expression in which the amount that a gene is transcribed and/or translated is reduced

GENE SILENCING: any form of genetic regulation in which the expression of a gene is completely repressed, either by preventing transcription (pre-transcriptional gene silencing) or after a messenger RNA (mRNA) has been transcribed (post-transcriptional gene silencing

RNA INTERFERENCE (RNAI): Sequence-specific degradation of messenger RNA (mRNA) caused by complementary double-stranded RNA

UP-REGULATION: a process of gene expression in which the amount that a gene is transcribed and/or translated is increased

Discovery

In addition to the three main types of RNA—messenger RNA (mRNA), transfer RNA (tRNA), and ribosomal RNA (rRNA)—there are numerous other types of RNA molecules. Some have an effect, through complementary binding, on mRNA molecules. When this kind of RNA binds to an mRNA, it effectively blocks translation of the mRNA and can therefore be described as having an antisense action (that is, it blocks the expression of the message in the mRNA). Antisense RNA was first discovered in 1981 as a mechanism regulating copy number of bacterial plasmids. Other RNAs, such as small nuclear and small nucleolar RNAs (snRNA and snoRNA), act in RNA splicing and editing, with a catalytic effect guided by complementary base pairing.

Various forms of gene down-regulation were discovered throughout the 1990's, including plant post-transcriptional gene silencing (that is, preventing the mRNA from being translated), gene silencing in fungi (that is, preventing transcription of a gene), and RNA interference in the nematode *Caenorhabditis elegans*. The importance of noncoding RNA molecules, including antisense RNA, is becoming clear. They add a previously unknown level of genetic complexity, and the extent of their influence is yet to be determined fully.

Natural Function

Antisense RNA is utilized in a number of ways by bacterial plasmids. Replication of ColE1 plasmids requires an RNA preprimer, called RNA II, that interacts with the origin of replication and forms a particular secondary structure. This allows an enzyme to cut and form the mature primer needed for DNA replication. Antisense RNA I can bind to RNA II, preventing the formation of the necessary structure. In the R1 plasmid, the CopA antisense RNA can bind and prevent the translation of the RNA transcript for replication initiation protein RepA. Thus, change in plasmid number is controlled by changing levels of antisense RNA, modifying the ability of plasmids to replicate.

Many plasmids use antisense RNA to ensure their maintenance within bacteria. The R1 plasmid transcribes Hok toxin mRNA, but interaction with antisense Sok RNA prevents its translation. Sok RNA is less stable than Hok RNA, so plasmid loss leads to Sok degradation but leaves some Hok transcripts, which are translated into a toxin that kills the cell. This is an ingenious way of selecting for plasmid propagation. Antisense regulation has also been found in some transposons and bacteriophages.

Bacteria use antisense RNA to regulate particular genes. Such RNA is often encoded in a region different from that of the target and may affect multiple genes. For example, the OxyS RNA, induced by oxidative stress, inhibits translation of fhlA mRNA (involved in formate metabolism). In conjunction with the protein Hfq, OxyS RNA binds near the ribosome-binding site in fhlA mRNA, preventing translation. MicF RNA is induced under cellular stress and binds to the mRNA of membrane pore protein ompF to prevent its translation.

One of the first examples of antisense regulatory mechanisms in eukaryotes came from the nematode *Caenorhabditis elegans*. Small antisense RNA molecules lin-4 and let-7 show imperfect base-pairing to the 3′ untranslated region of their target gene mRNAs. This results in translational inhibition and is important for normal development. These small antisense RNAs are members of the microRNA (miRNA) class of molecules, which are single-stranded RNA molecules, about 21 nucleotides long, found throughout eukaryotes. They are produced by cleavage of longer molecules (about 60-100 nucleotides) which contain partial self-complementarity that produces a hairpin structure. The function of most miRNAs is unknown.

Antisense RNA has been implicated in other

Dr. Phillip A. Sharp, who with Richard J. Roberts won the 1993 Nobel Prize in Physiology or Medicine. Sharp acknowledges that the discovery of antisense RNA and RNA interference has changed his cancer research. The process could theoretically offer ways of "silencing" the genes associated with cancer. (AP/Wide World Photos)

processes. Imprinted genes are often associated with antisense transcripts from the same locus, although their role is unclear. Double-stranded RNA may be capable of affecting DNA chromatin structure through methylation of homologous sequence.

RNA Interference

RNA interference (RNAi) causes sequence-specific gene silencing in response to the presence of double-stranded RNA. The process is proposed to have evolved as a mechanism of avoiding viral infection and limiting transposable element replication, since both can involve double-stranded RNA.

The process is present in a wide variety of eukaryotes (including mice and probably hu-

mans), and the steps involved are likely to be similar. RNAi begins by the recognition of long double-stranded RNA molecules by the conserved protein Dicer, which cuts these long RNAs to produce small interfering RNAs (siRNA, double-stranded RNA about 21-23 nucleotides long with overhanging 3′ ends). These become part of the RNA-induced silencing complex (RISC), which uses the antisense strand of the siRNA to recognize complementary RNA sequences. These sequences are then cut and degraded. Some organisms show amplification, where siRNA stimulates production of more double-stranded RNA to be processed by Dicer. *Caenorhabditis elegans* and plants show evidence of a systemic response, whereby initial silencing in one cell spreads to other cells.

Therapeutic Applications

Many disease states are caused by abnormal gene expression and are therefore potential targets for gene therapy. One approach is to use antisense RNA or RNAi. For example, cancer cells often show overexpression of genes involved in growth and proliferation. These genes could be targeted by antisense RNA to decrease the amount of gene product produced, in hopes of preventing tumor growth. RNAi could be used to target specific mutations. The mRNA from a mutant allele could be targeted for degradation in a heterozygous patient, allowing production of the correct protein from the wild-type allele alone. Antisense techniques could be used to target viruses and prevent their replication, or could be used to correct aberrant splicing. Many of the processes described above have been successfully demonstrated in experimental systems such as cell culture or mouse models.

Many issues need to be addressed before antisense RNA therapeutics become feasible. A delivery system is needed to produce a long-lasting effect in the correct cell type. The fact that RNA interference appears to spread systemically in some organisms may facilitate its application. The safety and effectiveness of such approaches also need to be tested. Nevertheless, the approach is promising and potentially very useful.

—*Peter J. Waddell and Michael J. Mclachlan*

See also: Gene Regulation: Eukaryotes; Human Genetics; Model Organism: *Caenorhabditis elegans*; Noncoding RNA Molecules; RNA Structure and Function; RNA Transcription and mRNA Processing; RNA World; Viral Genetics.

Further Reading

Brantl, S. "Antisense-RNA Regulation and RNA Interference." *Biochimica et Biophysica Acta* 1575 (2002): 15-25. A detailed survey of the wide variety of ways in which antisense RNAs operate.

Shuey, D. J., D. E. McCallus, and T. Giordano. "RNAi: Gene Silencing in Therapeutic Intervention." *Drug Discovery Today* 7 (2002): 1040-1046. A look at challenges ahead for using RNAi in a medical setting.

Archaea

Field of study: Cellular biology
Significance: *Archaea are diverse prokaryotic organisms distinct from the historically familiar bacteria. Archaea have certain molecular properties previously thought to occur only in eukaryotes. Many archaea require severe conditions for growth, and their genetic processes have adapted to these extreme conditions in ways that are not fully understood.*

Key terms

CONJUGATION: the process by which one bacterial cell transfers DNA directly to another

DOMAIN: the highest-level division of life, sometimes called a superkingdom

EXTREME HALOPHILES: microorganisms that require extremely high salt concentrations for optimal growth

INSERTION SEQUENCE: a small, independently transposable genetic element

METHANOGENS: microorganisms that derive energy from the production of methane

PROKARYOTES: unicellular organisms with simple ultrastructures lacking nuclei and other intracellular organelles

SMALL SUBUNIT RIBOSOMAL RNA (SSU rRNA): the RNA molecule found in the small subunit of the ribosome; also called 16S rRNA (in prokaryotes) or 18S rRNA (in eukaryotes)

Gene Sequences Measure the Diversity of Prokaryotes

Prokaryotic microorganisms have been on earth for as many as 3.8 billion years and have diverged tremendously in genetic and metabolic terms. Unfortunately, the magnitude of this divergence has made it difficult to measure the relatedness of prokaryotes to one another. In the 1970's, Carl R. Woese addressed this problem using a method of reading short sequences of ribonucleotides from a highly conserved RNA molecule, the small subunit ribosomal RNA (ssu rRNA). Because this RNA is present in all organisms and has evolved very slowly, any two organisms have at least a few of these short nucleotide sequences in common.

The proportion of shared sequences thus provided a quantitative index of similarity by which all cellular organisms could, in principle, be compared.

When the nucleotide sequence data were used to construct an evolutionary tree, eukaryotes (plants, animals, fungi, and protozoa) formed a cluster clearly separated from the common bacteria. Unexpectedly, however, a third cluster emerged that was equally distinct from both eukaryotes and common bacteria. This cluster consisted of prokaryotes that (1) lacked biochemical features of most bacteria (such as a cell wall composed of peptidoglycan), (2) possessed other features not found in any other organisms (such as membranes composed of isoprenoid ether lipids), and (3) occurred in unusual, typically harsh, environments. Woese and his co-workers eventually designated the three divisions of life represented by these clusters "domains," naming the nonbacterial prokaryotes the domain *Archaea.*

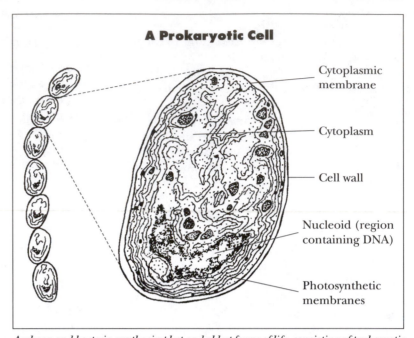

A Prokaryotic Cell

Cytoplasmic membrane

Cytoplasm

Cell wall

Nucleoid (region containing DNA)

Photosynthetic membranes

Archaea and bacteria are the simplest and oldest forms of life, consisting of prokaryotic cells, which differ from the cells that form higher organisms (fungi, algae, protozoa, plants, and animals), called eukaryotic cells. Based on an electron microscope image of one cell in a string forming a cyanobacterium, this depiction shows the basic features of a prokaryote. Note the lack of a defined nucleus and organelles (there are no plastids or mitochondria)—the components that house genetic information in eukaryotes. Instead, genetic material in prokaryotes is located in an unbound region called the nucleoid. (Kimberly L. Dawson Kurnizki)

Biology of the Domain *Archaea*

Archaea tend to require unusual conditions for growth, which has made it challenging to determine their genetic properties. The methanogens, for example, live by converting hydrogen (H_2) and carbon dioxide (CO_2) or other simple carbon compounds into methane and are killed by even trace amounts of oxygen. The extreme halophiles, in contrast, normally live in brine lakes and utilize oxygen for growth. However, they require extremely high concentrations of salt to maintain their cellular structure. A third class of archaea, the extreme thermophiles, occur naturally in geothermal springs and grow best at temperatures ranging from 60 to 105 degrees Celsius (140–

221 degrees Fahrenheit). Many derive energy from the oxidation or reduction of sulfur compounds. Sequencing of DNA fragments recovered from "moderate" environments, such as ocean water or soil, has revealed many additional archaeal species that presumably do not require unusual environmental conditions but have never been cultured in the laboratory.

The Genetic Machinery of Archaea

Because bacteria and eukaryotes differ greatly with respect to gene and chromosome structure and the details of gene expression, molecular biologists have examined the same properties in archaea and have found a mixture of "bacterial" and "eukaryotic" features. The organization of DNA within archaeal cells is bacterial, in the sense that archaeal chromosomes are circular DNAs of between 2 million and 4 million base pairs having single origins of replication. As in bacteria, the genes are densely packed and often grouped into clusters of related genes transcribed from a common promoter. The promoters themselves, however, resemble the TATA box/BRE element combination of eukaryotic DNA polymerase II (Pol II) promoters, and the RNA polymerases have the complex subunit composition of eukaryotes rather than the simple composition found in bacteria. Furthermore, archaea initiate transcription by a simplified version of the process seen in eukaryotic cells: Transcription factors (TATA-binding protein and a TFIIB) first bind to regions ahead of the promoter, then recruit RNA polymerase to attach and begin transcription. Introns are rare in archaea, however, and do not interrupt protein-encoding genes. Also, the regulation of transcription in archaea seems to depend heavily on the types of repressor and activator proteins found in bacteria.

Genomes of Archaea

The availability of complete DNA sequences now enables archaeal genomes to be compared to the genomes of bacteria and eukaryotes. One pattern that emerges from these comparisons is that most of the archaeal genes responsible for the processing of information (synthesis of DNA, RNA, and proteins) resemble their eukaryotic counterparts, whereas most of the archaeal genes for metabolic functions (biosynthetic pathways, for example) resemble their bacterial counterparts. The genomes of archaea also reveal probable cases of gene acquisition from distant relatives, a process called lateral gene transfer.

A third pattern to emerge from genome comparisons is that some archaea are missing genes thought to be important or essential. For example, the genomes of at least two methanogenic archaea do not encode an enzyme that charges transfer RNA (tRNA) with cysteine. These archaea instead use a novel strategy for making cysteinyl-tRNA: Some of the seryl tRNA made by these cells is converted to cysteinyl tRNA by a specialized enzyme. Even more intriguing is the much longer list of archaea—all of which happen to be hyperthermophiles, which grow optimally at 80 degrees Celsius (176 degrees Fahrenheit) or above—that lack genes for the DNA mismatch repair proteins found in all other organisms.

Unique Genetic Properties?

This last observation raises an important question: Has an evolutionary history distinct from all conventional genetic systems, combined with the special demands of life in unusual environments, resulted in unique genetic properties in archaea? Although basic genetic assays can be performed in only a few species, the results help identify which genetic properties of cellular organisms are truly universal and which ones may have unusual features in archaea.

The methanogen *Methanococcus voltae* transfers short pieces of chromosome from one cell to another, using particles that resemble bacterial viruses (bacteriophages). This means of gene transfer has been seen in only a few bacteria. In other methanogens, researchers have used more conventional genetic phenomena, such as antibiotic-resistance genes, plasmids, and transposable elements, to develop tools for cloning or inactivating genes. As a result, new details about the regulation of gene expression in archaea and the genetics of methane formation are now coming to light.

The extreme halophile *Halobacterium salinarum* exhibits extremely high rates of spontaneous mutation of the genes for its photosynthetic pigments and gas vacuoles. This genetic instability reflects the fact that insertion sequences transpose very frequently into these and other genes. A distantly related species, *Haloferax volcanii*, has the ability to transfer chromosomal genes by means of conjugation. Although many bacteria engage in conjugation, the mechanism used by *H. volcanii* does not resemble the typical bacterial system, since no plasmid seems to be involved, and there is no apparent distinction between donor strain and recipient strain in the transfer of DNA.

Genetic tools for the archaea from geothermal environments are less well developed, but certain selections have made it possible to study spontaneous mutation and homologous recombination in some species of *Sulfolobus*. At the normal growth temperatures of these aerobic archaea, 75-80 degrees Celsius (167-176 degrees Fahrenheit), spontaneous chemical decomposition of DNA is calculated to be about one thousand times more frequent than in the organisms previously studied by geneticists. In spite of this, *Sulfolobus acidocaldarius* exhibits the same frequency of spontaneous mutation as *Escherichia coli* and significantly lower proportions of base-pair substitutions and deletions. This indicates especially effective mechanisms for avoiding or accurately repairing DNA damage, including mismatched bases, despite the fact that no mismatch repair genes have been found in *Sulfolobus* species. Also, *S. acidocaldarius*, like *H. volcanii*, has a mechanism of conjugation that does not require a plasmid or distinct donor and recipient genotypes. The transferred DNA recombines efficiently into the resident chromosome, as indicated by frequent recombination between mutations spaced only a few base pairs apart.

Finding two similar and unusual mechanisms of conjugation in two dissimilar and distantly related archaea (*H. volcanii* and *S. acidocaldarius*) raises questions regarding the possible advantages of this capability. Population genetic theory predicts that organisms that reproduce clonally (as bacteria and archaea do) would benefit from occasional exchange and recombination of genes, because this accelerates the production of beneficial combinations of alleles. Such recombination may be particularly important for archaea such as *Haloferax* and *Sulfolobus* species, whose extreme environments are like islands separated by vast areas that cannot support growth. For these organisms, frequent DNA transfer between cells of the same species may provide an efficient way to enhance genetic diversity within small, isolated populations.

—*Dennis W. Grogan*

See also: Antisense RNA; Bacterial Genetics and Cell Structure; Gene Regulation: Bacteria; Lateral Gene Transfer; Noncoding RNA Molecules.

Further Reading

Madigan, Michael T., John M. Martinko, and Jack Parker. *Brock Biology of Micro-organisms.* 10th ed. Upper Saddle River, N.J.: Prentice Hall, 2003. Chapter 13 of this popular microbiology text provides an accurate and well-illustrated overview of the biological diversity of the archaea.

Olsen, Gary, and Carl R. Woese. "Archaeal Genomics: An Overview." *Cell* 89 (1997): 991-994. This mini-review article, along with several accompanying articles, summarizes for specialists numerous molecular differences and similarities between archaea and bacteria or eukaryotes, based on the first archaeal genomes to be sequenced.

Woese, Carl. R. "Archaebacteria." *Scientific American* 244 (1981): 98-122. A clear, though somewhat dated, description of the archaea and the various lines of evidence for their status as a "third form of life."

Artificial Selection

Fields of study: Evolutionary biology; Population genetics

Significance: *Artificial selection is the process through which humans have domesticated and improved plants and animals. It continues to be the primary means whereby agriculturally important plants and animals are modified to improve their desirability. However, artificial selection is also a threat to genetic diversity of agricultural organisms as uniform and productive strains replace the many diverse, locally produced varieties that once existed around the globe.*

Key terms

GENETIC MERIT: a measure of the ability of a parent to contribute favorable characteristics to its progeny

GENETIC VARIATION: a measure of the availability of genetic differences within a population upon which artificial selection has potential to act

HERITABILITY: a proportional measure of the extent to which differences among organisms within a population for a particular character result from genetic rather than environmental causes (a measure of nature versus nurture)

Natural vs. Artificial Selection

Selection is a process through which organisms with particular genetic characteristics leave more offspring than do organisms with alternative genetic forms. The process may occur because the genetic characteristics confer upon the organism a better ability to survive and ultimately produce more offspring than individuals with other characteristics (natural selection), or it may be caused by selective breeding of individuals with characteristics valuable to humans (artificial selection). Natural and artificial selection may act in concert, as when a genetic characteristic confers a disadvantage directly to the organism. Dwarfism in cattle, for example, not only directly reduces the survival of the affected individuals but also reduces the value of the animal to the breeder. Conversely, natural selection may act in opposition to artificial selection. For example, a genetic characteristic that results in the seed being held tightly in the head of a grass such as wheat is an advantage to the farmer, as it makes harvesting easier. That same characteristic would be a disadvantage to wild wheat because it would limit seed dispersal.

Early Applications

Artificial selection was probably conducted first by early farmers who identified forms of crop plants that had characteristics that favored cultivation. Seeds from favored plants were preferentially kept for replanting. Any characteristics that were to some degree heritable would have had the tendency to be passed on to the progeny through the selected seeds. Some favored characteristics may have been controlled by a single gene and were therefore quickly established, whereas other favored characteristics may have been controlled by a large number of genes with individually small effects, making them more difficult to estab-

The beefalo is created by breeding a cow and a bison, and then breeding the offspring again to a cow. Such hybridization, in both plants and animals, is a form of artificial selection that has been practiced by humans for thousands of years to meet agricultural needs. (AP/Wide World Photos)

lish. Nevertheless, seeds selected from the best plants would tend to produce offspring that were better than average, resulting in gradual improvement in the population. It would not have been necessary to have knowledge of the mechanisms of genetics to realize the favorable effects of selection. Likewise, individuals who domesticated the first animals for their own use would have made use of selection to capture desirable characteristics within their herds and flocks. The first of those characteristics was probably docile behavior, a trait known to be heritable in contemporary livestock populations.

From Pedigrees to Genome Maps

Technology to improve organisms through selective breeding preceded an understanding of its genetic basis. Recording of pedigrees and performance records began with the formal development of livestock breeds in the 1700's. Some breeders, notably Robert Bakewell, began recording pedigrees and using progeny testing to determine which sires had superior genetic merit. Understanding of the principles of genetics through the work of Gregor Mendel enhanced but did not revolutionize applications to agricultural plant and animal improvement.

Development of reliable methods for testing the efficiency of artificial selection dominated advances in the fields of plant and animal genetics during the first two-thirds of the twentieth century. Genetic merit of progeny was expected to be equal to the average genetic merit of the parents. More effective breeding programs are dependent on identifying potential parents with superior genetic merit. Computers and large-scale databases have greatly improved selection programs for crops and livestock. Genetic change on the order of 2 percent per generation became possible. However, selection to improve horticultural species and companion animals continued to rely largely on the subjective judgment of the breeder to identify superior stock. Plant and animal genome-mapping programs would facilitate the next leap forward in genetic improvement of agricultural organisms. Selection among organisms based directly on their gene sequences promised to allow researchers to bypass the time-consuming data-recording programs upon which genetic progress of the 1990's relied.

Diversity vs. Uniformity

The ultimate limit to what can be achieved by selection is the exhaustion of genetic variants. One example of the extremes that can be accomplished by selection is evident in dog breeding: The heaviest breeds weigh nearly one hundred times as much as the lightest breeds. Experimental selection for body weight in insects and for oil content in corn has resulted in variations of similar magnitudes.

However, most modern breeding programs for agricultural crops and livestock seek to decrease variability while increasing productivity. Uniformity of the products enhances the efficiency with which they can be handled mechanically for commercial purposes. As indigenous crop and livestock varieties are replaced by high-producing varieties, the genetic variation that provides the source of potential future improvements is lost. Widespread use of uniform varieties may also increase the susceptibility to catastrophic losses or even extinction from an outbreak of disease or environmental condition. The lack of biodiversity in the wake of such species loss could threaten entire ecosystems and human beings themselves.

—*William R. Lamberson*

See also: Eugenics; Eugenics: Nazi Germany; Evolutionary Biology; Gene Therapy; Genetic Engineering: Agricultural Applications; Genetic Engineering: Historical Development; Genomes; Genome Libraries; Hardy-Weinberg Law; High-Yield Crops; Inbreeding and Assortative Mating; Natural Selection; Polyploidy; Population Genetics; Pedigree Analysis; Punctuated Equilibrium; Quantitative Inheritance; Sociobiology; Speciation.

Further Reading

Lurquin, Paul F. *The Green Phoenix: A History of Genetically Modified Plants.* New York: Columbia University Press, 2001. Gives equal weight to the science behind developing improved crop strains and the multinational corporations marketing the results.

Rissler, Jane, and Margaret Mellon. *The Ecological Risks of Engineered Crops.* Cambridge, Mass.: MIT Press, 1996. A scientific and policy assessment of the dangers. The authors—both of whom work for the Union of Concerned Scientists—not only outline the risks but also make suggestions for ways to minimize them.

Tudge, Colin. *The Engineer in the Garden: Genes and Genetics, from the Idea of Heredity to the Creation of Life.* New York: Hill and Wang, 1995. A British science journalist follows the history of human manipulation of genetics to explore the social ramifications of artificial selection, which has resulted in genetic "advances" that have not been adequately exposed to ethical and policy considerations. *Library Journal* praised the book as "a good balance between overall breadth of coverage and the intelligent, readable synthesis of the myriad issues" of genetic research.

Zohary, Daniel, and Maria Hopf. *Domestication of Plants in the Old World: The Origin and Spread of Cultivated Plants in West Asia, Europe, and the Nile Valley.* 3d ed. New York: Oxford University Press, 2001. A thorough review of information on the beginnings of agriculture, particularly utilizing new molecular biology findings on the genetic relations between wild and domesticated plant species.

Autoimmune Disorders

Fields of study: Diseases and syndromes; Immunogenetics

Significance: *Autoimmune disorders are chronic diseases that arise from a breakdown of the immune system's ability to distinguish between the body's own cells and foreign substances. Autoimmune disorders can be caused by both genetic and environmental factors and cause an individual's immune system to react against the organs or tissues of the individual's own body.*

Key terms

ANTIBODIES: molecules in blood plasma responsible for recognizing and binding to antigens

ANTIGENS: foreign substances recognized by the immune system that result in the production of antibodies and lymphocytes directed specifically against them

IMMUNE SYSTEM: the system that normally responds to foreign agents by producing antibodies and stimulating antigen-specific lymphocytes, leading to destruction of these agents

LYMPHOCYTES: sensitized cells of the immune system that recognize and destroy harmful agents via antibody and cell-mediated responses that include B lymphocytes from the bone marrow and T lymphocytes from the thymus

MAJOR HISTOCOMPATIBILITY COMPLEX (MHC): a system of protein markers on a cell's outer membrane following infection with a virus, malignant cell, or foreign cell that signals the immune system to destroy the cell

Autoimmune Disorders and Immune System Dysfunction

Autoimmune disorders involve a large group of chronic and potentially life-threatening diseases that are initiated by an individual's own immune system attacking the organs or tissues of the individual's own body. The main function of the immune system is to defend against invading microorganisms such as bacteria, fungi, viruses, protozoa, and parasites by producing antibodies or lymphocytes that recognize and destroy the harmful agent. The ability to distinguish normal body constituents (self) from foreign substances (nonself) is crucial to appropriate immune functioning. Loss of this ability to distinguish between self and nonself can lead to serious damage to the affected organs and tissues.

Autoimmunity—the inaccurate recognition of a normal body component as foreign, followed by the mounting of an autoimmune response—results when normal autoantigens on body cells stimulate the development of autoantibodies. These autoantibodies are most often the result of a genetic defect during viral and bacterial infections or from environmental or chemical factors; they may also, according to some researchers, be a natural consequence of the aging process.

History and Classification of Autoimmune Disorders

The concept of autoimmune disorders was first proposed in 1901, but it was not until the 1950's that autoimmunity was experimentally created in animals via immunization. By the 1960's, it was recognized that autoimmunity was a direct or indirect cause of numerous human ailments. Many diseases formerly classified as collagen-vascular diseases (collagenoses) were later classified as autoimmune disorders.

Autoimmune disorders are generally categorized as organ-specific diseases and non-organ-specific (also called systemic) diseases. Organ-specific autoimmune diseases involve an attack directed against one main organ and have been documented for essentially every organ in the body. Common examples include multiple sclerosis (brain), insulin-dependent diabetes mellitus (pancreas), Graves' disease (thyroid), Addison's disease (adrenal glands), pernicious anemia (stomach), myasthenia gravis (muscles), autoimmune hemolytic anemia (blood), primary biliary cirrhosis (liver), pemphigus vulgaris (skin), and glomerulonephritis (kidneys). Non-organ-specific autoimmune diseases involve an attack by the immune system on several body areas, potentially causing diseases such as systemic lupus erythematosus, rheumatoid arthritis, polyarteritis nodosa, scleroderma, ankylosing spondylitis, and rheumatic fever.

Some evidence suggests that other conditions (such as certain types of eye inflammation and male and female infertility) may be autoimmune related. Allergies involve hypersensitivity reactions that result in immune reactions that can lead to inflammation and tissue damage. Environmental antigens such as pollen, dust mites, food proteins, and bee venom may cause allergic reactions such as hay fever, asthma, and food intolerance in sensitive individuals via the antibody class known as immunoglobulin E (IgE). Medications such as antibiotics may also be recognized as chemical antigens, causing adverse allergic reactions.

Immunologists do not know the precise origin of most autoimmune diseases. What researchers have shown is that most autoimmune diseases occur more frequently in females than in males and that the development of autoimmune disorders often requires both a genetic susceptibility and additional stimuli such as exposure to a toxin. Of the numerous theories proposed for the cause of autoimmunity development, three models have received the most consideration by clinical researchers. The clonal deletion theory suggests that autoimmunity develops if autoreactive T or B cell clones are not eliminated during the fetal period or very soon after birth. The body normally does not react to its own fetal or neonatal antigens, which are recognized because the corresponding T and B cell clones are eliminated from the immune system. In the unfortunate event that "forbidden clones" of autoreactive cells remain active, antibodies are produced that are directed against its own antigens, and autoimmunity develops, frequently involving the loss of the helper T cells' ability to regulate B-cell function. A second theory suggests that some antigens that are normally nonimmunogenic (hidden antigens) somehow become autoimmunogenic and stimulate the immune system to react against itself. A third theory suggests that autoimmunity can be initiated by an exogenous antigen, assuming that the antibodies produced to fight it cross-react with a similar determinant on the body's own cells.

Diagnosis and Treatment

Diagnosis of autoimmune disorders generally begins with the often difficult task of documenting autoantibodies and autoreactive T cells. A condition suspected to be caused by autoimmunity can also be confirmed by a number of other direct and indirect methods, such as a favorable response to immunosuppressive, corticosteroid, or anti-inflammatory drug treatment along with several other immunologic techniques.

Treatment strategies lag behind the ability to diagnose autoimmune disorders. Initial management involves the control and reduction of both pain and loss of function. Correction of deficiencies in hormones such as insulin or thyroxin caused by autoimmune damage to glands is often performed first. Replacing blood components by transfusion is also considered, but treatment effectiveness is often limited by the lack of knowledge of the precise disease mechanisms. Suppression of the immune system is also often attempted, but achieving a delicate balance between controlling the autoimmune disorder and maintaining the body's ability to fight disease in general is critical.

Medication therapy commonly includes corticosteroid drugs, with more powerful immunosuppressant drugs such as cyclophosphamide, methotrexate, azathioprine, chloroquine derivatives, and small doses of antimetabolic or anticancer drugs often required. A majority of these medications can rapidly damage dividing tissues such as the bone marrow and thus must be used with caution. Plasmapheresis (removal of toxic antibodies) is often helpful in diseases such as myasthenia gravis, while other treatments involve drugs that target immune system cells such as the cyclosporines. Fish oil and antioxidant supplementation have been shown to be an effective anti-inflammatory intervention and may help suppress autoimmune diseases such as rheumatoid arthritis and systemic lupus erythematosus.

—Daniel G. Graetzer, updated by Bryan Ness

See also: Aging; Allergies; Antibodies; Developmental Genetics; Diabetes; Hybridomas and Monoclonal Antibodies; Immunogenetics; Organ Transplants and HLA Genes; Stem Cells; Steroid Hormones.

Further Reading

Abbas, Abul K., and Richard A. Flavell, eds. *Genetic Models of Immune and Inflammatory Diseases*. New York: Springer, 1996. Individual chapters provide descriptions of research results or mini-reviews of transgenic and targeted gene disruption models used to study autoimmune and inflammatory diseases. Illustrations.

Bona, Constantin A., et al., eds. *The Molecular Pathology of Autoimmune Diseases*. 2d ed. New York: Taylor and Francis, 2002. A review of the latest research carried out in immunology, particularly at the molecular level. Examines developments in the diagnosis and treatment of such conditions as hemolytic anemia, diabetes, Graves' disease, Addison's disease, multiple sclerosis, inflammatory bowel disease, and autoimmune hepatitis.

Fernandes, Gabriel, and Christopher A. Jolly. "Nutrition and Autoimmune Diseases." *Nutrition Reviews* 56 (January, 1998). Summarizes several topics from a conference on nutrition and immunity and relates striking benefits of fish oil and antioxidant supplementation on gene and T-cell subsets.

Lachmann P. J., et al., eds. *Clinical Aspects of Immunology*. Malden, Mass.: Blackwell, 1993. Provides an excellent overview of immunology for the clinician, covering new areas of immunologic diagnosis, mechanisms, and techniques.

Theofilopoulos, A. N., ed. *Genes and Genetics of Autoimmunity*. New York: Karger, 1999. Examines a broad spectrum of topics related to autoimmunity, including the chapters "Immunoglobulin Transgenes in B Lymphocyte Development, Tolerance, and Autoimmunity," "The Role of Cytokines in Autoimmunity," "Genetics of Human Lupus," and "Genetics of Multiple Sclerosis." Illustrated.

Vyse, Timothy J., and John A. Todd. "Genetic Analysis of Autoimmune Disease." *Cell* 85 (May, 1996). Describes how ongoing study of the entire human genetic code will assist in the isolation and correction of aberrant genes that cause immunological disease and presents evidence that environmental factors have only minor effects on immune system abnormalities.

Web Site of Interest

American Autoimmune Related Diseases Association. http://www.aarda.org. Site of the only national organization devoted specifically to autoimmune disorders and chronic illness. Includes a medical glossary, research articles, information on the science of autoimmune-related diseases, and links to related resources.

Bacterial Genetics and Cell Structure

Fields of study: Cellular biology; Bacterial genetics

Significance: *The study of bacterial structure and genetics has made tremendous contributions to the fields of genetics and medicine, leading to the development of drugs for the treatment of disease, the discovery of DNA as the master chemical of heredity, and knowledge about the regulation of gene expression in other organisms, including humans.*

Key terms

CLONING: the generation of many copies of DNA by replication in a suitable host

EUKARYOTE: an organism made up of cells having a membrane-bound nucleus that contains chromosomes

MUTATION: the process by which a DNA base-pair change or a change in a chromosome is produced; the term is also used to describe the change itself

PROKARYOTE: an organism lacking a membrane-bound nucleus

RECOMBINANT DNA: a DNA sequence that has been constructed or engineered from two or more distinct DNA sequences

Bacteria and Their Structure

The old kingdom Monera contained what has now been classified into the domains *Bacteria* and *Archaea*. Organisms in these domains are unicellular (one-celled) and prokaryotic (lacking a membrane-bound nucleus). Bacteria are among the simplest, smallest, and most ancient of organisms, found in nearly every environment on earth. While some bacteria are autotrophic (capable of making their own food), most are heterotrophic (forced to draw nutrients from their environment or from other organisms). For most of human history, the existence of bacteria was unknown. It was not until the late 1800's that bacteria were first identified. Their role in nature is that of decomposers: They break down organic molecules into their component parts. Along with fungi, they are the major recyclers in nature.

They are also capable of changing atmospheric nitrogen to a form that is usable by plants and animals.

It has long been known that some bacteria are pathogens, or causers of disease. Scientists have expended tremendous effort in describing the role bacteria play in disease and in creating agents that could kill them. Other bacteria, such as *Escherichia coli*, may be part of a mutualistic relationship with another organism, such as humans. Bacteria have been used extensively in genetics research because of their small size and because they reproduce rapidly; some bacteria produce a new generation every twenty minutes. Since they have been so thoroughly studied, a great deal is known about their structure and genetics.

Most bacteria are less than one micron (one-millionth of a meter) in length. They do not contain mitochondria (organelles that produce the energy molecule adenosine triphosphate, or ATP), chloroplasts (plant organelles in which the reactions of photosynthesis take place), lysosomes (organelles that contain digestive enzymes), or interior membrane systems such as the endoplasmic reticulum or Golgi bodies. They do, however, contain RNA, ribosomes (organelles that serve as the sites of protein synthesis), and DNA, which is organized as part of a single, circular chromosome. The circular chromosome is centrally located within the cell in a region called the nucleoid region and is capable of supercoiling. Bacteria often have additional genes carried on small circular DNA molecules called plasmids, which have been used extensively in genetic research. Some plasmids carry genes that impart antibiotic resistance to the cells that contain them.

Bacteria have three basic morphologies, or cell shapes. Bacteria that are spherical are called cocci. Some coccus bacteria form clusters (staphylococcus), while others may form chains (streptococcus). Bacteria that have a rodlike appearance are called bacilli. Spiral or helical bacteria are called spirilla (sometimes called spirochetes).

Classification of Bacteria

Bacteria fall into three basic types: those that lack cell walls, those with thin cell walls, and those with thick cell walls. Mycoplasmas lack cell walls entirely. The bacteria that cause tuberculosis, *Mycobacterium tuberculosis*, do have cell walls and, unlike *Archaea*, their cell walls are composed of peptidoglycan, a complex organic molecule made of two unusual sugars held together by short polypeptides (short chains of amino acids). In 1884, Hans Christian Gram, a Danish physician, found that certain bacterial cells absorbed a stain called crystal violet, while others did not. Those cells that absorb the stain are called gram-positive, and those that do not are called gram-negative. It has since been found that gram-positive bacteria have thick walls of peptidoglycan, while gram-negative bacteria have thin peptidoglycan walls covered by a thick outer membrane. It is this thick outer membrane that prevents crystal violet from entering the bacterial cell. Distinguishing between gram-positive and gram-negative bacteria is an important step in the treatment of disease since some antibiotics are more effective against one class than the other.

By contrast with bacteria, members of *Archaea* have cell walls that do not contain peptidoglycan. Members of *Archaea* are usually found in extreme environments, such as hot springs, extremely saline environments, and hydrothermal vents. Methanogens are the most common and are strict anaerobes, which means that they are killed by oxygen. They live in oxygen-free environments, such as sewers and swamps, and produce methane gas as a waste product of their metabolism. Halobacteria live in only those environments that have a high concentration of salt, such as salt ponds. Thermoacidophiles grow in very hot or very acidic environments.

Bacteria can be further differentiated by the presence or absence of certain surface structures. Some strains produce an outer slime layer called a "capsule." The capsule permits the bacterium to adhere to surfaces (such as human teeth, for example, where the build-up of such bacteria causes dental plaque) and provides some protection against other microorganisms. Some strains display pili, which are fine, hairlike appendages that also allow the bacterium to adhere to surfaces. Some pili,

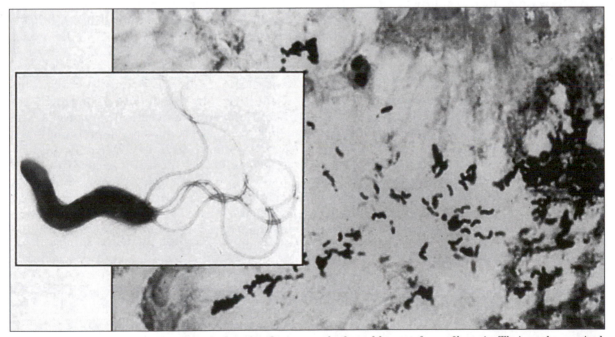

Helicobacter pylori, *which causes stomach ulcers, is only one example of one of the many forms of bacteria. The inset shows a single bacterim.* (AP/Wide World Photos)

such as F pili in *E. coli*, are involved in the exchange of genetic material from one bacterium to another in a process called conjugation. Some bacterial strains have one or more flagella, which allow them to be motile (capable of movement). Any bacterium may have one or more of these surface structures.

Research in molecular genetics is continuing to expand insight into bacterial classification and gene function. Many researchers have been actively sequencing the genomes of bacteria from a broad spectrum. The number of species that have been sequenced is now in the hundreds and includes many human pathogens, such as those that cause tuberculosis, bacterial pneumonia, ulcers, bacterial influenza, leprosy, and Lyme disease. The genomes of a wide range of nonpathogenic bacteria have also been sequenced. Comparisons among the genomes that have been sequenced is beginning to show extensive evidence that bacteria of different species have transferred genes back and forth many times in the past, thus making it difficult to trace their evolutionary lineages.

Bacterial Reproduction

Bacteria reproduce in nature by means of binary fission, wherein one cell divides to produce two daughter cells that are genetically identical. As bacteria reproduce, they form clustered associations of cells called colonies. All members of a colony are genetically identical to one another, unless a mutagen (any substance that can cause a mutation) has changed the DNA sequence in one of the bacteria. Changes in the DNA sequence of the chromosome often lead to changes in the physical appearance or nutritional requirements of the colony. While a bacterium is microscopic, bacterial colonies can be seen with the naked eye; changes in the colonies are relatively easy to perceive. This is one of the reasons bacteria have been favored organisms for genetic research.

For the most part, there is very little genetic variation between one bacterial generation and the next. Unlike higher organisms, bacteria do not engage in sexual reproduction, which is the major source of genetic variation within a population. In laboratory settings, however,

bacteria can be induced to engage in a unidirectional (one-way) exchange of genetic material via conjugation, first observed in 1946 by biochemists Joshua Lederberg and Edward Tatum. The unidirectional nature of the gene transfer was discovered by William Hayes in 1953. He found that one bacterial cell was a donor cell while the other was the recipient. In the 1950's, molecular biologists François Jacob and Elie Wollman used conjugation and a technique called "interrupted mating" to map genes onto the bacterial chromosome. By breaking apart the conjugation pairs at intervals and analyzing the times at which donor genes entered the recipient cells, they were able to determine a correlation between time and the distance between genes on a chromosome. The use of this technique led to a complete map of the sequence of genes contained in the chromosome. It also led to a surprise: It was use of interrupted mating with *E. coli* that first demonstrated the circularity of the bacterial chromosome. The circular structure of the chromosome was in striking contrast to eukaryotic chromosomes, which are linear.

Transformation and Transduction

The bacterium *Streptococcus pneumoniae* was used in one of the early studies that eventually led to the identification of DNA as the master chemical of heredity. Two strains of *S. pneumoniae* were used in a study conducted by microbiologist Frederick Griffith in 1928. One strain (S) produces a smooth colony that is virulent (infectious) and causes pneumonia. The other strain (R) produces a rough colony that is avirulent (noninfectious). When Griffith injected mice with living type R bacteria, the mice survived and no bacteria were recovered from their blood. When he injected mice with living type S, the mice died, and type S bacteria were recovered from their blood. However, if type S was heat-killed before the mice were injected, the mice did not die, and no bacteria were recovered from their blood. This confirmed what Griffith already knew: Only living type S *S. pneumoniae* caused lethal infections. Something interesting happened when Griffith mixed living type R with heat-killed type S, however: Mice injected with this mixture died, and virulent type S bacteria were recovered from their blood. An unknown agent apparently transformed avirulent type R into virulent type S. Griffith called the agent the "transforming principle." It was his belief that the transforming principle was a protein.

Sixteen years later, in 1944, bacteriologists Oswald Avery, Colin MacLeod, and Maclyn McCarty designed an experiment that showed conclusively that the transforming principle was DNA rather than protein. They showed that R bacteria could be transformed to S bacteria in a test tube. They then progressively purified their extract until only proteins and the two nucleic acids, RNA and DNA, remained. They placed some of the mixture onto agar plates (glass dishes containing a gelatin growth medium). At this point, transformation still occurred; therefore, it was clear that one of these three molecules was the transforming agent. They treated their extract with protein-degrading enzymes, which denatured (destroyed) all the proteins in the extract. Despite the denaturing of the proteins, transformation still occurred when some of the extract was plated; had protein been the transforming agent, no transformation could have occurred. Protein was eliminated as the transforming agent. The next step was to determine which of two nucleic acids was responsible for the transformation of the R strain into the S strain. They introduced RNase, an enzyme that degrades RNA, to the extract. The RNA was destroyed, yet transformation took place. RNA was thus eliminated. At this point, it was fairly obvious that DNA was the transforming agent. To conclusively confirm this, they introduced DNase to the extract. When the DNA was degraded by the enzyme, transformation did not take place, showing that DNA was the transforming agent.

Another way that genetic material can be exchanged between bacteria is by transduction. Transduction requires the presence of a bacteriophage (a virus that infects bacteria). A virus is a simple structure consisting of a protein coat called a capsid that contains either RNA or DNA. Viruses are acellular, nonliving, and extremely small. To reproduce, they must infect living cells and use the host cell's internal structures to replicate their genetic material and manufacture viral proteins. Bacteriophages, or phages, infect bacteria by attaching themselves to a bacterium and injecting their genetic material into the cell. Sometimes, during the assembly of new viral particles, a piece of the host cell's DNA may be enclosed in the viral capsid. When the virus leaves the host cell and infects a second cell, that piece of bacterial DNA enters the second cell, thus changing its genetic makeup. Generalized transduction (the transfer of a gene from one bacterium to another) was discovered by Joshua and Esther Lederberg and Norton Zinder in 1952. Using *E. coli* and a bacteriophage called *P1*, the Lederbergs and Zinder were able to show that transduction could be used to map genes to the bacterial chromosome.

Hershey-Chase Bacteriophage Experiments

The use of bacteriophages has been instrumental in confirming DNA as the genetic material of living cells. Alfred Hershey and Martha Chase devised a series of experiments using *E. coli* and the bacteriophage *T2* that conclusively established DNA as genetic material in 1953. Bacteria are capable of manufacturing all essential macromolecules by utilizing material from their environment. Hershey and Chase grew cultures of *E. coli* in a growth medium enriched with a radioactive isotope of phosphorus, phosphorus 32. DNA contains phosphorus; as the succeeding generations of bacteria pulled phosphorus from the growth medium to manufacture DNA, each DNA strand also carried a radioactive label. *T2* phages were used to infect the cultures of *E. coli*. When the new *T2* viruses were assembled in the bacterial cells, they too carried the radioactive label phosphorus 32 on their DNA. A second culture of *E. coli* was grown in a medium enriched with radioactive sulfur 35. Proteins contain sulfur (but no phosphorus). *T2* viruses were used to infect this culture. New viruses contained the sulfur 35 label on their protein coats.

Since the *T2* phage consists of only protein and DNA, one of these two molecules had to be the genetic material. Hershey and Chase infected unlabeled *E. coli* with both types of radioactive *T2* phages. Analysis has shown that the

An example of a bacterial colony formed by a pathogenic strain of Escherichia coli, *displayed by microbiologist Jay Lewis shortly after an outbreak of food poisoning in Washington state in 1996.* (AP/Wide World Photos)

phosphorus 32 label passed into the bacterial cells, while the sulfur 35 label was found only in the protein coats that did not enter the cells. Since the protein coat did not enter the bacterial cell, it could not influence protein synthesis. Therefore, protein could not be the genetic material. The Hershey-Chase experiment confirmed DNA as the genetic material.

Restriction Enzymes and Gene Expression

Using the aforementioned methods, it has been possible to construct a complete genetic map showing the order in which genes occur on the chromosome of *E. coli* and other bacteria. Certain genes are common to all bacteria. There are also several genes that are shared by bacteria and higher life-forms, including humans. Further research showed that genes can be either inserted into or deleted from bacte-

rial DNA. In nature, only bacteria contain specialized enzymes called restriction enzymes. Restriction enzymes are capable of cutting DNA at specific sites called restriction sites. The function of restriction enzymes in bacteria is to protect against invading viruses. Bacterial restriction enzymes are designed to destroy viral DNA without harming the host DNA. Hundreds of different restriction enzymes have been isolated from bacteria, and each is named for the bacterium from which it comes. The discovery and isolation of restriction enzymes led to a new field of biological endeavor: genetic engineering.

Use of these enzymes has made gene cloning possible. Cloning is important to researchers because it permits the detailed study of individual genes. Restriction enzymes have also been used in the formation of genomic libraries (a collection of clones that contains at least one copy of every DNA sequence in the genome). Genomic libraries are valuable because they can be searched to identify a single DNA recombinant molecule that contains a particular gene or DNA sequence.

Bacterial studies have been instrumental in understanding the regulation of gene expression, or the translation of a DNA sequence first to a molecule of messenger RNA (mRNA) and then to a protein. Bacteria live in environments that change rapidly. To survive, they have evolved systems of gene regulation that can either "turn on" or "turn off" a gene in response to environmental conditions. François Jacob and Jacques Monod discovered the *lac* operon, a regulatory system that permits *E. coli* to respond rapidly to changes in the availability of lactose, a simple sugar. Other operons, such as the tryptophan operon, were soon discovered as well. An operon is a cluster of genes whose expression is regulated together and involves the interaction of regions of DNA with regulatory proteins. The discovery of operons in bacteria led to searches for them in eukaryotic cells. While none has been found, several other methods of regulating the expression of genes in eukaryotes have been described.

Impact and Applications

Diabetes mellitus is a disease caused by the inability of the pancreas to produce insulin, a protein hormone that is part of the critical system that controls the body's metabolism of sugar. Prior to 1982, people who suffered from diabetes controlled their disease with injections of insulin that had been isolated from other animals, such as cows. In 1982, human insulin became the first human gene product to be manufactured using recombinant DNA. The technique is based on the knowledge that genes can be inserted into the bacterial chromosome; that once inserted, the gene product, or protein, will be produced; and that once produced, the protein can be purified from bacterial extracts. Human proteins are usually produced by inserting a human gene into a plasmid vector, which is then inserted into a bacterial cell. The bacterial cell is cloned until large quantities of transformed bacteria are produced. From these populations, human proteins, such as insulin, can be recovered.

Many proteins used against disease are manufactured in this manner. Some examples of recombinant DNA pharmaceutical products that are already available or in clinical testing include atrial natriuretic factor, which is used to combat heart failure and high blood pressure; epidermal growth factor, which is used in burns and skin transplantation; factor VIII, which is used to treat hemophilia; human growth hormone, which is used to treat dwarfism; and several types of interferons and interleukins, which are proteins that have anticancer properties.

Bacterial hosts produce what are called the "first generation" of recombinant DNA products. There are limits to what can be produced in and recovered from bacterial cells. Since bacterial cells are different from eukaryotic cells in a number of ways, they cannot process or modify most eukaryotic proteins, nor can they add sugar groups or phosphate groups, additions that are often required if the protein is to be biologically active. In some cases, human proteins produced in prokaryotic cells do not fold into the proper three-dimensional shape; since shape determines function in proteins, these proteins are nonfunctional. For this reason, it may never be possible to use bacteria to manufacture all human proteins. Other organisms are used to produce what are called the second generation of recombinant DNA products.

The impact of the study of bacterial structures and genetics and the use of bacteria in biotechnology, cannot be underestimated. Bacterial research has led to the development of an entirely new branch of science, that of molecular biology. Much of what is currently known about molecular genetics, the expression of genes, and recombination comes from research involving the use of bacteria. Moreover, bacteria have had and will continue to have applications in the production of pharmaceuticals and the treatment of disease. The recombinant DNA technologies developed with bacteria are now being used with other organisms to produce medicines and vaccines.

—Kate Lapczynski, updated by Bryan Ness

See also: Archaea; Bacterial Resistance and Super Bacteria; Biopharmaceuticals; Cholera; Chromosome Walking and Jumping; Cloning; Diabetes; Diphtheria; Gene Regulation: Bacteria; Gene Regulation: *Lac* Operon; Gene Regulation: Viruses; Genetic Code, Cracking of; Lateral Gene Transfer; Model Organism: *Escherichia coli*; Molecular Genetics; Plasmids; Restriction Enzymes; Transposable Elements.

Further Reading

Birge, Edward A. *Bacterial and Bacteriophage Genetics.* 4th ed. New York: Springer, 2000. Examines how genetic investigations and manipulations of bacteria and bacteriophages have made vital contributions to the basic understanding of living cells and to the development of genetic engineering and biotechnology.

Drlica, Karl. *Understanding DNA and Gene Cloning: A Guide for the Curious.* New York: John Wiley & Sons, 2003. A basic overview designed to help lay readers understand molecular biology and recombinant DNA technology. Good illustrations and graphics.

Goldberg, Joanna B., ed. *Genetics of Bacterial Polysaccharides.* Boca Raton, Fla.: CRC Press, 1999. Gives background on the field's history, polysaccharide diversity, research gaps, and nomenclature issues. Nine chapters by international researchers present the genetic analysis of polysaccharides from various bacteria pathogens to humans and one symbiotic with legumes.

Hacker, J., and J. B. Kaper, eds. *Pathogenicity Islands and the Evolution of Pathogenic Microbes.* 2 vols. New York: Springer, 2002. Explores pathogenicity islands, plasmids, and bacteriophages, which are able to carry genes whose products are involved in pathogenic processes. Shows how such elements and their products play an important role in pathogenesis due to the intestinal *E. coli* as well to *Shigellae.*

Hatfull, Graham F., and William R. Jacobs, Jr., eds. *Molecular Genetics of Mycobacteria.* Washington, D.C.: ASM Press, 2000. Surveys all aspects of mycobacterial genetics in the context of new genomic information, starting with the development of mycobacterial genetics and then presenting the molecular genetics of mycobacteria in sections on genomes and genetic exchange, gene expression, metabolism, and genetic strategies.

Russell, Peter J. *Fundamentals of Genetics.* 2d ed. San Francisco: Benjamin Cummings, 2000. Introduces the three main areas of genetics: transmission genetics, molecular genetics, and population and quantitative genetics. Reflects advances in the field, such as the structure of eukaryotic chromosomes, alternative splicing in the production of mRNAs, and molecular screens for the isolation of mutants.

Schumann, Wolfgang, S. Dusko Ehrlich, and Naotake Ogasawara, eds. *Functional Analysis of Bacterial Genes: A Practical Manual.* New York: Wiley, 2001. Follows two teams of laboratories that analyze thousands of newly discovered bacterial genes to try to discover their functions. Addresses the biology of *Bacillus subtilis.*

Thomas, Christopher M., ed. *The Horizontal Gene Pool: Bacterial Plasmids and Gene Spread.* Amsterdam: Harwood Academic, 2000. International geneticists, biologists, and biochemists discuss the various contributions plasmids make to horizontal gene pools: replication, stable inheritance, and transfer modules; the phototypic markers they carry; how they evolve; how they contribute to their host population; and approaches for studying and classifying them.

Watson, James D., et al. *Recombinant DNA: A Short Course.* New York: Scientific American Books, 1983. A classic account by one of three men who shared a Nobel Prize in Physiology or Medicine for describing the molecular structure of DNA.

Web Site of Interest

E. coli Genome Project, University of Wisconsin. http://www.genome.wisc.edu. The genome research center that sequenced the organism's complete K-12 genome now maintains and updates that sequence as well as those of other strains and other pathogenic *Enterobacteriaceae.*

Bacterial Resistance and Super Bacteria

Field of study: Bacterial genetics
Significance: *Antibiotic-resistant bacteria have become a significant worldwide health concern. Some strains of bacteria (called super bacteria) are now resistant to most, if not all, of the available antibiotics and threaten to return health care to a preantibiotic era. Understanding how and why bacteria become resistant to antibiotics may aid treatment, the design of future drugs, and efforts to prevent other bacterial strains from becoming resistant to antibiotics.*

Key terms

ANTIMICROBIAL DRUGS: chemicals that destroy disease-causing organisms without damaging body tissues; chemicals made naturally by bacteria and fungi are also known as antibiotics

PLASMIDS: small, circular pieces of DNA that can exist separately from the bacterial chromosome; plasmids can be transferred among bacteria, and they may carry more than one R factor

RESISTANCE FACTOR (R FACTOR): a piece of DNA that carries a gene encoding for resistance to an antibiotic

TRANSPOSONS: also known as jumping genes, transposons are pieces of DNA that carry R factors and can integrate into a bacterial chromosome; they are also responsible for the spread of drug resistance in bacteria and fungi, and, like plasmids, each transposon may carry more than one R factor

History of Antibiotics

Throughout history, illnesses such as cholera, pneumonia, and sexually transmitted diseases have plagued humans. However, it was not until the early twentieth century that antibiotics were discovered. Until then, diseases such as diphtheria, cholera, and influenza were serious and sometimes deadly. With the advent of the antibiotic era, it appeared that common infectious diseases would no longer be a serious health concern. A laboratory accident led to the discovery of the first mass-produced antibiotic. In 1928, Scottish bacteriologist Alexander Fleming grew *Staphylococcus aureus* in petri dishes, and the plates became contaminated with a mold. Before Fleming threw out the plates, he noticed that there was no bacterial growth around the mold. The mold, *Penicillium notatum*, produced a substance that was later called penicillin, which was instrumental in saving the lives of countless soldiers during World War II. From the 1950's until the 1980's, antibiotics were dispensed with great regularity for most bacterial infections, for earaches, for colds, and as a preventive measure.

As the twentieth century progressed, however, it became apparent that the initial promise of antibiotics was mitigated by the ability of microorganisms to evolve quickly, given their relatively short life spans. Emerging infectious diseases such as multidrug-resistant tuberculosis, vancomycin-resistant enterococci, and penicillin-resistant gonorrhea became serious global health care concerns. The problem was exacerbated by the seemingly haphazard dispensing of antibiotics for viral infections (against which antibiotics are ineffective, although often prescribed as a hedge against secondary infections or simply to palliate patients).

The Rise of Bacterial Resistance

On average, bacteria can replicate every twenty minutes. Several generations of bacteria can reproduce in a twenty-four-hour period. This quick generation time leads to a rapid adaptation to changes in the environment. English naturalist Charles Darwin's *On the Origin of Species by Means of Natural Selection* (1859) first explained the theory of natural selection, the process whereby this adaptation occurs. If an organism has an advantage over other organisms (such as the ability to grow in the presence of a potentially harmful substance), that organism will survive to pass that characteristic on to its offspring while the other organisms die. The emergence of antibiotic-resistant bacteria is an excellent example of Darwin's theory of natural selection at work.

In the early twentieth century, German microbiologist Paul Ehrlich coined the term "magic bullet" in reference to chemotherapy (the treatment of disease with chemical compounds). For a drug such as an antibiotic to be a "magic bullet," it must have a specific target that is unique to the disease-causing agent and cannot harm the host in the process of curing the disease. In 1910, Ehrlich discovered that arsphenamine (Salvarsan), a derivative of arsenic, could be used to treat syphilis, a common sexually transmitted disease in the early twentieth century. Until that time syphilis had no known cure. The use of Salvarsan did cure some patients of syphilis, but, since it was a rat poison, it killed other patients. Generally speaking, antimicrobials have specific targets (or modes of action) within bacteria. They target the following structures or processes: synthesis of the bacterial cell wall, injury to the plasma membrane, and inhibition of synthesis of proteins, DNA, RNA, and other essential metabolites (all of these substances are building blocks for the bacteria). A good antibiotic will have a target that is unique to the bacteria so the host (the patient) will not be harmed by the drug.

Bacteria and fungi are, of course, resistant to the antibiotics they naturally produce. Other bacteria have the ability to acquire resistances to antimicrobials, and this drug resistance occurs either through a mutation in the DNA or resistance genes on plasmids or transposons. Plasmids are small, circular pieces of DNA that can exist within or independently of the bacterial chromosome. Transposons, or "jumping genes," are pieces of DNA that can jump from one bacterial species to another and be integrated into the bacterial chromosome. The spread of plasmids and transposons that carry antibiotic resistance genes has led bacteria to become resistant to many, if not all, currently available antibiotics.

Several antimicrobial resistance mechanisms allow bacteria to become drug resistant. The first mechanism does not allow the drug to enter the bacterial cell. A decrease in the permeability of the cell wall will inhibit the antimicrobial drug from reaching its target. An alteration in a penicillin-binding protein (pbp), a protein found in the bacterial cell wall, will allow the cell to "tie up" the penicillin. Also, the pores in the cell wall can be altered so the drug cannot pass through. A second strategy is to pump the drug out of the cell after it has entered. Such systems are found in pathogenic *Escherichia coli*, Pseudomonas aeruginosa, and *Staphylococcus aureus*. These pumps are usually nonspecific and can cause bacteria to become resistant to more than one antibiotic at a time. Another method of resistance is through chemical modification of the drug. Penicillin is inactivated by breaking a chemical bond found in its ring structure. Other drugs are inactivated by the addition of other chemical groups. Finally, the target of the drug can be altered in such a manner that it is no longer affected by the drug. For example, *Mycobacterium tuberculosis*, which causes tuberculosis, became resistant to the drug rifampin by altering the three-dimensional structure of a specific protein.

Antibiotic Misuse and Drug Resistance

The misuse of antibiotics over several decades has caused many strains of bacteria to become resistant. For some bacterial infections, only one or no effective drugs are available for treatment. Many different factors of misuse, overuse, and abuse of antibiotics have led to drug-resistant diseases. Perhaps one of the most important factors in the emergence of drug-resistant bacteria is the overprescription or inappropriate use of antibiotics. Another major factor is misuse by the patient. After several days of taking an antibiotic, a patient may begin to feel better and decide not to finish all of the prescription. By not completing the full course of treatment, the patient merely kills the bacteria that are sensitive to the antibiotic, leaving the resistant bacteria to grow, multiply, pass on their resistant genes, and cause the same infection. This time, another antibiotic (if there is one available that is effective) must be used.

Another contributing factor is the ease with which the newest and best antibiotics may be obtained in many countries. In several countries in Central America, for example, one can walk into the local pharmacy and receive any antibiotic without a prescription. Another factor in the worldwide spread of drug-resistant infectious diseases is the ease of travel. Infected people can carry bacteria from one continent to another in a matter of hours and infect anyone with whom they come in contact.

The use of antibiotics is not limited to humans. They also play an important role in agriculture. Antibiotics are added to animal feed on farms to help keep herds healthy, and they are also used on fish farms for the production of fish for market. Antibiotics are used to treat domestic animals such as cats, dogs, birds, and fish and are readily available in pet stores to clear up fish aquariums. This widespread use of antibiotics allows bacteria in all environmental niches the possibility of becoming resistant to potentially useful drugs.

Emerging Resistant Infections and Super Bacteria

The misuse of antibiotics over the decades has led to more infectious diseases becoming resistant to the current arsenal of drugs. Some diseases that could be treated effectively in the 1970's and 1980's can no longer be controlled with the same drugs. Two very serious problems have emerged: vancomycin-resistant enterococci and multidrug-resistant tuberculosis. The enterococcus is naturally resistant to many types of antibiotics, and the only effective treat-

Multiple-Resistant Bacteria

They lurk in schools, nursing homes, and hospitals—perhaps even in your home. Often, you cannot see them to avoid them. Increasingly, they are a global health problem. What are these unseen purveyors of disease? Antibacterial soaps.

Antibacterial soaps contain antibacterials, a subclass of antimicrobials, which kill or inhibit the growth of bacteria and other microorganisms. Antiseptics are antimicrobial agents that are sufficiently nontoxic to be applied to human tissue. Antibiotics are chemicals that inhibit a specific pathway or enzyme in a bacterium and are critical to the treatment of a bacterial infection. When bacteria are exposed to sublethal concentrations of an antibiotic, resistance can develop through the elimination of normal bacteria, allowing the resistant ones to survive and reproduce. The question has been whether exposure to antibacterial products can promote antibiotic resistance. The answer is that the use of antibacterial products may actually *increase* the prevalence of antibiotic-resistant bacteria.

Antibiotic resistance is irreversible and unavoidable, due to the selective pressure on bacteria to become resistant. This selection is in large part a result of the widespread use of antibiotics to increase growth rates in livestock, as well as unnecessary and improper use of antibiotics to restore and maintain human health. The indiscriminate use or overuse of antibiotics has been widely blamed for the appearance of so-called super bacteria—bacteria that are unaffected by more than one antibiotic. In addition, a widely used antibacterial agent found in toothpaste, kitchen utensils and appliances, clothing, cat litter, and toys could cause resistant strains of bacteria to develop.

Triclosan is a good example of the potent antibacterial and antifungal agents that are increasingly used to produce "germ-free" consumer products. Until recently, triclosan was considered a broad-spectrum antiseptic rather than a true antibiotic. As a general biocide, triclosan was not expected to have a specific target in the bacterial cell. However, Stuart Levy and his colleagues at Tufts University School of Medicine determined that triclosan specifically interferes with an enzyme important in the synthesis of plasma membrane lipids. As triclosan kills off normal bacteria, it could make way for the growth of strains with triclosan-insensitive enzymes. More troubling, one of the front-line antibiotics commonly used to treat tuberculosis, isoniazid, targets the same enzyme, raising the possibility that the use of triclosan will lead to new drug-resistant strains of *Mycobacterium tuberculosis*.

Consumers are convinced that use of products with antimicrobial chemicals will lower their risk of infection. While this has not been demonstrated scientifically, effective handwashing has been demonstrated to prevent illness. However, the key to effective handwashing is the length of time (15-30 seconds) spent scrubbing, not the inclusion of antibacterials in the soap. Regular soap, combined with scrubbing action, physically dislodges and removes microorganisms. The constant exposure of bacteria to sublethal concentrations of triclosan promotes development of resistance; the substitution of antibacterial soap for proper handwashing techniques will eventually render triclosan ineffective. In the battle of the soaps, "plain" wins.

—*Laurie F. Caslake*

ment has been vancomycin. With the appearance of vancomycin-resistant enterococci, however, there are no reliable alternative treatments. The fear that vancomycin resistance will spread to other bacteria such as *staphylococci* seems well founded: A report from Japan in 1997 indicated the existence of a strain of staphylococcus that had become partially resistant to vancomycin. If a strain of methicillin-resistant *Staphylococcus aureus* (MRSA) also becomes resistant to vancomycin, there will be no effective treatment available against this super bacterium.

A second problem is the appearance of multidrug-resistant tuberculosis. *Mycobacterium tuberculosis* is a slow-growing bacterium that requires a relatively long course of antibiotic therapy. Tuberculosis (TB) is spread easily, and it is a deadly disease. In the United States in 1900, tuberculosis was the number-one cause of death. In the 1990's, it was still a leading cause of death worldwide. Treatment of multidrug-resistant tuberculosis requires several antibiotics taken over a period of at least six months, with a success rate of approximately 50 percent; on the other hand, susceptible strains of TB have a cure rate of nearly 100 percent.

Another contributing factor to the emergence of drug-resistant infectious diseases is the lack of basic knowledge about some bacteria. Funding for basic genetic research on tuberculosis was reduced dramatically in the mid-twentieth century when it appeared that TB would be eradicated just as smallpox had been. The appearance of multidrug-resistant tuberculosis caught scientists and physicians unprepared. Little was known about the genetics of tuberculosis or how drug resistance occurred.

Another concern about drug-resistant infections is how to control them. Hospitals are vigilant, and, in some cases, very proactive in screening for drug-resistant infections. People can be asymptomatic carriers (that is, they carry the disease-causing organism but are still healthy) of a disease such as methicillin-resistant *Staphylococcus aureus* and could infect other people without knowing it. The role of the infection-control personnel is to find the source of the infection and remove it.

Impact and Applications

There is little encouraging news about the availability of new antibiotics. The crisis of super bacteria has altered the view that few new antibiotics would be needed. Pharmaceutical companies are scrambling to discover new antimicrobial compounds and modify existing antibiotics. Policy decisions of the 1970's and 1980's requiring more and larger clinical trials for antibiotics before they are approved for use by the Food and Drug Administration have increased the price of antibiotics and the amount of time it takes to market them. It may take up to ten years from the time of "discovery" for an antibiotic to be approved for use. The scientific community has therefore had to meet the increase of drug-resistant bacterial strains with fewer and fewer new antibiotics.

The emergence of antibiotic-resistant bacteria and super bacteria is a serious global health concern that will lead to a more prudent use of available antibiotics. It has also prompted pharmaceutical companies to search for potentially new and novel antibiotics in the ocean depths, outer space, and other niches. "Rational" drug design—or RDD, drug design based on knowledge of how bacteria become drug resistant—will also be important. Exactly how scientists and physicians will be able to combat super bacteria is a question that remains to be answered. Until a more viable solution is found, prudent use of antibiotics, surveillance of drug-resistant infections, and well-orchestrated worldwide monitoring and containment of emerging diseases appear to be the answers.

—*Mary Beth Ridenhour*

See also: Archaea; Bacterial Genetics and Cell Structure; Chromosome Walking and Jumping; DNA Replication; Emerging Diseases; Gene Regulation: Bacteria; Gene Regulation: *Lac* Operon; Lateral Gene Transfer; Mutation and Mutagenesis; Model Organism: *Escherichia coli*; Natural Selection; Plasmids; Transposable Elements.

Further Reading

Franklin, T. J., and G. A. Snow. *Biochemistry of Antimicrobial Action.* New York: Chapman and Hall, 2001. Provides an explanation of the chemistry of antimicrobials and how bacteria may become resistant to their effects.

Levy, Stuart. *The Antibiotic Paradox: How the Misuse of Antibiotics Destroys Their Curative Powers.* Cambridge, Mass.: Perseus, 2002. Provides an overview of antibiotic resistance in bacteria. Levy also discusses mechanisms of resistance, reasons for the spread of antimicrobial resistance, and ways to combat this spread.

Murray, Patrick, ed. *Manual of Clinical Microbiology.* Washington, D.C.: ASM Press, 2003. Presents a direct approach to organizing information with thorough but concise treatments of all the major areas of microbiology, including new microbial discoveries, changing diagnostic methods, and emerging therapeutic challenges facing clinicians.

Tortora, Gerard. *Microbiology: An Introduction.* 7th ed. San Francisco: Benjamin Cummings, 2001. An accessible introduction to the basic principles of microbiology, the interaction between microbe and host, and human diseases caused by microorganisms. Provides an overview of antibiotics and how bacterial resistances to antibiotics occur.

Web Site of Interest

National Center for Infectious Diseases. http://www.cdc.gov/ncidod. Searchable on "antibiotic resistance" and other keywords to provide access to research articles.

Behavior

Field of study: Population genetics

Significance: *One of the long-standing questions pondered by biologists is, to what extent do genes control the way we behave? By the mid-1990's, researchers had identified human genes that had been linked to such behavioral characteristics as depression, homosexuality, schizophrenia, and alcoholism; however, such findings were complicated by methodological questions and by the problem of distinguishing between the effects of genetic and environmental factors.*

Key terms

EUGENICS: a process in which negative genetic traits are removed from the population and positive genetic traits are encouraged, by controlling, in some manner, who is allowed to reproduce

GENOME: the entire set of genes required by an organism; a set of chromosomes

HERITABILITY: the probability that a specific gene or trait will be passed from parent to offspring, rendered as a number between zero and 100 percent, with zero percent being not heritable and 100 percent being completely heritable

LINKAGE: a relation of gene loci on the same chromosome; the more closely linked two loci are, the more often the specific traits controlled by these loci are expressed together

NEUROTRANSMITTER: a chemical messenger that transmits a nerve impulse between neurons

Brain Biology

As the first organ system to begin development and the last to be completed, the vertebrate nervous system—brain, spinal cord, and nerves—with the brain at the control, remains something of an enigma to biologists. It is based on neurons, special cells that generate and transmit bioelectrical impulses. The vertebrate brain consists of as many as three major areas: the brain stem, the cerebellum, and the cerebrum. A reptilian brain consists of only the brain stem, while the mammalian brain has all three, including a well-developed cerebrum (the two large hemispheres on top). The brain stem controls basic body functions such as breathing and heart rate, while the cerebrum is the ultimate control center. Consisting of billions of neurons (commonly called brain cells), the cerebrum controls functions such as memory, speech, hearing, vision, and analytical skills.

The brain is an exceedingly complex network of billions of neurons. As messages enter the brain stem from the spinal cord, groups of neurons either respond directly or transfer information to higher levels. In order to communicate with other neurons, each individual neuron generates impulses much like the impulse that carries a voice over a telephone line, and this message travels from the beginning to the end of each neuron. At the end of one neuron and the beginning of the next in line, a small open space occurs. This space is filled with fluids, and the message is carried across to the next neuron by a chemical known as a neurotransmitter. Neurotransmitters may be of several biochemical classifications, including acetylcholines, amines, amino acids, and peptides. An individual neuron and an entire neuronal circuit may fire or not fire an impulse based on the messages carried by these neurotransmitters. For example, the signal for pain is transmitted from neuron to neuron by a peptide-based neurotransmitter known as substance P, while another peptide transmitter (endorphin) acts as a natural painkiller. Thought, memory, and behavior, then, are produced by the activity along neuronal circuits. A genetic link occurs here, since neurotransmitters are expressed either directly or indirectly based on information in genes.

By birth, the collection of approximately 21,200 genes in humans has directed the development of the nervous system. At birth, the brain consists of approximately 100 billion neurons and trillions of supporting glial cells to protect and nourish neurons. However, the intricate wiring between these neurons is yet to be determined. Studies from the 1980's and 1990's suggested that the critical networking and circuit formation between these billions of neurons that control later brain function are determined not from genes but from environmental input and experiences from birth until the brain is fully developed around age seven.

Genes and Behavior

Genes make proteins, and proteins cause biochemical responses in cells. The behavior of an animal takes place under the combined influences of its genes, expressed through the actions of proteins, and its environment. A good example is the phenomenon of mating seasons in many animals. As day length gradually increases toward spring and summer, a critical length is reached that signals the release of hormones that result in increased sexual activity, with the ultimate goal of seasonal mating. The production and activity of hormones involve genes or gene products. If the critical number of daylight hours is not reached, the genes will not be activated, and sexual behavior will not increase.

Each neuron making up the intricate networks and circuits throughout the cerebrum (80 percent of the human brain) has protein receptors (chemoreceptors) that respond to specific signaling molecules. The production of the receptors and signaling molecules used for any type of brain activity is directly tied to genes. A slightly different gene may lead to a slightly different signaling molecule or receptor and thus a slightly different cell (neuron) response. A larger difference among genes may lead to a larger difference among signaling molecules or receptors and thus a larger variation in cell response. Since human behavior involves the response of neurons and neuron networks in the brain to specific signals, and since the response of neurons occurs because of the interaction between a signaler and a receptor built by specific genes, the genetic link seems straightforward: input, signal, response, behavior. However, when the slight variations between genes are added to the considerable variation among noncoding or regulatory sequences of DNA, the genetic connection to behavior becomes much less direct. Since a gene is under the control of one or several regulatory sequences that in turn may be under the control of various environmental inputs, the amount of genetic variation among individuals is compounded by two other critical factors: the environmental variations under which the brain develops and the daily environmental variations to which the individual is exposed. A convenient way to think of genetics and behavior is to consider that genes simply allow humans to respond to a specific stimulus by building the pathway required for a response, while behavior is defined by the degree and the manner of human response.

Eugenics

The concept of eugenics was born during the evolution and study of basic genetics in the early twentieth century. Eugenics is the categorization of a specific human behavior to an underlying genetic cause. Human characteristics such as alcoholism and laziness were thought to be caused entirely by inherited genes. Since then, research has provided a much clearer picture of a genetic-behavior link. People inherit specific genes to build specific pathways that allow them to respond in certain ways to environmental input. With variations possible—from the gene to gene regulators to the final cellular response—it is virtually impossible to disconnect the "nature vs. nurture" tie that ultimately controls human behavior. Genes are simply the tools by which the environment shapes and reshapes human behavior. There is a direct correlation between gene and protein: Change the gene, change the protein. However, there is no direct correlation between gene and behavior: Changing the gene does not necessarily change the behavior. Behavior is a multifaceted, complex response to environmental influences that is only partially related to genetic makeup. Most studies conducted on humans based on twin and other relative data suggest that most behavioral characteristics have between a 30 and 70 percent genetic basis, leaving considerable room for environmental influence. For example, studies of twins indicate that homosexuality may be as much as 50 percent genetic, leaving 50 percent under environmental control.

Another important fact is that almost no behaviors are controlled by a single gene locus, and the more complex the behavior, the more likely that it is controlled by several to many genes. Hence, not only do environmental effects cloud the picture; each gene involved in more complex behavioral traits represents just a small part of the genetic basis for the trait. The study of the genetic basis for complex traits, therefore, involves the search for quantitative trait loci (QTLs), rather than for single genes. Searching for QTLs requires that a large number of genetic markers be identified in the human genome, and the Human Genome Project has provided numerous such markers. A QTL is identified by looking for "linkage" between a specific genetic marker and the trait being studied. Linkage occurs when a marker is close to one of the genes that control the trait. Practically speaking, this means that individuals with the behavioral trait have the marker, and those who do not have the trait lack the marker. Thus, geneticists are not directly identifying the genes involved, but are identifying the approximate locations of the genes. Unfortunately, the more genes control a trait, the harder it is to identify QTLs. Environmental effects can also mask the existence of QTLs, causing some people to have the trait that lack a QTL and others to lack the trait but have a QTL. In spite of these difficulties, QTLs have been identified for a number of behavioral traits, such as aggression, depression, and a number of other mental disorders.

Single-Gene Behavioral Traits

Although behavioral traits controlled by a single gene have been identified, they probably require interaction with other genes in order to produce the specific characteristics of the behavior. On top of this are laid environmental effects. The most dramatic case of a single gene that controls a complex behavior was the discovery in 2002 of the gene that controls honeybee social status. This same gene is found in fruit flies and affects how actively fruit flies seek food. Bees with a more actively expressed form of the gene (called the *for* gene) were much more likely to forage than bees with a less active *for* gene. Not surprisingly, the *for* gene produces a protein that acts as a cell-signaling molecule.

In humans, only a few behavioral traits are clearly controlled by a single gene. The best examples include Huntington's disease (a rare, autosomal dominant gene), early-onset Alzheimer's disease (also a rare, autosomal dominant gene), and fragile X syndrome (actually involves two genes). The remaining traits, discussed below, actually represent multigene traits where one primary QTL has been identified as primarily responsible.

Several genes were identified during the late 1980's and early to mid-1990's with possible direct behavioral links. A gene has been identified that seems to be involved in neurotic behaviors associated with anxiety, depression, hostility, and impulsiveness. This gene produces a protein that transports a chemical called serotonin, across neuronal membranes. Serotonin is a neurotransmitter and is the chemical that is affected by the antidepressant drug Prozac and other serotonin reuptake inhibitors (SSRIs). Scientists have also identified a gene that may be related to schizophrenia and a gene that may determine how well alcohol is cleared from the brain after overindulgence.

One of the more recent, and in some ways controversial, discoveries involved a gene for antisocial behavior (ASB). The study, started in 1972, followed the lives of 1,037 boys from birth. Children who grew up in abusive environments were more likely to display antisocial behavior later, which is not a surprise. However, about half of the boys were found to have lower levels of an enzyme called monoamine oxidase A (MAOA), which is involved in the metabolism of several neurotransmitters. The boys with the lower MAOA activity were twice as likely to have been diagnosed with conduct disorder and were three times more likely to have been convicted of a violent crime by age twenty-six. It should be noted that lower MAOA activity alone was not enough; the boys also had to be exposed to abusive upbringings. Although the link seems strong, it has not been proved, and further study is being conducted.

A better understanding of single-gene behavioral traits could open the way to better treatment and more accurate diagnoses, but it also opens the potential for discrimination. This is especially the case for traits like antisocial behavior. Those who test for low MAOA activity might be incarcerated more readily by society or be punished differently if they are considered beyond rehabilitation. To avoid such dangers, society needs to be educated better about the interplay between genes and environment. Instead of punishing offenders more severely, earlier intervention, as early as childhood, might prevent later problems. It may also be possible to stabilize MAOA activity to near-normal levels in those who have inherently low activity levels. For most behavioral traits, though, such concerns are probably unwarranted. Because most behavioral traits are controlled by many genes, all interacting with the environment, diagnosis will probably never be possible. Yet, when genetic details of the QTLs that have been discovered are learned, therapies might be developed to offset their potential harm.

Multiple-Gene Behavioral Traits

Most geneticists concede that for many behavioral traits it may never be possible to sort out the details of the underlying genetic causes. Some genes may play such a minor role that the search for some QTLs will be fruitless. Nevertheless, geneticists have been able to discover QTLs for some important behavioral traits, and the heritability of a number of traits has been determined. The better data available from the Human Genome Project has spawned a new field of study called behavioral genomics.

Some traits, such as IQ, may never be fully understood from a genetic perspective. Heritability of IQ is high, but environment also plays an important role, and numerous genes are likely involved. More success has come from focusing on specific disorders. For example, four genes have been identified so far that are involved in attention deficit hyperactivity disorder. Other QTLs have been identified in some studies but have not been found in others. This shows one of the frustrating aspects of studying the genetics of behavior. QTLs identified using one set of data will not be supported by another set of data. This may be true because such QTLs play such a small part in developing the trait that they are undetectable under certain environmental conditions. Genes and QTLs for dyslexia, schizophrenia, and homosexuality have also been discovered. The study on homosexuality has been controversial, because the QTLs discovered by one set of researchers have never been successfully identified by anyone else.

For the most complex human traits QTLs still await discovery, but twin studies have given some insights. Twin studies involve comparing the traits of identical twins that were separated from birth. The assumption is that, because they have been raised in different environments, any traits they share will be primarily due to genetics rather than environment. A study of Swedish men showed that heritability of cognitive (thinking) ability was 62 percent, while spatial ability was 32 percent. Heritability of other personality traits fell somewhere between these values. Although these kinds of studies are interesting, they may be misleading.

Consequently, a number of geneticists criticize such research, especially twin studies, which have some inherent statistical problems. Such studies can also lead to misunderstandings, especially by nonscientists, who often interpret the numbers incorrectly. Saying that cognitive ability has a 62 percent heritability does not mean that a child has a 62 percent chance of being as intelligent as his or her parents, but rather that, of the factors involved in determining a person's intelligence, genetics accounts for approximately 62 percent of the observed variation in the population.

The Future of Behavioral Genetics

Researchers are actively seeking additional and stronger links between behavior and genetics, but even when such links are found, the degree to which a particular gene is involved and the amount of variation among humans may be hard to uncover. The Human Genome Project has greatly accelerated the search for the genetic bases of behavior, but with these new data has come an even clearer realization of the complexity of the connections between genes and human behavior. If nothing else, the future should hold more precise answers to the long-standing questions about what makes human beings who they are. The current understanding makes it clear that behavior is determined neither solely by genes nor solely by the environment. Continuing studies should make the relative contributions of genes and environment more understandable.

—*W. W. Gearheart, updated by Bryan Ness*

See also: Aggression; Alcoholism; Altruism; Biological Clocks; Biological Determinism; Criminality; Developmental Genetics; Eugenics; Gender Identity; Genetic Engineering: Medical Applications; Genetic Engineering: Social and Ethical Issues; Genetic Screening; Genetic Testing; Genetic Testing: Ethical and Economic Issues; Heredity and Environment; Homosexuality; Human Genetics; Inbreeding and Assortative Mating; Intelligence; Klinefelter Syndrome; Knockout Genetics and Knockout Mice; Miscegenation and Antimiscegenation Laws; Natural Selection; Sociobiology; Steroid Hormones; Twin Studies; XYY Syndrome.

Further Reading

Avital, Eytan, and Eva Jablonka. *Animal Traditions: Behavioural Inheritance in Evolution.* New York: Cambridge University Press, 2000. Broadens the evolutionary approach to behavior by arguing that the transfer of learned information across generations is indispensable.

Badcock, C. R. *Evolutionary Psychology: A Critical Introduction.* Malden, Mass.: Polity Press in association with Blackwell, 2000. An introductory text that addresses such topics as selection and adaptation, survival of the fittest, the benefits and costs of brain evolution, psychological conflict between parent and child, language, and development and conflict.

Benjamin, Jonathan, Richard P. Ebstein, and Robert H. Belmaker, eds. *Molecular Genetics and the Human Personality.* Washington, D.C.: American Psychiatric Association, 2002. Provides a comprehensive overview of the genetic basis for human personality. Eighteen chapters, each of which ends with a reference section. Index.

Briley, Mike, and Fridolin Sulser, eds. *Molecular Genetics of Mental Disorders: The Place of Molecular Genetics in Basic Mechanisms and Clinical Applications in Mental Disorders.* Malden, Mass.: Blackwell, 2001. Explores the role of molecular genetics in the understanding of mental disorders and how molecular genetics might help in the development of new drugs for mental illness. Illustrations.

Burnham, Terry, and Jay Phelan. *Mean Genes: From Sex to Money to Food—Taming Our Primal Instincts.* Cambridge, Mass.: Perseus, 2000. In examining the issues that most affect people's lives—body image, money, addiction, violence, and relationships, friendship, love, and fidelity—Burnham argues that struggles for self-improvement are, in fact, battles against one's own genes.

Carson, Ronald A., and Mark A. Rothstein. *Behavioral Genetics: The Clash of Culture and Biology.* Baltimore: Johns Hopkins University Press, 1999. Experts from a range of disciplines—genetics, ethics, neurosciences, psychiatry, sociology, and law—address the cultural, legal, and biological underpinnings of behavioral genetics.

Cartwright, John. *Evolution and Human Behav-*

ior: Darwinian Perspectives on Human Nature. Cambridge, Mass.: MIT Press, 2000. Offers an overview of the key theoretical principles of human sociobiology and evolutionary psychology and shows how they illuminate the ways humans think and behave. Argues that humans think, feel, and act in ways that once enhanced the reproductive success of our ancestors.

Clark, William R., and Michael Grunstein. *Are We Hardwired? The Role of Genes in Human Behavior.* New York: Oxford University Press, 2000. Explores the nexus of modern genetics and behavioral science, revealing that few elements of behavior depend upon a single gene; instead, complexes of genes, often across chromosomes, drive most of human heredity-based actions. Asserts that genes and environment are not opposing forces but work in conjunction.

DeMoss, Robert T. *Brain Waves Through Time: Twelve Principles for Understanding the Evolution of the Human Brain and Man's Behavior.* New York: Plenum Trade, 1999. Provides an accessible examination on what makes humans unique and delineates twelve principles that can explain the rise of humankind and the evolution of human behavior.

Plomin, Robert, et al. *Behavioral Genetics.* 4th ed. New York: Worth, 2001. Introductory text that explores the basic rules of heredity, its DNA basis, and the methods used to find genetic influence and to identify specific genes.

Rosen, David H., and Michael C. Luebbert, eds. *Evolution of the Psyche.* Westport, Conn.: Praeger, 1999. Surveys a range of scientific and theoretical approaches to understanding some of the most important markers connected with the evolution of the psyche, including sex and mating, evolution and creativity and humor, the survival value of forgiveness, and the evolutionary significance of archetypal dreams.

Wright, William. *Born That Way: Genes, Behavior, Personality.* New York: Knopf, 1998. Uses twin and adoption studies to trace the evolution of behavioral genetics and discusses the corroborating research in molecular biology that underlines the links between genes and personality.

Web Sites of Interest

American Association for the Advancement of Science, Behaviorial Genetics Project. http://www.aaas.org/spp/bgenes/meetings.shtml.

Human Genome Project Information, Behavioral Genetics. http://www.ornl.gov/techresources/human_genome/elsi/behavior.html. This site includes information on the basics of behavioral genetics and links to related resources.

National Institute of Mental Health, Center for Genetic Studies. http://zork.wustl.edu/nimh. A technical site on the collecting of clinical data to help determine the possible genetic bases of certain mental disorders.

Biochemical Mutations

Fields of study: Human genetics and social issues; Molecular genetics

Significance: *The study of the biochemistry behind a particular phenotype is often necessary to understand the modes of inheritance of mutant genes. Knowledge of the biochemistry of mutant individuals is especially useful in determining treatments for genetic diseases.*

Key terms

ALLELE: a form of a gene at a specific gene locus; a locus in an individual organism typically has two alleles

BIOCHEMICAL PATHWAY: the steps in the production or breakdown of biological chemicals in cells; each step usually requires a particular enzyme

GENOTYPE: the genetic characteristics of a cell or organism, expressed as a set of symbols representing the alleles present

HETEROZYGOUS: a genotype in which a locus has two alleles that are different

HOMOZYGOUS: a genotype in which a locus has two alleles that are the same

PHENOTYPE: expressed or visible characteristics of a genotype; different genotypes often are expressed as different phenotypes but may have the same phenotype

Proteins and Simple Dominant and Recessive Alleles

In order to understand how certain genotypes are expressed as phenotypes, knowledge of the biochemistry behind gene expression is essential. It is known that the various sequences of nitrogenous bases in the DNA of genes code for the amino acid sequences of proteins. How the proteins act and interact in an organism determines that organism's phenotype.

Simple dominant and recessive alleles are the easiest to understand. For example, in the genetic disease phenylketonuria (PKU), two alleles of the PKU locus exist: p^+, which codes for phenylalanine hydroxylase, an enzyme that converts phenylalanine (a common amino acid in proteins) to tyrosine (another common amino acid); and p, which is unable to code for the functional form of the enzyme. Individuals with two normal alleles, p^+p^+, have the enzyme and are able to perform this conversion. However, individuals with two abnormal alleles, pp, do not have any of this enzyme and are unable to make this conversion. Since phenylalanine is not converted to tyrosine, the phenylalanine accumulates in the organism and eventually forms phenylketones, which are toxic to the nervous system and lead to mental retardation. The heterozygote, p^+p, has one normal and one abnormal allele. These individuals have phenylalanine and tyrosine levels within the normal range, since the enzyme can be used over and over again in the conversion. In other words, even when there is only one normal allele present, there is enough enzyme produced for the conversion to proceed at the maximum rate.

Many other inborn errors of metabolism follow this same pattern. In the case of albinism, for example, afflicted individuals are missing the enzyme necessary to produce the brown-black melanin pigments. Galactosemics are missing an essential enzyme for the breakdown of galactose.

Other Single-Gene Phenomena

Many other genetic phenomena can be explained by looking at the biochemistry behind them. For example, the "chinchilla coat" mutation in rabbits causes a gray appearance in the homozygous state, $c^{ch}c^{ch}$. This occurs because the c^{ch} allele codes for a pigment enzyme that is partially defective. The partially defective enzyme works much more slowly than the normal enzyme, and the smaller amount of pigment produced leads to the gray phenotype. When this allele is heterozygous with the fully defective c allele, $c^{ch}c$, there is only half as much of an enzyme that works very slowly. As one might expect, there is less pigment produced, and the phenotype is an even lighter shade of gray called light chinchilla. The enzyme concentration does affect the rate of the reaction and, ultimately, the amount of product made. This phenomenon is known as incomplete, or partial, dominance. Genes for the red pigments in such flowers as four-o'clocks and snapdragons show incomplete dominance, as do the hair, skin, and eye pigment genes of humans and the purple pigment genes of corn kernels.

Sometimes a mutation occurs that creates an enzyme with a different function instead of creating a defective enzyme. The B allele in the ABO blood-group gene codes for an enzyme that adds galactose to a short sugar chain that exists on the blood cell's surface forming the B antigen. The A allele codes for an enzyme that adds N-acetylgalactosamine to the same previously existing sugar chain, forming the A antigen. Anyone with two B alleles, I^BI^B, makes only the B antigen and is type B. Those with two A alleles, I^AI^A, make only the A antigen and are type A. Heterozygotes, I^AI^B, have the enzymes to make both antigens, and they do. Since they have both antigens on their blood cells, they are classified as type AB. This phenomenon is known as codominance and is also seen in other blood-type genes.

Biochemistry can also explain other single-gene phenomena such as the pigmentation pattern seen in Siamese cats and Himalayan rabbits. The Siamese-Himalayan allele codes for an enzyme that is so unstable that it falls apart and is completely nonfunctional at the normal body temperature of most mammals. Only at cooler temperatures can the enzyme retain its stability and function. Since mammals have lower temperatures at their extremities, it is there that the enzyme produces pigment; at more centrally located body areas, it cannot

Many genetic phenomena can be explained by looking at the biochemistry behind them. For example, the chinchilla coat in rabbits such as this one at the Dallas Zoo is caused by a mutant allele that, in the homozygous state, codes for a pigment enzyme that is partially defective. This partially defective enzyme works much more slowly than the normal enzyme, and the smaller amount of pigment produced leads to the gray phenotype. (AP/Wide World Photos)

function. This leaves a pattern of dark pigmentation on the tail, ears, nose, feet, and scrotum, with no pigmentation at other areas.

Multiple-Gene Phenomena

Few genes act completely independently, and biochemistry can be used to explain gene interactions. One simple interaction can be seen in fruit-fly eye pigmentation. There are two separate biochemical pathways to make pigment. One produces the red pteridines, and the other produces the brown omochromes. If *b* is an allele that cannot code for an enzyme necessary to make red pigments, a *bbr$^+$r$^+$* fly would have brown eyes. If *r* is an allele that cannot code for an enzyme necessary to make

brown pigment, a *b$^+$b$^+$rr* fly would have red eyes. When mated, the resulting progeny would be *b$^+$br$^+$r.* They would make both brown and red pigments and have the normal brick-colored eyes. Interbreeding these flies would produce some offspring that were *bbrr.* Since these offspring make neither brown nor red pigments, they would be white-eyed.

Another multigene phenomenon that is seen when looking at the genes of enzymes that are in the same biochemical pathway is epistasis. Consider the following pathway in dogs:

colorless → brown → black

The *a$^+$* allele codes for the enzyme that converts colorless to brown, but the *a* allele cannot, and the *b$^+$* allele codes for the enzyme that converts brown to black, but the *b* allele cannot. The phenotype of an organism that is *aab$^+$b$^+$* depends only on the *aa* genotype, since an *aa* individual produces no brown and the *b$^+$b$^+$* enzyme can make black only by converting brown to black. The cross *a$^+$ab$^+$b × a$^+$ab$^+$b* would be expected to produce the normal 9*a$^+$_b$^+$_* (black) : 3*a$^+$_bb* (brown) : 3*aab$^+$_* (white) : 1*aabb* (white) phenotypic ratio of the classic dihybrid cross, but this is more appropriately expressed as 9 black : 3 brown : 4 white ratio. (The symbol "_" is used to indicate that the second gene can be either dominant or recessive; for example, *A_* means that both *AA* and *Aa* will result in the same phenotype.) Other pathways give different epistatic ratios such as the following pathway in peas:

white → white → purple

If *A* codes for the first enzyme, *B* codes for the second enzyme, and *a* and *b* are the nonfunctional alleles, both *AAbb* and *aabb* are white. Their progeny when they are crossed, *AaBb*, is purple because it has both of the enzymes in the pathway. Interbreeding the *AaBb* progeny gives a ratio of 9 purple to 7 white.

Human pigmentation is another case in which many genes are involved. In this case, the various genes determine how much pigment is produced by nonalbino individuals. Several gene loci are involved, and the contributions of each allele of these loci is additive. In other words, the more functional alleles one has, the darker the pigmentation; the fewer one has, the lighter. Since many of the genes involved for skin, eye, and hair color are independent, ranges of color in all three areas are seen that may or may not be the same. In addition, there are genes that code for enzymes that produce chemicals that modify the expression of the pigment genes (for example, to change blue eyes to gray, convert hazel eyes to green, or change brown hair to auburn). This gives rise to the great diversity of pigmentation seen in humans today. Add to these many possible expression patterns at the biochemical level the effect of the environment, and it is clear why such great variation in phenotypic expression is possible.

—*Richard W. Cheney, Jr.*

See also: Chemical Mutagens; Chromosome Mutation; Classical Transmission Genetics; Complete Dominance; Dihybrid Inheritance; Epistasis; Inborn Errors of Metabolism; Incomplete Dominance; Monohybrid Inheritance; Mutation and Mutagenesis; Oncogenes; Phenylketonuria (PKU); Tumor-Suppressor Genes.

Further Reading

Neumann, David, et al. *Human Variability in Response to Chemical Exposures: Measures, Modeling, and Risk Assessment.* Boca Raton, Fla.: CRC Press, 1998. Addresses genetic evidence for variability in the human response to chemicals associated with reproductive and developmental effects, the nervous system and lungs, and cancer.

Strachan, Tom, and Andrew Read. *Human Molecular Genetics.* 2d ed. New York: Wiley, 1999. Provides introductory material on DNA and chromosomes and describes principles and applications of cloning and molecular hybridization. Surveys the structure, evolution, and mutational instability of the human genome and human genes, and examines mapping of the human genome, study of genetic diseases, and dissection and manipulation of genes.

Bioethics

Field of study: Bioethics; Human genetics and social issues

Significance: *Bioethics is the practice of helping society and, more specifically, families, patients, and medical teams, make tough health care decisions. This branch of philosophy focuses on helping individuals decide what is right for them while addressing the needs of families, health care providers, and society.*

Key terms

GENETIC TESTING: the use of the techniques of genetics research to determine a person's risk of developing, or status as a carrier of, a disease or other disorder

INFORMED CONSENT: the right of patients to know the risks of medical treatment and to determine what is done to their bodies

The Emergence of Bioethics

As early as the mid-1960's, advances in genetics and reproduction, life support, and transplantation technologies spurred an increased focus on ethical issues in medicine and scientific research. From the late 1960's through the mid-1970's, bioethicists were preoccupied with the moral difficulties of obtaining voluntary, informed consent from human subjects in scientific research. They concentrated on the development of ethical guidelines in research that would ensure the protection of individuals vulnerable to exploitation, including mentally or physically handicapped individuals, prisoners, and children. Beginning in the mid-1970's and continuing through the mid-1980's, bioethicists became increasingly involved in discussions of the definitions of life, death, and what it means to be human. In the mid-1980's, practitioners began to focus on cost contain-

President's Council on Bioethics

President George W. Bush established the President's Council on Bioethics by executive order on November 28, 2001. Its mission was to advise the chief executive on bioethical issues emerging from advances in biomedical science and technology. Specifically mentioned in the council's mission were embryo and stem cell research, assisted reproduction, cloning, and end-of-life issues. Other ethical and social issues identified for discussion included the protection of human research subjects and the appropriate use of biomedical technologies. The council, chaired by Leon Kass, consisted of eighteen members appointed by the president, who were eligible for reappointment. Included in that group were scientists, physicians, ethicists, social scientists, lawyers, and theologians. The council was scheduled to terminate two years after its creation unless extended.

Deeply controversial issues constituted the subject matter of the inquiries undertaken by the council. Debate among its members as well as discussions on the floors of the Senate and House of Representatives were strongly divisive, producing heated argument and disagreement. The council's members were particularly divided on the issue of human cloning, producing two recommendations for national policy. Both recommendations would ban cloning to produce children, and ten of the eighteen council members recommended a four-year moratorium on human cloning for biomedical research while the issue continued to be studied. Declining to call for an outright ban on cloning, the divided council stated that "prudent and sensible" regulation was the best way to advance research while guarding against abuse. The minority favored regulating cloned embryos used in biomedical research, including federal licensing, oversight, and time limits on the length of time for development of cloned embryos.

President Bush stated his strong opposition to human cloning in a speech in August, 2001. The Human Cloning Prohibition Act of 2003, which banned all forms of human cloning, including cloning to create a pregnancy and cloning for medical research, passed the House of Representatives in February of 2003 by a vote of 241 to 155. It also made it a crime to "receive or import a cloned human embryo or any product derived from a cloned human embryo," punishable by $1 million in fines and ten years' imprisonment. This part of the law essentially made it illegal to harvest embryonic stem cells for medical research.

Stem cells—undifferentiated cells that have the potential to grow into any type of tissue—are created in the first days of pregnancy. Scientists hope to direct stem cells to grow a variety of tissues for use in transplantation to treat serious illnesses such as cancer, heart disease, and diabetes. Embryos have been valued in research for their ability to produce these stem cells, but the harvesting process requires the destruction of days-old embryos (a procedure condemned by the Catholic Church, President Bush, anti-abortion activists, and women's rights organizations). Other research, however, points to similar promise using stem cells harvested from adults, so that no embryos are destroyed.

—*Marcia J. Weiss*

ment in health care and the allocation of scarce medical resources.

Bioethicists worry about such matters as the guarantee of privacy, especially when compulsory testing for genetic disorders is involved, and about the limits of a person's right to threaten the health of others versus the personal right to freedom of choice. For example, the dissemination of information about genetic predispositions to chronic, costly, or incapacitating conditions can result in the denial of insurance coverage, job opportunities, and admittance to educational programs. Bioethicists also debate such matters as the use of people's reproductive materials—their eggs or sperm—to create embryos or fetuses without their explicit consent.

Beginning in 1992, the Joint Commission on Accreditation of Health Care Organizations, the U.S. agency that accredits hospitals and health care institutions, required these organizations to establish committees to formulate ethics policies and address ethical conflicts and issues. Centers for the study of biomedical ethics such as the Society for Health and Human Values and the Park Ridge Center for the Study of Health, Faith, and Ethics became important forums for public debate and research.

The overriding principle of bioethics and U.S. law is to respect each person's right to decide, free of coercion, what treatments or procedures he or she will undergo, except when the person making the decision is not competent because of youth, mental retardation, or medical deterioration. Other important rights discussed by bioethicists include a patient's right to know that medical practitioners are telling the truth and the right to know the risks of proposed medical treatment.

Impact and Applications

Advances in genetics and genetic testing have created a host of dilemmas for bioethicists, patients, and the health care establishment. For example, as the ability to forecast and understand the genetic code progresses, people will have to decide whether knowing the future, even if it cannot be altered or changed, is a good thing for them or their children.

Bioethicists help people to decide whether genetic testing can be valuable for them. Factors typically considered before a person undergoes genetic testing include the nature of the test, the timing of the test, and the options that having the test results will bring. Testing can be done prenatally to detect disorders in fetuses; it can also be done before conception to determine whether a prospective parent is a carrier of a gene for a particular disorder or disease. Tests can also provide information about whether an adult is susceptible to or even in a presymptomatic state for a genetic disorder.

Practicing bioethicists help patients to focus on whether genetic testing will help them with the nature and severity of any disorders they or their children may have, the degree of disability or discomfort they may face, the costs and rigors of treatment, and the options that might be opened or closed as a result of testing. The key for consumers of genetic testing is whether the information obtained can be provided in time and at a time when it can help to guide treatments or family planning. Some affected persons need only to make lifestyle changes or

Leon Kass of the University of Chicago was appointed head of the President's Council on Bioethics in November, 2001. Professor Kass and a panel of scientists, doctors, lawyers, and ethicists advised the Bush administration on policy issues surrounding stem cell and other research in biology, medicine, and genetics. (AP/Wide World Photos)

take medications to help prevent or manage a disease; others learn that they or their offspring are at risk for, or even likely to develop, serious and often untreatable disorders. Knowing one's genetic fate may be more of a burden than a person wants, particularly if there is nothing that can be done to change or alter the risks the person faces. Bioethicists act as guides through the complicated and often wrenching decision process.

—*Fred Buchstein*

See also: Bioinformatics; Biological Determinism; Cloning: Ethical Issues; Criminality; DNA Fingerprinting; Eugenics; Eugenics: Nazi Germany; Forensic Genetics; Gene Therapy; Gene Therapy: Ethical and Economic Issues; Genetic Counseling; Genetic Engineering: Social and Ethical Issues; Genetic Screening; Genetic Testing; Genetic Testing: Ethical and Economic Issues; Human Genetics; In Vitro Fertilization and Embryo Transfer; Insurance; Miscegenation and Antimiscegenation Laws; Patents on Life-Forms; Paternity Tests; Prenatal Diagnosis; Race; Stem Cells; Sterilization Laws.

Further Reading

Bulger, Ruth Ellen, Elizabeth Heitman, and Stanley Joel Reiser, eds. *The Ethical Dimensions of the Biological and Health Sciences.* 2d ed. New York: Cambridge University Press, 2002. Designed for graduate students who will be conducting research in the medical and biological sciences. Provides essays, readings, and questions to stimulate thinking about ethical issues and implications.

Caplan, Arthur. *Due Consideration: Controversy in the Age of Medical Miracles.* New York: Wiley, 1997. A leading bioethicist analyzes the moral questions regarding scientific advancements, among them cloning, assisted suicide, genetic engineering, and treating illnesses during fetal development.

Charon, Rita, and Martha Montello, eds. *Stories Matter: The Role of Narrative in Medical Ethics.* New York: Routledge, 2002. Explores the narrative interaction of the medical field— the written and verbal communication involved in doctors' notes, patients' stories, the recommendations of ethics committees, and insurance justifications—and the way in which this interaction profoundly affects decision making, patient health, and treatment.

Comstock, Gary L., ed. *Life Science Ethics.* Ames: Iowa State Press, 2002. Introduces ethical reasoning in the area of humankind's relationship with nature and presents twelve fictional case studies as a means to show the application of ethical reasoning.

Danis, Marion, Carolyn Clancy, and Larry R. Churchill, eds. *Ethical Dimensions of Health Policy.* New York: Oxford University Press, 2002. The three authors, from varied professions within the medical field, attempt to identify the goals of health care, examine how to connect ethical considerations with the making of health policy, and discuss specific areas of ethical controversy such as resource allocation, accountability, the needs of vulnerable populations, and the conduct of health services research.

Evans, John Hyde. *Playing God? Human Genetic Engineering and the Rationalization of Public Bioethical Debate.* Chicago: University of Chicago Press, 2002. Provides a framework for understanding the public debate, and details the various positions of the debate's players, including eugenicists, theologians, and bioethicists.

Kass, Leon R. *Life, Liberty, and the Defense of Dignity: The Challenge for Bioethics.* San Francisco: Encounter Books, 2002. Examines genetic research, cloning, and active euthanasia, and argues that biotechnology has left humanity out of its equation, often debasing human dignity rather than celebrating it.

Kristol, William, and Eric Cohen, eds. *The Future Is Now: America Confronts the New Genetics.* Lanham, Md.: Rowman & Littlefield, 2002. Brings together classic writings (George Orwell, Aldous Huxley) as well as more recent essays and congressional testimony about human cloning, genetic engineering, stem cell research, biotechnology, human nature, and American democracy.

May, Thomas. *Bioethics in a Liberal Society: The Political Framework of Bioethics Decision Making.* Baltimore: Johns Hopkins University Press, 2002. Takes the debate about biomedical ethics into the political realm, analyzing how the political context of liberal constitutional democracy shapes the rights and obligations of both patients and health care professionals.

O'Neill, Onora. *Autonomy and Trust in Bioethics.* New York: Cambridge University Press, 2002. Examines issues surrounding reproductive and principled autonomy, trust, consent, and the media and bioethics.

Singer, Peter. *Unsanctifying Human Life: Essays on Ethics.* Edited by Helga Kuhse. Malden, Mass.: Blackwell, 2002. Singer is one of today's major bioethicists. Here he examines the role of philosophers and philosophy in such questions as the moral status of the embryo, animal rights, and how we should live.

Veatch, Robert M. *The Basics of Bioethics.* 2d ed. Upper Saddle River, N.J.: Prentice Hall, 2003. In a textbook designed for students, Veatch presents an overview of the main theories and policy questions in biomedical ethics. Includes diagrams, case studies, and definitions of key concepts.

Web Sites of Interest

American Journal of Bioethics Online. http://www.bioethics.net. Provides sections on cloning basics, animal cloning, stem cells, U.S. federal and state laws, the cloning debate, news, and more.

Kennedy Institute of Ethics, Georgetown University. http://www.georgetown.edu/research/kie. Links to many resources on bioethics as well as a "bioethics library" that in turn leads to resources on human genetics and ethics.

National Information Resource on Ethics and Human Genetics. http://www.georgetown.edu/research/nrcbl/nirehg. Supports links to databases, annotated bibliographies, and articles about the ethics of genetic testing and human genetics.

President's Council on Bioethics. http://bioethics.gov. Government arm that advises on ethical issues surrounding biomedical science and technology. Includes links to bioethics literature and other resources on ethics and human genetics.

The Hastings Center. http://www.thehastingscenter.org. This independent nonprofit or-

ganization specializes in bioethics, and its site contains news postings, articles on bioethics and different aspects of genetic science, and announcements of events and publications.

Biofertilizers

Field of study: Genetic engineering and biotechnology

Significance: *Biofertilizers were used in agriculture long before chemical fertilizers became prevalent during and after the Industrial Revolution. The depleted soil fertility and contamination of ecosystems caused by the extensive use of chemical fertilizers, however, has prompted the redevelopment of biofertilizers, which are designed to work according to basic principles at work in nature, taking advantage of plants and other organisms to maintain healthy soil.*

Key terms

ALGAE: minute plants that live in fresh water; they are used as biofertilizers because of their high productivity and ability to fix atmospheric nitrogen

NODULE: a symbiotic relationship between bacteria and plant roots that causes the conversion of nitrogen gas into a form readily accessible by plants

SYMBIOSIS: a mutually beneficial association between two living organisms

Overview

Biofertilizers are living microorganisms that work either alone or in association with matter and other organisms to enhance the fertility of soil. For many centuries, biofertilizers were used in organic farming in countries such as China, India, and Egypt until modernization resulted in a move toward the use of environmentally destructive chemical fertilizers. Organic agriculture integrates livestock, aquatic organisms, plants, and the scientific enhancement of natural processes to maintain ecological equilibrium, thus maximizing the production of foods and goods through complete recycling of all resources. Biofertilizers may include microorganisms, nitrogen-fixing algae, green manure, plant residues, and sewer sludge. Biofertilizers not only provide an alternative method of farming but may also provide the only way to reduce environmental contamination and soil fertility depletion caused by chemical fertilizers.

Microorganisms and Algae

Microorganisms must be applied to the soil or mixed with seeds and other ingredients before they can establish a symbiotic relationship with a plant's root system. They have been shown to stabilize manure, increase the amount of nitrogen in the soil, increase root surface area for absorption, and control the leaking of nitrogen into the groundwater. Another type of association between plant roots (particularly legumes) and nitrogen-fixing bacteria is a symbiotic relationship called a "nodule." Nodules are natural "factories" that produce ample fertilizers by converting nitrogen into ammonium, a form of nitrogen that may be used directly by plants or deposited slowly into the soil, thus enhancing its fertility. While continuous application of chemical fertilizers depletes the soil's natural fertility and destroys beneficial microorganisms, natural fertilizers produced by nodules alleviate the contamination to the ecosystem. Unfortunately, nodules are formed only between legumes and bacteria, thus limiting their use. Genetically engineered bacteria are being used to improve the efficiency of nitrogen fixation, and some researchers have attempted to transfer genes responsible for forming such symbiotic relationships from legumes to other plant species.

Algae are minute plants that are almost entirely aquatic. They grow and reproduce very rapidly during the growing season but die off during the nongrowing season. Such a "boom and bust" life cycle provides the soil with substantial amounts of nutrients through the degradation of dead algae and the deposit of the nitrogen fixed by algae during the growing season. Two types of algae, azospirillum and azotobacter, have been used as biofertilizers in rice fields for centuries in Southeast Asia. Algae are the only plants that are able to fix nitrogen by themselves. The cost of raising algae is minimal, but if they are grown too densely, they may become weeds and suffocate aquatic animals.

Green Manure, Plant Residues, and Treated Sewer Sludge

Some crops, particularly legumes, are grown and harvested to be used as green manure to restore the soil's fertility. They may be used directly or mixed with microorganisms. Green manure releases nutrients slowly and provides long-term fertility for soil. After grains are harvested for food, plant residues are processed into fertilizer in one of two ways. They may be burned to extract energy, after which the ashes are applied to the soil. Alternatively, they may be fermented in a sealed underground tank to make methane to be used as "natural" gas or ethanol, which may become an increasingly important fuel source. Genetically engineered microbes convert fiber-rich crop residues such as wheat straw and corn stalks into ethanol. Solid residue left over after fermentation is used as a biofertilizer. The fermentation approach is promising because it is environmentally friendly and also creates alternative energy sources, additional income, and market opportunities for farmers.

Even though sewer sludge has been used as a fertilizer in developing countries for some time, people are more resistant to it because of the animal and human wastes present in the sludge. Proper treatments must be in place to get rid of possible heavy metals or pathogens. More research is needed on sewer sludge before it is considered for wide use as a fertilizer.

—*Ming Y. Zheng*

See also: Biopesticides; Genetic Engineering; Genetic Engineering: Agricultural Applications; Genetically Modified (GM) Foods; High-Yield Crops; Quantitative Inheritance; Transgenic Organisms.

Further Reading

Bethlenfalvay, G. J., and R. G. Linderman. *Mycorrhizae in Sustainable Agriculture.* Madison, Wis.: American Society of Agronomy, 1992. Provides excellent discussions on the knowledge and importance of mycorrhizae to plants and the soil.
Chrispeels, M. J., and D. E. Sadava. *Plants, Genes,*

and Crop Biotechnology. Boston: Jones and Bartlett, 2003. Offers a discussion on the potential genetic modifications necessary to increase the use of microorganisms to supply nitrogen fertilizer.

Legocki, Andrezej, Hermann Bothe, and Alfred Pühler, eds. *Biological Fixation of Nitrogen for Ecology and Sustainable Agriculture.* New York: Springer, 1997. From the proceedings of the NATO Advanced Research Workshop held in Poznań, Poland, in 1996. Includes bibliographical references.

Lynch, J. M. *Soil Biotechnology: Microbiological Factors in Crop Productivity.* Boston: Blackwell Scientific, 1983. Contains some excellent information on the potential for genetically engineered microorganisms to improve crop production.

Matson, P. A., et al. "Agricultural Intensification and Ecosystem Properties." *Science* 277 (July 25, 1997): 504-509. Provides useful information on agriculture intensification and its sustainability, a strategy that can help reduce negative consequences.

Salisbury, F. B., and C. W. Ross. *Plant Physiology.* 4th ed. Belmont, Calif.: Wadsworth, 1992. Contains excellent chapters on plant nutrition and nitrogen metabolism.

Web Site of Interest

U.S. Department of Agriculture, Biotechnology: An Information Resource. http://www.nal.usda.gov/bic. A government site that offers dozens of links to information on applications of genetic engineering to agriculture.

Bioinformatics

Fields of study: Bioinformatics; Molecular genetics; Techniques and methodologies

Significance: *Bioinformatics is the application of information technology to the management of biological information to organize data and extract meaning. It is a hybrid discipline that combines elements of computer science, information technology, mathematics, statistics, and molecular genetics.*

Key terms

ALGORITHM: a mathematical rule or procedure for solving a specific problem; in bioinformatics, a computer program is built to implement an algorithm, but different algorithms may be used to achieve the same result—that is, to align two sequences

DATABASE: an organized collection of information within a computer system that can be used for storage and retrieval as well as for complex searches and analyses

GENBANK: a comprehensive, annotated collection of publicly available DNA sequences maintained by the National Center for Biotechnology Information and available through its Web site

GENOMICS: the use of high-throughput technology to analyze molecular events within cells at the whole genome scale (for example, all of the genes, all of the messenger RNA, or all of the proteins)

HUMAN GENOME PROJECT: a publicly funded international project to determine the complete DNA sequence of human genomic (chromosomal) DNA and to map all of the genes, which produced a "final" sequence in April, 2003

MICROARRAY: a technology to measure gene expression using nucleic acid hybridization of messenger RNA to a miniature array of DNA probes for many genes

PROTEOMICS: a collection of technologies that examine proteins within a cell in a holistic fashion, identifying or quantitating a large number of proteins within a single sample, identifying many protein-protein or protein-DNA interactions, and so on

The Need for Bioinformatics

The sequencing of cloned DNA molecules has become a routine, automated task in the modern molecular genetics laboratory, and large, publicly funded genome projects have determined the complete genomic sequences for humans, mice, fruit flies, dozens of bacteria, and many other species of interest to geneticists. All of this information is now freely available in online databases. Computational molecular biology tools allow for the design of polymerase chain reaction (PCR) primers, restriction enzyme cloning strategies, and even entire *in silico* experiments. This greatly accelerates the work of researchers but also changes the daily lives of many biologists so that they spend more time working with computers and less time working with test tubes and pipettors. The rapid accumulation of enormous amounts of molecular sequence data and their cryptic and subtle patterns have created a need for computerized databases and analysis tools.

Bioinformatics provides essential support services to modern molecular genetics for organizing, analyzing, and distributing data. As DNA sequencing and other molecular genetic technologies become more automated, data are generated ever more rapidly, and computing systems must be designed to store the data and make them available to scientists in a useful fashion. The use of these vast quantities of data for the discovery of new genes and genetic principles relies on the development of sophisticated new data-mining tools. The challenge of bioinformatics is in finding new approaches to deal with the volume and complexity of the data, and in providing researchers with access both to the raw data and to sophisticated and flexible analysis tools in order to advance researchers' understanding of genetics and its role in health and disease.

Database Design

The DNA sequence data collected by automated sequencing equipment can be represented as a simple sequence of letters: G, A, T, and C—which stand for the four nucleotide bases on one strand of the DNA molecule (guanine, adenine, thymine, and cytosine). These letters can easily be stored as plain text files on a computer. Similarly, protein sequences can also be stored as text files using the twenty single-letter abbreviations for the amino acids.

There is a significant advantage to storing DNA and protein sequence as plain text files, also known as flat files. Text files take up minimal amounts of hard-drive space, can be used on any type of computer and operating system, and can easily be moved across the Internet. However, a text file with a bunch of letters representing a DNA or protein sequence is essentially meaningless without some basic descrip-

tive information, such as the organism from which it comes, its location on the genome, the person or organization that produced the sequence, and a unique identification number (accession number) so that it can be referenced in scientific literature. This additional annotation information can also be stored as text—even in the same file with the sequence information—but there must be a consistent format, a standard.

In addition to maintaining basic flat-file structures for text data, it is useful to maintain sequence data in relational databases, which allow for much faster searching across multiple query terms and the linkage of sequence data files with other relevant information. The most sophisticated and widely used relational database system for bioinformatics is the Entrez system at the National Center for Biotechnology Information (NCBI). Entrez is a relational database that includes cross-links between all of the DNA sequences in GenBank. GenBank exchanges data with the DNA DataBank of Japan and the European Molecular Biology Laboratory on a daily basis to ensure that all three centers maintain the same set of data, and all peer-reviewed journals require the submission of sequence data to GenBank prior to publication of research articles; publicly funded sequencing projects, such as the Human Genome Project, submit new sequence data to GenBank as it is collected, so that the scientific community can have immediate access to it. Entrez also includes all of the derived protein sequences (translations from cDNAs and predicted coding sequences in genomic DNA), the scientific literature in MedLine/PubMed, three-dimensonal protein structures from the Protein Data Base (PDB), and human genetic information from the Online Mendelian Inheritance in Man (OMIM) database. Relational databases are even more important for more complex types of genomic data, such as gene expression microarrays and genetic variation and genotyping data sets.

Key Algorithms

Some of the key algorithms used in bioinformatics include sequence alignment (dynamic programming), sequence similarity

Steven Brenner, of the University of California at Berkeley, next to a computer running bioinformatics software in November, 2001. He advocates distributing information freely as "open source code," claiming that this is the best way to debug bioinformatics software and advance research. (AP/Wide World Photos)

(word matching from hash tables), assembly of overlapping fragments, clustering (hierarchical, self-organizing maps, principal components, and the like), pattern recognition, and protein three-dimensonal structure prediction. Bioinformatics is both eclectic and pragmatic: Algorithms are adopted from many different disciplines, including linguistics, statistics, artificial intelligence and machine learning, remote sensing, and information theory. There is no consistent set of theoretical rules at the core of bioinformatics; it is simply a collection of whatever algorithms and data structures have been found to work for the current data-management problems being faced by biologists. As new types of data become important in the work of molecular geneticists, new algorithms for bioinformatics will be invented or adopted.

New Types of Data

In addition to DNA and protein sequences, bioinformatics is being called upon to organize many other types of biological information that are being collected in ever greater amounts. Gene expression microarrays collect information on the amounts of mRNA produced from tens of thousands of different genes in a single tissue sample. Proteomics technologies are automating the process of mass spectroscopy, which allows investigators to identify and measure thousands of proteins in a single cell extract sample. Genes and proteins can also be organized into gene families based on sequence similarity, homology across organisms (comparative genomics), and function in metabolic or regulatory pathways. Many new technologies are being developed to measure genetic variation: genetic tests either for alleles of well-studied genes or for anonymous single nucleotide polymorphisms (SNPs) identified from genome sequence data. As these genotyping technologies are improved, it is becoming possible to collect data in an automated fashion for many genetic loci from a single DNA sample, or to test a single genetic locus on many thousands of DNA samples in parallel. These new data types require new database designs and the inclusion of new types of algorithms (from statistics, population genetics, and other disciplines) in bioinformatics data-management solutions.

Integration

The biggest challenge facing bioinformatics is the integration of various types of data in a form that allows scientists to extract meaningful insights into biology from the masses of information in molecular genetic databases. Genome browsers are one example of this challenge. It is extremely difficult to provide a display that allows someone to view all of the relevant information about a gene or a chromosomal region, including the identity of encoded proteins, protein structure and functional information, involvement in metabolic and regulatory pathways, developmental and tissue-specific gene expression, evolutionary relationships to proteins in other organisms, DNA motifs bound by regulatory proteins, genetic synteny with other species (that is, having genes with loci on the same chromosome), phenotypes of mutations, and known alleles and SNPs and their frequency in various populations.

Another, much more modest, goal would be simply to alert a person viewing a DNA or protein sequence in one database of the existence of additional information about that entity in other databases. At the present time, such cross-database links are inconsistent and unreliable. The NCBI cross-references its own databases—from DNA to proteins to three-dimensional structures to PubMed articles to genomes. Most special subject databases, such as those that focus on a particular species or on a particular type of molecule, link DNA and protein sequences back to the corresponding "reference"

entries in GenBank; however, these links are not reciprocal. Someone looking at a GenBank cDNA sequence in the Entrez browser would have no way of knowing that a corresponding protein entry is present in a database dedicated to *Drosophila* genetics or to G-protein coupled receptor mutants. It is never possible for scientists to be certain that they have collected all of the relevant information about a molecule of interest from all online databases.

—*Stuart M. Brown*

See also: cDNA Libraries; DNA Fingerprinting; DNA Sequencing Technology; Forensic Genetics; Genetic Testing: Ethical and Economic Issues; Genetics, Historical Development of; Genomic Libraries; Genomics; Human Genome Project; Icelandic Genetic Database; Linkage Maps; Proteomics.

Further Reading

Baxevanis, Andreas D., and B. F. Francis Ouellette. *Bioinformatics: A Practical Guide to the Analysis of Genes and Proteins.* 2d ed. Hoboken, N.J.: John Wiley & Sons, 2003. This book provides a sound foundation of basic concepts of bioinformatics, with practical discussions and comparisons of both computational tools and databases relevant to biological research. The standard text for most graduate-level bioinformatics courses.

Claverie, Jean-Michel, and Cedric Notredame. *Bioinformatics for Dummies.* Hoboken, N.J.: John Wiley & Sons, 2003. A practical introduction to bioinformatics: computer technologies that biochemical and pharmaceutical researchers use to analyze genetic and biological data. This reference addresses common biological questions, problems, and projects while providing a UNIX/Linux overview and tips on tweaking bioinformatic applications using Perl.

Krawetz, Stephen A., and David D. Womble. *Introduction to Bioinformatics: A Theoretical and Practical Approach.* Totowa, N.J.: Humana Press, 2003. Aimed at undergraduates, graduate students, and researchers. Four sections: "Biochemistry: Cell and Molecular Biology," "Molecular Genetics," "Unix Operating System," and "Computer Applications."

Mount, David W. *Bioinformatics: Sequence and Genome Analysis.* Cold Spring Harbor, N.Y.: Cold Spring Harbor Laboratory Press, 2001. A textbook written for the biologist who wants to acquire a thorough understanding of popular bioinformatics programs and molecular databases. It does not teach programming but does explain the theory behind each of the algorithms.

Nucleic Acids Research 31, no. 1 (2003). This widely respected journal produces a special issue in January of each year devoted entirely to online bioinformatics databases. The articles represent the definitive statement by the directors of each of the major public databases of molecular biology data regarding the types of information and analysis tools in their databases and plans for development in the immediate future.

Web Sites of Interest

Bioinformatics Organization. http://www.bio informatics.org. Provides a helpful tutorial on bioinformatics.

European Bioinformatics Institute. http://www .ebi.ac.uk. Maintains databases concerning nucleic acids, protein sequences, and macromolecular structures, as well as postings of news and events and descriptions of ongoing scientific projects.

Biological Clocks

Field of study: Human genetics and social issues

Significance: *Biological clocks control those periodic behaviors of living systems that are a part of their normal function. The rhythms may be of a daily, monthly, yearly, or even longer periodicity. In some cases, the clocks may be "programmed" to regulate processes that may occur at some point in the lifetime of the individual, such as those processes related to aging. Altered or disturbed rhythms may result in disease.*

Key terms

ALZHEIMER'S DISEASE: a disorder characterized by brain lesions leading to loss of memory, personality changes, and deterioration of higher mental functions

CIRCADIAN RHYTHM: a cycle of behavior, approximately twenty-four hours long, that is expressed independent of environmental changes

FREE-RUNNING CYCLE: the rhythmic activity of an individual that operates in a constant environment

HUNTINGTON'S DISEASE: an autosomal dominant genetic disorder characterized by loss of mental and motor functions in which symptoms typically do not appear until after age thirty

SUPRACHIASMATIC NUCLEUS (SCN): a cluster of several thousand nerve cells that contains a central clock mechanism that is active in the maintenance of circadian rhythms

Types of Cycles

Biological clocks control a number of physiological functions, including sexual behavior and reproduction, hormonal levels, periods of activity and rest, body temperature, and other activities. In humans, phenomena such as jet lag and shift-work disorders are thought to result from disturbances to the innate biological clock.

The most widely studied cycles are circadian rhythms. These rhythms have been observed in a variety of animals, plants, and microorganisms and are involved in regulating both complex and simple behaviors. Typically, circadian rhythms are innate, self-sustaining, and have a cyclicity of nearly, but not quite, twenty-four hours. Normal temperature ranges do not alter them, but bursts of light or temperature can change the rhythms to periods of more or less than twenty-four hours. Circadian rhythms are apparent in the activities of many species, including humans, flying squirrels, and rattlesnakes. They are also seen to control feeding behavior in honeybees, song calling in crickets, and hatching of lizard eggs.

What is known about the nature of the biological clock? The suprachiasmatic nucleus (SCN) consists of a few thousand neurons or specialized nerve cells that are found at the base of the hypothalamus, the part of the brain that controls the nervous and endocrine systems. The

SCN appears to play a major role in the regulation of circadian rhythms in mammals and affects cycles of sleep, activity, and reproduction. The seasonal rhythm in the SCN appears to be related to the development of seasonal depression and bulimia nervosa. Light therapy is effective in these disorders. Blind people, whose biological clocks may lack the entraining effects of light, often show free-running rhythms.

Genetic control of circadian rhythms is indicated by the findings of single-gene mutations that alter or abolish circadian rhythms in several organisms, including the fruit fly (*Drosophila*) and the mouse. A mutation in *Drosophila* affects the normal twenty-four-hour activity pattern so that there is no activity pattern at all. Other mutations produce shortened (nineteen-hour) or lengthened (twenty-nine-hour) cycles. The molecular genetics of each of these mutations is known.

A semidominant autosomal mutation, CLOCK, in the mouse produces a circadian rhythm one hour longer than normal. Mice that are homozygous (have two copies) for the CLOCK mutation develop twenty-seven- to twenty-eight-hour rhythms when initially placed in darkness and lose circadian rhythmicity completely after being in darkness for two weeks. No anatomical defects have been seen in association with the CLOCK mutation.

Biological Clocks and Aging

Genes present in the fertilized egg direct and organize life processes from conception until death. There are genes whose first effects may not be evident until middle age or later. Huntington's disease (also known as Huntington's chorea) is such a disorder. An individual who inherits this autosomal dominant gene is "programmed" around midlife to develop in-

A squirrel hibernates in the hands of University of Minnesota biochemist Matt Andrews. Because the ground squirrel possesses the ability to put its body into this form of stasis, it is nearly immune to strokes. The genetics of such biological clocks may one day lead to better treatments for strokes in humans. (AP/Wide World Photos)

voluntary muscle movement and signs of mental deterioration. Progressive deterioration of body functions leads to death, usually within fifteen years. It is possible to test individuals early in life before symptoms appear, but such tests, when no treatment for the disease is available, are controversial.

Alzheimer's disease (AD) is another disorder in which genes seem to program processes to occur after middle age. AD is a progressive, degenerative disease that results in a loss of intellectual function. Symptoms worsen until a person is no longer able to care for himself or herself, and death occurs on an average of eight years after the onset of symptoms. AD may appear as early as forty years of age, although most people are sixty-five or older when they are diagnosed. Age and a family history of AD are clear risk factors. Gene mutations associated with AD have been found on human chromosomes 1, 14, 19, and 21. Although these genes, especially the apolipoprotein *e4* gene, increase the likelihood of a person getting AD, the complex nature of the disorder is underscored when it is seen that the mutations account for less than half of the cases of AD and that some individuals with the mutation never get AD.

Impact and Applications

Evidence has accumulated that human activities are regulated by biological clocks. It has also become evident that many disorders and diseases, and even processes that are associated with aging, may be affected by abnormal clocks. As understanding of how genes control biological clocks develops, possibilities for improved therapy and prevention should emerge. It may even become possible to slow some of the harmful processes associated with normal aging.

—*Donald J. Nash*

See also: Aging; Alzheimer's Disease; Biological Determinism; Cancer; Developmental Genetics; Huntington's Disease; Telomeres.

Further Reading

Finch, Caleb Ellicott. *Longevity, Senescence, and the Genome*. Reprint. Chicago: University of Chicago Press, 1994. Provides a comparative review of research on organisms from algae to primates, expanding traditional gerontological and geriatric issues to intersect with behavioral, developmental, evolutionary, and molecular biology. Illustrated.

Hamer, Dean, and Peter Copeland. *Living with Our Genes: Why They Matter More than You Think*. New York: Doubleday, 1998. Links DNA and behavior and contains a good chapter on biological clocks and aging.

Medina, John J. *The Clock of Ages: Why We Age, How We Age—Winding Back the Clock*. New York: Cambridge University Press, 1996. A book written especially for the general reader. Covers aging on a system-by-system basis and includes a large section on the genetics of aging.

Nelson, James Lindemann, and Hilde Lindemann Nelson. *Alzheimer's: Answers to Hard Questions for Families*. New York: Main Street Books, 1996. Reviews Alzheimer's disease for the lay reader, guides caregivers through the difficult moral and ethical problems associated with the disease, and discusses support services.

Zallen, Doris Teichler. *Does It Run in the Family? A Consumer's Guide to DNA Testing for Genetic Disorders*. New Brunswick, N.J.: Rutgers University Press, 1997. Focuses on the practical aspects of obtaining genetic information, clearly explaining how genetic disorders are passed along in families. Provides useful information on genetic disorders, including Huntington's disease and Alzheimer's disease.

Web Sites of Interest

National Institute of Mental Health, How Biological Clocks Work. http://www.nimh.nih .gov/publicat/bioclock.cfm. An introductory site that focuses on the molecular basis of the biological clock. Includes references for further study.

National Science Foundation, Center for Biological Timing. http://www.cbt.virginia .edu/cbtdocs. This site covers the science of biological clocks and provides a biological timing tutorial.

Biological Determinism

Field of study: Human genetics and social issues

Significance: *Biological determinists argue that there is a direct causal relationship between the biological properties of human beings and their behavior. From this perspective, social and economic differences between human groups can be seen as a reflection of inherited and immutable genetic differences. This contention has been used by groups in power to claim that stratification in human society is based on innate biological differences. In particular, biological determinism has been used to assert that certain ethnic groups are biologically defective and thus intellectually, socially, and morally inferior to others.*

Key terms

DETERMINISM: the doctrine that everything, including one's choice of action, is determined by a sequence of causes rather than by free will

INTELLIGENCE QUOTIENT (IQ): performance on a standardized test, often assumed to be indicative of an individual's level of intelligence

REDUCTIONISM: the explanation of a complex system or phenomenon as merely the sum of its parts

REIFICATION: the oversimplification of an abstract concept such that it is treated as a concrete entity

The Use of Inheritance to Promote Social Order

The principle of biological determinism lies at the interface between biology and society. A philosophical extension of the use of determinism in other sciences, such as physics, biological determinists view human beings as a reflection of their biological makeup and hence simple extensions of the genes that code for these biological processes. Long before scientists had any knowledge of genetics and the mechanisms of inheritance, human societies considered certain groups to be innately superior by virtue of their family or bloodlines (nobility) while others were viewed as innately infe-

rior (peasantry). Such views served to preserve the social order. According to evolutionary biologist Stephen Jay Gould, Plato himself circulated a myth that certain citizens were "framed differently" by God, with the ranking of groups in society based on their inborn worth.

As science began to take a more prominent role in society, scientists began to look for evidence that would justify the social order. Since mental ability is often considered to be the most distinctive feature of the human species, the quantification of intelligence was one of the main tactics used to demonstrate the inferiority of certain groups. In the mid-1800's, measurements of the size, shape, and anatomy of the skull, brain, and other body features were compiled by physician Samuel George Morton and surgeon Paul Broca, among others. These measurements were used to depict races as separate species, to rank them by their mental and moral worth, and to document the subordinate status of various groups, including women. In the first decades of the twentieth century, such measurements were replaced by the intelligence quotient (IQ) test. Although its inventor, Alfred Binet, never intended it to be used in this way, psychologists such as Lewis M. Terman and Robert M. Yerkes promoted IQ as a single number that captured the complex, multifaceted, inborn intelligence of a person. IQ was soon used to restrict immigration, determine occupation, and limit access to higher education. Arthur Jensen, in 1979, and Richard Herrnstein and Charles Murray, in 1994, reasserted the claim that IQ is an inherited trait that differs among races and classes.

Problems with the Principle of Biological Determinism

Geneticists and sociobiologists (who study the biological basis of social behavior) have uncovered a variety of animal behaviors that are influenced by biology. However, the genetic makeup of an organism ("nature") is expressed only within the specific context of its environment ("nurture"). Thus genes that are correlated with behavior usually code for predispositions rather than inevitabilities. For such traits, the variation that occurs within a group is usually greater than the differences that occur between groups. In addition, the correlation between two entities (such as genes and behavior) does not necessarily imply a causal relationship (for example, the incidences of ice cream consumption and drowning are correlated only because both increase during the summer). Complex, multifaceted behaviors such as intelligence and violence are often reified, or treated as discrete concrete entities (as IQ and impulse control, respectively), in order to make claims about their genetic basis. Combined with the cultural and social bias of scientific researchers, reification has led to many misleading claims regarding the biological basis of social structure.

Biological and cultural evolution are governed by different mechanisms. Biological evolution occurs only between parents and offspring (vertically), while cultural evolution occurs through communication without regard to relationship (horizontally) and thus can occur quickly and without underlying genetic change. Moreover, the socially fit (those who are inclined to reproduce wealth) are not necessarily biologically fit (inclined to reproduce themselves). The reductionist attempt to gain an understanding of human culture through its biological components does not work well in a system (society) shaped by properties that emerge only when the parts (humans) are put together. Cultures cannot be understood as biological behaviors any more than biological behaviors can be understood as atomic interactions.

Impact and Applications

Throughout history, biological determinism has been used to justify or reinforce racism, genocide, and oppression, often in the name of achieving the genetic improvement of the human species (for example, the "racial health" of Nazi Germany). Gould has noted that claims of biological determinism tend to be revived during periods when it is politically expedient to do so. In times of economic hardship, many find it is useful to adopt an "us against them" attitude to find a group to blame for social and economic woes or to free themselves from the responsibility of caring for the "biologically inferior" underprivileged. As advances in molecular genetics lead to the identification of additional genes that influence behavior, society must guard against using this information as justification for the mistreatment or elimination of groups that are perceived as "inferior" or "undesirable" by the majority.

—*Lee Anne Martínez*

See also: Aggression; Aging; Alcoholism; Altruism; Behavior; Bioethics; Biological Clocks; Cloning: Ethical Issues; Criminality; Developmental Genetics; Eugenics; Eugenics: Nazi Germany; Gender Identity; Genetic Engineering: Social and Ethical Issues; Genetic Screening; Genetic Testing: Ethical and Economic Issues; Heredity and Environment; Human Genetics; Intelligence; Miscegenation and Antimiscegenation Laws; Natural Selection; Race; Sociobiology; Twin Studies; XYY Syndrome.

Further Reading

Begley, Sharon. "Gray Matters." *Newsweek,* March 27, 1995. Discusses the differences between the brains of males and females.

Gould, Stephen Jay. *The Mismeasure of Man.* New York: Norton, 1996. Refutes Richard Herrnstein and Charles Murray's argument and presents an engaging historical overview of how pseudoscience has been used to support racism and bigotry.

Herrnstein, Richard, and Charles Murray. *The Bell Curve: Intelligence and Class Structure in American Life.* New York: Simon and Schuster, 1994. Asserts that IQ plays a statistically important role in the shaping of society by examining such sociological issues as school dropout rates, unemployment, work-related injury, births out of wedlock, and crime.

Moore, David S. *Dependent Gene: The Fallacy of Nature vs. Nurture.* New York: W. H. Freeman, 2001. Few books examine the ways the genes and the environment interact to produce everything from eye color to behavioral tendencies. This book lays to rest the popular myth that some traits are purely genetic and others purely a function of environment; rather, all traits are the result of complex, dependent interactions of both—interactions that occur at all stages of biological and psychological development. An informed argument against simplistic determinism.

Rose, Steven. "The Rise of Neurogenetic Determinism." *Nature* 373 (February, 1995). Discusses how advances in neuroscience have led to a resurgence of the belief that genes are largely responsible for deviant human behavior.

Sussman, Robert, ed. *The Biological Basis of Human Behavior: A Critical Review.* 2d ed. New York: Simon and Schuster, 1998. Fifty-nine essays examine genetics, the various interpretations of the early evolution of human behavior, new attempts to link human physical variation to behavioral differences between people, modern evolutionary psychology, and the influences of hormones and the brain on behavior.

Biological Weapons

Fields of study: Genetic engineering and biotechnology; Human genetics

Significance: *Just as twentieth century discoveries in chemistry and physics led to such devastating weapons as poison gases and nuclear bombs, so humanity in the twenty-first century faces the prospect that the biotechnological revolution will lead to the development and use of extremely deadly biological weapons.*

Key terms

ANTHRAX: an acute bacterial disease that affects animals and humans and that is especially deadly in its pulmonary form

BIOLOGICAL WEAPON (BW): the military or terrorist use of such organisms as bacteria and viruses to cause disease and death in people, animals, or plants

BIOTERRORIST: an individual or group that coercively threatens or uses biological weapons, often for ideological reasons

ETHNIC WEAPONS: genetic weapons that target certain racial groups

GENETIC ENGINEERING: the use of recombinant DNA to alter the genetic material in an organism

IMMUNE SYSTEM: the biological defense mechanism that protects the body from disease-causing microorganisms

RECOMBINANT DNA: DNA prepared by transplanting and splicing genes from one species into the cells of another species

SMALLPOX: an acute, highly infectious, often fatal disease characterized by fever followed by the eruption of pustules

Early History

Biological warfare antedates by several centuries the discovery of the gene. Just as the history of genetics did not begin with Gregor Mendel, whose pea-plant experiments eventually helped found modern genetics, the history of biological warfare began long before the Japanese dropped germ-filled bombs on several Chinese cities during World War II. For example, the Assyrians, six centuries before the common era, knew enough about rye ergot, a fungus disease, to poison their enemies' wells. The ancient Greeks also used disease as a military weapon, and the Romans catapulted diseased animals into enemy camps. A famous medieval use of biological weapons occurred during the Tatar siege of Kaffa, a fortified Black Sea port, then held by Christian Genoans. When Tatars started dying of the bubonic plague, the survivors catapulted cadavers into the walled city. Many Genoans consequently died of the plague, and the remnant who sailed back to Italy contributed to the spread of the Black Death into Europe.

Once smallpox was recognized as a highly contagious disease, military men made use of it in war. For example, the conquistador Francisco Pizarro presented South American natives with smallpox-contaminated clothing, and, in an early case of ethnic cleansing, the British and Americans used deliberately induced smallpox epidemics to eliminate native tribes from desirable land.

As scientists in the nineteenth and twentieth centuries learned more about the nature and modes of reproduction of such diseases as anthrax and smallpox, germ warfare began to become part of such discussions as the First International Peace Conference in The Hague (1899). The worldwide revulsion against the chemical weapons used in World War I, along with a concern that biological weapons would be more horrendous, led to the Geneva Proto-

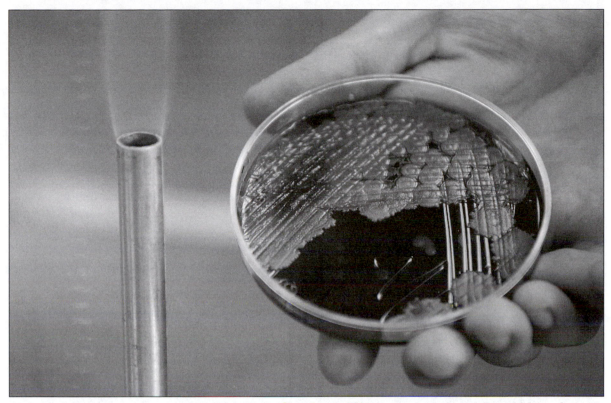

Anthrax colonies grow on culture in a petri dish in Mexico City, where in 2001 two germ banks housed dozens of these cultures virtually unguarded. (AP/Wide World Photos)

col (1925), which prohibited the first use of germ weapons, but not their development.

From Germ Warfare to Genetic Weapons

With the accelerating knowledge about the genetics of various disease-causing microorganisms, several countries became concerned with the threat to their security posed by the weaponizing of these pathogens. Although several states signed the Geneva Protocol in the late 1920's, others signed only after assurances of their right to retaliate. The United States, which did not ratify the treaty until 1975, did extensive research on germ weapons during the 1950's and 1960's. American scientists were able to make dry infectious agents that could be packed into shells and bombs, and estimates were made that ten airplanes with such bombs could kill or seriously disable tens of millions of people. Unknown to Congress and the American people, tests using apparently harmless microbes were performed on such large commu-

nities as San Francisco. When news of these secret tests was made public, many questioned their morality. Extensive criticism of the research and development of these weapons, together with the realization that these weapons posed a threat to the attackers as well as the attacked, led President Richard Nixon to end the American biological weapons program formally in 1969.

Abhorrence of biological weapons extended to the world community, and in 1972 the Biological and Toxin Weapons Convention (BTWC)—a treaty that prohibited the development, production, and stockpiling of bacteriological weapons—was signed in Washington, D.C., London, and Moscow and was put into force in 1975. Although it was eventually signed by most members of the United Nations, the nations that signed the pact failed to reach agreement on an inspection system that would control the proliferation of these weapons. A pivotal irony of the BTWC is that while most of the world was renouncing germ warfare, biologists were learning how to manipulate DNA, the molecule that carries genetic information, in powerful new ways. This knowledge made possible the creation of "superbugs," infectious agents for which there are no cures.

Some scientists warned the public and international agencies about these new germ weapons. Other investigators discovered that American researchers were creating infectious agents that would confuse diagnosticians and defeat vaccines. Similarly, Soviet researchers on an island in the Aral Sea, described as the world's largest BW test site, were producing germ weapons that could be loaded on missiles. When Boris Yeltsin became president of Russia, he discovered that the secret police and military officials had misinformed him about BW programs, in which deadly accidents had occurred. Also troubling was the spread of biological agents to such countries as Iraq. American and French companies legally shipped anthrax and botulinum bacteria to Iraq, whose scientists later acknowledged that they had used these microbes to make tons of biological weapons during the 1980's.

With the demise of the Soviet Union and increasing violence in the Balkans and Middle East, politicians became fearful that experts who had dedicated their careers to making biological weapons would now sell their knowledge to rogue nations or terrorist groups. Indeed, deadly pathogens were part of world trade, since the line separating legitimate and illegitimate research, defensive and offensive BWs, was fuzzy. In the 1980's members of a religious cult spread salmonella, a disease-causing bacterium, in an Oregon town, causing more than seven hundred people to become very ill. The same company that sold salmonella to this religious cult also sold pathogens to the University of Baghdad. Bioterrorism had become both a reality and a threat.

In 1996, U.N. inspectors supervise the destruction of growth media for biological weapons in Iraq. (AP/Wide World Photos)

The Future of Genetic Weapons

Some scientists and politicians believe that a nation's best defense against bioterrorism is advanced genetic knowledge, so that vaccines can be tailored to respond to traditional and new BWs. For example, the Human Genome Project, which succeeded in mapping the human genetic material, has the potential for revealing both the vulnerabilities and defenses of the immune system. (The human genome sequence contains 3.2 billion bases and approximately 34,000 genes. These data freely are available on the Internet in a variety of forms, including text files and graphical "genome browsers.") On the other hand, such knowledge could prove dangerous if the genetic vulnerabilities of certain ethnic groups could be targeted by bioengineered microbes. Some scientists find these speculations about genocidal BWs unevidenced and unsubstantial. Genetic similarities between different ethnic groups are more significant than their differences. Other scientists point out that dramatic genetic differences between ethnic groups are a reality. For example, milk is a poison for certain Southeast Asian populations. Other genetic differences could therefore be exploited to create BWs to attack group-specific gene clusters. Believers in ethnic BWs point to existing techniques for selectively killing certain cells and for inactivating certain DNA sequences. These techniques, developed with the hope of curing genetic diseases, could also be used to cause harm. Knowledge of the structure of the human genome will increasingly lead to knowledge of its function, and this knowledge will make it possible to manipulate, in both benign and malign ways, these information-laden molecules. Modern biotechnology thus presents humanity with both a great promise, to better health and life in peace, and a great peril, to multiply sickness and death in war. The hope of many scientists, politicians, and ordinary people is that humanity will choose the path of promise.

—*Robert J. Paradowski*

See also: Anthrax; Bioethics; Biopesticides; Biopharmaceuticals; Emerging Diseases; Eugenics; Eugenics: Nazi Germany; Genetic Engineering: Risks; Genetic Engineering: Social and Ethical Issues; Smallpox.

Further Reading

Alibeck, Ken, with Stephen Handelman. *Biohazard: The Chilling True Story of the Largest Covert Biological Weapons Program in the World, Told from the Inside by the Man Who Ran It.* New York: Random House, 1999. Alibeck was a Kazakh physician who helped create the Soviet Union's advanced biological warfare program. For some, this autobiographical account is the best and most comprehensive overview of the BW controversy; for others, Alibeck's treatment is tarnished by his long association with the Soviet system.

British Medical Association. *Biotechnology, Weapons, and Humanity.* Amsterdam, Netherlands: Harwood Academic, 1999. Concerned that genetic engineering and biotechnology will be used to produce germ weapons, the physicians of the British Medical Association produced this helpful guide to facilitate public debate. Contains a glossary of technical terms and an excellent set of references.

Cole, Leonard A. *The Eleventh Plague: The Politics of Biological and Chemical Warfare.* New York: W. H. Freeman, 1996. Cole, who has published other books on chemical and biological weapons, examines various programs in the United States and Iraq, while emphasizing that morality is important in responding to the challenges posed by these weapons.

Miller, Judith, Stephen Engelberg, and William Broad. *Germs: Biological Weapons and America's Secret War.* New York: Simon & Schuster, 2001. This book, written by three *New York Times* reporters, explores the ideas and actions of scientists and politicians involved in the past, present, and future of germ warfare. Forty-two pages of notes and a select bibliography.

Piller, Charles, and Keith R. Yamamoto. *Gene Wars: Military Control over the New Genetic Technologies.* New York: William Morrow, 1988. A journalist teamed with a molecular biologist to write this book in order to demystify new biological technologies for the nonscientist and to alert scientists of their special responsibility to enlighten public debates about BW research. Appendices on recombinant DNA and BW treaties. Selected bibliography.

Web Site of Interest

Federation of American Scientists, Chemical and Biological Arms Control Program. http://www.fas.org/bwc. An organization dedicated to educating the public about biological weapons. The site posts position papers, information on bioterrorism, and links to related documents, including the Joint Concept for Non-Lethal Weapons statement from the U.S. Marine Corps' Non-Lethal Weapons Program.

Biopesticides

Field of study: Genetic engineering and biotechnology

Significance: *As an alternative to chemical pesticides, agricultural scientists have begun using ecologically safer methods such as biopesticides to protect plants from insects.*

Key terms

Agrobacterium tumefaciens: a species of bacteria that is able to transfer genetic information into plant cells

Bacillus thuringiensis: a species of bacteria that produces a toxin deadly to caterpillars, moths, beetles, and certain flies

BACULOVIRUS: a strain of virus that is capable of causing disease in a variety of insects

TRANSFORMATION: the process of transferring a foreign gene into an organism

TRANSGENIC ORGANISM: an organism synthesizing a foreign protein; the gene of which was obtained from a different species of organism

Bacillus thuringiensis

Hungry insects are the bane of gardeners. This problem is worsened for farmers, whose livelihoods depend on keeping fields free of destructive insects. Although effective, chemical pesticides have a variety of drawbacks. The increasing popularity of organically grown produce that is untreated by chemicals suggests that consumers are wary that human-made pesticides may hold hidden dangers. In response to consumers' worries over chemical-pesticide safety, agricultural biologists have turned to nature to solve pest problems. Biopesticides are insecticides taken from nature. They are designed not in a laboratory but through evolution, making them very specific and effective. A biopesticide may be sprayed directly on crops or may be genetically engineered to be produced by a crop itself.

Since the 1950's, the bacterial pesticide *Bacillus thuringiensis* (*Bt*) has been used on crops susceptible to destruction by insect larvae. Upon sporulation, *Bt* produces a crystallized protein that is toxic to many forms of larvae. The protein is synthesized by the bacteria as an

Although the biopesticide Bt *has been genetically integrated into crops to make them resistant to insects and other pests, there is evidence that new pests are emerging and old pests, such as the bollworm—against which the* Bt *crops were engineered to protect—are evolving their own resistance to the* Bt *toxin. Here a farmer in India displays the bollworm damage to his* Bt *cotton. (AP/Wide World Photos)*

Biopesticides and Nontarget Species

Researchers have long had a concern as to the effect of chemical insecticides on nontarget species. Target species frequently display resistance to chemical controls due to large effective population sizes and prior histories of exposure to chemical agents which favors the increase in resistance alleles in a population. The exact opposite is typically true for nontarget species that occupy the treatment area. When biopesticides such as the Cry1ab endotoxins, derived from the soil bacterium *Bacillus thuringiensis* (*Bt*), were first proposed as control agents, many scientists believed that the collateral effects on nontarget species would be significantly limited.

Initially these toxins were sprayed on crops, thus potentially increasing the exposure of nontarget species. Even with the development of transgenic crops such as corn, it was possible for *Bt* to move from the treatment area to the feeding grounds of nontarget species through pollen dispersal. One of the first documented accounts of *Bt*-induced mortality in a nontarget species was provided in 1999 by researchers at Cornell University. They demonstrated that the pollen from *Bt*-treated corn increased mortality among monarch butterflies (*Danaus plexippus*) when applied to the surfaces of milkweed plants, the butterfly's primary food source. In this study, monarchs exposed to *Bt* had a slower rate of growth and increased mortality. It was suggested that field monarchs could also be exposed to corn pollen containing *Bt* endotoxins. Given the popularity of the monarch and the noticeable decline in North American populations during the 1990's, it appeared that the future of biopesticides was dim.

Since then, additional studies indicate a less significant effect of *Bt* toxins on nontarget species. The dispersal of *Bt* pollen is not believed to occur more than a few meters from the edges of the treatment area, and even at these distances the levels have been shown to be sublethal. Research involving monarchs and swallowtail butterflies (*Papilio* species) has indicated that lethality is not elevated at low-level *Bt* exposure, although there is evidence of reduced growth rates. Furthermore, only a fraction of the nontarget organism's population would be exposed at a given time, and frequently the larval periods of the target and nontarget organisms do not overlap. This evidence suggests that biopesticides are not producing the observed decrease in nontarget populations.

It is likely that there may be a limited effect of biopesticides on nontarget species, and most researchers agree that additional research needs to be conducted. The genetics of *Bt* resistance have been determined for a number of insects, although for others the exact mechanism has remained elusive. However, the greatest threat to the nontarget organisms rests with habitat destruction. A decrease in the population size due to reduced resources may serve to weaken the population and enhance the sublethal effects of biopesticide production. The physiological effect and population genetics of *Bt* susceptibility in nontarget species will need to be examined in some detail to prove to the public the value of biopesticides.

—*Michael Windelspecht*

inactive proenzyme. After it is digested, enzymes in the insect's gut cleave the protein into an active, toxic fragment. The active toxin binds to receptors in the insect's midgut cells and blocks those cells from functioning. Only caterpillars (tobacco hornworms and cotton bollworms), beetles, and certain flies have the gut biochemistry to activate the toxin. The toxin does not kill insects that are not susceptible, nor does it harm vertebrates in any way.

The drawbacks of *Bt* are its expense and its short-lived effect. In the early 1990's, scientists overcame these two drawbacks through applied genetic engineering. They produced transgenic cotton plants that generated their own *Bt* toxin. The toxin gene was first isolated from *Bt* cells and ligated (enzymatically attached) into a Ti plasmid. A Ti plasmid is a circular string of double-stranded DNA that originates in the *Agrobacterium tumefaciens* bacteria. The *A. tumefaciens* has the ability to take a portion of that Ti plasmid, called the T-DNA, and transfer it and whatever foreign gene is attached to it into a plant cell. Cotton plants were exposed to the *A. tumefaciens* carrying the toxin gene and were transformed. The transgenic plants synthesized the Bt toxin and became resistant to many forms of larvae.

At the University of Florida's Institute of Food and Agricultural Sciences, Dov Borovsky has developed a "diet pill" for mosquitoes that causes them to starve to death. It may help eradicate mosquitoes and the diseases they transmit. (AP/Wide World Photos)

Many crystal toxins have been isolated from various strains of *Bt*. These toxins make up a large collection of proteins active against pests from nematodes to aphids. Researchers are in the process of reengineering the toxin genes to improve upon their characteristics and to design better methods of transporting genes from one *Bt* strain to another.

Other Biopesticides

Several species of fungi have been found to be toxic to insects, including *Verticillium lecanii* and *Metarhizium anisopliae*. Scientists have just begun to study these fungi, which are not yet commercially available to farmers.

In the mid-1990's, a viral biopesticide called baculovirus became widely popular. Baculoviruses are sprayed onto high-density pest populations just like chemical pesticides. Baculoviruses have several advantages over conventional pesticides. The most important advantage is their strong specificity against moths, sawflies, and beetles but not against beneficial insects. Also, viruses, unlike bacteria, tend to persist in the environment for a longer period. Finally, baculoviruses are ideal for use in developing countries because they can be produced cheaply and in great quantity with no health risks to workers. One limitation of baculovirus is that it must be administered in a precise temporal and spatial framework to be effective. Knowledge of insect behavior after hatching, the insect population's distribution within the crop canopy, and the volume of foliage ingested by each larva is essential. For example, moths usually do the most damage at the late larval stage. To minimize crop damage from moths, one must spray baculovirus as early as possible before the insects reach that late stage.

One final biopesticide approach has been to make transgenic plants that manufacture proteins isolated from insect-resistant plant species. Tomatoes naturally make an enzyme inhibitor that deters insects by keeping their digestive enzymes (trypsin and chymotrypsin) from functioning. These inhibitors were isolated by Clarence Ryan at the University of Washington. Ryan transformed tobacco plants with two different forms of inhibitor (inhibitors I and II from tomato). The tomato proteins were effectively produced in tobacco and made the transgenic plants resistant to tobacco hornworm larvae.

Biopesticide Resistance

As with chemical pesticides, over time insect populations grow resistant to biopesticides. *Bt*-resistant moths can now be found around the world. Resistance arises when pesticides are too effective and destroy more than 90 percent of a pest population. The few insects left are often very resistant to the pesticide, breed among themselves, and create large, resistant populations.

Entomologists have suggested strategies for avoiding pesticide-resistant insect populations. One strategy suggests mixing biopesticide–producing and nonproducing plants in the same field, thereby giving the pesticide-susceptible part of the insect population places of refuge. These refuges would allow resistant and nonresistant insects to interbreed, making the overall species less resistant. Other strategies include synthesizing multiple types of *Bt* toxin in a single plant to increase the toxicity range and reduce resistance, making other biological toxins besides *Bt* in a single plant, and reducing the overall exposure time of insects to the biopesticides.

—*James J. Campanella*

See also: Biofertilizers; Genetic Engineering: Agricultural Applications; Genetic Engineering: Risks; Genetic Engineering: Social and Ethical Issues; Genetically Modified (GM) Foods; High-Yield Crops; Population Genetics.

Further Reading

Carozzi, Nadine, and Michael Koziel, eds. *Advances in Insect Control: The Role of Transgenic Plants*. Bristol, Pa.: Taylor & Francis, 1997. Presents technologies that have proved successful for engineering of insect-tolerant crops as well as an overview of new technologies for future genetic engineering.

Cory, Jenny, and David Bishop. "Use of Baculoviruses as Biological Insecticides." *Molecular Biotechnology*, March, 1997. Presents a complete discussion of baculoviruses.

Deacon, J. W. *Microbial Control of Plant Pests and Diseases*. Washington, D.C.: American Society for Microbiology, 1983. Good if brief monograph for a broad perspective on biocontrol agents.

Entwhistle, Philip F., Jenny S. Cory, Mark J. Bailey, and Steven R. Higgs. *Bacillus Thuringiensis, an Environmental Biopesticide: Theory and Practice*. New York: Wiley, 1994. Addresses the use of one of the oldest and best-known biopesticides, including its manufacture and use in developing nations.

Gibbons, Anne. "Moths Take the Field Against Biopesticides." *Science*, November 1, 1991. Discusses resistance to biopesticides and its implications.

Hall, Franklin R., and Julius J. Menn, eds. *Biopesticides: Use and Delivery*. Totowa, N.J.: Humana Press, 1999. A guide to development, application, and use of biopesticides as a complementary or alternative treatment to chemical pesticides. Reviews their development, mode of action, production, delivery systems, and market prospects and discusses current registration requirements for biopesticides as compared with conventional pesticides.

Koul, Opender, and G. S. Dhaliwal, eds. *Microbial Biopesticides*. New York: Taylor & Francis, 2002. International experts on biopesticides explore developments in using those based on bacteria, fungus, viruses, and nematodes, discussing their advantages and disadvantages and their role in genetic engineering.

_____. *Phytochemical Biopesticides*. Amsterdam, Netherlands: Harwood Academic, 2001. Addresses advances made on phytochemical biopesticides, covering behavioral, chemical, biochemical, and molecular aspects. Topics include the role of phytochemical biopesticides in integrated pest management (IPM), the potential uses of phytochemical pesticides in deciduous temperate fruit crops, and prospects and problems with phytochemical biopesticides.

Web Site of Interest

U.S. Environmental Protection Agency. http://www.epa.gov/pesticides. Government site with a link to information on biopesticides, including the Federal Insecticide, Fungicide, and Rodenticide Act, enacted to monitor the harmful effects of toxic pesticides on humans and the environment and ensure industry compliance.

Biopharmaceuticals

Fields of study: Diseases and syndromes; Genetic engineering and biotechnology

Significance: *Biopharmaceuticals encompass a class of drugs that are designed by combining genetics with biotechnology. Biopharmaceutical products are typically derived from proteins, such as enzymes or antibodies, and are genetically engineered in an attempt to treat a specific disease. They differ from traditional pharmaceuticals, which are usually simpler compounds that are produced by chemical synthesis.*

Key terms

CLINICAL TRIAL: an experimental research study used to determine the safety and effectiveness of a medical treatment or drug

HUMANIZED ANTIBODY: a human antibody that has been engineered to contain a portion of a nonhuman variable region with known therapeutic activity

PHARMACOGENOMICS: the field of science that examines how variations in genes alter the metabolism and effectiveness of drugs

History of Biopharmaceuticals

Drugs have been used by humans for thousands of years. The Sumerians are the first people known to have compiled medical information in a written form that outlined symptoms and treatments for different diseases more than three thousand years ago. Most ancient cultures used medicines derived from plants and animals. These drugs are different from modern biopharmaceuticals in many ways, but the most significant difference is that there was no engineering used to shape the drugs for a particular disease. Since there was no real understanding of the underlying problem, a rational approach to drug selection and design was difficult if not impossible. One philosophy of medicine that developed to address this problem was called the doctrine of similitudes, in which treatments were based on similarities of structure with disease manifestation. For example, the leaves of St John's wort look similar to damaged skin, so it was thought that extracts from this plant would be beneficial for treating cuts and burns.

It was not until the twentieth century that the underlying genetic basis for disease was discovered. The discovery that DNA is the genetic material and provides the instructions to make proteins was revolutionary. In the mid-1900's it was demonstrated that sickle-cell disease was caused by a single nucleotide change from an A (adenine) to a T (thymine) in the hemoglobin beta-chain gene. This small change alters the shape of a red blood cell from a biconcave disc to a sharply pointed crescent. Even though this finding showed it was possible to identify genetic mutations, there was still no way to manipulate or make changes to the genetic information itself.

The advent of recombinant DNA technology in the 1970's provided the first chance to engineer, or manipulate, genes. Restriction enzymes became an important tool of this technology. Restriction enzymes were first found in bacteria, where they function to protect the cell from foreign DNA by cutting it up. Restriction enzymes cut DNA at specific sequences, which are usually palindromes of the letters that signify the four nucleotides that make up DNA: guanine (G), adenine (A), thymine (T), and cytosine (C). Most restriction enzymes cut the DNA in such a way that an overhang, called a sticky end, is created. Since DNA readily recombines with complementary strands, these sticky ends can be used to splice different pieces of DNA together. The resulting sequence is called recombinant DNA.

With the ability to engineer DNA now possible, scientists looked again to bacteria to provide a way to convert that DNA into protein. Bacteria are ideal for protein production because they reproduce quickly, are easy to manipulate, and can be grown in large quantities. Many bacteria contain circular pieces of DNA apart from their genome, called plasmids. These plasmids can be readily transferred between bacteria and are also inherited by the daughter cells when a bacterium divides. With the use of restriction enzymes, plasmids can be taken from bacteria and engineered to contain a foreign gene. The resulting recombinant plasmid can be put back in bacteria, transforming them into protein factories that work nonstop transcribing and translating the recombinant gene. The first biopharmaceutical produced in bacteria was recombinant human insulin, which was marketed in 1982.

The future for biopharmaceuticals looks bright. In 1991, there were only fourteen biopharmaceuticals approved for use by the U.S. Food and Drug Administration (FDA). By 2001, nearly three hundred had been approved for use, with an additional fifty in phase III clinical trials, and by 2003 more than 330 major companies in the United States were working to produce and develop biopharmaceuticals.

Design of Biopharmaceuticals

A popular method for identifying disease-related genes is called genomics. This method uses gene chip analysis to screen thousands of genes in a single experiment. This approach is dramatically faster and more efficient than traditional methods and can be used for any disease, even those that are not hereditary.

Once the genomic information is obtained, it is used to build a broad understanding of how a disease gene functions and what role it plays in the cell. This information is gathered through the use of experimental models, genetic analysis, biochemical analysis, and structural analysis. Experimental models can range from cell culture to transgenic mice and provide physiological information about the disease. Genetic analysis provides information about where and when the gene is expressed. At the molecular level, biochemical analysis provides information about protein-protein interactions, post-translational modifications, and enzymatic activity. Structural analysis yields extremely detailed information about the physical arrangement of the atoms that make up the protein. All of these approaches can provide clues as to what may be important targets for treatment of the disease.

A better understanding of the disease at the genetic and molecular levels facilitates an attempt at designing a biopharmaceutical to treat the disease. Once a disease is understood, it becomes possible to target a key pathway or protein involved in the disease. The resultant drug and the way that it is used clinically will vary from disease to disease. For example, type I diabetes is caused by a deficiency in the hormone insulin. Without insulin, the body is not able to regulate the level of glucose in the blood. Lack of insulin is easily corrected by a simple injection of recombinant human insulin, the first biopharmaceutical. Another example of a biopharmaceutical that is currently in use is the enzyme tissue plasminogen activator (tPA). Most heart attacks are caused by a blood clot blocking the flow of blood through a coronary artery. Formation and removal of blood clots is a highly regulated and well-understood process. The enzyme tPA is known to be one of the key players in blood clot removal. This knowledge led to the development of recombinant tPA, which can be provided by injection or infusion to heart attack patients. Once in the bloodstream, the tPA breaks up the coronary artery clots and restores blood flow to the heart, preventing any further muscle damage.

Clinical Trials

Before a biopharmaceutical can be used to treat disease, it must undergo a clinical trial to test its safety and effectiveness. There are four phases to a clinical trial. Phase I trials involve studies on a small number of patients (fewer than one hundred) to determine drug safety and dosage. Phase II trials involve more patients (up to five hundred) to determine effectiveness and additional safety information, such as potential side effects. Phase III trials are the most extensive and involve large numbers of people (between one thousand and three thousand) to establish risk-benefit information and comparisons with other currently used treatments. Phase IV trials determine the drug's optimal use in a clinical setting. In 2003, the entire process of drug design—from discovery to clinical trials—cost approximately $802 million and took an average of twelve years. Many years of research and millions of dollars are wasted, because only one in five thousand potential drugs actually makes it to market.

Biopharmaceuticals Today

Two examples of approved biopharmaceuticals are Aralast and Campath. Aralast is marketed by Baxter and was approved for use by the FDA in 2003. Aralast is the trade name for the recombinant human protein known as alpha-1 proteinase inhibitor (A1PI). A1PI deficiency results in the destruction of lung tissue, which can lead to emphysema. Aralast is given to patients intravenously each week and helps protect them against future lung damage. The second drug, Campath, is marketed by Millennium Pharmaceuticals and was approved by the FDA in 2001. Campath is the trade name for a humanized antibody against the CD52 antigen found on lymphocytes. The antibody is used to treat chronic lymphocytic leukemia and works by destroying lymphocytes through agglutination and complement activation.

The current trend in pharmaceutical research is the production of designer drugs through the new field of pharmacogenomics. Instead of giving a patient a drug that works for the average person with the average form of a disease, a patient will be given a drug that is specifically matched to his or her genetic profile and to his or her particular form of the disease. Currently 100,000 people die each year because of adverse drug reactions. With the use of pharmacogenomics it will be possible to determine a patient's genetic profile prior to treatment and avoid adverse drug reactions. Research in pharmacogenomics will also increase the pool of drugs available to treat disease. Currently, many drugs never make it to market because they work for only a small subset of patients. Pharmacogenomic research will allow these specific patients to be identified and enable previously abandoned drugs to be used to treat disease.

—Matthew J. F. Waterman

See also: Cloning; Cloning: Ethical Issues; Cloning Vectors; Gene Therapy; Gene Therapy: Ethical and Economic Issues; Genetic Engineering; Genetic Engineering: Medical Applications; Genetic Engineering: Risks; Genetic Engineering: Social and Ethical Issues; Genetically Modified (GM) Foods; Molecular Genetics; Synthetic Genes; Transgenic Organisms; Xenotransplants.

Further Reading

Collins, F., and V. McKusick. "Implications of the Human Genome Project for Medical Science." *JAMA* 285, no. 5 (2001): 540-541. Overviews the significant impact of the Human Genome Project on medical research, including specific examples of drug design.
Wu-Pong, S., and Y. Rojanasakul. *Biopharmaceutical Drug Design and Development.* Totowa, N.J.: Humana Press, 1999. Outlines the process of biopharmaceutical design, including basic molecular biology, major classes of biopharmaceuticals, and clinical trials.

Web Site of Interest

American Chemical Society, The Pharmaceutical Century. http://pubs.acs.org/journals/pharmcent. Posts articles about the science of biopharmaceuticals, including the role played by genetics and the Human Genome Project in the development of new drugs.

Blotting: Southern, Northern, and Western

Field of study: Techniques and methodologies

Significance: *Blotting is a technique that allows identification of a specific nucleic acid or amino acid sequence even when it is mixed in with all of the other material from a cell. This allows the rapid identification of the changes associated with mutant alleles.*

Key terms

BLOTTING: the transfer of nucleic acids or proteins separated by gel electrophoresis onto a filter paper, which allows access by molecules that will interact with only one specific sequence hybridization: incubation of a target sequence with an identifying probe, which allows the formation of annealed hybrids

northern blot: a blot designed to detect messenger RNA

probe: a nucleic acid sequence or antibody that can attach to a specific DNA or RNA sequence or protein; the probes are often labeled with radioactive compounds or enzymes so their position can be determined

southern blot: a blot designed to detect specific DNA fragments

western blot: a blot that uses antibodies to detect specific proteins

Limitations of Gel Electrophoresis

Using gel electrophoresis to separate proteins and nucleic acids has been an invaluable tool in analyzing living systems. Changes in these molecules—such as a mobility shift in a mutant protein or the change in the size of a plasmid that has received a DNA insert—can be easily detected using this technique. However, the ability to differentiate between types of molecules is quite limited. An extract of red blood cell proteins run through an acrylamide gel might show one major band for hemoglobin which can be discerned from the many other proteins in the cell. However, the hundreds of different proteins that might be produced in a liver extract will produce a tight ladder of bands that are impossible to tell apart.

The situation can be even worse with DNA. A restriction enzyme digest of a plasmid or simple virus might yield fewer than six pieces of DNA that could be easily separated on an agarose gel. If one were to digest the total genomic DNA of even a simple organism, such as *Escherichia coli*, with a typical restriction enzyme such as *Eco*RI, the result would be a thousand bands of numerous sizes (4×10^6 base pairs of DNA, since *Eco*RI recognizes a six-base-pair site, which should occur, on average, every 4^6 or 4,096 bp). After separation on a gel, the result would be a smear with no individual bands visible. Working with an even more complex genome, such as the human genome, would result in millions of bands. The only way to study a specific protein or nucleic acid sequence using gel electrophoresis, therefore, would be to find

a way to label it specifically so that it could be differentiated from the general background.

Basic Blotting Techniques

In 1975, Ed Southern developed a method that allowed the detection of specific DNA sequences after they had been separated by agarose gel electrophoresis. What makes a piece of DNA unique is the sequence of the nucleotides. This is most efficiently detected by the hybridization of the antiparallel strand. This can only occur if the two strands are separated into single strands. Therefore, the first step is to soak the agarose gel in a strong base, such as 1 molar sodium hydroxide, and high salt, which stabilizes the single-stranded form. The base is then neutralized with a strong buffer, such as tris-hydrochloride, again in high salt. The DNA can now be analyzed by its ability to hybridize to a radioactive piece of single-stranded DNA. Since this radioactive DNA can "explore" the different sequences to find the one matching sequence, it is also known as a probe (an instrument or device that can be used to explore and send back information).

Although the agarose is porous, it would be very slow and inefficient to try to perfuse the gel with radioactive probe and then remove the pieces that did not hybridize. Southern realized that he needed to move the DNA to a thin material to be able to probe it efficiently. The material chosen was nitrocellulose, consisting of a variant of paper (cellulose) with reactive nitro groups attached. The treated gel is placed on a sponge soaked with a high-salt solution. The nitrocellulose sheet is placed onto the gel and then a stack of dry paper towels is laid on top. The salt solution is drawn through the gel to the dry towels and carries the DNA from the gel up into the paper. The positively charged nitro groups on the nitrocellulose stick to the negatively charged DNA, thereby holding the DNA in a pattern matching the band locations in the gel. The nitrocellulose is removed from the gel and baked at 80 degrees Celsius (176 degrees Fahrenheit) or treated with ultraviolet light, both of which covalently cross-link the DNA to the paper, locking it in its position. The filter is soaked in a solution that promotes reassociation of single-stranded DNA, and radioactive, single-stranded DNA is added. Since the added DNA could stick nonspecifically to the nitrocellulose, the paper is pretreated with unrelated DNA, such as sheared salmon DNA, which will bind the available nitro groups but not react with the probe.

A large molar excess of probe must be used to drive the hybridization reaction (reforming the "hybrid" of two matching antiparallel strands together), which means that it is necessary to make sure that enough probe is available in the solution to randomly run into the correct sequence on the paper and reanneal to it. The hybridization is done at an elevated temperature—often 50-65 degrees Celsius (122-149 degrees Fahrenheit), so that only strands that match exactly will stay together and those with short, random matches will come apart. After overnight hybridization, the paper is washed multiple times with a detergent-salt solution, which removes the DNA that did not hybridize. The paper is placed against a piece of X-ray film, and the radioactive emissions from the probe darken the film next to them. When the film is developed, a pattern of bands appears that corresponds to the position in the original gel of the DNA piece for which the researcher was probing.

Expanded Techniques to Study RNA and Proteins

The basic method of blotting has been expanded to include the study of RNA and proteins. James Alwine developed a very similar method to transfer messenger RNA (mRNA) that had been separated on an agarose gel. Since the mRNA started as single-stranded, there was no need to treat the gel with denaturant. However, to block the formation of internal double-stranded regions, which could alter the migration during electrophoresis, the gel contained an organic solvent. Other than that, the two methods are very similar. Although the DNA transfer system was named the Southern blot in honor of Ed Southern, Alwine decided to defer the credit and called his system the Northern blot to indicate that it was related but in a different direction.

Similarly, when W. N. Burnette developed a system for transferring and detecting specific proteins, he named the system Western blotting. This system of naming has been expanded: A technique for detecting viral DNA in tree leaves was named the Midwestern blot and a variant of the Northern blot developed in Israel was named the Middle Eastern blot.

Since proteins are generally smaller than DNA fragments, they are usually separated on polyacrylamide gels, which have a much smaller pore size than agarose gels. It is therefore necessary to use electrical current to pull the proteins out of the gel. The nitrocellulose is pressed onto the gel with a porous plastic pad. The gel is then placed in a buffer tank and electrodes are placed on either side. When a voltage is applied, the current that flows through the gel carries the proteins onto the nitrocellulose. The reactive side chains of the nitrocellulose also bind proteins very effectively, so they are all retained on the paper. The specific probe used to detect a protein is an antibody that either can be radioactively labeled or can have an enzymatic side chain attached, which will produce light or a colored dye when the appropriate chemicals are added. Since the antibody is a protein, it could also stick nonspecifically to the paper, so the blot is pretreated with a general protein such as serum albumin before the antibody is added.

Blotting in Genetic Analysis

The ability to detect individual molecules in a large background has been very important for genetic analyses. For instance, restriction fragment length polymorphism (RFLP) analysis is a method that uses the change in the size of a DNA fragment in the genome, generated by restriction enzyme digestion as a genetic marker. The isolation of many disease genes, including the one causing Huntington's disease, depended on RFLP mapping to localize the gene. It would not be possible to detect the changes in a single DNA fragment out of the millions generated by digesting the human genome without having the Southern blot to pick out the correct piece. Many other mutations that change a specific region of DNA—such as deletions, inversions, and duplications—are often detected by changes in a Southern blot pattern. The sensitivity of hybridization can be tuned to a level where probes that differ by only a single nucleotide will not attach efficiently. This allows the rapid identification of the positions of point mutations. When polymerase chain reaction (PCR) is used to amplify DNA from a crime scene or to detect human immunodeficiency virus (HIV) in the bloodstream, the presence of DNA pieces on a gel is not sufficient proof that the correct DNA has been found. The DNA must be blotted and probed with the expected sequence to confirm that it is the correct piece.

Northern blot analysis allows scientists to see how mRNA is altered in different mutants. Northern blots can indicate if a mutant allele is no longer transcribed or if the level of mRNA produced has been dramatically decreased or increased. Deletions or insertions will also show up as shortened or lengthened messages. Alternative splicing can be seen as multiple bands on a Northern blot which hybridize to the same probe. Point mutations that do not detectably alter the mRNA can still dramatically alter the protein product. Changes of a single amino acid can alter the electrophoretic mobility and the difference in apparent molecular weight can often only be detected by Western blot. These changes can also alter protein stability, which can be detected as decreased protein levels showing up on the Western. The ability to detect changes at the DNA, RNA, and protein level through blotting techniques has greatly increased the ability of scientists to study genetic alterations.

Future Directions

Blotting techniques are the most generally efficient methods for detecting specific proteins or nucleic acids. Most improvements in the past years have been aimed at speeding up the transfer process using vacuums or pressure or the hybridization process by changing the conditions. The next step will be developing silicon chips that can interact with specific nucleic acid or amino acid sequences and produce an electrical output when they "hybridize" with the correct sequence. This will diminish the time required to confirm a sequence from several hours to minutes.

—*J. Aaron Cassill*

See also: Antibodies; DNA Sequencing Technology; Gel Electrophoresis; Genetic Testing; Huntington's Disease; Immunogenetics; Model Organisms; Polymerase Chain Reaction; Repetitive DNA; RFLP Analysis.

Further Reading

Alwine, J. C., D. J. Kemp, and G. R. Stark. "Method for Detection of Specific RNAs in Agarose Gels by Transfer to Diazobenyloxymethyl-Paper and Hybridization with DNA Probes." *Proceedings of the National Academy of Sciences* 74 (1977): 5350. The original description of Northern blotting.

Southern, E. M. "Detection of Specific Sequences Among DNA Fragments Separated by Gel Electrophoresis." *Journal of Molecular Biology* 98, no. 3 (1975): 503-517. The original description of Southern blotting and of blotting in general. This is one of the most often cited articles in biology research.

Breast Cancer

Field of study: Diseases and syndromes
Significance: *While the majority of breast cancers are caused by acquired mutations, about 5 percent of all breast cancers are caused by inherited mutations that greatly increase the chances of developing the disease. Germ-line mutations in the* BRCA1 *and* BRCA2 *genes are associated with most of these inherited breast cancers.*

Key terms

BRCA1 AND *BRCA2* GENES: the genes associated with most inherited breast cancers

CELL CYCLE: the sequence of events of a dividing cell

EXON: the coding sequence (part of a messenger RNA, or mRNA) that specifies the amino acid sequence of the protein produced during translation

GERM-LINE MUTATION: a heritable change in the genes of an individual's reproductive cells, often linked to hereditary diseases

p53 GENE: a tumor-suppressor gene, the first gene identified in an inherited breast cancer

TUMOR-SUPPRESSOR GENE: a gene that produces a protein product that limits cell division and therefore acts to inhibit the uncontrolled cell growth of cancers

Genes Associated with Breast Cancer

Approximately one in eight women develops breast cancer over the course of her lifetime. In the United States there are approximately 180,000 new cases of breast cancer yearly. By 2002, more than forty different genes had been found to be altered in breast cancers. Those breast cancers that are not familial (inherited)

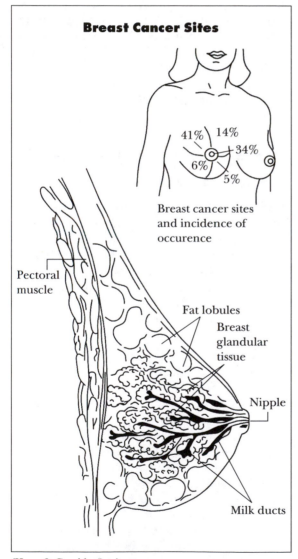

Breast Cancer Sites

41% 14%
34%
6%
5%

Breast cancer sites and incidence of occurence

Pectoral muscle

Fat lobules

Breast glandular tissue

Nipple

Milk ducts

(Hans & Cassidy, Inc.)

are termed "sporadic." It is estimated that about 5 to 10 percent of all breast cancers are familial. Approximately 80 to 85 percent of these can be attributed to mutations in the *BRCA1* or *BRCA2* gene.

The first gene identified in an inherited breast cancer was *p53*, which is mutated in Li-Fraumeni syndrome. It is a tumor-suppressor gene that encodes a protein transcription factor that stops the cell cycle until DNA repair has occurred; a defective *p53* gene no longer stops cell division, and unrepaired DNA can be replicated, resulting in accumulated mutations in the cell. About 1 percent of women who develop breast cancer before the age of thirty have germ-line mutations in *p53*. Families with this syndrome have extremely high rates of brain tumors and other cancers in both children and adults.

Some gene mutations may predispose an individual to develop breast cancer. For example, there is an increased incidence of breast cancer associated with the ataxia telangiectasia (*AT*) gene and the *HRAS1* gene. A mutated form of the *AT* gene is found in the rare recessive hereditary disorder ataxia telangiectasia, which has a very wide range of symptoms, including cerebellar degeneration, immunodeficiency, balance disorder, high risk of blood cancers, extreme sensitivity to ionizing radiation, and an increased risk of breast cancer. Individuals with one mutated copy of the *AT* gene have an increased risk of cancer. The *AT* gene was identified as a phosphatidylinositol-3 kinase (an enzyme that adds a phosphate group to a lipid molecule) that transmits growth signals and other signals from the cell membrane to the cell interior. The *AT* gene was found to be similar in sequence to other genes that are known to have a role in blocking the cell cycle in cells whose DNA is damaged by ultraviolet light or X rays. It is possible that the mutated *AT* gene does not stop the cell from dividing, and the damaged DNA may lead to cancers. It is disturbing to note that individuals with a mutated *AT* gene may be more sensitive to ionizing radiation. It must be determined if these individuals should then avoid low X-ray doses, such as those received from a mammogram used to detect the early stages of breast cancer.

Discoveries of Breast Cancer Genes

Prior to the discoveries of *BRCA1* and *BRCA2*, there were many hints that susceptibility to at least some breast cancers was inherited. The time line below shows some of the discoveries leading up to the discoveries of *BRCA1* and *BRCA2* as well as later discoveries about breast cancer genes.

1966   Henry Lynch began the first studies on inherited cancers.

1970   The first cancer-causing gene (oncogene) was reported in chickens by Peter Vogt.

1976   J. Michael Bishop and Harold Varmus reported the discovery of oncogenes in the DNA of normal chromosomes.

1978   M. H. Bronstein et al. see a link between Cowden disease, an inherited tumorogenic syndrome, and breast cancer.

1979   Arnold Levine and David Baltimore discover *p53*, a gene mutated in approximately half of all known cancers, including breast cancer.

1985   The mutant *p53* gene is cloned by Arnold Levine.

1987   Michael Swift et al. report a hereditary link between ataxia telangiectasia mutated (*ATM*) and many cancers, including breast cancer.

1988   Dennis Slamon reports that the *HER-2/neu* growth factor gene is overexpressed in 30 percent of the most aggressive breast cancers.

1990   Mary Claire King and coworkers report the discovery of *BRCA1* in Ashkenazi Jewish women and locate it on chromosome 17.

1990   David Malkin et al. report a link between the *p53* gene product and breast cancer.

1991   Elizabeth Claus et al. do a statistical analysis of familial breast cancer and predict a dominant breast cancer gene will be found.

1993   Theodore Krontiris et al. report an association between *HRAS1* (Harvey rat sarcoma oncogene 1) and breast cancer.

1994   Yoshio Miki et al. announce the cloning of *BRCA1* on chromosome 17.

1995   Richard Wooster et al. announce the discovery and cloning of *BRCA2* on chromosome 13.

1996   Prasanna Athma et al. report that heterozygotes for the recessive allele *ATM* are more susceptible to breast cancer.

1997   Danny Liaw et al. find that germ-line mutations in the *PTEN* gene lead to Cowden disease and associated breast cancer.

1998   Dennis Slamon tests Herceptin, a monoclonal antibody that targets the product of *HER-2/neu*, against aggressive breast cancers.

1999   François Ugolini et al. implicate *FGFR1* (fibroblast growth factor receptor gene 1) in some breast cancers.

2000   Tommi Kainu et al. propose a *BRCA3* gene to explain non-*BRCA1/BRCA2* hereditary breast cancers in several families.

2001   Paul Yaswen reports that multiple copies of the gene *ZNF217* are seen in 40 percent of breast cancers.

2001   Minna Allinen et al. find a mutation in *CHEK2* gene that leads to hereditary breast cancers. This is proposed as *BRCA3*.

2002   Alan D'Andrea et al. report that the same inherited mutations in the six genes that cause Fanconi anemia also increase the susceptibility to breast cancer.

—Richard W. Cheney, Jr.

Possible Functions of Breast Cancer Genes

The *BRCA1* gene is on chromosome 17 and encodes a protein that is 1,863 amino acids long. Germ-line mutations of *BRCA1* are associated with 50 percent of hereditary breast cancers and with an increased risk of ovarian cancer. The *BRCA2* gene is on chromosome 13q12-13 and encodes a protein of 3,418 amino acids. Germ-line mutations of *BRCA2* are thought to account for approximately 35 percent of families with multiple-case, early-onset female breast cancer. Mutations of *BRCA2* are also associated with an increased risk of male breast cancer, ovarian cancer, prostate cancer, and pancreatic cancer.

Although *BRCA1* was cloned in 1994 and *BRCA2* in 1995, the function of these genes has been difficult to identify. Part of the difficulty has been that the proteins coded by these genes do not resemble any proteins of known function. In 1997, David Livingston and coworkers of the Dana-Farber Cancer Institute found that the *BRCA1* gene product associates with repair protein RAD51. A few months later, Allan Bradley of Baylor College of Medicine and Paul Hasty of Lexicon Genetics reported that the *BRCA2* protein binds to the RAD51 repair protein. This work suggests that both genes may be in the same DNA-repair pathway. Bradley and Hasty also showed that embryonic mouse cells with inactivated mouse *BRCA2* genes are unable to survive radiation damage, again suggesting that the *BRCA* genes are DNA-repair genes. Initially, it was thought that the breast cancer genes were typical tumor-suppressor genes that normally function to control cell growth. The 1997 work suggests that the breast cancer gene mutations act indirectly to disrupt DNA repair and allow cells to accumulate mutations, including mutations that allow cancer development. In 2002 the detailed structure of the *BRCA2* protein was determined. It has structural motifs that show it to be capable of binding to DNA. Although the specific role of the *BRCA2* protein is uncertain, it is now clear that it does play a role in repairing double-stranded breaks in DNA. The understanding of the function of *BRCA1* and *BRCA2* is incomplete, but what is known will encourage additional studies.

Social Implications of Genetic Screening

With the cloning of the *BRCA1* and *BRCA2* genes, it became possible to test them for mutations. Such testing has been controversial, raising a number of social and psychological issues. There is a concern that the technical ability to test for genetic conditions is ahead of the ability to predict outcomes or risks, prescribe the

Myriad Genetic Laboratories president Gregory Critchfield in 2002, posing in front of walls lined with the DNA data on BRCA1 *and* BRCA2 *genes from just one woman. A mutation on either of these genes increases the risk of developing breast cancer significantly, but the ability to identify the mutation in individuals may also dramatically increase the chances for early detection and survival.* (AP/Wide World Photos)

most effective treatment, or counsel individuals. Part of the dilemma about testing is the uncertainty about the meaning of the test results. If a test confirms the presence of a mutation in a breast cancer gene in a woman with a family history of breast cancer, there is a high risk, but not a certainty, that the woman will develop breast cancer. Even if a test is negative, it does not mean the woman is not at risk for breast cancer, because the large majority of breast cancers are not inherited. If a test is positive, it is not clear what the best course for the woman would be. Increased monitoring with mammography and even removal of both breasts as a preventive measure should reduce the chances of developing cancer but do not guarantee a cancer-free life. Even if a woman does not yet have cancer, she may feel the additional psychological stress of knowing she has a high risk of developing cancer.

There is also concern that test results may be misused by employers or insurers. A number of states have passed laws that prevent health insurance companies from using genetic test results to discriminate against patients. In 1996, the National Cancer Institute established the Cancer Genetics Network as a means for individuals with a family history of cancer to enroll in research studies and learn of their genetic status while receiving counseling.

—Susan J. Karcher, updated by Bryan Ness

See also: Aging; Cancer; Cell Cycle, The; DNA Repair; Genetic Counseling; Genetic Screening; Genetic Testing; Genetic Testing: Ethical and Economic Issues; Hereditary Diseases; Human Genome Project; Model Organism: *Mus musculus*; Mutation and Mutagenesis; Oncogenes; Tumor-Suppressor Genes.

Further Reading

Bowcock, Anne M., ed. *Breast Cancer: Molecular Genetics, Pathogenesis, and Therapeutics.* Totowa, N.J.: Humana Press, 1999. Detailed information geared toward researchers and health professionals. The chapter "Hereditary Breast Cancer Genes" discusses *BRCA1* and *BRCA2* mutations among Ashkenazi women. Also addresses surgery, chemotherapy, drug resistance, and the *MDR* gene.

Dickson, Robert B., and Marc E. Lipman, eds. *Genes, Oncogenes, and Hormones: Advances in Cellular and Molecular Biology of Breast Cancer.* Boston: Kluwer Academic, 1992. Contains papers on the genetics and molecular biology of breast cancer, including the role of suppressor genes, the role of the retinoblastoma gene, oncogenes and stimulatory growth factors, and much more. Index.

Kemeny, Mary Margaret, and Paula Dranov. *Beating the Odds Against Breast and Ovarian Cancer: Reducing Your Hereditary Risk.* Reading, Mass.: Addison-Wesley, 1992. Designed for women with a family history of breast or ovarian cancer who are motivated to evaluate risk factors, nutrition, warning signs, and options for treatment.

Love, Susan M., with Karen Lindsey. *Dr. Susan Love's Breast Book.* Illustrations by Marcia Williams. 3d ed. New York: Perseus, 2000. Perhaps the most comprehensive book on breast health, including breast cancer. One chapter discusses the genetic risks for breast cancer.

Lynch, Henry T. *Genetics and Breast Cancer.* New York: Van Nostrand Reinhold, 1981. One of the seminal works on genetic breast cancer, by a pioneer in the investigation of hereditary breast-ovarian cancer syndrome.

National Cancer Institute. *Genetic Testing for Breast Cancer: It's Your Choice.* Bethesda, Md.: Author, 1997. One of the National Cancer Institute's large number of pamphlets and monographs on various cancers, including genetic risks for cancer, designed to provide responsible and detailed information to the public.

Yang, Haijuan, et al. "BRCA2 Function in DNA Binding and Recombination from a BRCA2-DSS1-ssDNA Structure." *Science* 297 (September 13, 2002): 1837-1848. This study presents evidence that the failure of *BRCA2* in DNA repair through homologous recombination may account for unsuppressed tumor growth.

Web Sites of Interest

American Cancer Society, All About Breast Cancer. http://www.cancer.org. Searchable information on breast cancer, including an overview, a detailed guide, and practical resources.

National Cancer Institute, National Institutes of Health. http://www.nci.nih.gov/breast. Provides information on the genetics of breast cancer and useful links.

National Institutes of Health, National Library of Medicine. Genetics Home Reference. http://www.nlm.nih.gov. This site includes information on breast cancer genetics.

Burkitt's Lymphoma

Field of study: Diseases and syndromes

Significance: *Burkitt's lymphoma, a cancer of B lymphocytes, is the most common tumor among children and young adults in Central Africa and New Guinea. It is one of the most aggressive malignancies known. However, early clinical and laboratory diagnosis usually leads to effective treatment and survival.*

Key terms

B LYMPHOCYTE: an antibody-producing lymphocyte

ONCOGENE: a mutated or improperly expressed gene that can cause cancer; the normal form of an oncogene, called a proto-oncogene, is involved in regulating the cell cycle

RECIPROCAL TRANSLOCATION: a chromosomal abnormality in which there is an exchange of chromosome segments between nonhomologous chromosomes

SARCOMA: a cancer arising from cells of mesodermal origin

The Discovery of Burkitt's Lymphoma

Burkitt's lymphoma was first described by Denis Burkitt, a surgeon working in Uganda in the 1950's, as a sarcoma of the jaw in African children. Males are affected more commonly than females. The mean age for affected children in Africa is seven years, whereas the mean age in the United States is eleven years. Tumor infiltration usually occurs in abdominal sites such as bowels, kidneys, ovaries, or other organs. Rare cases occur as acute leukemia with circulating Burkitt tumor cells. The acute lymphocytic leukemia (ALL) presentation is particularly common in cases associated with acquired immunodeficiency syndrome (AIDS). Burkitt's lymphoma grows very rapidly, with a doubling time of approximately twenty-four hours, and thus prompt diagnosis is essential. A healthy child may become critically ill in about four to six weeks. These children often exhibit a head or neck mass or a large abdominal mass with fluid (ascites) accumulation. Other symptoms include vomiting, pain, anemia, and increased bleeding.

Diagnosis

The diagnosis of Burkitt's lymphoma is usually made by a needle biopsy from a suspected disease site such as the bone marrow, ascites, or a lymph node. Microscopic analysis is used to determine if the disease is present and, if so, its stage of development. Early clinical and laboratory diagnosis spares the child any life-threatening complications from the rapid tumor growth. Other common diagnostic tests may include a complete blood count (CBC), a platelet count, a bone marrow aspiration, a biopsy, and a lumbar puncture. Further tests may include specialized radiographic exams such as a computer-assisted tomography (CAT) scan to look for hidden tumor masses. The National Cancer Institute (NCI) stages Burkitt's lymphoma according to the amount of the disease present. The less disease, the better the outlook for improvement after treatment. Patients who remain free of disease for more than one year from the time of diagnosis are considered to be cured.

Culprit or Consort?

As with many other cancers, the exact cause of Burkitt's lymphoma is not known. In patients from equatorial Africa, however, there is a close correlation with the Epstein-Barr virus (EBV). More than 97 percent of lymphomas from equatorial Africa carry the EBV genome. By contrast, only 15 to 20 percent of sporadic cases of Burkitt's lymphoma in Europe and North America are positive for EBV. EBV has a single, linear, double-stranded DNA genome and was the first herpesvirus to be completely sequenced. EBV infection is not limited to areas where Burkitt's lymphoma is found. It infects people worldwide without producing symptoms. EBV is also the causative agent of infectious mononucleosis, a common disease in which B cells are infected.

At least two EBV subtypes have been identified in human populations: EBV-1 is detected more commonly in Western societies, whereas EBV-1 and EBV-2 subtypes seem to be equally distributed in Africa. Although EBV is identified as a possible causative agent of African Burkitt's lymphoma, it appears that non-African Burkitt's lymphoma EBV may be just one factor in a multistep process of development. Burkitt's lymphoma is a monoclonal proliferation of B lymphocytes. The lymphocytes have membrane receptors for EBV. African children who develop Burkitt's lymphoma are thought to be unable to mount an appropriate immune response to primary EBV infection, possibly because of coexistent malaria, which is immunosuppressive. As time passes, excessive B-cell proliferation occurs. The precise role of EBV in the development of Burkitt's lymphoma remains unclear, but much research in this area continues to be done.

Good Genes and Bad Genes

The cells of Burkitt's lymphoma are characterized by a specific chromosomal defect known as a balanced reciprocal translocation. Observation of fresh Burkitt's lymphomas and cultured cells has revealed an additional DNA fragment at the end of the long arm of chromosome 14, while the end of one chromosome, 8, was consistently absent. Researchers suggested that the missing part of chromosome 8 was translocated to chromosome 14. This 8/14 translocation has also been observed in sporadic Burkitt's tumors from America, Japan, and Europe and has been observed in Burkitt's tumors with or without EBV markers.

The part of chromosome 8 involved in the translocation is known as the c-*myc* proto-oncogene. Proto-oncogenes normally help control the cell cycle by regulating the number of cell divisions. They are especially active when high rates of cell division are needed, as in embryonic development, wound healing, or regeneration. The proto-oncogene is transformed into an oncogene when the chromosomes break

and reunite, resulting in a reciprocal translocation. The rearrangement of genes in this kind of translocation causes the c-*myc* gene to become an oncogene by forming an abnormal fusion protein that triggers the onset of cancer. More than sixty human proto-oncogenes have now been localized to a specific chromosome or chromosome region. The new location of the c-*myc* gene results in deregulation and subsequent overexpression. A normal-acting proto-oncogene is transformed into an abnormally active oncogene.

Ninety percent of Burkitt's tumors are associated with a reciprocal translocation involving

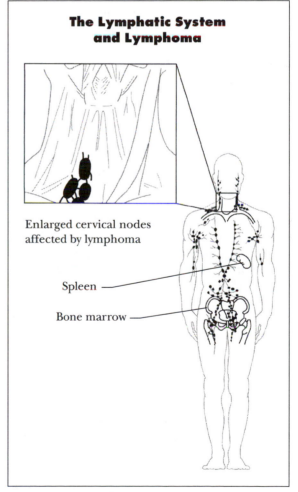

Anatomy of the lymphatic system, showing major lymph nodes. Enlarged lymph nodes may occur for a wide variety of reasons, including but not limited to lymphoma (cancer). (Hans & Cassidy, Inc.)

chromosomes 8 and 14. As additional tumors have been examined, two other related translocations involving chromosome 8 have been observed. The variant translocations involved chromosomes 2 or 22. However, no unified theory exists to explain the role of chromosome abnormalities in the activation of oncogenes. The Epstein-Barr virus has been implicated in Burkitt's lymphoma and is known to be a B-cell mitogen (a substance that stimulates cell division). As a mitogen, it stimulates inactive cells to transform into actively dividing cells. Perhaps EBV plays a role in the origin of 8/14 translocation abnormalities simply by increasing the number of B cells undergoing DNA replication. This could increase the chances for developing a chromosome abnormality with the potential to become cancerous.

—*Phillip A. Farber*

See also: Cancer; Oncogenes; Tumor-Suppressor Genes.

Further Reading

Burkitt, Denis Parsons, and D. H. Wright. *Burkitt's Lymphoma.* Foreword by Sir Harold Himsworth. Edinburgh: Livingstone, 1970. The classic description by the discoverer of this type of sarcoma. Illustrations (some color), maps.

Cotran, R. S., et al. *Robbins Pathologic Basis of Disease.* 6th ed. Philadelphia: Saunders, 1999. A leading textbook that presents comprehensive information accompanied by more than 170 newly created computer-generated diagrams, including schematics, flow charts, and diagrammatic representations of the disease.

Heim, S., and Felix Mitelman. *Cancer Cytogenetics.* 2d ed. New York: J. Wiley, 1995. Provides excellent correlation of molecular and chromosomal findings in Burkitt's lymphoma.

Lenoir, G. M., G. T. O'Conor, and C. L. M. Olweny, eds. *Burkitt's Lymphoma: A Human Cancer Model.* New York: Oxford University Press, 1985. Proceedings of a symposium organized by the International Agency for Research on Cancer and co-sponsored by the WHO Regional Office for Africa, the General Motors Cancer Research Foundation, and the Association Pour le Développement de la Recherche sur le Cancer.

Weinberg, R. A. "How Cancer Arises." *Scientific American* (September, 1996). A general discussion for the layperson, which applies to Burkitt's as well as other cancers. Provides excellent illustrations, tables, and discussions of important research.

Web Sites of Interest

American Cancer Society. http://www.cancer.org. Comprehensive and searchable site covering all aspects of cancer.

Lymphoma Research Foundation, Getting the Facts on Non-Hodgkin's Lymphoma. http://www.lymphoma.org. Site has searchable information on Burkitt's lymphoma (non-Hodgkin's lymphoma), including online guides and information on obtaining guides by mail.

The Leukemia and Lymphoma Society. http://www.leukemia-lymphoma.org. Searchable site that includes a free, sixty-four-page downloadable guide on all types of lymphomas, including Burkitt's.

Cancer

Field of study: Diseases and syndromes

Significance: *At its root cancer is a genetic disease. It is characterized by unrestrained growth and reproduction of cells, loss of contact inhibition, and, eventually, metastasis (the wandering of cancer cells from a primary tumor to other parts of the body). All of these changes represent underlying mutations or inappropriate expression of genes involved in the control of the cell cycle and related processes.*

Key terms

CARCINOGEN: a substance or other environmental factor that produces or encourages cancer

ONCOGENES: genes that cause cancer but that, in their normal form, called proto-oncogenes, are important in controlling the cell cycle and related processes

TUMOR: a mass formed by the uncontrolled growth of cells, which may be malignant (considered cancerous) or benign (nonmalignant)

TUMOR-SUPPRESSOR GENES: genes involved in regulating the cell cycle and preventing cell division until an appropriate time; when mutated, these genes can cause cancer

The Problem of Cancer

Cancer is characterized by abnormal cell growth that leads to the invasion and destruction of healthy tissue by cells that meet certain criteria. Normal cells in the human body are continuously growing but are under normal cell control mechanisms. Cancer cells begin as normal cells that, due to genetic mutations, start to grow uncontrollably, escaping from the normal rules regulating cell growth and behavior. Contact inhibition, in which cells contacting other cells prevent unrestrained growth, is lost in cancer cells. Normal cells also remain in one location, or at least in the same tissue, but malignant tumors, in their later stages, metastasize, allowing their cells to wander freely in the body, leading to the development of tumors in other organs. A final common feature is that cancer cells lose their normal cell shape.

The area where cancer begins to form a tumor is called the primary site. Most types of cancer begin in one place (the breast, lung, or bowel, for example) from which the cells invade neighboring areas and form secondary tumors. To make matters more complicated, some types of cancer, such as leukemia, lymphoma, and myeloma, begin in several places at the same time, usually in the bone marrow or lymph nodes. Primary tumors begin with one abnormal cell. This cell, as is true of all cells, is extremely small, no more than 0.002 or 0.003 millimeter across (about one-twentieth the width of a human hair). Therefore early cancer is very difficult to locate. Even if there are more than 100,000 cancer cells in a tumor, it is barely visible except under a microscope.

Cancer cells divide and reproduce about every two to six weeks. If they divide on the average of once per month, a single cell will multiply into approximately four thousand cells by the end of a year. After twenty months, there will be one million cells, which would form a tumor about the size of a pinhead and would still be undetectable. A tumor can be discovered only when a lump of approximately one billion cells is present. This would be about the size of a small grape. It would take about two and one-half years for a single cancer cell to reach this size. Within seven months, the one billion cells would grow to more than 100 billion cells, and the tumor would weigh about four ounces. By the fortieth month of growth, the lump of cancer cells would weigh about two pounds. By the time a tumor has reached this size, death often occurs. Death normally occurs about three and one-half years after the first cancer cell begins to grow. It takes about forty-two cell doublings to reach the lethal stage. The problem is that, in most cases, tumors are detectable only after thirty doublings. By this time, cancer cells may have invaded many other areas of the body beyond the primary site.

How Cancer Cells Grow and Invade

Cancer cells are able to break down the barriers that normally keep cells from invading other groups of cells. With the aid of a microscope, cancer cells can be observed breaking through the boundary between cells, called the basement membrane. Cancer cells can make substances that break down the intercellular matrix, the "glue" that holds cells together. The intercellular matrix is a complex mixture of substances, including collagen, a strong, fibrous protein that gives strength to tissues. Cancer cells produce collagenase, an enzyme that breaks down collagen. Cancer cells also produce hyaluronidase, which further breaks down the intercellular matrix. This causes cancer cells to lose their normal shape and allows them to push through normal boundaries and establish themselves in surrounding tissues. Cancer cells have jagged edges, are irregular in shape, and have hard-to-detect borders, making them relatively easy to identify microscopically. Normal cells, on the other hand, have a regular, smooth edge and shape.

There are many steps involved in the process of metastasizing, not all of which are understood by researchers. First is the entry into a blood vessel or lymph channel. Lymph channels, or lymphatics, comprise a network of vessels that carry lymph from the tissues to the bloodstream. Lymph is a colorless liquid that drains from spaces between cells. It consists mainly of water, salts, and proteins and eventually enters the bloodstream near the heart. The function of lymph is to filter out bacteria and other foreign particles that might enter the blood and cause infections. A mass of lymph vessels is called a lymph node. In the human body, lymph nodes are found in the neck, under the arms, and in several other places. Every body tissue has a network of lymph and blood vessels running through it.

Once a malignant tumor develops and metastasizes, the cells often travel through the body using the lymphatic system, a network of vessels that filter pathogens and transport lymph, a fluid similar to blood plasma. Cancer cells may gain entry into a nearby lymph vessel by breaking down defensive enzymes. Once in the lymph system, they can travel to nodes (gland-like masses of cells that produce white blood cells) and eventually into the bloodstream. Whatever route they take, groups of cancer cells can break away from the primary site of the tumor and float along whatever vessel they have invaded, forming numerous secondary tumors along the way. Because cancer cells are not considered foreign substances, such as bacteria or viruses, they are able to evade the body's immune system. Because of their overall resemblance to normal cells, cancer cells fool the body into thinking they are normal and therefore not dangerous.

Cancer cells eventually enter narrow blood vessels called capillaries and stay there for a brief period before they enter tissues such as lungs, bones, skin, and muscle. The secondary tumors then capture their own territory. As a tumor establishes itself, its cells often secrete signal proteins that stimulate new blood vessels to form (a process called angiogenesis) to increase blood supply to the growing tumor. The body thus not only fails to destroy developing tumors, but unwittingly helps establish them as well.

The Genetics of Cancer

Cancer has been known since antiquity, but it was not until the twentieth century that the underlying causes of cancer began to be explored. In 1910, Peyton Rous discovered a type of cancer in chickens called a sarcoma (a cancer of connective tissue) that could be passed on to other chickens. He demonstrated this by removing tumors from affected chickens, grinding the tumors up, filtering the grindate, and then injecting the filtrate into healthy chickens. Injected chickens invariably developed sarcoma tumors, suggesting that something smaller than the tumor cells was being passed on and was stimulating cancer development in otherwise normal cells. It is now known that the filtrate contained a cancer-causing virus, now called the Rous sarcoma virus. Similar types of viruses were discovered to be responsible for cancers in a variety of animals, but none was discovered in humans initially.

As the genetic material of some of the tumor viruses was later analyzed, all of them were discovered to contain genes called oncogenes, because they promoted oncogenesis (tumor development). Even more surprising was the discovery that humans have genes in their genome that are homologous (having a high degree of similarity) to viral oncogenes. The human genes did not seem to cause cancer under normal circumstances and were called proto-oncogenes. In cancer cells, some of these proto-oncogenes were discovered to have mutations or, in some cases, were simply overexpressed. In recognition of their abnormal state, these genes were called cellular oncogenes, to distinguish them from viral oncogenes. It is now known that proto-oncogenes are important in controlling the cell cycle by stimulating cell division only at the appropriate time. When they are transformed into oncogenes, uncontrolled cell growth and division occurs, two of the hallmarks of cancer.

A second type of cancer-causing gene, called a tumor-suppressor gene, was discovered to be the cause of retinoblastoma, a cancer of the retina, most often occurring in children. Tumor-suppressor genes have an effect opposite to that of proto-oncogenes; they suppress cell division and thus prevent unrestrained cell proliferation. If both alleles of a tumor-suppressor gene have a mutation that makes them nonfunctional, then cell division can occur un-checked. Retinoblastoma occurs in children when they inherit one faulty copy from a parent. If the other copy experiences a mutation, which frequently occurs, then retinoblastoma develops.

How Cancer Develops

The development of cancer is typically more complicated than implied above. It generally requires mutations in more than a single proto-oncogene or tumor-suppressor gene. Any factors that increase mutation rates or decrease a cell's ability to repair mutations will increase the likelihood that cancer will develop. Inheritance of already mutated genes can also greatly increase a person's chance of developing cancer, which accounts for the above-normal occurrence of cancer in some families.

One of the best-studied cases of oncogenesis involves colorectal cancer, which takes years to develop from a small cluster of abnormal cells into life-threatening cancer. It involves the loss or mutation of three tumor-suppressor genes

Leland H. Hartwell, co-winner of the 2001 Nobel Prize in Physiology or Medicine with R. Timothy Hunt and Paul M. Nurse, at the Fred Hutchinson Cancer Research Center in Seattle, Washington, shortly after the Nobel Foundation's announcement. The three men won for their work on cell division and its implications for cancer research. (AP/Wide World Photos)

and one proto-oncogene. Often colorectal cancer runs in families, because the loss of the first gene, the *APC* tumor-suppressor gene, is often inherited, resulting in an increased chance of developing colorectal cancer. Loss of this gene causes increased cell growth and some other genetic changes. In the next step, the *ras* oncogene is mutated, causing even more cell growth. Two more tumor-suppressor genes are lost, *DCC* and *p53*, at which point a tumor called a carcinoma has developed. Additional gene loss, which occurs much more easily in tumor cells, leads to metastasis, and the cancer then spreads to other organs and tissues.

Inheritance of a gene loss or mutation does not mean a person will get cancer; it simply means they have a higher chance of developing cancer. Although development of cancer is ultimately genetically based, environmental factors also play a part. In the case of colorectal cancer, a diet low in roughage is often considered to increase colorectal cancer rates. Exposure to carcinogens, chemicals, or other factors, such as radiation, can also increase the likelihood of cancer. Exposure can occur in the diet, as a result of skin exposure, or inhalation. For example, smoking cigarettes is known to increase the occurrence of lung cancer, as well as a variety of other cancers. Excess exposure to damaging UV rays in sunlight or other sources is known to significantly increase the occurrence of skin cancer. Carcinogens promote cancer because they cause damage to DNA, and if the damage happens to occur to a tumor-suppressor gene or oncogene, then cancer may occur.

Inheritance of some mutations is particularly potent in increasing the chances of developing cancer. One example is the genetic disease xeroderma pigmentosa. Individuals with this disease develop skin cancer in response to even relatively brief exposure to UV radiation and must therefore avoid exposure to sunlight. A small percentage of breast cancers are also highly heritable and are now known to involve mutations in the *BRCA1* and *BRCA2* genes. In these types of highly heritable cancers, it appears that the mutations cause some kind of deficiency in the cellular DNA repair systems. As a result of a decreased ability to repair mutations, it is just a matter of time before mutations occur in proto-oncogenes or tumor-suppressor genes, so that the only way to prevent cancer is to control exposure to as many environmental carcinogens as possible and to aggressively screen for tumors.

Cancer Treatment

Cancers vary in their severity and rate of growth, which means that proper treatment depends on correctly diagnosing the type of cancer. For example, some forms of prostate cancer grow extremely slowly, and metastasis is rare until very late stages in the disease, sometimes many years after initial diagnosis. Treatment may comprise simply monitoring the tumor, avoiding carcinogenic exposure as much as possible, and possibly changing lifestyle. On the other hand, some types of skin cancer progress so rapidly that aggressive treatment may be required, unless it is caught very early. Although survival rates for many types of cancer have risen, treatment for most cancers is still only partially successful, and the later a tumor is detected, the greater chance that it will be untreatable.

New therapies are constantly being developed, but most cancers are still treated using surgery (removal of tumors), chemotherapy, and radiation therapy, either singly or, more often, in combination. More important than the specific treatment used is detecting tumors in their earliest stages, before they have extensively invaded surrounding tissues or metastasized. Survival rates are high for most cancers when treated very early.

The very nature of cancer makes treatment difficult. Because the cells involved are difficult for the immune system to recognize as dangerous, the body is typically inefficient at destroying them. Many of the treatments, other than surgical removal, rely on the fact that cancer cells divide much faster and more frequently than normal cells. Therefore, any agent that can cause higher mortality in rapidly dividing cells has potential as a cancer treatment. Chemotherapeutic agents are essentially toxic chemicals that are most toxic to dividing cells. Thus, they kill cancer cells much more readily than most other body cells, but any other body cells undergoing cell division are susceptible, so chemotherapy also kills some normal cells. Cancer patients often feel very ill during chemotherapy because of this.

Radiation therapy works similarly, being more damaging to dividing cells. An added advantage of radiation therapy, if the tumor has not yet metastasized, is that it can be focused more intensely in the vicinity of the tumor, preventing damage to other tissues. If the tumor has metastasized, then more widespread exposure to radiation may be used, with the obvious drawback that many other normal cells will also be damaged. Radiation therapy is often used to treat leukemia. Radiation is used to kill the patient's bone marrow, and then new bone marrow is transplanted from a compatible donor. The new bone marrow can then restore normal function to production of blood cells.

Genetics has played a part in improving chemotherapy. It has long been known that some people will respond better than others to certain chemotherapeutic drugs. It is now known that some of these differences are genetic, and the underlying genetic differences have been uncovered in some cases. Therefore, as part of cancer treatment for some kinds of cancer, a person may be tested genetically to make more intelligent choices about which drugs to use. As more genetic data become available, it is anticipated that more effective and personalized treatments will be developed.

Innovations and Future Treatments

Although the immune system cannot normally identify cancer cells accurately, there has been some success in immunological approaches. A recent new treatment involves Herceptin, a specially designed monoclonal antibody that attacks cells that overproduce a protein called HER-2. It has shown some promise in treating breast cancer and is being studied as an agent for treating other forms of cancer. Unfortunately, some of the possible side effects include damage to the heart and lungs and serious allergic reactions. Careful monitoring is required to prevent life-threatening damage. Research is also progressing on development of vaccines against cancer, but so far this approach is still in its early experimental stages.

Angiogenesis inhibitors are also at the experimental stage. As noted above, many cancer cells secrete agents that encourage angiogenesis (the formation of blood vessels) in the region of the tumor to increase blood supply to the rapidly dividing cells. It is hoped that angiogenesis inhibitors might counteract this activity and essentially "starve" the cancer cells. Shrinkage of tumors in laboratory animals has been observed using this approach.

Photodynamic therapy also shows promise. It is based on the observation that certain chemicals, when ingested by single-celled organisms, release damaging oxygen radicals when exposed to light, thus killing the organisms. It has been observed that cancer cells retain these chemicals longer than normal cells. Treatment involves administering the chemical by injection, then waiting for a specified period for it to be retained by cancer cells and flushed out of normal cells. Then the tissue in which the cancer cells are located is exposed to laser light. This method works on any tissues that can be exposed to laser light, which includes any part of the body accessible to endoscopy.

The ultimate treatment for cancer would be replacement or repair of the mutated genes responsible. Currently such treatment is not possible. There are many hurdles to overcome, including designing safe methods for inserting corrected gene copies. There is danger that improper gene therapy methods could actually make things worse, causing additional tumors or other diseases. A much better understanding of the genetics of cancer and future improvements in gene therapy techniques hold promise of someday being able to cure or prevent most kinds of cancer.

—*Leslie V. Tischauser, updated by Bryan Ness*
See also: Aging; Breast Cancer; Burkitt's Lymphoma; Cell Culture: Animal Cells; Cell Cycle, The; Cell Division; Chemical Mutagens; Chromosome Mutation; Developmental Genetics; DNA Repair; Gene Therapy; Genetic Engineering: Medical Applications; Genetic Testing: Ethical and Economic Issues; Hereditary Diseases; Homeotic Genes; Human Genome Project; Huntington's Disease; Hybridomas and Monoclonal Antibodies; Hypercholesterolemia; Insurance; Model Organism: *Caenorhab-*

ditis elegans; Model Organism: *Mus musculus*; Mutation and Mutagenesis; Nondisjunction and Aneuploidy; Oncogenes; Proteomics; Reverse Transcriptase; RNA Transcription and mRNA Processing; Signal Transduction; Stem Cells; Steroid Hormones; Telomeres; Tumor-Suppressor Genes.

Further Reading

Bowcock, Anne M., ed. *Breast Cancer: Molecular Genetics, Pathogenesis, and Therapeutics.* Totowa, N.J.: Humana Press, 1999. Detailed information geared toward researchers and health professionals. The chapter "Hereditary Breast Cancer Genes" discusses *BRCA1* and *BRCA2* mutations among Ashkenazi women. Also addresses surgery, chemotherapy, drug resistance, and the *MDR* gene.

Bradlow, H. Leon, Jack Fishman, and Michael P. Osborne, eds. *Cancer Prevention: Novel Nutrient and Pharmaceutical Developments.* New York: New York Academy of Sciences, 1999. Examines several classes of nutrients and pharmaceutical agents believed to be important for tumor inhibition. Reviews novel preclinical models that facilitate analysis of chemopreventive agent efficacy and mechanisms of gene-nutrient interaction and provides information on clinical trials studying chemopreventive regimens.

Coleman, William B., and Gregory J. Tsongalis, eds. *The Molecular Basis of Human Cancer.* Totowa, N.J.: Humana Press, 2002. Surveys the known molecular mechanisms governing neoplastic transformation in the breast, prostate, lung, liver, colon, skin, leukemias, and lymphomas. Illuminates both recent developments and established concepts in epidemiology, molecular techniques, oncogenesis, and mutation mechanisms.

Cowell, J. K., ed. *Molecular Genetics of Cancer.* 2d ed. San Diego: Academic Press, 2001. Focuses on tumors of tissues such as liver, lung, bladder, and brain and surveys research in the cloning and analysis of genes central to the development and progression of human cancers.

Davies, Kevin, and Michael White. *Breakthrough: The Race to Find the Breast Cancer Gene.* New York: John Wiley, 1996. A history of the research into the genetic causes of breast cancer and other types of cancer.

Ehrlich, Melanie, ed. *DNA Alterations in Cancer: Genetic and Epigenetic Changes.* Natick, Mass.: Eaton, 2000. A comprehensive overview of the numerous and varied genetic alterations leading to the development and progression of cancer. Topics include oncogenes, tumor-suppressor genes, cancer predisposition, DNA repair, and epigenetic alteration such as methylation.

Fitzgerald, Patrick J. *From Demons and Evil Spirits to Cancer Genes: The Development of Concepts Concerning the Causes of Cancer and Carcinogenesis.* Washington, D.C.: American Registry of Pathology, Armed Forces Institute of Pathology, 2000. Traces the history, epidemiology, and genetics of neoplasms, cancer, and medical oncology.

Greaves, Mel F. *Cancer: The Evolutionary Legacy.* New York: Oxford University Press, 2000. Presents a Darwinian explanation for cancer that includes historical anecdotes and scientific findings.

Habib, Nagy A., ed. *Cancer Gene Therapy: Past Achievements and Future Challenges.* New York: Kluwer Academic/Plenum, 2000. Reviews forty-one preclinical and clinical studies in cancer gene therapy, organized into sections on the vectors available to carry genes into tumors, cell cycle control, apoptosis, tumor-suppressor genes, antisense and ribozymes, immuno-modulation, suicidal genes, angiogenesis control, and matrix metallo proteinase.

Hanski, C., H. Scherübl, and B. Mann, eds. *Colorectal Cancer: New Aspects of Molecular Biology and Immunology and Their Clinical Applications.* New York: New York Academy of Sciences, 2000. Explores the immunological and molecular aspects of colon and rectal cancer.

Hodgson, Shirley V., and Eamonn R. Maher. *A Practical Guide to Human Cancer Genetics.* 2d ed. New York: Cambridge University Press, 1999. Gives a general overview of the underlying molecular genetic basis of cancer, the genetics of human cancers by site of origin, and a review of inherited cancer-predisposing syndromes.

Liotta, L. A. "Cancer Cell Invasion and Metastasis." *Scientific American* (1992). Provides a basic description of cancer genetics.

Maruta, Hiroshi, ed. *Tumor-Suppressing Viruses, Genes, and Drugs: Innovative Cancer Therapy Approaches.* San Diego, Calif.: Academic, 2002. An international field of experts address a number of innovative approaches to treating cancer, such as viral therapy using specific viral mutants, gene therapy using a variety of tumor-suppressor genes, and drug therapy targeted to block oncogenic signal pathways.

Mendelsohn, John, et al. *The Molecular Basis of Cancer.* 2d ed. Philadelphia: Saunders, 2001. Surveys the principles that constitute the scientific basis for understanding the pathogenesis of cancer and emphasizes clinical implications for treatment. Covers recent advances and current research, with descriptions of the basic mechanisms of malignant cells and molecular abnormalities, as well as new approaches to therapy.

Schneider, Katherine A. *Counseling About Cancer: Strategies for Genetic Counseling.* 2d ed. New York: Wiley-Liss, 2002. A reference guide to assist genetic counselors and other health care providers help patients and families through the emotional difficulties of managing hereditary cancer.

Vogelstein, Bert, and Kenneth W. Kinzler, eds. *The Genetic Basis of Human Cancer.* 2d ed. New York: McGraw-Hill, 2002. Introduces the fundamentals of genetics and human phenotypes, gene mutation, the Human Genome Project, and gene imprinting and covers advances in the field.

Wilson, Samuel, et al. *Cancer and the Environment: Gene-Environment Interaction.* Washington, D.C.: National Academy Press, 2002. Includes "The Links Between Environmental Factors, Genetics, and the Development of Cancer," "Gene-Environment Interaction in Special Populations," and "Gene-Environment Interaction in Site-Specific Cancers."

Web Sites of Interest

American Cancer Society. http://www.cancer.org. Comprehensive and searchable site covering all aspects of cancer.

National Cancer Institute. http://www.cancer.gov. Site links to comprehensive information on genetics and cancer, including a cancer-basics tutorial.

cDNA Libraries

Fields of study: Bioinformatics; Techniques and methodologies

Significance: *A cDNA library is a set of cloned DNA copies of the RNAs found in a specific cell type at a specific time. This library can be used to construct probes for mapping these genes, to study the changing expression of genes over time (during development, for example), or to clone genes into organisms for further study or production of proteins.*

Key terms

COMPLEMENTARY DNA (cDNA): also known as copy DNA, a form of DNA synthesized by reverse transcribing RNAs (usually messenger RNAs) into DNA

DNA LIBRARY: a collection of DNA fragments cloned from a single source, such as a genome, chromosome, or set of mRNAs

IN SITU HYBRIDIZATION: a technique that uses a molecular probe to determine the chromosomal location of a gene

INTRONS: noncoding segments of DNA within a gene that are removed from mRNA copies of the gene before polypeptide translation

REVERSE TRANSCRIPTASE: an enzyme, isolated from retroviruses, that synthesizes a DNA strand from an RNA template

Gene Cloning and DNA Libraries

In order to study and map genes, researchers need to take potentially very large sections of DNA (such as a chromosome or whole genome), break them into smaller, manageable fragments, and clone these fragments to construct a DNA library. A genomic or chromosome library may contain many thousands of cloned fragments, many of which will represent stretches of noncoding DNA between genes. If the researcher is interested is studying the protein-coding regions, or genes, of the DNA, it is better to start with the messenger RNAs (mRNAs) of the cell, which represent the genes being actively transcribed in the cell at that time. By constructing and cloning complementary DNA (cDNA) copies of these mRNAs, researchers can create a library that contains copies of only the active genes.

cDNA Library Construction

DNA copies of mRNAs are synthesized using the enzyme reverse transcriptase. This enzyme was independently discovered by Howard Temin and David Baltimore in 1970 in retroviruses, which "reverse transcribe" their RNA genomes into DNA after infecting their host cells. In the late 1970's, researchers began using the enzyme to make DNA copies of mRNAs, and later to construct cDNA libraries.

To create a cDNA library from a sample of cells, mRNAs from the cells are isolated and purified. Reverse transcriptase is used to synthesize a complementary DNA strand using each mRNA strand as a template, resulting in a collection of double-stranded RNA-DNA hybrids. To obtain double-stranded cDNAs suitable for cloning, the enzyme RNase H is used to digest the RNA strand, and DNA polymerase I is used to synthesize the second DNA strand using the first as a template. If desired, "sticky ends" can be added to the cDNAs for cloning into a vector. The set of recombinant vectors are inserted into bacterial cells in the process of transformation, resulting in a cloned cDNA library. The library is maintained as a collection of bacterial colonies, each colony containing a different cloned DNA fragment.

Applications

A cDNA library represents the coding sequences of genes that were actively expressed in the original cell sample at the time the sample was taken. In effect, it can represent a snapshot of active genes in the cells at that time. Comparing the cDNAs of different tissues from the same organism can reveal the differences in gene expression of these tissues. Also, comparing cDNAs of cells in the same tissue over time can show how gene expression changes in the same cells. This approach has been especially fruitful in developmental genetic re-

search, because the developmental pattern of an organism can be correlated with the activity of specific genes.

Cloned cDNAs can also be used to find the chromosomal location of an expressed gene. One strand of the cDNA clone is labeled with a fluorescent tag and used as a molecular probe. In the technique of in situ hybridization, the probe will base pair, or hybridize, to the complementary sequence in a preparation of partially denatured chromosomes, and the chromosomal location of the original gene will be visible because of the fluorescent label. Such a probe can also be used to screen a chromosome or genomic library for the cloned fragment containing the target gene. Using the entire cDNA library to probe a genome will generate a cDNA map that suggests the most biologically and medically important parts of the genome, aiding researchers in the search for disease genes.

Genes of eukaryotes (nonbacterial organisms) usually contain introns, noncoding segments that are transcribed but removed from mRNAs before translation, but bacterial genes do not. Often, a eukaryotic gene put into a bacterial cell will not produce a functional polypeptide because the cell does not have the biochemical machinery for removing introns. If the goal of the research is to have a bacterium make the protein product of a gene, it may be necessary to clone a cDNA version of the gene, which lacks introns, using a special expression vector that allows the cell to transcribe the inserted gene and translate it to the proper polypeptide.

Advantages and Disadvantages

Because cDNA libraries contain only DNA of expressed genes, they are much smaller and more easily managed and studied than chromosome or genomic libraries that have all coding and noncoding regions. The cDNA versions of genes have only the protein-coding sequence, without introns, so that cloning them in bacteria allows expression of the protein products of the genes. In contrast to other DNA libraries, cDNA libraries can be used to study variable patterns of gene expression among cell types or over time. In eukaryotes, cDNA copies of genes are not identical to the original sequences of the genes and also lack the promoter region necessary for proper transcription of the gene. However, using cDNA as a molecular probe can lead to the identification of the original gene.

—*Stephen T. Kilpatrick*

See also: Bioinformatics; DNA Fingerprinting; DNA Sequencing Technology; Forensic Genetics; Genetic Testing: Ethical and Economic Issues; Genetics, Historical Development of; Genomic Libraries; Genomics; Human Genome Project; Icelandic Genetic Database; Linkage Maps; Proteomics; Reverse Transcriptase.

Further Reading

Sambrook, Joseph, and David Russell. *Molecular Cloning: A Laboratory Manual.* 3d ed. Cold Spring Harbor, N.Y.: Cold Spring Harbor Laboratory Press, 2001. Contains detailed protocols for mRNA isolation, cDNA synthesis, and library construction.

Watson, James D., John Tooze, and David T. Kurtz. *Recombinant DNA: A Short Course.* New York: W. H. Freeman, 1983. An introduction to techniques for cloning genes, including construction of cDNA libraries.

Cell Culture: Animal Cells

Fields of study: Cellular biology; Techniques and methodologies

Significance: *The ability to grow and maintain cells or tissues in laboratory vessels has provided researchers with a means to study cell genetics and has contributed to the understanding of what differentiates "normal" cells from cancer cells. The technology involved in growing viruses in cell culture has proved vital both to understanding virus replication and for development of viral vaccines.*

Key terms

CELL LINES: cells maintained for an indeterminate time in culture

HeLa CELLS: the first human tumor cells shown to form a continuous cell line

MICROPROPAGATION: removal of small pieces of plant tissue for growth in culture

PRIMARY CELLS: explants removed from an animal

TRANSFORMATION: any physical change to a cell, but generally the change of a normal cell into a cancer cell

Early History

Methodology for maintaining tissues in vitro (in laboratory vessels) began in 1907 with Ross Harrison at Yale College. Harrison placed tissue extracts from frog embryos on microscope slides in physiological fluids such as clotted frog lymph. The material was sealed with paraffin and observed; specimens could be maintained for several weeks. In 1912, Alexis Carrel began the maintenance of cardiac tissues from a warm-blooded organism, a chicken, in a similar manner. The term "tissue culture" was originally applied to the cells maintained in the laboratory in this manner, reflecting the origin of the technique. More appropriate to modern techniques, the proper terminology is "cell culture," since it is actually individual cells which are grown, developing as explants from tissue. Nevertheless, the terms tend to be used interchangeably for convenience.

Types of Cell Culture

The most common form of mammalian cell culture is that of the primary explant. Cells are removed from the organism, preferably at the embryonic stage, treated with an enzyme such as trypsin, which serves to disperse the cells, and placed in a laboratory growth vessel. Most vessels used today are composed of polystyrene or similar forms of plastic.

Most forms of cells are anchorage-dependent, meaning they will attach and spread over a flat surface. Given sufficient time, such cells will cover the surface in a layer one cell thick, known as a monolayer.

A few forms of cells, mainly hematopoietic (blood-forming) or transformed (cancer) cells, are anchorage-independent and will grow in suspension as long as proper nutrients are supplied.

Similar procedures are used in preparation of nonmammalian cell lines such as those from poikilotherms (cold-blooded organisms such as fish) or insects. Insect lines have become particularly important as techniques were developed for cloning genes in insect pathogens known as baculoviruses. Such cells can often be maintained at room temperature in suspension.

Development of Cell Lines

A characteristic of primary cells is that of a finite life span; normal cells will replicate approximately fifty times, exhibit symptoms of "aging," and die. When primary cells are removed from a culture and cultured separately, they become known as a cell strain.

A few rare cells may enter "crisis" and begin to exhibit characteristics of abnormal cells such as anchorage-independence or unusual chromosome numbers. If these cells survive, they represent what is called a "cell line." Cell lines express characteristics of cancer cells and are often immortal.

During the first half century of work in cell culture, only nonhuman cells were grown in culture. In 1952, George Gey, a physician at Johns Hopkins Hospital, demonstrated that human cells could also be grown continuously in culture. Using cervical carcinoma explants from a woman named Henrietta Lacks, Gey prepared a continuous line from these cells. Known as HeLa cells, these cultures became standard in most laboratories studying the growth of animal viruses. Ironically, growth of HeLa cells was so convenient and routine that the cells frequently contaminated other cultures found in the same laboratories.

Nutrient Requirements

Particular cells may have more stringent requirements for growth than other types of cells; in addition, primary cells have greater requirements than cell lines. However, certain generalities apply to the growth requirements for all cells. All cells must be maintained in a physiological salt solution. Required vitamins and amino acids are included in the mixture. Antibiotics such as penicillin and streptomycin are routinely added to suppress the growth of unwanted microorganisms. Nevertheless, sterility

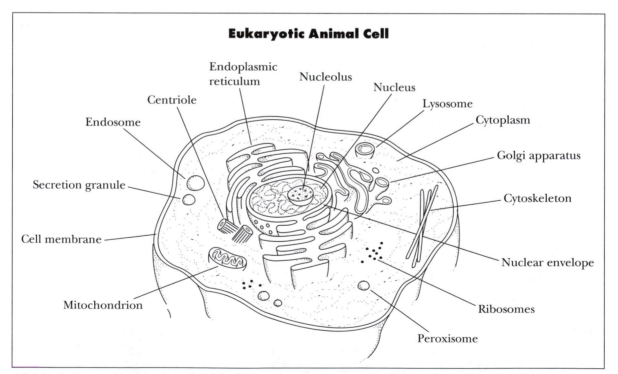

Eukaryotic Animal Cell

All animal cells are eukaryotic cells, which differ from more primitive prokaryotic cells in having a nucleus that houses the primary genetic material. This drawing depicts the basic features of a eukaryotic animal cell. (Electronic Illustrators Group)

is of utmost importance since some organisms are unaffected by these antibiotics. Depending upon the type of cell, the particular pH, or acid content, of the culture may be variable. Most mammalian cells grow best at a pH of 7-7.2. For this reason, cells are generally grown in special incubators which utilize a relatively high CO_2 atmosphere.

Replication of animal cells requires the presence of certain growth factors to be present in the medium. Historically, the source of such factors has been serum, usually obtained from fetal bovines. Genetic engineering techniques have resulted in production of commercially available growth factors, eliminating the requirement for expensive serum for growth of some forms of cells in culture.

Genetics of Cells in Culture

Study of cultured animal cells has resulted in significant advancement in understanding many areas of cell regulation. For example, the role played by cell receptors in response to the presence of extracellular ligands such as hormones and other metabolites was clarified by studying the response of cells to such stimulation. Intracellular events, including the role of enzymes in cell activities, was clarified and remains a primary area of research.

The ability to transform mammalian cells using isolated DNA has allowed for significant applications in genome analysis. Such genetic manipulation has led to a greater understanding in the role specific genes play in cell regulation. In particular, use of cultured cells was instrumental in clarifying the role played by specific gene products in intracellular trafficking, the movement of molecules to specific sites within the cell. Similar techniques continue to be used to further understand the regulatory process.

Mammalian Cells and Oncogenesis

During the 1960's, Leonard Hayflick at the Wistar Institute in Philadelphia, Pennsylvania, observed that primary cells in culture exhibit a finite life span; normal cells generally divide no more than approximately fifty times (a phenomenon now called the Hayflick limit). Any cells that survive generally take on the characteristics of cancer cells.

During the same period, Howard Temin at the University of Wisconsin, while studying the growth of RNA tumor viruses in cultured cells, reported the apparent requirement for DNA production by these viruses in transforming normal cells into cancer cells. Temin's and Hayflick's investigations contributed significantly to the question of how cancer cells differ from normal cells, and the understanding of genes involved in such changes. Eventually, this led to the discovery of oncogenes.

The term "oncogene" is somewhat misleading. Its definition was originally based on the fact that mutations in such genes may contribute to transformation of cells from normal to cancerous. The study of these genes in cultured cells clarified their role: Most oncogene products can be classified as growth factors, which stimulate cell growth; receptors, which respond to such stimulation; or intracellular molecules, which transfer such signals to the cell DNA. In other words, the normal function of the oncogene is to regulate replication of normal cells; only when these proteins are inappropriately expressed do they result in transformation of the cell.

Application of Cell Culture to Virology

The use of mammalian cells for the study of viruses represented among the earliest, and arguably among the most important, applications of the technique of cell culture. Prior to the 1940's, study of most animal viruses, including those that cause disease in humans, was confined to in vivo studies in animals. For example, the study of poliovirus required inoculation of the virus directly into the brains of suitable monkeys.

In 1949, John Enders and his co-workers demonstrated the growth of poliovirus in human embryonic cells, eliminating the requirement for monkeys. Their work played a critical role in the later development of poliovirus vaccines by Jonas Salk and Albert Sabin. The ability to grow viruses in cells maintained in the laboratory opened the field to nearly all virologists and biochemists, rather than restricting such studies to those with access to animal facilities.

—*Richard Adler*

See also: Cancer; Cell Culture: Plant Cells; Cell Cycle, The; Cell Division; Gene Regulation: Eukaryotes; Gene Regulation: Viruses; Mitosis and Meiosis; Oncogenes; Stem Cells; Totipotency; Tumor-Suppressor Genes; Viral Genetics.

Further Reading

Freshney, R. Ian. *Culture of Animal Cells.* New York: Wiley-Liss, 2000. Basically a how-to text on the science and art of tissue culture. Useful as a source of recipes and techniques, as well as an extensive bibliography.

Gold, Michael. *A Conspiracy of Cells.* Albany: State University of New York Press, 1986. A full account of the history behind development of the HeLa cell line. Much of the account deals with the (literal) spread of these cells throughout the field of cell culture.

Hayflick, L., and P. Moorhead. "The Serial Cultivation of Human Diploid Cell Strains." *Experimental Cell Research* 25 (1961): 585-621. The classic work that first reported the limited life span of human cells in culture.

Pollack, Robert, ed. *Readings in Mammalian Cell Culture.* 2d ed. Cold Spring Harbor, N.Y.: Cold Spring Harbor Press, 1981. A collection of reprints consisting of nearly all classical papers in the field of cell culture.

Cell Culture: Plant Cells

Fields of study: Cellular biology; Techniques and methodologies

Significance: *Plant cell culture is the establishment and subsequent growth of various plant cells, tissues, or organs in vitro, using an artificial nutritional medium usually supplemented by various plant growth regulators. It has become a tool that plant geneticists use for purposes ranging from the basic study of plant development to the genetic improvement of economically important agricultural plant species.*

Key terms

CALLUS: a group of undifferentiated plant cells growing in a clump

MORPHOGENESIS: the induction and formation of organized plant parts or organs

PLANT GROWTH REGULATORS: hormonelike substances that profoundly affect plant growth and development

SOMATIC EMBRYOS: asexual embryoid structures derived from somatic cells

TOTIPOTENCY: the ability of a plant cell or part to regenerate into a whole plant

Culturing Plant Cells

Plant cell cultures are typically initiated by taking explants—such as root, stem, leaf, or flower tissue—from an intact plant. These explants are surface-sterilized and then placed in vitro on a formulated, artificial growth medium containing various inorganic salts, a carbon source (such as sucrose), vitamins, and various plant growth regulators, depending on the desired outcome. There are many commercially available media formulations; the two most common include MS (murashige and skoog) and WPM (woody plant media). Alternatively, customized formulations may be necessary for culturing certain plant species. One of the most important uses of plant tissue culture has been for the mass propagation of economically important agricultural and horticultural crops. Since the 1980's, however, plant cell culture has become an important tool allowing for direct genetic manipulations of several important agricultural crops, including corn, soybeans, potatoes, cotton, and canola, to name only a few.

Appearance in Culture

The underlying basis for the prevalent and continued use of plant cell culture is the remarkable totipotent ability of plant cells and tissues. They are able to dedifferentiate in culture, essentially becoming a nondifferentiated clump of meristematic, loosely connected cells termed callus. Callus tissue can be systematically subcultured and then, depending on exposure to various plant growth regulators incorporated in the growth media, induced to undergo morphogenesis. Morphogenesis refers to the redifferentiation of callus tissue to form specific plant organs, such as roots, shoots, or subsequent whole plants. Many plant species can also be manipulated in culture to form somatic embryos, which are asexual embryoid

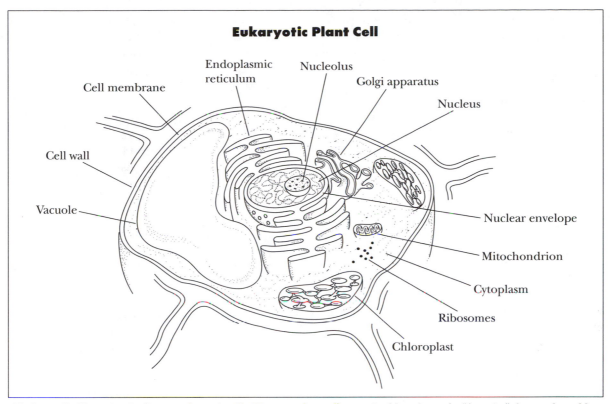

Eukaryotic Plant Cell

All plant cells, like animal cells, are eukaryotic cells. However, plant cells contain chloroplasts, the "factories" that produce chlorophyll during photosynthesis. This drawing depicts the basic features of a eukaryotic plant cell. (Electronic Illustrators Group)

structures that can then develop into plantlets. The totipotency of plant cells thus allows for a single cell, such as a plant protoplast, to be able to regenerate into a complete, whole plant. An analogous comparison of the totipotency of plant cells would be that of stem cells in animals. Genetic manipulation of individual plant cells coupled with their totipotency makes plant cell culture a powerful tool for the plant geneticist.

Role of Plant Growth Regulators

Hormones or plant growth regulators (PGRs) are naturally occurring or synthetic compounds that, in small concentrations, have tremendous regulatory influence on the physiological and morphological growth and development of plants. There are several established classes of PGRs, including auxins, cytokinins, gibberellins, abscisic acid (ABA), and ethylene. Additionally, several other compounds, such as polyamines, oligosaccharides, and sterols,

exert hormonelike activity in plant cell cultures. While each class has a demonstrative and unique effect on overall whole plant growth and development, auxins and cytokinins continue to be the most widely used in manipulating plant growth in vitro. Auxins (such as IAA, NAA, and 2,4-D) and cytokinins (such as zeatin, kinetin, and BAP) are frequently used in combination in plant tissue culture. Generally, a high auxin-to-cytokinin ratio results in the induction of root tissue from callus, while a high cytokinin-to-auxin ratio results in the induction of shoot formation. For many plant species, an intermediate ratio of auxin to cytokinin results in continued callus formation.

There are also specific uses of certain PGRs in plant cell culture. For example, 2,4-D is typically used to induce somatic embroygenesis in cultures but then must be removed for subsequent embryoid development. Gibberellins, such as GA_4 and GA_7, can be inhibitory to morphogenesis. Some PGRs may even elicit opposite morphogenic effects in two different plant species. Nevertheless, the use of PGRs remains essential in plant cell culture to direct morphological development.

Applications and Potential

Plant cell culture as a tool has greatly enhanced the ability of the plant geneticist in the area of crop improvement. Haploid cell cultures initiated from pollen can result in homozygous whole plants, which are very useful as pure lines in breeding programs. In such plants, recessive mutations are easily identified.

The enzymatic removal of the plant cell wall yields naked plant protoplasts, which are more amenable to genetic manipulation. Protoplasts of different species can be chemically or electrically fused to give somatic hybrids that may not be obtained through traditional sexual crossing due to various types of sexual incompatibility. As they divide and regenerate cell walls, these somatic hybrids can then be selected for desired agriculture characteristics, such as insect or disease resistance.

The isolation of plant protoplasts from leaves results in millions of individual cells. As they divide, grow, and differentiate into whole plants, some may contain spontaneous mutations or other changes which can be selected for. Screening for such characteristics, such as salt tolerance or disease resistance, can be done in vitro, thereby saving time and space.

A relatively recent use of plant cell culture in crop improvement involves directed genetic transformation. Genes from other species, including bacteria, animals, and other plants, have been introduced into cell cultures, resulting in genetically modified (GM) plants. The most common technique used to transfer desired genes uses the bacterium *Agrobacterium tumefaciens*. Other techniques include electroporation, microinjection, and particle bombardment with "gene guns." As genetic engineering of plants proceeds and is refined, plant cell culture will continue to play a vital role as a tool in this effort.

—*Thomas J. Montagno*

See also: Cancer; Cell Culture: Animal Cells; Cell Cycle, The; Cell Division; Gene Regulation: Eukaryotes; Genetic Engineering: Agricultural Applications; Genetically Modified (GM) Foods; High-Yield Crops; Mitosis and Meiosis; Oncogenes; Shotgun Cloning; Stem Cells; Totipotency.

Further Reading

George, Edwin F. *Plant Propagation by Tissue Culture.* 2d ed. 2 vols. Edington, Wiltshire, England: Exegetics, 1993. An exhaustive presentation of nutritional media components and discussion of PGR effects in culture. Also contains specifics on the culture of several hundred species. Illustrations, photographs.

Trigiano, Robert, and Dennis Gray, eds. *Plant Tissue Culture Concepts and Laboratory Exercises.* 2d ed. Boca Raton, Fla.: CRC Press, 2000. A concise historical presentation of plant cell culture along with current trends. Also includes detailed student experiments and procedures. Illustrations, photographs.

The Cell Cycle

Field of study: Cellular biology
Significance: *During the phases of the cell cycle, cells divide (mitosis and cytokinesis), grow (G_1), replicate their DNA (S), and prepare for another cell division (G_2). Protein signals regulate progress through these phases of the cell cycle. Mutations that alter signal structure, time of synthesis, or how it is received can cause cancer.*

Key terms

CHECKPOINT: the time in the cell cycle when molecular signals control entry to the next phase

CYCLINS: proteins whose levels rise and fall during the cell cycle

KINASE: an enzyme that catalyzes phosphate addition to molecules

ONCOGENE: a gene whose products stimulate inappropriate cell division, causing cancer

TUMOR SUPPRESSOR: a gene whose product normally prevents or slows cell division; when mutated, these genes can lead to uncontrolled cell division

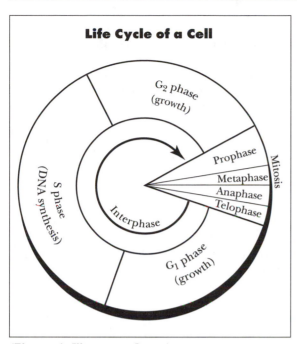

Life Cycle of a Cell

G_2 phase (growth)

Mitosis

Prophase
Metaphase
Anaphase
Telophase

Interphase

S phase (DNA synthesis)

G_1 phase (growth)

(Electronic Illustrators Group)

Defining Cell Cycle Phases

The eukaryotic cell cycle is defined by five phases. Two of these, mitosis and cytokinesis, do not last long. Mitosis itself has five phases:

(1) prophase, when duplicated attached chromatids with replicated DNA condense and become visible as chromosomes, each composed of two sister chromatids

(2) metaphase, when the chromasomes attach to spindle fibers and move to the middle of the cell

(3) anaphase, when sister chromatids separate

(4) telophase, when the separated sister chromatids, now chromosomes, move to opposite poles of the cell, during which cytokinesis often starts

(5) interphase, a time between successive mitoses when cells approximately double in size

An early experiment showed that DNA replicates long before mitosis. After a short exposure (pulse) of cells to radioactive thymidine to allow synthesis of radioactive DNA, the "hot" nucleotide was removed and cells were allowed to grow for different chase times in a medium containing nonradioactive nucleotides. (The term "chase" refers to the second part of what is called a radioactive pulse-chase experiment.) Autoradiography showed that after the pulse period, 40 percent of cells were labeled, but only interphase nuclei were radioactive. This established that DNA is not actually synthesized during mitosis. Labeled mitotic chromosomes were seen only in cells chased for 4 to 14 hours. After longer chase times, label was again confined to interphase nuclei. From this kind of study, the cell cycle could be divided into five major phases (times listed are typical of a cell in an adult organism):

(1) mitosis (M), one hour

(2) cytokinesis (C), thirty minutes

(3) gap 1 (G_1), a time of cell growth, which lasts the generation time minus the times of the other phases

(4) synthesis (S), nine hours of DNA synthesis

(5) gap 2 (G_2), four to five hours of preparation for the next mitosis

Identification of Cell Cycle Switches

Cells reproduce at different rates. Embryonic cells divide hourly or more often, while neurons stop dividing altogether shortly after birth. Cells divide, or stop dividing, in response to chemical signals. When mitotic cells are fused with G_1, S, or G_2 cells, fused cells at first contain the chromosomes of the mitotic cell alongside the intact nucleus of the other cell. After a few minutes, the intact nucleus disintegrates and its chromosomes also condense, suggesting that some chemical signal is causing the intact nucleus to respond as if it were undergoing mitosis. This suggests that cells in mitosis contain a substance that induces nondividing cells to become "mitotic."

The first chemical signal controlling the cell cycle was discovered in studies of amphibian oogenesis. Oocyte maturation begins with the first meiotic division, when the germinal vesicle (the oocyte nucleus) breaks down and chromosomes and spindle fibers first appear at one pole of the oocyte. In a key experiment, cytoplasm taken from oocytes during germinal vesicle breakdown was injected into immature oocytes. Condensed chromosomes quickly appeared in the injected oocytes. A protein called

MPF (maturation promoting factor) was purified from the older oocytes. MPF was later found in developing frog embryos, where its levels fluctuated, peaking just before the embryonic cells began mitosis. Thus, MPF also controls mitosis as well as meiosis and is often called mitosis-promoting factor. MPF consists of cyclin and cyclin-dependent protein kinase (cdk). Cyclin-bound cdk catalyzes phosphorylation of other cellular proteins. Levels of cdk were shown to be constant in the cell, while cyclin levels rose and peaked late in G_2, explaining why MPF activity is highest during mitosis and why mitotic cells induce nuclear breakdown and chromosome condensation when fused to nonmitotic cells.

To study cell cycle regulation further, researchers turned to yeast, a model single-cell eukaryote easily subject to genetic manipulation. Mutagenized yeast was screened for temperature-sensitive mutations that reproduced at lower temperatures but were blocked at one or another point in the cell cycle when grown at higher temperatures. One such temperature-sensitive mutant was arrested in G_2 at the higher temperature. These cells had a defective cell-division-cycle-2 (*cdc2*) gene encoding a yeast version of the frog cdk in MPF. Cellular cdc2 levels are stable, but its kinase activity depends on a yeast cyclin whose levels peak at the end of G_2. The active yeast MPF triggers passage through the G_2 checkpoint, committing the cell to mitosis. Other mutants were found encoding separate G_1 cyclin and G_1 cdk proteins that together form an active kinase that triggers passage through a G_1 checkpoint into the S phase of the cell cycle. Among higher eukaryotes, different combinations of cyclins and cdk's act at still other checkpoints in the cell cycle.

How MPF and G_1 cdk's Work

The proteins phosphorylated by yeast MPF and G_1 cyclin-cdk catalysis function in pathways that promote mitosis and cytokinesis, on one hand, and DNA replication, on the other. How are cdk's activated, what proteins do they phosphorylate, and what do these phosphorylated proteins do?

Yeast MPF made by joining late G_2 cyclin and cdk is not active until it is itself phosphorylated. MPF first receives two phosphates. Then the addition of a third phosphate causes the first two to come off in a peculiar MPF activation pathway. In fact, MPF remains unphosphorylated and inactive in cells experimentally prevented from replicating. In normal cells, blocking premature activation of MPF could prevent condensing chromosomes from damaging DNA that is still replicating. When properly activated, MPF phosphorylates (a) proteins that break down the nuclear membrane, (b) histones and other chromatin proteins thought to start chromosome condensation, and (c) microtubule-associated proteins associated with mitotic spindle formation.

G_1 cyclin and cdk production occur when cells reach a suitable size during G_1 and when they are stimulated by a growth factors. For example, EGF (epidermal growth factor) stimulates embryonic cell growth by binding to cell membrane receptors. EGF-receptor binding converts the intracellular domain of the receptor into an active protein kinase that catalyzes self-phosphorylation. The auto-phosphorylated receptor activates a G-protein encoded by the *ras* gene, which binds GTP. Then, ras-GTP activates the first in a series of protein kinases, setting off an intracellular kinase cascade. Sequential phosphorylations finally stimulate synthesis of G_1 cyclin and G_1 cdk. Active cyclin-bound G_1 cdk then phosphorylates the Rb protein, causing it to detach from protein EF2, which becomes an active transcription factor that stimulates synthesis of proteins needed for replication in the S phase.

To summarize, MPF is activated by a phosphorylation pathway in which the kinase itself becomes phosphorylated, while G_1 cdk is made in response to growth factors like EGF that initiate phosphorylation cascades, resulting in the eventual synthesis of cyclin and cdk. MPF phosphorylates other proteins, permitting transition across the G_2 checkpoint, while the G_1 cdk allows progress through the G_1 checkpoint.

The Cell Cycle and Cancer

With the discovery of the first MPF, scientists had already begun to suspect that mutations in genes encoding proteins involved in cell cycling might cause the uncontrollable cell divisions associated with cancer. Many cancers are associated with oncogenes (called proto-oncogenes when they function and are expressed correctly) encoding proteins involved in cell cycle control. Some oncogenes are carried by viruses, but most arise by mutation of their normal counterparts, resulting in inappropriate activity of the protein encoded by the gene. Representative human oncogenes include *neu* (a growth-factor-receptor oncogene associated with breast and ovarian cancers), *trk* (a receptor oncogene associated with colon cancer), *ras* (a G-protein oncogene), *L-myc* (a transcription-factor oncogene causing small-cell lung cancer), *cdk-4* (a cyclin-dependent kinase oncogene causing a muscle sarcoma), and *CYCD1* (a cyclin oncogene associated with lymphoma). Each of these oncogenes produces proteins that promote unrestricted passage through the cell cycle. In contrast, retinoblastoma is a rare eye cancer in which the *Rb* oncogene product is not made, so that *EF2* transcription factor is always active and genes involved in replication are continuously on. Because *Rb* restrains unwanted cell divisions, it is called a tumor-supressor gene. Unfortunately, the *Rb* oncogene is also associated with more common human adult lung, breast, and bladder cancers. Another tumor-suppressor gene, *p53*, is also implicated in several human cancers; a defective *p53* gene allows cells with damaged DNA to replicate, increasing the chances of cancer development.

In the brief history of cell cycle studies, the discovery of an oncogene identifies the cause of a cancer while the newfound actor in a phosphorylation cascade is a candidate for an oncogene. The study of the cell cycle is an excellent example of the synergy between basic and applied science: The fundamental drive to know how cells grow and divide has merged with a fervent desire to conquer a group of human diseases increasingly prevalent in our aging population.

—Gerald K. Bergtrom

See also: Cancer; Cell Division; Chemical Mutagens; Chromosome Mutation; Cytokinesis; DNA Repair; Gene Regulation: Eukaryotes; Mitosis and Meiosis; Mutation and Mutagenesis; Oncogenes; Stem Cells; Telomeres; Totipotency; Tumor-Suppressor Genes.

Further Reading

Becker, W. M., L. J. Kleinsmith, and J. Hardin. *The World of the Cell.* 5th ed. San Fransisco, Calif.: Benjamin Cummings, 2003. Provides an excellent overview of cell components regulating the cell cycle and how their dysfunction can cause cancer.

Campbell, N., and J. B. Reece. *Biology.* 6th ed. San Fransisco, Calif.: Benjamin Cummings, 2002. A standard, periodically updated introductory biology textbook for undergraduate majors that includes a detailed account of meiosis.

McCormick, F. "Signaling Networks That Cause Cancer." *Trends in Cell Biology* 9 (1999): M53-M56. A review of signaling pathway components whose inappropriate activity causes cancer.

Murray, A. W., and Tim Hunt. *The Cell Cycle: An Introduction.* New York: W. H. Freeman, 1993. An informative overview for both students and general readers, without too much scientific jargon. Bibliographical references, index.

Orr-Weaver, T. L., and R. A. Weinberg. "A Checkpoint on the Road to Cancer." *Nature* 392 (1998): 223-224. Describes a mutation in a gene regulating the cell cycle in cancer cells.

Cell Division

Field of study: Cellular biology

Significance: *Eukaryotic cell division (mitosis and cytokinesis) are a short part of the cell cycle. In the longer time between successive cell divisions cells grow and replicate their DNA. Molecular signals tell cells when to enter each stage of the cycle.*

Key terms

ASEXUAL REPRODUCTION: a form of reproduction wherein an organism's cell DNA doubles and is distributed equally to progeny cells

BINARY FISSION: cell division in prokaryotes in which the plasma membrane and cell wall grow inward and divide the cell in two

CHROMATID: one-half of a replicated chromosome

CHROMATIN: the material that makes up chromosomes; a complex of fibers composed of DNA, histone proteins, and nonhistone proteins

CHROMOSOME: a self-replicating structure, consisting of DNA and protein, that contains part of the nuclear genome of a eukaryote; also used to describe the DNA molecules comprising the prokaryotic genome

CYCLINS: proteins whose levels rise and fall during the cell cycle

CYCLIN-DEPENDENT PROTEIN KINASES (CDK'S): proteins that regulate progress through the eukaryotic cell cycle

CYTOKINESIS: movements of and in a cell resulting in the division of one eukaryotic cell into two

DNA REPLICATION: synthesis of new DNA strands complementary to parental DNA

GENOME: the species-specific, total DNA content of a single cell

MEIOSIS: a type of cell division that leads to production of gametes (sperm and egg) during sexual reproduction

MITOSIS: nuclear division, a process of allotting a complete set of chromosomes to two daughter nuclei

PHASES OF MITOSIS AND MEIOSIS: periods—including prophase, metaphase, anaphase, telophase—characterized by specific chromosomal events during cell division

PHASES OF THE CELL CYCLE: mitosis, cytokinesis, G_1 (gap 1), S (DNA synthesis), and G_2 (gap 2)

PHOSPHORYLATION: a chemical reaction in which a phosphate is added to a molecule, common in the control of cell activity, including the regulation of passage through different stages of the cell cycle

Asexual vs. Sexual Reproduction

A cell's genetic blueprint is encoded in genes written in the four-letter alphabet of DNA, which stands for the four nucleotides that make up the strands of DNA: guanine (G), adenine (A), thymine (T), and cytosine (C).

Reproduction of this blueprint is an essential property of life. Prokaryotes (cells without nuclei) contain a single chromosome in the form of a circular double helix. They replicate their DNA and reproduce asexually by binary fission. Eukaryotic cells, with two or more pairs of linear, homologous chromosomes in a nucleus, replicate their DNA and reproduce asexually by mitosis. In sexual reproduction in higher organisms, special cells called germ cells are set aside to form gametes by meiosis. During meiosis, the germ cells duplicate their chromosomes and separate the homologs into gametes. After mitosis, new cells have a copy of all of the chromosomes originally present in the parent cell; after meiosis, gametes (sperm or egg) contain only one of each homologous chromosome originally present in the parent cell. Though their chromosomal outcomes are quite different, the cellular events of mitosis and meiosis share many similar features, discussed below mostly in the context of mitosis. The focus here is on when cells replicate their DNA, when they physically divide, and how they partition duplicate sets of genetic information into progeny cells.

Binary Fission vs. Meiosis, Mitosis, and Cytokinesis

During binary fission, which occurs in prokaryotic cells (cells that have no nucleus—primary bacteria), these small cells grow larger, become pinched in the middle, and eventually produce two new cells. A specific base sequence in the circular bacterial DNA molecule attaches to the cell membrane. When this sequence replicates during DNA synthesis it also attaches to the cell membrane, but on the opposite side of the cell. As the bacterial cell grows and divides, the two DNA attachment points become separated into the progeny cells, ensuring that each gets a copy of the original circular DNA molecule. DNA replication and cell division in prokaryotes are therefore simultaneous processes.

Mitosis (and meiosis) and cytokinesis, by contrast, are processes well separated in time from DNA replication. When first observed in the microscope in the 1880's, mitosis seemed to be a busy time in the life of a cell. During prophase (the initial phase of mitosis), nuclei seem to disintegrate in a matter of minutes at the same time that chromosomes take shape from nondescript nuclear substance. Spindle fibers form at opposite poles and grow toward the center of the cell. After about thirty minutes, cells are in metaphase. The spindle fibers extend across the cell, attaching to fully formed chromosomes lined up at the metaphase plate in the middle of the cell. Each chromosome is actually composed of two attached strands, or chromatids.

During anaphase the chromatids of each chromosome pull apart and move toward opposite poles of the cell. Telophase is characterized by the re-formation of nuclei around the chromosomes and the de-condensation of the chromosomes back to the shapeless substance now called chromatin.

Cytokinesis, meaning "cell movement," begins during telophase, lasts about thirty minutes, and is the actual division of the parent cell into two cells, each of which gets one of the newly forming nuclei. The processes of mitosis and cytokinesis, which together typically last about 1.5 hours, ensures that duplicated pairs of chromosomes are partitioned correctly into progeny cells.

Meiosis actually consists of two cell divisions, each progressing through prophase, metaphase, anaphase, and telophase. In the first division, homologous chromosomes with their chromatids are separated into progeny cells; in the second, chromatids are pulled apart into the cells that will become gametes. The result is to produce haploid eggs or sperm, rather than the diploid progeny with paired homologous chromosomes that result from mitosis.

The Cell Cycle

Early histologists studying mitosis noted that it often took cells about twenty hours to double, implying a long period between successive cell divisions. This period was called interphase, meaning simply "between" the mitotic phases. An interphase also separates the first meiotic division from a prior mitosis, though there is not always an interphase between the first and second meiotic divisions. One might have suspected that cells were not just biding their time

between mitoses, but it was only in the middle of the twentieth century that the cell cycle was fully characterized, showing interphase to be a long and very productive time in the life of a cell.

In an elegant experiment, cultured cells were exposed to radioactive thymidine, a DNA precursor. After a few minutes, radioactive DNA was detected in the nuclei of some cells. However, no cells actually in mitosis were radioactive. This meant that DNA is not synthesized during mitosis. Radioactive condensed mitotic chromosomes were detected only four to five hours after cells had been exposed to the radioactive DNA precursor, suggesting that replication had ended four to five hours before the beginning of mitosis. Studies like this eventually revealed the five major intervals of the cell cycle: mitosis, cytokinesis, gap 1 (the G_1 phase, a time of cell growth), DNA synthesis (the S phase of DNA synthesis), and gap 2 (the G_2 phase, during which a cell continues growing and prepares for the next mitosis).

The overall length of the cell cycle differs for different cell types. Human neurons stopped dividing shortly after birth, never to be replaced. Many other differentiated cells do not divide but are replaced periodically by stem cells that have the capacity to continue to divide and differentiate. Clearly, human genes must issue instructions telling cells when and when not to reproduce.

Controlling the Cell Cycle

Sometimes cells receive faulty instructions (for example, from environmental carcinogens) or respond inappropriately to otherwise normal commands from other cells. Cancer is a group of diseases in which normal regulation of the cell cycle has been lost and cells divide out of control. In research published in the 1970's, cells synchronized in mitosis were mixed with others synchronized in other phases of the cell cycle in the presence of polyethylene glycol (the main ingredient in automobile antifreeze). The antifreeze caused cells to fuse. Right after mixing, chromosomes and a mitotic spindle could be seen alongside an intact nucleus in the fused cells. Later, the intact nucleus broke down and chromosomes condensed. The conclusion from studies like this is that mitosing cells contain a substance that causes nuclear breakdown and chromosome condensation in nonmitosing cells. Similar results were seen when cells in meiosis were fused with non-meiotic cells. When purified, the substances from meiotic and mitotic cells could be injected into nonmitosing cells, where they caused nuclear breakdown and the appearance of chromosomes from chromatin. The substance was called maturation (or mitosis) promoting factor (MPF). MPF contains one polypetide called cyclin and another called cyclin-dependent protein kinase (cdk). The kinase enzyme catalyzes transfer of a phosphate to other proteins; it is active only when bound to cyclin—hence the name. The kinase is always present in cells, while cyclin concentrations peak at mitosis and then fall. This explains why MPF activity is highest during mitosis and why mitotic cells fused to G_1 cells, for example, can cause the G_1 cell nucleus to disappear and chromosomes to emerge from chromatin.

Since the initial discovery of MPF, studies of eukaryotic cells, from yeast cells to human cells, have revealed many different cyclin-dependent kinases and other regulatory proteins that exert control at different checkpoints on the cell cycle, determining whether or not cells progress from one stage to another. Scientists remain ignorant of the exact causes of most cancers, but because of the compelling need to know, researchers are beginning to understand the normal controls on cellular reproduction.

A final word on the cyclin-dependent protein kinase: This enzyme is one of a large number of kinases that participate in regulating cell chemistry and behavior in response to many extracellular signals (such as hormones). The phosphorylation of cellular proteins has emerged as a major theme in the regulation of many cellular activities, including cell division.

—*Gerald K. Bergtrom*

See also: Cell Culture: Animal Cells; Cell Culture: Plant Cells; Cell Cycle, The; Cytokinesis; Gene Regulation: Eukaryotes; Mitosis and Meiosis; Polyploidy; Totipotency.

Further Reading

Baringa, M. "A New Twist to the Cell Cycle." *Science*, August 4, 1995. Addresses how periodic changes in cyclin concentrations regulate the cell cycle.

Campbell, N., and J. B. Reece. *Biology*. 6th ed. San Fransisco, Calif.: Benjamin Cummings, 2002. Includes a detailed account of meiosis (pp. 239-243) in a standard, periodically updated textbook for undergraduate majors.

Karp, Gerald. "Cell and Molecular Biology." 3d ed. Hoboken, N.J.: John Wiley & Sons, 2002. Detailed accounts of mitosis and events and regulation of the cell by cyclins and kinases (pp. 580-608) in a standard, periodically updated textbook for professionals and undergraduate majors.

Murray, A. W., and Tim Hunt. *The Cell Cycle: An Introduction*. New York: W. H. Freeman, 1993. An informative overview for both students and general readers, without too much scientific jargon. Bibliographical references, index.

Orr-Weaver, T. L., and R. A. Weinberg. "A Checkpoint on the Road to Cancer." *Nature*, March 9, 1998. Describes a mutation in a gene regulating the cell cycle in cancer cells.

Central Dogma of Molecular Biology

Field of study: Molecular genetics

Significance: *The central dogma states precisely how DNA is processed to produce proteins. Originally thought to be a unidirectional process proceeding from DNA to RNA and then to protein, it is now known to include reverse transcription and the enzymatic activity of certain RNA molecules. The central dogma lies at the core of molecular genetics, and understanding it, and particularly reverse transcription, is key to comprehending both the way viruses cause disease and methods that have revolutionized biology.*

Key terms

CODON: three nucleotides in DNA or RNA that correspond with a particular amino acid or stop signal

COLINEARITY: the exact correspondence between DNA or RNA codons and a protein amino acid sequence

COMPLEMENTARY BASES: the nucleic acid bases in different strands of nucleic acid in RNA and DNA that pair together through hydrogen bonds: guanine-cytosine and adenine-thymine (in DNA and RNA) and adenine-uracil (in RNA)

EXON: the part of the coding sequence of mRNA that specifies the amino acid sequence of a protein

HYDROGEN BOND: a weak chemical bond that forms between atoms of hydrogen and atoms of other elements, including oxygen and nitrogen

INTRON: a noncoding intervening sequence present in many eukaryotic genes that is transcribed but removed before translation

RETROVIRUS: a virus that carries reverse transcriptase that converts its RNA genome into a DNA copy that integrates into the host chromosome

REVERSE TRANSCRIPTION: the conversion of RNA into DNA catalyzed by the enzyme reverse transcriptase

RIBOZYME: catalytic RNA

SUBUNIT: a polypeptide chain of a protein

Francis Crick in 2003, who with James Watson won the 1962 Nobel Prize in Physiology or Medicine for their discovery of the double helix structure of DNA. Crick articulated the "central dogma" of molecular biology and coined the term. (AP/Wide World Photos)

Original Central Dogma

Nobel Prize winner Francis Crick, who was co-discoverer with James Watson of the double helical structure of DNA, coined the term "central dogma" in 1958 to describe the fact that the processing of genetic information contained in DNA proceeded unidirectionally by its conversion first to an RNA copy, called messenger RNA (mRNA), in a molecular process called transcription. Then the genetic information contained in the sequence of bases in the mRNA was read in the ribosome, and the appropriate amino acids carried by transfer RNAs (tRNAs) were assembled into protein according to the genetic code in a process called translation. The basis of these reactions stemmed from the properties of DNA, particularly its double helical structure. The fact that the two strands of DNA were held together by hydrogen bonds between specific nucleic acid bases (guanine-cytosine, adenine-thymine) on the two strands clearly suggested how the molecule could be duplicated. Watson and Crick postulated that if they split the double-stranded structure at the

hydrogen bonds, attaching new complementary bases, and reforming the hydrogen bonds, precise copies identical to the original DNA would result. In an analogous manner, RNA was produced by using one DNA strand as a template and adding the correct complementary bases according to what came to be called Watson Crick base pairing. Thus the original dogma stated that transfer of genetic information proceeded unidirectionally, that is, only from DNA to RNA to protein. The only exception was the duplication of DNA in a process called replication.

Modified Central Dogma

Several discoveries made it necessary to change the central dogma. The first and most heretical information came from the study of retroviruses, including the human immunodeficiency virus (HIV). Howard Temin reported that viruses of this group contained an enzyme called reverse transcriptase, which was capable of converting RNA to DNA and thus challenging the whole basis of molecular reactions and the central dogma. Temin and David Baltimore were subsequently awarded Nobel Prizes for their work describing this new enzyme. They were able to show that it synthesizes a DNA strand complementary to the RNA template, and then the DNA-RNA hybrid is converted to a DNA-DNA molecule, which inserts into the host chromosome. Only then can transcription and translation take place.

The second significant change was finding that RNA can act as a template for its own synthesis. This situation occurs in RNA bacteriophages such as MS2 and QB. These phages are very simple, with genomes specifying only three proteins, a coat and attachment proteins and an RNA replicase subunit. This subunit combines with three host proteins to form the mature RNA replicase that catalyzes the replication of the single-stranded RNA. Thus translation to form the protein subunit of RNA replicase occurs using the RNA genome as mRNA upon viral infection without transcription taking place. Only then is the RNA template successfully replicated.

The third natural modification of the original dogma also concerned the properties of RNA. Thomas Cech in 1982 discovered that introns could be spliced out of eukaryotic genes without proteins catalyzing the process. For the discovery and characterization of catalytic RNA, Cech and Sidney Altman were awarded Nobel Prizes for their work in 1989. Their experiments demonstrated that RNA introns, also called ribozymes, had enzymatic activity that could produce a functional mRNA. This process occurred by excising the introns and combining the exons, thus restoring colinearity of DNA and amino acid sequence. RNA processing thus demonstrates another needed modification of the central dogma: The colinearity of gene and protein in prokaryotes predicts that gene expression results directly from the sequence of bases in its DNA. In the case of eukaryotic genes with multiple introns, however, colinearity does not result until the RNA pro-

cessing has taken place. Therefore, the correspondence of the codons in the original DNA sequence containing the introns does not correspond to the order of amino acids in the protein product.

Numerous examples also exist of DNA rearrangements occurring before final gene expression takes place. Examples include the formation of antibodies, the expression of different mating types in yeast, and the expression of different surface antigens in parasites, such as the trypanosome protozoan parasite, which causes sleeping sickness. All of these gene products are produced as a result of gene rearrangements, and the original DNA sequences are not colinear with the amino acid sequences in the protein.

Importance and Applications

The theoretical importance of the central dogma is unquestioned. For example, one modern-day scourge, the human immunodeficiency virus (HIV), replicates its genetic material by reverse transcription (central dogma modification), and one of the drugs shown to contain this virus, azidothymidine (AZT), targets the reverse transcriptase enzyme. Perhaps even more important is the use of the reverse transcription polymerase chain reaction (RT-PCR), one application of the polymerase chain reaction originally devised in 1983 by Kary B. Mullis, formerly of Cetus Corporation. RT-PCR employs reverse transcriptase to form a double-stranded molecule from RNA, resulting in a revolutionary technique that can generate usable amounts of DNA from extremely small quantities of DNA or from poor-quality DNA. Also of practical importance is the laboratory modification of hammerhead ribozymes (central dogma modification), found naturally in plant pathogens, for clinical uses, such as to target RNA viruses infecting patients, including HIV and papillomavirus.

—Steven A. Kuhl

See also: DNA Structure and Function; Gene Regulation: Viruses; Genetic Code; Genetic Code, Cracking of; Molecular Genetics; Protein Synthesis; Reverse Transcriptase; RNA Structure and Function; RNA World.

Further Reading

Cech, T. R. "RNA as an Enzyme." *Scientific American,* November, 1986, pp. 64-75. A Nobel Prize winner describes his revolutionary discovery that RNA can catalyze reactions. Includes both charts and color illustrations.

Crick, F. "Central Dogma of Molecular Biology." *Nature* 227 (1970): 561-563. The seminal paper in which Nobel laureate Crick, co-discoverer of DNA's double helical structure, proposed his theory of how molecular reactions occur.

O'Connell, Joe, ed. *RT-PCR Protocols.* Totowa, N.J.: Humana Press, 2002. Collects several papers on the use of reverse transcription polymerase chain reaction in analysis of mRNA, quantitative methodologies, detection of RNA viruses, genetic analysis, and immunology. Tables, charts, index.

Varmus, H. "Retroviruses." *Science* 240 (1988): 1427-1435. Describes properties of different retroviruses, including the mechanism of reverse transcription.

Watson, James D., et al. *Molecular Biology of the Gene.* 5th ed. 2 vols. Menlo Park, Calif.: Benjamin Cummings, 2003. An eminently readable discussion of the subject, by the other co-discoverer of DNA's double helical structure. Contains numerous illustrations.

Chemical Mutagens

Field of study: Molecular genetics

Significance: *Mutagens are naturally occurring or human-made chemicals that can directly or indirectly create mutations or changes in the information carried by the DNA. Mutations may cause birth defects or lead to the development of cancer.*

Key terms

DEAMINATION: the removal of an amino group from an organic molecule

TAUTOMERIZATION: a spontaneous internal rearrangement of atoms in a complex biological molecule which often causes the molecule to change its shape or its chemical properties

The Discovery of Chemical Mutagens

The first report of mutagenic action of a chemical occurred in 1946, when Charlotte Auerbach showed that nitrogen mustard (a component of the poisonous "mustard" gas widely used in World War I) could cause mutations in fruit flies (*Drosophila melanogaster*). Since that time, it has been discovered that many other chemicals are also able to induce mutations in a variety of organisms. This led to the birth of genetic toxicology during the last half of the twentieth century, dedicated to identifying potentially mutagenic chemicals in food, water, air, and consumer products. Continued research has identified two modes by which mutagens cause mutations in DNA: (1) by interacting directly with DNA, and (2) indirectly, by tricking the cell into mutating its own DNA.

Chemical Mutagens with Direct Action on DNA

Base analogs are chemicals that structurally resemble the organic bases purine and pyrimidine and may be incorporated into DNA in place of the normal bases during DNA replication. An example is bromouracil, an artificially created compound extensively used in research. It resembles the normal base thymine and differs only by having a bromine atom instead of a methyl (CH_3) group. Bromouracil is incorporated into DNA by DNA polymerase, which pairs it with an adenine base just as it would thymine. However, bromouracil is more unstable than thymine and is more likely to change its structure slightly in a process called tautomerization. After the tautomerization process, the new form of bromouracil pairs better with guanine rather than adenine. If this happens to a DNA molecule being replicated, DNA polymerase will insert guanine opposite bromouracil, thus changing an adenine-thymine pair to guanine-cytosine by way of the two intermediates involving bromouracil. This type of mutation is referred to as a transition, in which a purine is replaced by another purine and a pyrimidine is replaced by another pyrimidine.

Another class of chemical mutagens are those that alter the structure and the pairing properties of bases by reacting chemically with them. An example is nitrous acid, which is formed by digestion of nitrite preservatives found in some foods. Nitrous acid removes an amino (NH_3) group from the bases cytosine and adenine. When cytosine is deaminated, it becomes the base uracil, which is not a normal component of DNA but is found in RNA. It is able to pair with adenine. Therefore, the action of nitrous acid on DNA will convert what was a cytosine-guanine base pair to uracil-guanine, which, if replicated, will give rise to a thymine-adenine pair. This is also a transition type of mutation.

Alkylating agents are a large class of chemical mutagens that act by causing an alkyl group (which may be methyl, ethyl, or a larger hydrocarbon group) to be added to the bases of DNA. Some types of alkylation cause the base to become unstable, resulting in a single-strand break in the DNA; this type of event can cause a mutation if the DNA is replicated with no base present or can lead to more serious breaks in the DNA strand. Other alkylation products will change the pairing specificity of the base and create mutations when the DNA is replicated.

Intercalating agents such as acridine orange, proflavin, and ethidium bromide (which are used in labs as dyes and mutagens) have a unique mode of action. These are flat, multiple-ring molecules that interact with bases of DNA and insert themselves between them. This insertion causes a "stretching" of the DNA duplex, and the DNA polymerase is "fooled" into inserting an extra base opposite an intercalated molecule. The result is that intercalating agents cause frame-shift mutations in which the "sense" of the DNA message is lost, just as if an extra letter were inserted into the phrase "the fat cat ate the hat" to make it "the ffa tca tat eth eha t." This occurs because genes are read in groups of three bases during the process of translation. This type of mutation always results in production of a nonfunctional protein.

Chemical Mutagens with Indirect Action

Aromatic amines are large molecules that bind to bases in DNA and cause them to be unrecognizable to DNA polymerase or RNA polymerase. An example is N-2-acetyl-2-amino-fluorine (AAF), which was originally used as an insecticide. This compound and other aromatic amines are relatively inactive on DNA until they react with certain cellular enzymes, after which they react readily with guanine. Mutagens of this type and all others with indirect action work by triggering cells to induce mutagenic DNA repair pathways, which results in a loss of accuracy in DNA replication.

One of the oldest known environmental carcinogens is the chemical benzo(α)pyrene, a hydrocarbon found in coal tar, cigarette smoke, and automobile exhaust. An English surgeon, Percivall Pott, observed that chimney sweeps had a high incidence of cancer of the scrotum. The reason for this was later found to be their exposure to benzo(α)pyrene in the coal tar and soot of the chimneys. Like the aromatic amines, benzo(α)pyrene is activated by cellular enzymes and causes mutations indirectly.

Another important class of chemical mutagens with indirect action are agents causing cross-links between the strands of DNA. Such cross-links prevent DNA from being separated into individual strands as is needed during DNA replication and transcription. Examples of cross-linking agents are psoralens (compounds found in some vegetables and used in treatments of skin conditions such as psoriasis) and cis-platinum (a chemotherapeutic agent used to fight cancer).

Another important class of chemical mutagens are those that result in the formation of active species of oxygen (oxidizing agents). Some of these are actually created in the body by oxidative respiration (endogenous mutagens), while others are the result of the action of chemicals such as peroxides and radiation. Reactive oxygen species cause a wide variety of damage to the bases and the backbone of DNA and may have both direct and indirect effects.

Detection of Chemical Mutagens

The Ames test, developed by biochemistry professor Bruce Ames and his colleagues, is one of the most widely used screening methods for chemical mutagens. It employs particular strains of the bacterium *Salmonella typhimurium* that require the amino acid histidine because of mutations in one of the genes controlling histidine production. The bacteria are exposed to the potential mutagen and then spread on an agar medium lacking histidine. The strains can grow only if they develop a mutation restoring function to the mutated gene required for histidine synthesis. The degree of growth indicates the strength of the mutagen; mutagens of different types are detected by using bacterial strains with different mutations. Mutagens requiring metabolic activation are detected by adding extracts of rat liver cells (capable of mutagen activation) to the tested substance prior to exposure of the bacteria. The Ames test and others like it involving microorganisms are rapid, safe, and relatively inexpensive ways to detect mutagenic chemicals, but it is not always clear how the results of the Ames test should be interpreted when determining the degree of mutagenicity predicted in humans.

Impact and Applications

Mutations can have serious consequences for cells of all types. If they occur in gametes, they can cause genetic diseases or birth defects. If they occur in somatic (body) cells of multicellular organisms, they may alter a growth-controlling gene in such a way that the mutated cell begins to grow out of control and forms a cancer. DNA is subject to a variety of types of damage by interaction with a wide array of chemical agents, some of which are ubiquitous in the environment, while others are the result of human intervention. Methods of detection of chemicals with mutagenic ability have made it possible to reduce the exposure of humans to some of these mutagenic and potentially carcinogenic chemicals.

—*Beth A. Montelone*

See also: Biochemical Mutations; Cancer; DNA Repair; DNA Replication; Mutation and Mutagenesis; Oncogenes; Repetitive DNA; Tumor-Suppressor Genes.

Further Reading

Ahern, Holly. "How Bacteria Cope with Oxidatively Damaged DNA." *ASM News* (March, 1993). Discusses oxidative damage to DNA.

Friedberg, Errol C., et al. *DNA Repair and Mutagenesis.* Washington, D.C.: ASM Press, 1995. Provides extensive descriptions of the mechanisms of chemical mutagenesis.

Hollaender, Alexander, and Frederick J. De Serres, eds. *Chemical Mutagens: Principles and Methods for Their Detection.* 10 vols. New York: Kluwer Academic, 1971-1986. This illustrated, multivolume work is the most comprehensive set of volumes on chemical mutagenesis.

Kuroda, Yukioki, et al., eds. *Antimutagenesis and Anticarcinogenesis Mechanisms II.* New York: Plenum Press, 1990. Part of the proceedings of the Second International Conference on Mechanisms of Antimutagenesis and Anticarcinogenesis, held December, 1988, in Ohito, Japan. Addresses topics such as antimutagens in food, environmental toxicology, free radicals, aspects of mammalian and human genetics, and molecular aspects of mutagenesis and antimutageneis.

Neumann, David, et al. *Human Variability in Response to Chemical Exposures: Measures, Modeling, and Risk Assessment.* Boca Raton, Fla.: CRC Press, 1998. Addresses genetic evidence for variability in the human response to chemicals associated with reproductive and developmental effects, the nervous system and lungs, and cancer.

Chloroplast Genes

Field of study: Molecular genetics

Significance: *Plants are unique among higher organisms in that they meet their energy needs through photosynthesis. The specific location for photosynthesis in plant cells is the chloroplast, which also contains a single, circular chromosome composed of DNA. Chloroplast DNA (cpDNA) contains many of the genes necessary for proper chloroplast functioning. A better understanding of the genes in cpDNA has improved the understanding of photosynthesis, and analysis of the DNA sequence of these genes has also been useful in studying the evolutionary history of plants.*

Key terms

CHLOROPLAST: the cell structure in plants responsible for photosynthesis

GENOME: all of the DNA in the nucleus or in one of the organelles such as a chloroplast

OPEN READING FRAMES: DNA sequences that contain all the components found in active genes, but whose functions have not yet been identified

PHOTOSYNTHESIS: the process in which sunlight is used to take carbon dioxide from the air and convert it into sugar

The Discovery of Chloroplast Genes

The work of nineteenth century Austrian botanist Gregor Mendel showed that the inheritance of genetic traits follows a predictable pattern and that the traits of offspring are determined by the traits of the parents. For example, if the pollen from a tall pea plant is used to pollinate the flowers of a short pea plant, all the offspring are tall. If one of these tall offspring is allowed to self-pollinate, it produces a mixture of tall and short offspring, three-quarters of them tall and one-quarter of them short. Similar patterns are observed for large numbers of traits from pea plants to oak trees. Because of the widespread application of Mendel's work, the study of genetic traits by controlled mating is often referred to as Mendelian genetics.

In 1909, German botanist Carl Erich Correns discovered a trait in four-o'clock plants (*Mirabilis jalapa*) that appeared to be inconsistent with Mendelian inheritance patterns. He discovered four-o'clock plants that had a mixture of leaf colors on the same plant: Some were all green, many were partly green and partly white (variegated), and some were all white. If he took pollen from a flower on a branch with all-green leaves and used it to pollinate a flower on a branch with all-white leaves, all the resulting seeds developed into plants with white leaves. Likewise, if he took pollen from a flower on a branch with all-white leaves and used it to pollinate a flower on a branch with all-green leaves, all the resulting seeds developed into plants with green leaves. Repeated pollen transfers in any combina-

tion always resulted in offspring whose leaves resembled those on the branch containing the flower that received the pollen—that is, the maternal parent. These results could not be explained by Mendelian genetics.

Since Correns's discovery, many other such traits have been discovered. It is now known that the reason these traits do not follow Mendelian inheritance patterns is that their genes are not on the chromosomes in the nucleus of the cell where most genes are located. Instead, the gene for the four-o'clock leaf color trait is located on the single, circular chromosome found in chloroplasts. Because chloroplasts are specialized for photosynthesis, many of the genes on the single chromosome produce proteins or RNA that either directly or indirectly affect synthesis of chlorophyll, the pigment primarily responsible for trapping energy from light. Because chlorophyll is green and because mutations in many chloroplast genes cause chloroplasts to be unable to make chlorophyll, most mutations result in partially or completely white or yellow leaves.

Carl Erich Correns, whose experiments with four-o'clock plants led to the discovery of chloroplast genes. (National Library of Medicine)

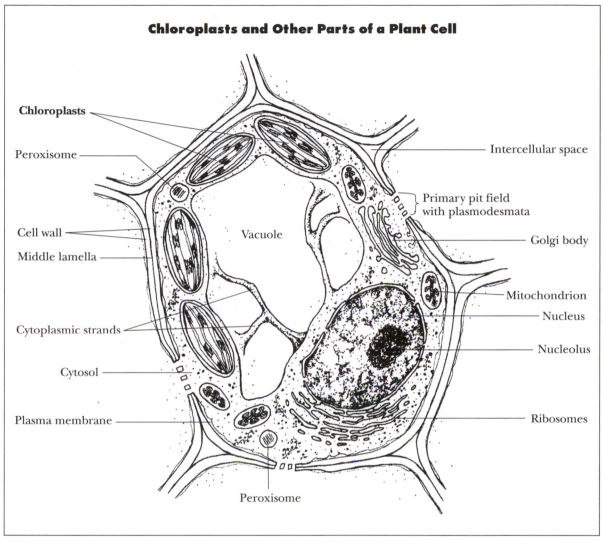

Chloroplasts and Other Parts of a Plant Cell

Chloroplasts

Peroxisome

Cell wall

Middle lamella

Cytoplasmic strands

Cytosol

Plasma membrane

Vacuole

Peroxisome

Intercellular space

Primary pit field
with plasmodesmata

Golgi body

Mitochondrion

Nucleus

Nucleolus

Ribosomes

(Kimberly L. Dawson Kurnizki)

Identity of Chloroplast Genes

Advances in molecular genetics have allowed scientists to take a much closer look at the chloroplast genome. The size of the genome has been determined for a number of plants and algae and ranges from 85 to 292 kilobase pairs (kb, or one thousand base pairs), with most being between 120 kb and 160 kb. The complete DNA sequence for many different chloroplast genomes of plants and algae have been determined. Although a simple sequence does not necessarily identify the role of each gene, it has allowed the identity of a number of genes to be determined, and it has al-

lowed scientists to estimate the total number of genes. In terms of genome size, chloroplast genomes are relatively small and contain a little more than one hundred genes.

Roughly half of the chloroplast genes produce either RNA molecules or polypeptides that are important for protein synthesis. Some of the RNA genes occur twice in the chloroplast genomes of almost all land plants and some groups of algae. The products of these genes represent all the ingredients needed for chloroplasts to carry out transcription and translation of their own genes. Half of the remaining genes produce polypeptides directly required for the biochemical reactions of photosynthesis. What is unusual about these genes is that their products represent only a portion of the polypeptides required for photosynthesis. For example, the very important enzyme ATPase—the enzyme that uses proton gradient energy to produce the important energy molecule adenosine triphosphate (ATP)—comprises nine different polypeptides. Six of these polypeptides are products of chloroplast genes, but the other three are products of nuclear genes that must be transported into the chloroplast to join with the other six polypeptides to make active ATPase. Another notable example is the enzyme ribulose biphosphate carboxylase (RuBP carboxylase, or Rubisco), which is composed of two polypeptides. The larger polypeptide, called *rbcL*, is a product of a chloroplast gene, whereas the smaller polypeptide is the product of a nuclear gene.

The last thirty or so genes remain unidentified. Their presence is inferred because they have DNA sequences that contain all the components found in active genes. These kinds of genes are often called open reading frames (ORFs) until the functions of their polypeptide products are identified.

Impact and Applications

The discovery that chloroplasts have their own DNA and the further elucidation of their genes have had some impact on horticulture and agriculture. Several unusual, variegated leaf patterns and certain mysterious genetic diseases of plants are now better understood. The discovery of some of the genes that code for polypeptides required for photosynthesis has helped increase understanding of the biochemistry of photosynthesis. The discovery that certain key chloroplast proteins such as ATPase and Rubisco carboxylase are composed of a combination of polypeptides coded by chloroplast and nuclear genes also raises some as yet unanswered questions. For example, why would an important plant structure like the chloroplast have only part of the genes it needs to function? Moreover, if chloroplasts, as evolutionary theory suggests, were once free-living bacteria-like cells, which must have had all the genes needed for photosynthesis, why and how

did they transfer some of their genes into the nuclei of the cells in which they are now found?

Of greater importance has been the discovery that the DNA sequences of many chloroplast genes are highly conserved—that is, they have changed very little during their evolutionary history. This fact has led to the use of chloroplast gene DNA sequences for reconstructing the evolutionary history of various groups of plants. Traditionally, plant systematists (scientists who study the classification and evolutionary history of plants) have used structural traits of plants such as leaf shape and flower anatomy to try to trace the evolutionary history of plants. Unfortunately, there are a limited number of structural traits, and many of them are uninformative or even misleading when used in evolutionary studies. These limitations are overcome when gene DNA sequences are used.

A DNA sequence several hundred base pairs in length provides the equivalent of several hundred traits, many more than the limited number of structural traits available (typically much fewer than one hundred). One of the most widely used sequences is the *rbcL* gene. It is one of the most conserved genes in the chloroplast genome, which in evolutionary terms means that even distantly related plants will have a similar base sequence. Therefore, *rbcL* can be used to retrace the evolutionary history of groups of plants that are very divergent from one another. The *rbcL* gene, along with a few other very conservative chloroplast genes, has already been used in attempts to answer some basic questions about the origins and evolution of some of the major flowering plant groups. Less conservative genes and ORFs show too much evolutionary change to be used at higher classification levels but are extremely useful in answering questions about the origins of closely related species, genera, or even families. As analytical techniques are improved, chloroplast genes show promise of providing even better insights into plant evolution.

—Bryan Ness

See also: Cell Culture: Plant Cells; DNA Isolation; Extrachromosomal Inheritance; Genomics; Hybridization and Introgression; Model Organism: *Chlamydomonas reinhardtii*.

Further Reading

Doyle, Jeff J. "DNA, Phylogeny, and the Flowering of Plant Systematics." *Bioscience* 43 (June, 1993). Introduces the reader to the basics of using DNA to construct plant phylogenies and discusses the future of using DNA in evolutionary studies in plants.

Palevitz, Barry A. "'Deep Gene' and 'Deep Time': Evolving Collaborations Parse the Plant Family Tree." *The Scientist* 15, no. 5 (2001). Describes the Deep Green Project, an attempt to use DNA sequence data to trace the evolutionary history of all plants.

Palmer, Jeffry D. "Comparative Organization of Chloroplast Genomes." *Annual Review of Genetics* 19 (1985). One of the best overviews of chloroplast genome structure, from algae to flowering plants.

Svetlik, John. "The Power of Green." *Arizona State University Research Magazine* (Winter, 1997). Provides a review of research at the Arizona State University Photosynthesis Center and provides good background for understanding the genetics of chloroplasts.

Cholera

Fields of study: Bacterial genetics; Diseases and Syndromes;

Significance: *Cholera is an extremely dangerous intestinal disease that has the potential to kill millions of people. Understanding of its genetic basis simplifies treatment and may lead to its eradication.*

Key terms

ENDEMIC: prevalent and recurring in a particular geographic region, specific to a particular region

PANDEMIC: an epidemic that occurs over a large area

PROPHYLAXIS: prevention or cure of a disease

Cholera, Its Symptoms, and Its Cure

Cholera arose centuries ago in India and was disseminated throughout Asia and Europe by trade and pilgrimage. It was devastating, causing epidemics that resulted in countless deaths. By the early twentieth century, cholera had been confined mostly to Asia. In 1961, however, a cholera pandemic beginning in Indonesia spread to Africa, the Mediterranean nations, and North America. In the poorer nations of the world, cholera is still widespread and occurs where sanitation is inadequate. In industrialized nations, where sanitation is generally good, only a few cases occur each year. These usually result from the return of afflicted travelers from regions where cholera is endemic. Because cholera has a 50 to 60 percent fatality rate when its symptoms are not treated quickly, occasional cases cannot be ignored; both the consequences to afflicted people and the potential for the outbreak of epidemics are great.

Cholera is an infection of the small intestine caused by the comma-shaped bacterium *Vibrio cholerae*. Infection is almost always caused by consumption of food or water contaminated with the bacterium. It is followed in one to five days by watery diarrhea that may be accompanied by vomiting. The diarrhea and vomiting may cause the loss of as much as a pint of body water per hour. This fluid loss depletes the blood water and other tissues so severely that if left unchecked it can cause death within a day. Treatment of cholera combines oral or intravenous rehydration of afflicted individuals with saline-nutrient solutions and chemotherapy with antibiotics, especially tetracycline. The two-pronged therapy replaces lost body water and destroys all *V. cholerae* in infected individuals. Antibiotic prophylaxis, which destroys the bacteria, leads to the cessation of production of cholera toxin, the substance that causes diarrhea, vomiting, and death.

Genetics and Cholera

The disease occurs when cholera toxin binds to intestinal cells and stimulates the passage of water from the blood into the intestine. This water depletion and resultant cardiovascular collapse are major causes of cholera mortality. Study of the genetics and the biochemistry of cholera has shown that the toxin is a protein composed of portions called A and B subunits, each produced by a separate gene. When a bacterium secretes a molecule of cholera toxin, it binds to a cell of the intestinal lining (an intestinal mucosa cell) via B subunits. Then the A subunits cause the mucosal cell to stimulate the secretion of water and salts from the blood to produce diarrhea. Lesser amounts of the watery mix are vomited and exacerbate dehydration.

The use of bacterial genetics to compare virulent *V. cholerae* and strains that did not cause the disease helped in the discovery of the nature of the cholera toxin and enabled production of vaccines against the protein. These vaccines are useful to those individuals who visit areas where cholera is endemic, ensuring that they do not become infected with it during these travels. Unfortunately, the vaccines are only effective for about six months.

The basis for the operation of cholera toxin is production of a hormone substance called cyclic adenosine monophosphate (cAMP). The presence of excess cAMP in intestinal mucosa cells causes movement of water and other tissue components into the intestine and then out of the body. The accumulation of cAMP is caused by the ability of the cholera toxin to modify an enzyme protein, adenyl cyclase, to make it produce excess cAMP via modification of a control substance called a G-protein. This modification, called adenine ribosylation, is a mechanism similar to that causing diphtheria, another dan-

Cholera in Marine Plankton

Outbreaks of cholera can occur in nonendemic areas when an infected person travels to another country or when infected water is carried in the ballast of ships to another country. These two processes alone, however, could not explain all of the outbreaks of cholera observed worldwide. In the late 1960's, *Vibrio cholerae* was found in the ocean associated with marine plankton. This association, along with climate change, helps to explain the spread of cholera.

Plankton are the small organisms suspended in the ocean's upper layers. Plankton can be divided into two groups, phytoplankton (small plants) and zooplankton (small animals). *Vibrio cholerae* is found associated with the surface and gut of copepods, which are members of the zooplankton group. These small crustaceans act as a reservoir for the cholera bacteria, allowing them to survive in the ocean for long periods of time. Then, a change in weather that causes the ocean temperature to rise could also cause currents that stir up nutrients from lower layers of the ocean to the upper layers. Numbers of phytoplankton, which live in the upper layers of ocean waters, increase in these periods as a result of the warmer temperatures and greater availability of nutrients. Zooplankton numbers increase as well, because of the increase in main food source, the phytoplankton. Consequently, the number of cholera bacteria increase to a level that can cause the disease. Thus, climate change can result in an outbreak of cholera in a region where cholera is endemic, or, if currents move the plankton to other coastal areas, in a new, nonendemic region. This scenario is believed to explain the 1991 cholera epidemic in Peru, when the oceanic oscillation known as El Niño caused a warming of ocean temperature.

Because of the association of *V. cholerae* with plankton, scientists believe they may be able to track or identify future epidemics by the use of satellite imagery. Increases in phytoplankton turn the ocean color from blue to green. Thus, changes in green areas in the ocean on satellite pictures show where the phytoplankton and, by association, zooplankton and cholera bacteria, are relocating or increasing in number.

The association of cholera with zooplankton has also helped reveal a new way to prevent the disease. People get cholera by ingesting several thousand cholera bacteria at one time. A single copepod can harbor ten thousand bacteria; therefore, the ingestion of one infected copepod can cause disease in a person. Researchers have found a simple and inexpensive way to reduce this risk from copepods dramatically. Filtering water through four layers of fabric used to make saris, which are commonly worn in regions plagued by cholera, removes 99 percent of copepods from water containing high levels of plankton.

Now that the entire genetic sequence of *V. cholerae* has been determined, scientists are armed with additional genetic data to elucidate the relationship of the bacterium with copepods, which may help them find more ways of controlling the spread of the disease.

—*Vicki J. Isola*

gerous disease that can be fatal, although in diphtheria other tissues and processes are affected.

Impact and Applications

Cholera has, for centuries, been a serious threat to humans throughout the world. During the twentieth century, its consequences to industrialized nations diminished significantly with the advent of sound sanitation practices that almost entirely prevented the entry of *V. cholerae* into the food and water supply. In poorer nations with less adequate sanitation, the disease flourishes and is still a severe threat.

It must be remembered that handling cholera occurs at three levels. The isolation and identification of cholera toxin, as well as development of current short-term cholera vaccines, were highly dependent on genetic methodology. Vaccine protects most travelers from the disease. However, wherever the disease afflicts individuals, its treatment depends solely upon rehydration and use of antibiotics. Finally, modern cholera prevention focuses solely on adequate sanitation. It is thus essential to produce a long-lasting vaccine for treatment of cholera to enable prolonged immunization at least at the ten-year level of tetanus shots. Efforts aimed at this goal are ongoing and utilize molecular genetics to define more clearly why long-term vaccination has so far been unsuccessful. Particularly useful will be fine genetic sequence analysis and the use of gene amplification followed by DNA fingerprinting.

—Sanford S. Singer

See also: Anthrax; Archaea; Bacterial Genetics and Cell Structure; Bacterial Resistance and Super Bacteria; Diphtheria; Emerging Diseases; Gene Regulation: Bacteria; Gene Regulation: *Lac* Operon; Transgenic Organisms.

Further Reading

Colwell, Rita R. "Global Climate and Infectious Disease: The Cholera Paradigm." *Science* 274 (1996). An analysis of the role climate change might have on the spread of cholera. Includes a good overview of the history of cholera as well.

Glenn, Gregory M., et al. "Skin Immunization Made Possible by Cholera Toxin." *Nature* 391 (1998). Describes a promising new approach to more safely and quickly immunizing people against cholera.

Heidelberg, John F., et al. "DNA Sequence of Both Chromosomes of the Cholera Pathogen *Vibrio cholerae.*" *Nature* 406 (2000). Publication of the complete genome sequence of the bacteria responsible for cholera.

Holmgren, John. "Action of Cholera Toxin and Prevention and Treatment of Cholera." *Nature* 292 (1981). Clearly describes both the composition and bioaction of cholera toxin.

Keusch, Gerald, and Masanobu Kawakami, eds. *Cytokines, Cholera, and the Gut.* Amsterdam: IOS Press, 1997. Surveys the role of peptide mediators in the intestinal responses to infectious and inflammatory challenges presented by diverse disease states, including inflammatory bowel disease and infectious diarrheas and dysenteries, and the epidemiology and pathogenesis of cholera and related diarrheal diseases.

Kudlick, Katherine. *Cholera in Post-revolutionary Paris: A Cultural History.* Berkeley: University of California Press, 1996. Explores the dynamics of class relations through an investigation of the responses to two cholera epidemics in Paris during 1832 and 1849.

Pennisi, Elizabeth. "Cholera Strengthened by Trip Through Gut." *Science* 296 (2002). Examines the effect that passing through a host's gut has on cholera bacteria.

Rakel, Robert E., ed. *Conn's Current Therapy: Latest Approved Methods of Treatment for the Practicing Physician.* Philadelphia: W. B. Saunders, 2003. Provides for general readers a succinct overview of cholera and its treatment.

Wachsmuth, Kate, et al., eds. *Vibrio Cholerae and Cholera: Molecular to Global Perspectives.* Washington, D.C.: ASM Press, 1994. A comprehensive guide to the disease and its genetics.

Web Site of Interest

Food and Drug Administration. http://vm.cfsan.fda.gov. The FDA's Bad Bug Book provides information on *Vibrio cholerae*, the bacterium that causes cholera. Links to the Centers for Disease Control and Prevention's "food illness fact sheet" on the disease.

Chromatin Packaging

Field of study: Molecular genetics
Significance: *The huge quantity of DNA present in each cell must be organized and highly condensed in order to fit into the discrete units of genetic material known as chromosomes. Gene expression can be regulated by the nature and extent of this DNA packaging in the chromosome, and errors in the packaging process can lead to genetic disease.*

Key terms

CHROMATIN: the material that makes up chromosomes; a complex of fibers composed of DNA, histone proteins, and nonhistone proteins

HISTONE PROTEINS: small, basic proteins that are complexed with DNA in chromosomes and that are essential for chromosomal structure and chromatin packaging

NONHISTONE PROTEINS: a heterogeneous group of acidic or neutral proteins found in chromatin that may be involved with chromosome structure, chromatin packaging, or the control of gene expression

NUCLEOSOME: the basic structural unit of chromosomes, consisting of 146 base pairs of DNA wrapped around a core of eight histone proteins

Chromosomes and Chromatin

Scientists have known for many years that an organism's hereditary information is encrypted in molecules of DNA that are themselves organized into discrete hereditary units called genes and that these genes are organized into larger subcellular structures called chromosomes. James Watson and Francis Crick elucidated the basic chemical structure of the DNA molecule in 1953, and much has been learned since that time concerning its replication and expression. At the molecular level, DNA is composed of two parallel chains of building blocks called nucleotides, and these chains are coiled around a central axis to form the well-known "double helix." Each nucleotide on each chain attracts and pairs with a complementary nucleotide on the opposite chain, so a DNA molecule can be described as consisting of a certain number of these nucleotide base pairs. The entire human genome consists of more than six billion base pairs of DNA, which, if completely unraveled, would extend for more than 2 meters (6.5 feet). It is a remarkable feat of engineering that in each human cell this much DNA is condensed, compacted, and tightly packaged into chromosomes within a nucleus that is less than 10^{-5} meters in diameter. What is even more astounding is the frequency and fidelity with which this DNA must be condensed and relaxed, packaged and unpackaged, for replication and expression in each individual cell at the appropriate time and place during both development and adult life. The essential processes of DNA replication or gene expression (transcription) cannot occur unless the DNA is in a more open or relaxed configuration.

Chemical analysis of mammalian chromosomes reveals that they consist of DNA and two distinct classes of proteins, known as histone and nonhistone proteins. This nucleoprotein complex is called chromatin, and each chromosome consists of one linear, unbroken, double-stranded DNA molecule that is surrounded in predictable ways by these histone and nonhistone proteins. The histones are relatively small, basic proteins (having a net positive charge), and their function is to bind directly to the negatively charged DNA molecule in the chromosome. Five major varieties of histone proteins are found in chromosomes, and these are known as H1, H2A, H2B, H3, and H4. Chromatin contains about equal amounts of histones and DNA, and the amount and proportion of histone proteins are constant from cell to cell in all higher organisms. In fact, the histones as a class are among the most highly conserved of all known proteins. For example, for histone H3, which is a protein consisting of 135 amino acid "building blocks," there is only a single amino acid difference in the protein found in sea urchins as compared with the one found in cattle. This is compelling evidence that histones play the same essential role in chromatin packaging in all higher organisms and that evolution has been quite intolerant of even minor sequence variations between vastly different species.

Nonhistones as a class of proteins are much more heterogeneous than the histones. They are usually acidic (carrying a net negative charge), so they will most readily attract and bind with the positively charged histones rather than the negatively charged DNA. Each cell has many different kinds of nonhistone proteins, some of which play a structural role in chromosome organization and some of which are more directly involved with the regulation of gene expression. Weight for weight, there is often as much nonhistone protein present in chromatin as histone protein and DNA combined.

Nucleosomes and Solenoids

The fundamental structural subunit of chromatin is an association of DNA and histone proteins called a "nucleosome." First discovered in the 1970's by Ada and Donald Olins and Chris Woodcock, each nucleosome consists of a core of eight histone proteins: two each of the histones H2A, H2B, H3, and H4. Around this histone octamer is wound 146 base pairs of DNA in one and three-quarter turns (approximately eighty base pairs per turn). The overall shape of each nucleosome is similar to that of a lemon or a football. Each nucleosome is separated from its adjacent neighbor by about 55 base pairs of "linker DNA," so that in its most unraveled state they appear under the electron microscope to be like tiny beads on a string. Portions of each core histone protein protrude outside of the wound DNA and interact with the DNA that links adjacent nucleosomes.

The next level of chromatin packaging involves a coiling and stacking of nucleosomes into a ribbon-like arrangement, which is twisted to form a chromatin fiber about 30 nanometers (nm) in diameter commonly called a "solenoid." Formation of solenoid fibers requires the interaction of histone H1, which binds to the linker DNA between nucleosomes. Each turn of the chromatin fiber contains about 1,200 base pairs (six nucleosomes), and the DNA has now been compacted by about a factor of fifty. The coiled solenoid fiber is organized into large domains of 40,000 to 100,000 base pairs, and these domains are separated by attached nonhistone proteins that serve to both organize and to control their packaging and unpackaging.

Long DNA Loops and the Chromosome Scaffold

Physical studies using the techniques of X-ray crystallography and neutron diffraction have suggested that solenoid fibers may be further organized into giant supercoiled loops. The extent of this additional looping, coiling, and stacking of solenoid fibers varies, depending on the cell cycle. The most relaxed and extended chromosomes are found at interphase, the period of time between cell divisions. Inter-

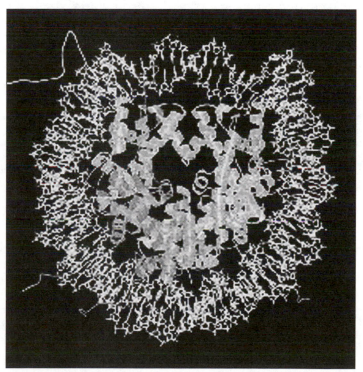

This image, captured through neutron crystallography, shows the molecular structure of the fundamental repeating unit of the chromosome, the nucleosome core complex: 146 base pairs of DNA wrapped around a core of eight histone proteins. (U.S. Department of Energy Genomes to Life Program, http://doegenomestolife.org)

phase chromosomes typically have a diameter of about 300 nm. Chromosomes that are getting ready to divide (metaphase chromosomes) have the most highly condensed chromatin, and these structures may have a diameter of up to 700 nm. One major study on the structure of metaphase chromosomes has shown that a skeleton of nonhistone proteins in the shape of the metaphase chromosome remains even after all of the histone proteins and the DNA have been removed by enzymatic digestion. If the DNA is not digested, it remains in long loops (10 to 90 kilobase pairs) anchored to this nonhistone protein scaffolding.

In the purest preparations of metaphase chromosomes, only two scaffold proteins are found. One of these forms the latticework of the scaffold, while the other has been identified as toposiomerase II, an enzyme that is critical in DNA replication. This enzyme cleaves double-stranded DNA and then rapidly reseals the cut after some of the supercoiling has been relaxed, thus relieving torsional stress and preventing tangles in the DNA. Apparently this same enzyme activity is necessary for the coiling and looping of solenoid fibers along the chromosome scaffold that occurs during the transition between interphase and metaphase chromosome structure. In the most highly condensed metaphase chromosomes, the DNA has been further compacted by an additional factor of one hundred.

Impact and Applications

Studies of chromatin packaging continue to reveal the details of the precise chromosomal architecture that results from the progressive coiling of the single DNA molecule into increasingly compact structures. Evidence suggests that the regulation of this coiling and packaging within the chromosome has a significant effect on the properties of the genes themselves. In fact, errors in DNA packaging can lead to inappropriate gene expression and developmental abnormalities. In humans, the blood disease thalassemia, several neuromuscular diseases, and even male sex determination can all be explained by the altered assembly of chromosomal structures.

Chromatin domains, composed of coiled so-lenoid fibers, may contain several genes, or the boundary of a domain can lie within a gene. These domains have the capacity to influence gene expression, and this property is mediated by specific DNA sequences known as locus control regions (LCRs). An LCR is like a powerful enhancer that activates transcription, thereby turning on gene expression. The existence of such sequences was first recognized from a study of patients with beta-thalassemia and a related condition known as hereditary persistence of fetal hemoglobin. In these disorders, there is an error in the expression of a cluster of genes, known as the beta-globin genes, that prevents the appearance of adult type hemoglobin. The beta-globin genes are linearly arrayed over a 50-kilobase-pair chromatin domain, and the LCR is found upstream from this cluster. Affected patients were found to have normal beta-globin genes, but there was a deletion of the upstream LCR that led to failure to activate the genes appropriately. Further investigation led to the conclusion that the variation in expression of these genes observed in different patients was caused by differences in the assembly of the genes into higher-order chromatin structures. In some cases, gene expression was repressed, while in others it was facilitated. Under normal circumstances, a nonhistone protein complex was found to bind to the LCR, causing the chromatin domain to unravel and making the DNA more accessible to transcription factors, thus enhancing gene expression.

DNA sequencing studies have demonstrated a common feature in several genes whose altered expression leads to severe human genetic disease. For example, the gene that causes myotonic dystrophy has a large number of repeating nucleotide triplets in the DNA region immediately adjacent to the protein-encoding segment. Physical studies have shown that this results in the formation of unusually stable nucleosomes, since these repeated sequences create the strongest naturally occurring sites for association with the core histones. It has been suggested that these highly stable nucleosomes are unusually resistant to the unwinding and denaturation of the DNA that must occur in order for gene expression to begin. RNA polymerase is the enzyme that makes an RNA transcript of the gene, and its movement through the protein-coding portion of the gene is inhibited if the DNA is unable to dissociate from the nucleosomes. Thus, although the necessary protein product would be normal and functional if it could be made, it is a problem with chromatin unpackaging that leads to reduced gene expression that ultimately leads to clinical symptoms of the disease. Both mild and severe forms of myotonic dystrophy are known, and an increase in the clinical severity correlates exactly with an increased number of nucleotide triplet repeats in the gene. Similar triplet repeats have been found in the genes responsible for Kennedy's disease, Huntington's disease (Huntington's chorea), spinocerebellar ataxia type I, fragile X syndrome, and dentatorubral-pallidoluysian atrophy.

Fascinating and unexpected recent research results have suggested that a central event in the determination of gender in mammals depends on local folding of DNA within the chromosome. Molecular biologists Peter Goodfellow and Robin Lovell-Badge successfully cloned a human gene from the Y chromosome that determines maleness. This *SRY* gene (named from the sex-determining region of the Y chromosome) encodes a protein that selectively recognizes a specific DNA sequence and helps assemble a chromatin complex that activates other male-specific genes. More specifically, binding of the SRY protein causes the DNA to bend at a specific angle and causes conformation that facilitates the assembly of a protein complex to initiate the cascade of gene activation leading to male development. If the bend is too tight or too wide, gene expression will not occur, and the embryo will develop as a female.

The unifying lesson to be learned from these examples of DNA packaging and disease is that DNA sequencing studies and the construction of human genetic maps will not by themselves provide all the answers to questions concerning human variation and genetic disease. An understanding of human genetics at the molecular level depends not only on the primary DNA sequence but also on the three-dimensional organization of that DNA within the chromosome. Compelling genetic and biochemical evidence has left no doubt that the packaging process is an essential component of regulated gene expression.

—Jeffrey Knight

See also: Cell Division; Central Dogma of Molecular Biology; Chromosome Structure; Developmental Genetics; Fragile X Syndrome; Gene Regulation: Eukaryotes; Huntington's Disease; Mitosis and Meiosis; Molecular Genetics.

Further Reading

Becker, Peter B. *Chromatin Protocols.* Totowa, N.J.: Humana Press, 1999. Western scientists provide step-by-step instructions for analyzing the relationship between chromatin structure and function and for elucidating the molecular mechanisms that control such vital cellular functions as transcription, replication, recombination, and DNA repair.

Elgin, Sarah C. R., and Jerry L. Workman, eds. *Chromatin Structure and Gene Expression.* 2d ed. New York: Oxford University Press, 2000. Examines numerous facets of chromatin structure, including its histones, nucleosomes, and fiber elements, its relationship to DNA structure, its replication and assembly, and its initiation of expression.

Kornberg, Roger, and Anthony Klug. "The Nucleosome." *Scientific American* 244 (1981). Provides a somewhat dated but highly readable summary of the primary association of DNA with histone proteins.

Lodish, Harvey, et al. *Molecular Cell Biology.* 4th ed. New York: W. H. Freeman, 2000. Covers chromatin structure from a cellular and biochemical perspective.

Russell, Peter. *Genetics.* 5th ed. Menlo Park, Calif.: Benjamin Cummings, 1998. A college-level textbook with an excellent discussion of chromatin structure and organization.

Turner, Bryan. *Chromatin and Gene Regulation: Mechanisms in Epigenetics.* Malden, Mass.: Blackwell, 2001. Explores the relationship between gene expression and DNA packaging by explaining the chromatin-based control mechanisms. An overview of transcription in bacteria, refined structures and the control mechanisms, and dosage compensation are covered.

Van Holde, Kensal. *Chromatin.* New York: Springer, 1988. Contemporary views on chromatin's functions and structure, addressing structures of DNA, proteins of chromatin (both histone and nonhistone), the nucleosome, higher-order structures, transcription, and replication. Bibliography.

Wolffe, Alan P. "Genetic Effects of DNA Packaging." *Scientific American,* November/December, 1995: 68-77. Excellent summary for the general reader of the relationship between gene expression and DNA packaging.

Chromosome Mutation

Fields of study: Cellular biology; Molecular genetics

Significance: *Unlike gene mutations, which alter individual genes, chromosome mutations delete, duplicate, or rearrange chromosome segments. Chromosome mutations may create gene mutations if they delete genes or if the breakpoints of rearranged segments disrupt gene structure or alter gene expression. Even when they do not create gene mutations, chromosome mutations may reduce fertility and are an important cause of inherited infertility in humans. They also play important roles in the evolution of species.*

Key terms

DELETION: a missing chromosome segment

DUPLICATION: a chromosome segment repeated in the same or in a different chromosome

FISSION: separation of a single chromosome into two chromosomes

FUSION: joining of two chromosomes to become a single chromosome

INVERSION: a chromosome segment with reversed orientation when compared to the original chromosome structure

TRANSLOCATION: a chromosome segment transferred from one chromosome to a nonhomologous chromosome

Discovery

As the fruit fly *Drosophila melanogaster* became a premier organism for genetic research in the early years of the twentieth century, geneticists who worked with it were the first to discover chromosome mutations. Calvin Bridges proposed deletions in 1917, duplications in 1919, and translocations in 1923 as explanations of phenomena he had observed in genetic experiments. Alfred Sturtevant proposed inversions in 1926 to explain experimental genetic data. Their proposals were directly confirmed as chromosome mutations when methods for microscopic examination of chromosomes were developed in the 1920's and 1930's.

Deletions

A deletion results when a chromosomal segment is lost. A deletion creates an imbalance in the genetic material because a relatively large segment of it is missing. Most deletions are lethal, even when heterozygous. Some small deletions persist in the heterozygous state but are usually lethal when homozygous. These small deletions are usually characterized by deleted portions of only one or two genes and behave genetically as recessive alleles when paired with a typical recessive allele of the affected gene.

Duplications

A duplication arises when a chromosomal segment is duplicated and inserted either into the same chromosome, as its parent segment, or into another chromosome. Duplications are present in most genomes. Genome projects (including the Human Genome Project) have revealed large duplicated segments containing multiple genes dispersed throughout the chromosomes in most species. Some duplications are repeated in tandem in the same chromosome and are subject to unequal crossing over, a process in which duplicated segments mispair with one another and a crossover takes place within the mispaired segment. Unequal crossing over increases the number of tandem duplications in one chromosome and decreases that number in the other.

Inversions

Two breaks within the same chromosome may liberate a chromosome segment. If the segment is reinserted into the same chromosome, but in reverse orientation, an inversion results. Also, rare crossing over between duplicated segments in the same chromosome may produce an inversion. If a breakpoint of the inversion lies within a gene, it disrupts the gene, causing a gene mutation. Additionally, an inversion may place a gene in another location in the chromosome, removing the gene from its regulatory elements and altering its expression, a phenomenon known as the position effect.

When one chromosome carries an inversion and its homologous partner does not, the individual carrying these two chromosomes is said to be an inversion heterozygote. The two homologous chromosomes in an inversion heterozygote cannot pair properly in meiosis; one of them must form a loop in the inverted region. A crossover within the inversion loop results in chromosomes that carry large deletions and duplications. Because of the imbalance of chromosomal material created by the deletions and duplications, progeny resulting from such crossovers usually do not survive. In genetic experiments, crossing over appears to be suppressed within an inversion, whereas, in reality, crossing over does take place within the inversion but crossover-type progeny fail to survive. For this reason, inversion heterozygotes may suffer a reduction in fertility that is proportional to the size of the inversion. An individual who is homozygous for an inversion, however, suffers no loss of fertility, because the chromosomes pair normally.

Translocations

A break in a chromosome may liberate a chromosome fragment, which if reattached to a different chromosome is called a translocation. Most translocations are reciprocal: Two chromosome breaks, each in a different chromosome, liberate two fragments, and each fragment reattaches to the site where the other fragment was originally attached; in other words, the two fragments exchange places. If the breakpoint of a translocation is within a gene, a gene mutation may result. Also, a gene at or near the breakpoint may undergo a change in its expression because of position effect.

Translocations alter chromosome pairing in meiosis. During meiosis in a reciprocal trans-location heterozygote, the two chromosomes with translocated segments pair with two other chromosomes without translocations. The pairing of these four chromosomes forms an X-shaped structure called a quadrivalent, so named because it contains four chromosomes paired with one another, instead of the usual two. Depending on the orientation of the quadrivalent during meiosis, some gametes may receive a balanced complement of chromosomes and others an unbalanced complement with large duplications and deletions. Typically, about half of all gametes in a reciprocal translocation heterozygote carry an unbalanced chromosome complement, a situation that significantly reduces the individual's fertility. However, translocation homozygotes suffer no loss of fertility, because the chromosomes pair normally with no quadrivalent.

Fusions

Very rarely, two chromosomes may fuse with one another to form a single chromosome. Chromosomes with centromeres at or very near the ends of the chromosomes may undergo breakage at the centromeres and fuse with each other in the centromeric region, resulting in a single chromosome with the long arms of the original chromosomes on either side of the fused centromere. Such a chromosome fusion is called a Robertsonian translocation. In other cases, two chromosomes may fuse with one another producing a dicentric chromosome (a chromosome with two centromeres). For the fused chromosome to persist, one of the centromeres ceases to function, leaving the other centromere as a single, functional centromere for the fused chromosome.

Fissions

A chromosome break produces two fragments, which may function as individual chromosomes if each has telomeres on both ends and a functional centromere. Typically, chromosome breakage produces one fragment with a telomere on one end and a centromere, and another fragment with a telomere on one end and no centromere. For both fragments to function as chromosomes, one must acquire a telomere and the other a centromere and a telomere. These events are highly unlikely, so fissions are rarer than fusions. However, complex translocations with other chromosomes may rarely produce functional chromosomes from a fission event, and cases of functional chromosomes arising from fissions have been documented.

Impact on Human Genetics and Medicine

Chromosome mutations are responsible for several human genetic disorders. For example, about 20 percent of hemophilia A cases result from a gene mutation caused by an inversion with a breakpoint in the *F8C* gene, which encodes blood clotting factor VIII. Cri du chat syndrome, a severe disorder characterized by severe mental retardation and distinctive physical features, is usually caused by deletion of a small chromosomal region near the end of chromosome 5. A few cases of this syndrome are associated with deletions that result from a translocation with a breakpoint near the end of chromosome 5 or crossovers within a small inversion in that chromosome region. Robertsonian translocations that fuse the long arm of chromosome 21 with the long arm of another chromosome (usually chromosome 14) are responsible for some inherited cases of Down syndrome. A reciprocal translocation between chromosomes 9 and 22, called the Philadelphia chromosome, causes increased susceptibility to certain types of cancer by altering the expression of a gene located at the breakpoint of the translocation. Other translocations are likewise associated with certain cancers. Chromosome mutations may also cause infertility in humans. Reciprocal translocations are especially notorious, although certain inversions are also associated with infertility.

Implications for Evolution

Heterozygous carriers of inversions, translocations, fusions, and fissions often suffer losses of fertility, but homozogotes do not. Thus, natural selection may disfavor heterozygotes while favoring homozygotes either for the original chromosome structure or for the mutation. Accumulation of different chromosome mutations in isolated populations of a species may eventually differentiate the chromosomes to such a degree that the isolated populations diverge into separate species. Their members can no longer produce fertile offspring when hybridized with members of another population because the chromosomes cannot properly pair with one another. Indeed, accumulated chromosome mutations are often evident when geneticists compare the chromosomes of closely related species. For example, the chromosomes of different *Drosophila* species are differentiated mostly by translocations and fusions. Comparison of human, chimpanzee, gorilla, and orangutan chromosomes reveals numerous inversions that distinguish the chromosomes of these species. One of the most striking cases of chromosome evolution is the origin of human chromosome 2. This chromosome matches two separate chromosomes in the great apes and apparently arose from a fusion of these two chromosomes after the divergence of the human and chimpanzee lineages. The presence in human chromosome 2 of DNA sequences corresponding to a nonfunctional centromere and telomere at sites corresponding to these structures in the great ape chromosomes is strong evidence of a chromosome fusion during evolution of the human lineage.

—Daniel J. Fairbanks

See also: Cell Cycle, The; Cell Division; Central Dogma of Molecular Biology; Chemical Mutagens; Chromosome Structure; Chromosome Theory of Heredity; Congenital Disorders; Cystic Fibrosis; Down Syndrome; Epistasis; Evolutionary Biology; Hemophilia; Hereditary Diseases; Huntington's Disease; Inborn Errors of Metabolism; Infertility; Mitosis and Meiosis; Molecular Genetics; Mutation and Mutagenesis; Punctuated Equilibrium.

Further Reading

Burnham, Charles R. *Discussions in Cytogenetics.* Minneapolis, Minn.: Burgess, 1962. A classic book on chromosome mutations written by one of the pioneers in the field. Though out of print, this book remains available in many libraries.

Calos, Michele. *Molecular Evolution of Chromosomes.* New York: Oxford University Press, 2003. Describes the role of chromosome mutations in evolution.

Miller, Orlando J., and Eeva Therman. *Human Chromosomes*. 4th ed. New York: Springer Verlag, 2001. A good textbook on human chromosomes, including common chromosome mutations.

Chromosome Structure

Fields of study: Cellular biology; Classical transmission genetics

Significance: *The separation of the alleles in the production of the reproductive cells is a central feature of the model of inheritance. The realization that the genes are located on chromosomes and that chromosomes occur as pairs that separate during meiosis provides the physical explanation for the basic model of inheritance. When chromosome structure is modified, changes in information transmission produce abnormal developmental conditions, most of which contribute to early miscarriages and spontaneous abortions.*

Key terms

HISTONES: a class of proteins associated with DNA

HOMOLOGOUS CHROMOSOMES: chromosomes that have identical physical structure and contain the same genes; humans have twenty-two pairs of homologous chromosomes and a pair of sex chromosomes that are only partially homologous

KARYOTYPING: an analysis or physical description of all the chromosomes found in an organism's cells; often includes either a drawing or photograph of the chromosomes

SPINDLE FIBERS: minute fibers composed of the protein tubulin that are involved in distributing the chromosomes during cell division

Discovery of Chromosomes' Role in Inheritance

The development of the microscope made it possible to study what became recognized as the central unit of living organisms, the cell. One of the most obvious structures within the cell is the nucleus. As study continued, dyes were used to stain cell structures to make them more visible. It became possible to see colored structures called chromosomes ("color bodies") within the nucleus that became visible when they condensed as the cell prepared to divide.

The association of the condensed, visible state of chromosomes with cell division caused investigators to speculate that the chromosomes played a role in the transmission of information. Chromosome counts made before and after cell division showed that the chromosome number remained constant from generation to generation. When it was observed that the nuclei of two cells (the egg and the sperm) fused during sexual reproduction, the association between information transport and chromosome composition was further strengthened. German biologist August Weismann, noting that the chromosome number remained constant from generation to generation despite the fusing of cells, predicted that there must be a cell division that reduced the chromosome number in the egg and sperm cells. The reductional division, meiosis, was described in 1900.

Following the rediscovery of Gregor Mendel's rules of inheritance in 1900, the work of Theodor Boveri and Walter Sutton led to the 1903 proposal that the character-determining factors (genes) proposed by Mendel were located on the chromosomes and that the factor segregation that was a central part of the model occurred because the like chromosomes of each pair separated during the reductional division that occurs in meiosis. This hypothesis, the "chromosome theory of heredity," was confirmed in 1916 by the observations of the unusual behavior of chromosomes and the determining factors located on them by Calvin Bridges.

Chromosome Structure and Relation to Inheritance

With the discovery of the nucleic acids came speculation about the roles of DNA and the associated proteins. During the early 1900's, it was generally accepted that DNA formed a structural support system to hold critical information-carrying proteins on the chromosomes. The identification of the structure of DNA in 1953 by American biologist James Watson and English physicist Francis Crick and the recogni-

External Structure of a Chromosome

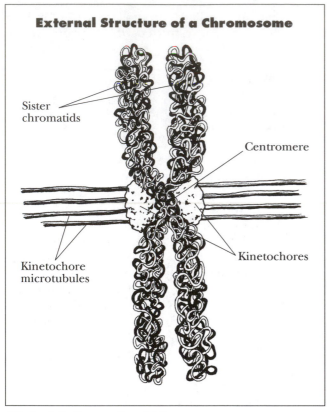

Sister chromatids

Centromere

Kinetochore microtubules

Kinetochores

(Kimberly L. Dawson Kurnizki)

tion that DNA, not the proteins, contained the genetic information led to study of chromosome structure and the relationships of the DNA and protein components.

It is now recognized that each chromosome contains one DNA molecule. Each plant and animal species has a specific number of chromosomes. Humans have twenty-three kinds of chromosomes, present as twenty-three pairs. Each chromosome can be recognized by its overall length and the position of constrictions, called centromeres, that are visible only when the cell is reproducing. At all other stages of the cell's life, the chromosome material is diffuse and is seen only as a general color within the nucleus. When the cell prepares for division, the fibrous DNA molecule tightly coils and condenses into the visible structures. Since there must be information for the two cells that result from the process of division, the chromosomes are present in a duplicated condition when they first become visible.

A major feature of the visible, copied chromosomes is the centromere. This constriction may be located anywhere along the chromosome, so its position is useful for identifying chromosomes. In karyotyping, the standard system used to identify human chromosomes, the numbering begins with the longest chromosome with the constriction nearest the center (chromosome 1) and are referred to as having a metacentric centromere placement. Chromosomes with nearly the same length but with the centromere constriction removed from the center position have higher numbers (chromosomes 2 and 3) and are referred to as acrocentric. Shorter chromosomes with a centromere near the middle are next, and the numbering proceeds based on the distance the centromere is removed from the central position. Short chromosomes with a centromere near one end have the highest numbers and are referred to as telocentric.

Most of the chromosomes have a centromere that is not centrally located, which results in arms of unequal length. The short arm is referred to as "petite" and is designated the p arm. The long arm is designated the q arm. This nomenclature is useful for referring to features of the chromosome. For example, when a portion of the long arm of chromosome 15 has been lost, the arm is shorter than normal. The loss, a deletion, is designated $15q^-$ (chromosome 15, long arm, deletion). Prader-Willi syndrome, in which an infant has poor sucking ability and poor growth, and later becomes a compulsive eater, results from this deletion. Cri du chat ("cry of the cat") syndrome results from $5p^-$, a deletion of a portion of the short arm of chromosome 5. The cry of these individuals is like that of a cat, and they are severely mentally retarded and have numerous physical defects.

Some chromosomes have additional constrictions referred to as secondary constrictions. The primary centromere constrictions are located where the spindle fibers attach to the chromosomes to move them to the appropriate poles during cell division. The secondary constrictions are sites of specific gene activity. Both of these regions contain DNA base sequence information that is specific to their functions.

Histones

The DNA of the chromosomes is wound around special proteins called histones. This results in an orderly structure that condenses the DNA mass so that the bulky DNA does not require as much storage space. The wrapped DNA units then fold into additional levels of compaction, by means of a process called condensation. The exact processes involved in these higher levels of folding are not fully understood, but the overall condensation reduces the bulk of the DNA nearly one thousandfold. If the DNA is removed from a condensed chromosome, the proteins remain and have nearly the same shape as the chromosome, indicating that it is the proteins that form the chromosome shape. The presence of these proteins and the fact that the DNA is wrapped around them raises many questions about how the DNA is copied in preparation for cell division and how the DNA information is read for gene activity. These are areas of active research.

The histone proteins form a structure called a "nucleosome" ("nuclear body"). There are four kinds of histones, and two of each kind join together to form a cylinder-shaped nucleosome structure. The fibrous DNA molecule wraps around each nucleosome approximately two and one-half times with a sequence of unwound DNA between each nucleosome along the entire length of the DNA molecule. The structure, called chromatin, looks like a string of beads when isolated sections are viewed with an electron microscope. When the chromatin is digested with enzymes that break the DNA backbone in the unwound regions, repeated lengths of chromatin are recovered, showing that the nucleosome wrapping is very regular. These nucleosome regions join together to form the additional folding as the chromosome condenses when the cell prepares for division.

In addition to the histone proteins, nonhistone proteins attach to the chromatin. With an electron microscope, chromatin loops can be seen extending from a protein matrix. There is evidence that these loops represent replication units along the chromosome, but how the DNA molecule is freed from the histone proteins to be replicated is a major unsolved puzzle.

The condensation of the chromatin is not uniform over the entire chromosome. In the regions immediately adjacent to the centromere, the chromatin is tightly condensed and remains that way throughout the visible cycle. All of the available evidence indicates that this chromatin does not contain actively expressed genes. It also replicates later than the remaining DNA. This more highly condensed chromatin is called heterochromatin ("the other chromatin"). The remaining chromatin is referred to as euchromatin ("true chromatin") because it contains actively expressed genes and it replicates as a unit.

Giemsa Stain and Chromosome Painting

When chromosomes are treated with a dye called Giemsa stain, regular banding patterns appear. The bands vary in width, but their positions on the individual chromosomes are consistent. This makes the bands useful in identifying specific chromosome regions. When a chromosome has a structural modification, such as an inversion—which results when two breaks occur and the region is reversed when the fragments are rejoined—the change in the banding pattern makes it possible to recognize where it has occurred. When a loss of a chromosomal region produces a deficiency disorder, changes in the banding patterns of a chromosome can identify the missing region. Karyotype analysis is a useful tool in genetic counseling because disorders caused by chromosome structure modifications can be identified. Associations between disorders and missing chromosome regions are useful in identifying which functions are associated with specific regions. Other stains produce different banding patterns and, when used in combination with the Giemsa banding patterns, allow diagnosis of structure modifications that can be quite complex.

It is also possible to use fluorescent dyes, in a process called chromosome painting, to identify the DNA of individual chromosomes, which allows the recognition of small regions

Internal Structure of a Chromosome

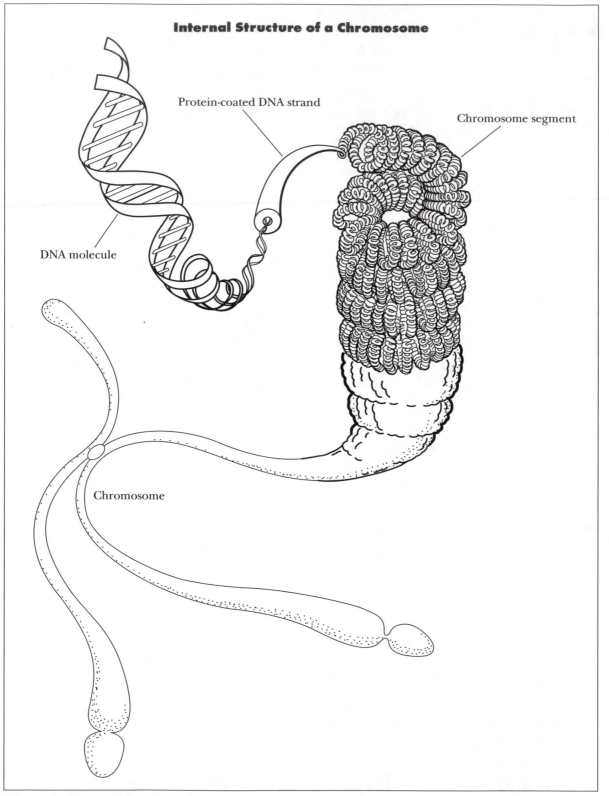

Protein-coated DNA strand

Chromosome segment

DNA molecule

Chromosome

(Electronic Illustrators Group)

that have been exchanged between chromosomes that are too small to be recognized otherwise. Color differences within chromosomes or at their tips clearly show which chromosomes have exchanged DNA, how much DNA each has been exchanged, and where on the chromosomes the exchanges have taken place. Many cancer cells, for example, have multiple chromosome modifications, with DNA from two or three chromosomes associated in one highly modified chromosome structure.

Chromosome Disorders

At the ends of the chromosomes are structures called telomeres, which are composed of specific repetitive DNA sequences that help protect the ends of chromosomes from damage and prevent DNA molecules from sticking together. Research that began in the early 1990's led to the discovery that the telomere regions of the chromosomes are shortened at each DNA replication. When the telomeres have been reduced to some critical point, the cell is no longer able to divide and often dies not long after, a phenomenon called apoptosis. Other observations indicate that the telomere is returned to its normal length in tumor cells, suggesting that this might contribute to the long life of tumor cells, possibly making them immortal. The relationship of cell age to telomere length and the mechanisms that lead to telomere shortening are not understood clearly, but this is an area of active research because it has implications for aging and cancer treatment.

The DNA of each chromosome carries a unique part of the information code in the sequence of the bases. The specific sequences are in linear order along the chromosome and form linked sequences of genes called linkage groups. When the like chromosomes pair and separate during meiosis, one copy of each chromosome is transmitted to the offspring. During meiosis, there may be an exchange of material between the paired chromosomes, but this does not change the information content because the information is basically the same for both chromosomes in any region. There may be differences in the coding sequences, but functionally the same informational content is transmitted. Extreme changes in chromosome structure that result in the moving of information to another chromosome may have consequences for how specific information is expressed; a change in position might result in different regulation or in changes in how the information is transmitted during meiosis.

Each chromosome has a specific arrangement of genes. Although homologous chromosomes exchange DNA during meiosis, as long as this process occurs normally, the gene arrangement on the chromosomes remains unchanged. Position affects result when genes are moved to different regions of the same chromosome or to another chromosome. A normal allele may show a mutant phenotype expression in a new position in the chromosome set. The best-known case occurs when a gene is placed adjacent to a heterochromatic region. The relocated DNA is condensed like the heterochromatic-region DNA and normally active genes now remain inactive. Ninety percent of patients with the disorder chronic myelogenous leukemia have an exchange of material called a translocation, between chromosomes 9 and 22. Chromosome 22 is shorter than normal and is called the Philadelphia chromosome, after the city in which it was discovered. The placing of a specific gene from chromosome 9 within the broken region adjacent to a gene on chromosome 22 causes the uncontrolled expression of both of the genes and uncontrolled cell reproduction, the hallmark of leukemia.

The separation of like chromosomes during meiosis occurs because the two chromosome arms are attached to a specific centromere. When the centromere is moved to one of the poles, the arms are pulled along, ensuring movement of all of the material of the paired chromosomes to the opposite poles and inclusion in the newly formed cells. Translocations occur when chromosomes are broken and material is placed in the wrong position by the repair system, causing a chromosome region to become attached to a different centromere. This leads to an inability to properly separate the regions of the arm, which can result in duplication of some of the chromosomal regions (when two copies of the same arm move to one cell) or deficiencies (when none of the material from a chromosome arm moves into a cell). This is a common outcome with translocation heterozygotes (individuals with both normal chromosomes and translocated chromosomes in the same cells). Pairing of like chromosome regions occurs, but rather than two chromosomes paired along their entire lengths, the arms of the two translocated chromosomes are paired with the arms of their normal pairing partners. The separation of the chromosomes produces duplications of material from one chromosome arm or a deficiency of that material 50 percent of the time. If these cells are involved in fertilization, the offspring will show duplication or deficiency disorders.

—*D. B. Benner, updated by Bryan Ness*

See also: Chromatin Packaging; Cell Cycle, The; Cell Division; Central Dogma of Molecular Biology; Chromosome Mutation; Chromosome Theory of Heredity; Chromosome Walking and Jumping; Classical Transmission Genetics; Dihybrid Inheritance; DNA Replication; DNA Structure and Function; Epistasis; Extrachromosomal Inheritance; Incomplete Dominance; Mendelian Genetics; Mitosis and Meiosis; Molecular Genetics; Monohybrid Inheritance; Multiple Alleles; Mutation and Mutagenesis; Nondisjunction and Aneuploidy; Parthenogenesis; Penetrance; Polygenic Inheritance; RNA Structure and Function; Transposable Elements.

Further Reading

Adolph, Kenneth W., ed. *Gene and Chromosome Analysis.* 3 vols. San Diego: Academic Press, 1993-1994. Covers a range of topics, including cDNA cloning methods, mammalian embryogenesis, microcell hybrids, tumor-suppressor genes, prenatal cytogenetics, and the polymerase chain reaction.

Bickmore, Wendy A. *Chromosome Structural Analysis: A Practical Approach.* New York: Oxford University Press, 1999. Describes research on chromatin and chromosome structure, specifically examining the mapping of protein, a study of whole chromosome structure and biochemical techniques for analyzing the substructure of mammalian nuclei, and the experimental manipulation of chromosome structure.

Campbell, Neil. *Biology.* 6th ed. San Francisco: Benjamin Cummings, 2001. A college-level biology textbook that provides introductory explanations of chromosomes.

Greider, Carol, and Elizabeth Blackburn. "Telomeres, Telomerase, and Cancer." *Scientific American* 274 (February, 1996). Contains a review of the nature of telomeres and their importance in the lives of cells.

Russell, Peter. *Genetics.* 5th ed. Menlo Park, Calif.: Benjamin Cummings, 1998. A college-level textbook with an excellent discussion of chromosome structure and function.

Sharma, Archana, and Sumitra Sen. *Chromosome Botany.* Enfield, N.H.: Science Publishers, 2002. Focuses on the chromosome as a vehicle of hereditary transmission, covering topics such as structural details, identification of gene sequences at the chromosome level, specific and genetic diversity in evolution, and the genome as affected by environmental agents.

Chromosome Theory of Heredity

Field of study: Classical transmission genetics; History of genetics

Significance: *The chromosome theory of heredity originated with American geneticist Walter Sutton, who first suggested that genes were located on chromosomes. This theory guided much of genetic research in the early twentieth century, including development of the earliest genetic maps based on linkage. In 1931, several experiments confirmed the chromosome theory by demonstrating that certain rearrangements of the heritable traits (or genes) were always accompanied by corresponding rearrangements of the microscopically observable chromosomes.*

Key terms

CROSSING OVER: the breakage of chromosomes followed by the interchange of the resulting fragments; also, the recombination of genes that results from the chromosomal rearrangement

GENETIC MAPPING: the locating of gene positions along chromosomes

INDEPENDENT ASSORTMENT: the inheritance of genes independently of one another when they are located on separate chromosomes

LINKAGE: the frequent inheritance of two or more genes together as a unit if they are located close together on the same chromosome

LINKAGE MAPPING: a form of genetic mapping that uses recombination frequencies to estimate the relative distances between linked genes

PHYSICAL MAPPING: a form of genetic mapping that associates a gene with a microscopically observable chromosome location

Mendel's Law of Independent Assortment

In a series of experiments first reported in 1865, Austrian botanist Gregor Mendel established the first principles of genetics. Mendel showed that the units of heredity were inherited as particles that maintained their identity across the generations; these units of heredity are now known as genes. These genes exist as pairs in all the body's cells except for the egg and sperm cells. When Mendel studied two traits at a time (dihybrid inheritance), he discovered that different genes were inherited independently of one another, a principle that came to be called the law of independent assortment. For example, if an individual inherits genes *A* and *B* from one parent and genes *a* and *b* from the other parent, in subsequent generations the combinations *AB*, *Ab*, *aB*, and *ab* would all occur with equal frequency. Gene *A* would go together with *B* just as often as with *b*, and gene *B* would go with *A* just as often as with *a*. Mendel's results were ignored for many years after he published his findings, but his principles were rediscovered in 1900 by Erich Tschermak von Seysenegg in Vienna, Austria, Carl Erich Correns in Tübingen, Germany, and Hugo de Vries in Amsterdam, Holland. Organized research in genetics soon began in various countries in Europe and also in the United States.

Sutton's Hypothesis

Mendel's findings had left certain important questions unanswered: Why do the genes exist in pairs? Why do different genes assort independently? Where are the genes located? Answers to these questions were first suggested in 1903 by a young American scientist, Walter Sutton, who had read about the rediscovery of Mendel's work. By this time, it was already well known that all animal and plant cells contain a central portion called the nucleus and a surrounding portion called the cytoplasm. Division of the cytoplasm is a very simple affair: The cytoplasm simply squeezes in two. The nucleus, however, undergoes mitosis, a complex rearrangement of the rod-shaped bodies called chromosomes, which exist in pairs. Sex cells (eggs or sperm) are "haploid," with one chromosome from each pair. All other body cells, called somatic cells, have a "diploid" chromosome number in which all chromosomes are paired. During mitosis, each chromosome becomes duplicated; then the two strands (or chromatids) split apart and separate. One result of mitosis is that the chromosome number of each cell is always preserved. Sutton also noticed that eggs in most species are many times larger than sperm because of a great difference in the amount of cytoplasm. The nuclei of egg and sperm are approximately equal in size, and these nuclei fuse during fertilization, a process in which two haploid sets of chromosomes combine to make a complete diploid set. From these facts, Sutton concluded that the genes are probably in the nucleus, not the cytoplasm, because the nucleus divides carefully and exactly while the cytoplasm divides inexactly. Also, if genes were in the cytoplasm, one would expect the mother's contribution to be much greater than the father's, contrary to the repeated observation that the parental contributions to heredity are usually equal.

Of all the parts of diploid cells, only the chromosomes were known to exist in pairs. If genes were located on the chromosomes, it would explain why they existed in pairs (except singly in eggs and sperm cells). In fact, the known behavior of chromosomes exactly paralleled the postulated behavior of Mendel's genes. Sutton's hypothesis that genes were located on chromosomes came to be called the chromosome theory of heredity. According to Sutton's hypothesis, Mendel's genes assorted independently because they were located on different chromosomes. However, there were only a limited number of chromosomes (eight in fruit flies, fourteen in garden peas, and forty-six in humans), while there were hundreds or thousands of genes. Sutton therefore predicted that Mendel's law of independent assortment would only apply to genes located on different chromosomes. Genes located on the same chromosome would be inherited together as a unit, a phenomenon now known as linkage.

In 1903, Sutton outlined his chromosomal theory of heredity in a paper entitled "The Chromosomes in Heredity." Many aspects of this theory were independently proposed by Theodor Boveri, a German researcher who had worked with sea urchin embryos at the Naples Marine Station in Italy.

Linkage and Crossing Over

Sutton had predicted the existence of linked genes before other investigators had adequately described the phenomenon. The subsequent discovery of linked genes lent strong support to Sutton's hypothesis. English geneticists William Bateson and Reginald C. Punnett described crosses involving linked genes in both poultry and garden peas, while American geneticist Thomas Hunt Morgan made similar discoveries in the fruit fly (*Drosophila melanogaster*). Instead of assorting independently, linked genes most often remain in the same combinations in which they were transmitted from prior generations: If two genes on the same chromosome both come from one parent, they tend to stay together through several generations and to be inherited as a unit. On occasion, these combinations of linked genes do break apart, and these rearrangements were attributed to "crossing over" of the chromosomes, a phenomenon in which chromosomes were thought to break apart and then recombine. Some microscopists thought they had observed X-shaped arrangements of the chromosomes that looked like the result of crossing over, but many other scientists were skeptical about this claim because there was no proof of breakage and recombination of the chromosomes in these X-shaped arrangements.

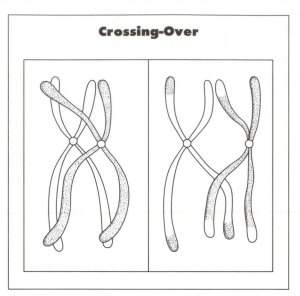

In the crossing-over process, chromosomes meet (left) and recombine (right). (Electronic Illustrators Group)

Genetic Mapping

Sutton had been a student of Thomas Hunt Morgan at Columbia University in New York City. When Morgan began his experiments with fruit flies around 1909, he quickly became convinced that Sutton's chromosome theory would lead to a fruitful line of research. Morgan and his students soon discovered many new mutations in fruit flies, representing many new genes. Some of these mutations were linked to one another, and the linked genes fell into four linkage groups corresponding to the four chromosome pairs of fruit flies. In fruit flies as well as other species, the number of linkage groups always corresponds to the number of chromosome pairs.

One of Morgan's students, Alfred H. Sturtevant, reasoned that the frequency of recombination of linked genes should be small for genes located close together and higher for genes located far apart. In fact, the frequency of crossing over between linked genes could serve as a rough measure of the distance between them along the chromosome. Sturtevant assumed that the frequency of recombination would be roughly proportional to the distance along the chromosome; recombination between closely linked genes would be a rare event, while recombination between genes further apart would be more common. Sturtevant first used this technique in 1913 to determine the relative positions of six genes on one of the chromosomes of *Drosophila*. For example, the genes for white eyes and vermilion eyes recombined about 30 percent of the time, and the genes for vermilion eyes and miniature wings recombined about 3 percent of the time. Recombination between white eyes and miniature wings took place 34 percent of the time, close to the sum of the two previously mentioned frequencies (30 percent plus 3 percent). Therefore, the order of arrangement of the genes was:

white ← 30 units → vermilion ← 3 units → miniature

Since the distances were approximately additive (the smaller distances added up to the larger distances), Sturtevant concluded that the genes were arranged along each chromosome in a straight line like beads on a string. In all, Sturtevant was able to determine such a linear arrangement among six genes in his initial study (an outgrowth of his doctoral thesis) and many more genes subsequently. Calvin Bridges, another one of Morgan's students, worked closely with Sturtevant. Over the next several years, Sturtevant and Bridges conducted numerous genetic crosses involving linked genes. They used recombination frequencies to determine the arrangement of genes along chromosomes and the approximate distances between these genes, thus producing increasingly detailed genetic maps of several *Drosophila* species.

The use of Sturtevant's technique of making linkage maps was widely copied. As each new gene was discovered, geneticists were able to find another gene to which it was linked, and the new gene was then fitted into a genetic map based on its linkage distance to other genes. In this way, geneticists began to make linkage maps of genes along the chromosomes of many different species. There are now more than one thousand genes in *Drosophila* whose locations have been mapped using linkage mapping. Extensive linkage maps have also been developed for mice (*Mus musculus*), humans (*Homo sapiens*), corn or maize (*Zea mays*), and bread mold

(*Neurospora crassa*). In bacteria such as *Escherichia coli*, other methods of genetic mapping were developed based on the order in which genes were transferred during bacterial conjugation. These mapping techniques reveal that the genes in bacteria are arranged in a circle or, more precisely, in a closed loop resembling a necklace. This loop can break at any of several locations, after which the genes are transferred from one individual to another in the order of their location along the chromosome. The order can be determined by interrupting the process and testing to see which genes had been transferred before the interruption.

Confirmation of the Chromosome Theory

The first confirmation of the chromosome theory was published in 1916 by Bridges, who studied the results of a type of abnormal cell division. When egg or sperm cells are produced by meiosis, only one chromosome of each chromosome pair is normally included in each of the resultant cells. In a very small proportion of cases, one pair of chromosomes fails to separate (or "disjoin"), so that one of the resultant cells has an extra chromosome while the other cell is missing that chromosome. This abnormal type of meiosis is called nondisjunction. In fruit flies, as in humans and many other species, females normally have two X chromosomes (XX) and males have two unequal chromosomes (XY). Bridges discovered some female fruit flies that had the unusual chromosome formula XXY; he suspected that these unusual females had originated from nondisjunction, in which two X chromosomes had failed to separate during meiosis. Bridges studied one cross using a white-eyed XXY female mated to a normal, red-eyed male. (The gene for white eyes was known to be sex-linked; it was carried on the X chromosome.) Bridges was able to predict both the genetic and chromosomal anomalies that would occur as a result of this cross. Among the unusual predictions that were verified experimentally was the existence of a chromosome configuration (XYY) that had never been observed before. Using the assumption that the gene for white eyes was carried on the X chromosome in this and other crosses, Bridges was able to make unusual predictions of both genetic and chromosomal results. These studies greatly strengthened the case for the chromosomal theory.

In 1931, Harriet Creighton and Barbara Mc-Clintock were able to confirm the chromosomal theory of inheritance much more directly. Creighton and McClintock used corn plants whose chromosomes had structural abnormalities on either end, enabling them to recognize the chromosomes under the microscope. One chromosome, for example, had a knob at one end and an attached portion of another chromosome at the other end, as shown in the figure headed "Creighton and McClintock's Cross." Creighton and McClintock then crossed plants differing in two genes located along this chromosome. One gene controlled the color of the seed coat while the other produced either a starchy or waxy kernel. The parental gene combinations (*C* with *wx* on the abnormal chromosome and *c* with *Wx* on the other chromosome) were always preserved in noncrossovers. However, a crossover between the two genes produced two new gene combinations: *C* with *Wx* and *c* with *wx*.

In this cross, Creighton and McClintock observed that the chromosomal appearance in the offspring could always be predicted from the phenotypic appearance: Seeds with colorless seed coats and starchy kernels had normal chromosomes, seeds with colored seed coats and waxy kernels had chromosomes with the knob at one end and the extra interchanged chromosome segment at the other end, seeds

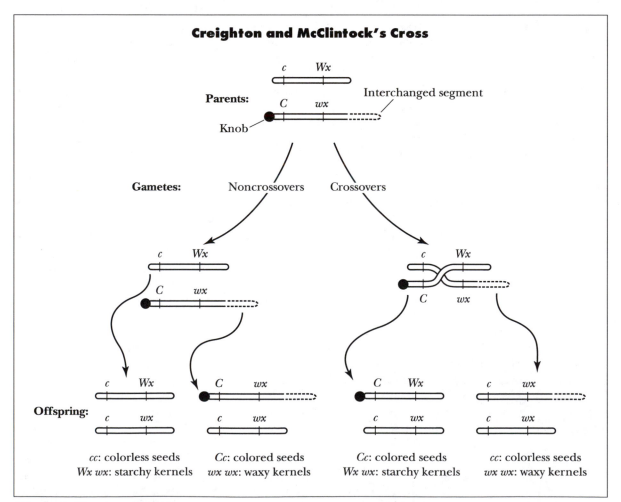

Creighton and McClintock's Cross

Parents:

Interchanged segment

Knob

Gametes: Noncrossovers   Crossovers

Offspring:

cc: colorless seeds
Wx wx: starchy kernels

Cc: colored seeds
wx wx: waxy kernels

Cc: colored seeds
Wx wx: starchy kernels

cc: colorless seeds
wx wx: waxy kernels

(Electronic Illustrators Group)

with colorless seed coats and waxy kernels had the interchanged segment but no knob, and seeds with colored coats and starchy kernels had the knob but not the interchanged segment. In other words, whenever the two genes showed rearrangement of the parental combinations, a corresponding switch of the chromosomes could be observed under the microscope. The interchange of chromosome segments was always accompanied by the recombination of genes, or, in the words of the original paper,

> cytological crossing-over . . . is accompanied by the expected types of genetic crossing-over. . . . Chromosomes . . . have been shown to exchange parts at the same time they exchange genes assigned to these regions.

In short, genetic recombination (the rearranging of genes) was always accompanied by crossing-over (the rearranging of chromosomes). This historic finding established firm evidence for the chromosomal theory of heredity. Later that same year, Curt Stern published a paper describing a very similar experiment using fruit flies.

Physical Mapping and Further Confirmation

Other evidence that helped confirm the chromosome theory came from the study of rare chromosome abnormalities. In 1933, Thomas S. Painter called attention to the large salivary gland chromosomes of *Drosophila*. Examination of these large chromosomes made structural abnormalities in the chromosomes easier to identify. When small segments of a chromosome were missing, a gene was often found to be missing also. These abnormalities, called chromosomal deletions, allowed the first physical maps of genes to be drawn. In all cases, the physical maps were found to be consistent with the earlier genetic maps (or linkage maps) based on the frequency of crossing over.

When Bridges turned his attention to the "bar eyes" trait in fruit flies, he discovered that the gene for this trait was actually another kind of chromosome abnormality called a "duplication." Again, a chromosome abnormality that could be seen under the microscope could be related to a genetic map based on linkage. Larger chromosome abnormalities included "inversions," in which a segment of a chromosome was turned end-to-end, and "translocations," in which a piece of one chromosome became attached to another. There were also abnormalities in which entire chromosomes were missing or extra chromosomes were present. Each of these chromosomal abnormalities was accompanied by corresponding changes in the genetic maps based on the frequency of recombination between linked genes. In cases in which the location of a chromosomal abnormality could be identified microscopically, this permitted an anchoring of the genetic map to a physical location along the chromosome. The correspondence between genetic maps and chromosomal abnormalities provided important additional evidence in support of the chromosomal theory. Other forms of physical mapping were developed decades later in mammals and bacteria. The increasingly precise mapping of gene locations led the way to the development of modern molecular genetics, including techniques for isolating and sequencing individual genes.

The discovery of restriction endonuclease enzymes during the 1970's allowed geneticists to cut DNA molecules into small fragments. In 1980, a team headed by David Botstein measured the sizes of these "restriction fragments" and found many cases in which the length of the fragment varied from person to person because of changes in the DNA sequence. This type of variation is generally called a "polymorphism." In this case, it was a polymorphism in the length of the restriction fragments (known as a restriction fragment length polymorphism, or RFLP). The use of the RFLP technique has allowed rapid discovery of the location of many human genes. The Human Genome Project (an effort by scientists worldwide to determine the location and sequence of every human gene) would never have been proposed had it not been for the existence of this mapping technique.

—Eli C. Minkoff

See also: Cell Division; Chromosome Mutation; Chromosome Structure; Classical Transmission Genetics; Genetic Code; Genetic Code,

Cracking of; Linkage Maps; Mendelian Genetics; Mitosis and Meiosis; Model Organism: *Drosophila melanogaster*; Restriction Enzymes; RFLP Analysis; Transposable Elements.

Further Reading

Botstein D., R. L. White, M. Skolnick, and R. W. Davis. "Construction of a Genetic Linkage Map in Man Using Restriction Fragment Length Polymorphisms." *American Journal of Human Genetics* 32, no. 3 (1980): 314-331. Botstein's initial paper on the RFLP technique.

Carlson, E. A. *The Gene: A Critical History.* 1966. Reprint. Ames: Iowa State University Press, 1989. A classic text that examines the major theories from the early to mid-twentieth century concerning the structure of the gene.

Cummings, Michael J. *Human Heredity: Principles and Issues.* 5th ed. Pacific Grove, Calif.: Brooks/Cole, 2000. College text that surveys topics such as genetics as a human endeavor; cells, chromosomes, and cell division; transmission of genes from generation to generation; cytogenics; the source of genetic variation; cloning and recombinant DNA; genes and cancer; genetics of behavior; and genes in populations.

Lewin, B. *Genes VII.* New York: Oxford University Press, 2001. Provides an integrated account of the structure and function of genes and incorporates all the latest research in the field, including topics such as accessory proteins (chaperones), the role of the proteasome, reverse translocation, and the process of X chromosome inactivation. More than eight hundred full color illustrations.

Peters, James A., ed. *Classic Papers in Genetics.* Englewood Cliffs, N.J.: Prentice-Hall, 1959. Includes many of the classic papers that contributed to chromosomal theory, including those in which Mendel established the principles of genetics, Sutton first proposed the chromosomal theory of heredity, Sturtevant produced the first genetic map based on linkage, and Creighton and McClintock confirmed that the recombination of linked genes always took place by a process that also rearranged the chromosomes.

Scriver, Charles, et al., eds. *The Metabolic and Molecular Bases of Inherited Disease.* 8th ed. 4 vols. New York: McGraw-Hill, 2001. An authority on heredity of disease and genetic inheritance, covering genetic perspectives, basic concepts, how inherited diseases occur, diagnostic approaches, and the effects of hormones.

Suzuki, D. T., et al. *An Introduction to Genetic Analysis.* 7th ed. New York: W. H. Freeman, 2000. A classic text updated to include discussions of the latest advances in genetics research.

Chromosome Walking and Jumping

Fields of study: Genetic engineering and biotechnology; Techniques and methodologies

Significance: *Chromosome walking and jumping are mapping methods used to find defective genes that cause hereditary diseases. Walking is a much slower process, whereas jumping has cut down the time it takes to locate defective genes sometimes by years. These techniques assist in curing diseases, seeking preventive measures, and detecting genetic carriers.*

Key terms

GENOMIC LIBRARY: a group of cloned DNA fragments representative of an organism's genome

KILOBASE PAIRS (kb): a measurement of 1,000 base pairs in DNA

MARKER: a unique DNA sequence with a known location with respect to other markers or genes

Gene Hunting

Several geneticists autonomously recognized the possibilities of chromosome walking and jumping to locate genes. Hans Lehrach suggested such techniques at the European Molecular Biology Laboratory, and Sherman Weissman posed similar methods at Yale University. Weissman's student Francis S. Collins elaborated his mentor's chromosome-jumping concepts. Interested in identifying disease-causing genes, Collins sought to examine sizable areas of genetic material for unknown genes believed to be responsible for triggering erratic biochemical behavior. His novel exploratory method enabled researchers to span chromosomes expeditiously and bypass repetitive or insignificant genetic information.

Investigators have adopted the chromosome-jumping procedure as a reliable, efficient molecular biology tool. Gene searching became less time-consuming and resulted in the identification of defective genes that code for abnormal proteins and cause such diseases as cystic fibrosis. Understanding the nature of such mutations makes the development of treatments and cures more likely and can lead to the ability to detect the presence of the mutated gene in carriers.

Procedure

Geneticists initiate chromosome walking and jumping by collecting genetic samples from people who have a specific disease and from their close relatives. For walking, researchers select a cloned DNA fragment from a genomic library that contains the marker closest to the gene being sought. A small part of the cloned DNA fragment that is on the end nearest the gene being sought is subcloned. The subcloned fragment is then used to screen the genomic library for a clone with a fragment closer to the gene. Then a small part of this new cloned fragment is subcloned to be used to screen for the next closer fragment. This series of steps is repeated as many times as needed, until a fragment is found that appears to contain a gene. This fragment is carefully analyzed, and if it does contain the gene of interest, the process is halted; if not, chromosome walking is continued. Chromosome walking is slow, and repetitive DNA sequences or regions that do not appear in the library can halt the process.

Geneticists choose chromosome jumping to maneuver to genes more quickly and to bypass troublesome regions of DNA. Using chromosome jumping, researchers can travel the same distance they can using chromosome walking—and farther—in one step requiring less time, because chromosome jumping uses much larger fragments. Chromosome jumping is achieved by selecting a large DNA segment from the area where geneticists believe the desired gene is located and joining the ends to form a circle. This moves DNA sequences together that naturally would occur at distances of several kilobases. Researchers cut out and clone these junctions to form libraries. They use probes from the DNA sample to seek clones with matching start and end sequences and jump along the chromosome. After each jump, bidirectional walking is often done in the new region. A combination of chromosome jumping and walking can be done until the gene is found.

Gene Discovery

Chromosome walking to find the gene for cystic fibrosis, *CF*, would have required eighteen years, while chromosome jumping reduced that time to four years. Collaborating with Lap-Chee Tsui and researchers at Toronto's Hospital for Sick Children, Collins examined DNA from patients suffering from cystic fibrosis. Tsui realized that the *CF* gene was located on chromosome 7. That chromosome consists of 150 million DNA base pairs. Using markers Tsui made from chromosome 7 library fragments, researchers, applying chromosome jumping, scanned the genetic material to target where they should use chromosome walking to find the *CF* gene.

They discovered the *CF* gene in 1989. Analysis revealed that mutation is a deletion of DNA base pairs. This gene codes the cystic fibrosis transmembrane conductance regulator (CFTR) protein. Tsui determined that the shape of CFTR and how it functions are affected by the mutated gene's coding. The abnormal CFTR is unable to create a release channel to remove chloride and sodium from cells. Mucus builds up, adhering to lungs and organs, and bacteria proliferate. Cystic fibrosis is the most frequent fatal hereditary disease in Caucasians. Geneticists estimate that one in twenty-five white Americans has a recessive *CF* gene, and one in two thousand white babies are born with cystic fibrosis. Internationally, researchers associated with Tsui's Toronto-based consortium continue to study DNA fragments for additional *CF* gene mutations and have detected at least one thousand distinct mutations.

Applications

Chromosome walking and jumping have been utilized to find other disease-causing genes. Collins and his team identified the tumor-producing neurofibromatosis gene in 1990. Three years later, they located the gene for Huntington's disease (Huntington's chorea), an extreme neurological disorder. This method also detected the location on the X chromosome of the choroideremia gene, which causes gradual blindness, mostly in males, as the retina and choroid coat degenerate. Investigating Duchenne muscular dystrophy, Louis Kunkel at the Harvard Medical School used chromosome walking to detect the absence of a gene on the X chromosome that codes the dystrophin protein for muscles. Not all genes found by these methods are linked to diseases. Andrew Sinclair and his team in London applied chromosome walking to seek the gene that signals development of testes in many embryonic mammals.

Although these techniques are useful, they raise ethical concerns. As soon as genes with disease-causing mutations are identified, people can undergo testing to determine whether they carry the mutations. This information can affect reproductive choices, particularly if both partners have a recessive allele for a potentially lethal disease. Fetal material can be genetically analyzed, resulting in complex decisions to continue or terminate a pregnancy if the fetus has the mutation.

Once the mapping of the human genome was completed in 2003, however, geneticists arrived at a time when they no longer needed to depend on chromosome walking and jumping as tools to seek human genes. However, investigators continue to use walking and jumping to locate genes of other organisms, particularly such agricultural plants as rice and wheat.

—*Elizabeth D. Schafer*

See also: Cystic Fibrosis; Genetic Screening; Genetic Testing; Genomic Libraries; Linkage Maps.

Further Reading

Gelehrter, Thomas D., Francis S. Collins, and David Ginsburg. *Principles of Medical Genetics.* 2d ed. Baltimore: Williams & Wilkins, 1998. Collins and his University of Michigan colleagues explore basic concepts and advances in genetics, including positional cloning, molecular genetics, genome mapping, and ethics, in a text comprehensible by readers unfamiliar with genetics. Glossary and illustrations.

Rommens, Johanna M., Michael C. Iannuzzi, et al. "Identification of the Cystic Fibrosis Gene: Chromosome Walking and Jumping." *Science* 245, no. 4922 (September 8, 1989): 1059-1065. This issue's cover story, cowritten by Francis S. Collins and his research team, announced one of chromosome jumping's first major discoveries.

Tsui, Lap-Chee, et al., eds. *The Identification of the CF (Cystic Fibrosis) Gene: Recent Progress and New Research Strategies.* New York: Plenum Press, 1991. This collection of technical papers represents the work of notable researchers on chromosome jumping who attended an international workshop seven months after the *CF* gene was identified.

Classical Transmission Genetics

Field of study: Classical transmission genetics; History of genetics

Significance: *In sexual reproduction, parents produce specialized cells (eggs and sperm) that fuse to produce a new individual. Each of these cells contains one copy of each of the required units of information, or genes, which provide the blueprint necessary for the offspring to develop into individual, functioning organisms. Transmission genetics refers to the passing of the information needed for the proper function of an organism from parents to their offspring as a result of reproduction.*

Key terms

CHROMOSOMES: structures in haploid cells (eggs and sperm) that carry genetic information from each parent

CROSS: the mating of parents to produce offspring during sexual reproduction

GENE: a sequence of base pairs that specifies a product (either RNA or protein); the average gene in bacteria is one thousand base pairs long

LINKAGE: a relation of gene loci on the same chromosome; the more closely linked two loci are, the more often the specific traits controlled by these loci are expressed together

MEIOSIS: the process of nuclear division during sexual reproduction that produces cells that contain half the number of chromosomes as the original cell

SEXUAL REPRODUCTION: reproduction that requires fusion of haploid gametes, each of which contains one copy of the respective parent's genes, as a first step

Discovery of Transmission Genetics

The desire to improve plant and animal production is as old as agriculture. For centuries, humans have been using selective breeding programs that have resulted in the production of thousands of varieties of plants and breeds of animals. The Greek philosopher-scientist Hippocrates suggested that small bits of the body of the parent were passed to the offspring during reproduction. These small bits of arms, heads, stomachs, and livers were thought to develop into a new individual. Following the development of the microscope, it became possible to see the cells, the small building blocks of living organisms. Study of the cell during the 1800's showed that sexual reproduction was the result of the fusion of specialized cells from two parents (eggs and sperm). It was also observed that these cells contained chromosomes ("color bodies" visible when the cells reproduced) and that the number and kind of these chromosomes was the same in both the parents and the offspring. This suggested that the chromosomes carried the genetic information and that each parent transmitted the same number and kinds of chromosomes. For example, humans have twenty-three kinds of chromosomes. The offspring receives one of each kind from each parent and so has twenty-three chromosome pairs. Since the parents and the offspring have the same number and kinds of chromosomes, and since each parent transmits one complete set of the chromosomes, it was thought that there must be a process of cell division that reduces the parent number from two sets of chromosomes to one set in the production of the egg or sperm cells. The parents would each have twenty-three pairs (forty-six) chromosomes, but their reproductive cells would each contain only one of each chromosome (twenty-three).

In the 1860's, the Austrian botanist Gregor Mendel repeated studies of inheritance in the garden pea and, using the results, developed a model of genetic transmission. The significance of Mendel's work was not recognized during his lifetime, but it was rediscovered in 1900. In that same year, the predicted reductional cell division during reproduction was fully described, and the science of genetics was born.

A Study of Variation

In many respects, genetics is the study of variation. It is recognized that a particular feature of an animal or plant is inherited because there is variation in the expression of that feature, and variation in expression follows a recognizable inheritance pattern. For example, it is known that blood types are inherited, both because there is variation (blood types A, B, and O) and because examination of family histories reveals patterns that show transmission of blood-type information from parents to children.

Variation in character expression may have one of two sources: environmental conditions or inherited factors. If a plant is grown on poor soil, it might be short. The same plant grown on good soil might be tall. A plant that is short because of an inherited factor cannot grow tall even if it is placed on richer soil. From this example, it can be seen that there may be two different ways to determine whether a specific character expression is environmentally or genetically determined: testing for environmental influences and testing for inherited factors. Many conditions are not so easily resolved as this example; there may be many complex environmental factors involved in producing a condition, and it would be impossible to test them all. Knowledge of inheritance patterns can, however, help in determining whether inherited factors play a role in a condition. Cancer-associated genes have been located using family studies that show patterns consistent with a genetic contribution to the disease. There are certainly environmental factors that influence cancer production, but those factors are not as easily recognized.

The patterns of transmission genetics were discovered because the experimenters focused their attention on single, easily recognized characteristics. Mendel carefully selected seven simple characteristics of the pea plant, such as height of the plant, color of the flower, and color of the seeds. The second reason for success was the use of carefully controlled crosses. The original parents were selected from varieties that did not show variations in the characteristic of interest. For example, plants from a pure tall variety were crossed with plants from a pure short variety. Control of the information passed by the parents allowed the experimenter to follow the variation of expression from parents to offspring through a number of generations.

Transmission Patterns

The classic genetic transmission pattern is the passing of information for each characteristic from each parent to each offspring. The offspring receives two copies of each gene. (The term "gene" is used to refer to a character-determining factor; Mendel's original terminology was "factor.") Each parent also had two copies of each gene, so in the production of the specialized reproductive cells, the number must be reduced. Consider the following example. A tall pea plant has two copies of the information for height, and both copies are for tall height (tall/tall). This plant is crossed with a plant with two genes for short height (short/short). The information content of each plant is reduced to one copy: The tall plant transmits one tall gene, and the short plant transmits one short gene. The offspring receive both genes and have the information content tall/short.

The situation becomes more complex and more interesting when one or both of the parents in a cross have two different versions of the gene for the same characteristic. If, for example, one parent has the height genes tall/short and the other has the genes short/short, the cells produced by the tall/short parent will be of two kinds: ½ carry the tall gene and ½ carry the short gene. The other parent has only one kind of gene for height (short), so all of its reproductive cells will contain that gene. The offspring will be of two kinds: ½ will have both genes (tall/short), and ½ will have only one kind of gene (short/short). Had it been known that the one parent had one copy for each version of the gene, it could have been predicted that the offspring would have been of two kinds and that each would have an equal chance of appearing. Had it not been known that one of the parents had the two versions of the gene, the appearance of two kinds of offspring would have revealed the presence of both genes. The patterns are repeatable and are therefore useful in predicting what might happen or revealing what did happen in a particular cross. For example, blood-type patterns or DNA variation patterns can be used to identify the children that belong to parents in kidnapping cases or in cases in which children are mixed up in a hospital.

In a second example, the pattern is more complex, because both parents carry both versions of the gene: a tall/short to tall/short cross. Each parent will produce ½ tall-gene-carrying cells and ½ short-gene-carrying cells. Any cell from one parent may randomly join with any cell from the other parent, which leads to the following patterns: ½ tall × ½ tall = ¼ tall/tall; ½ tall × ½ short = ¼ tall/short; ½ short × ½ tall = ¼ short/tall; ½ short × ½ short = ¼ short/short. Tall/short and short/tall are the same, yielding totals of ¼ tall/tall; ½ tall/short; and ¼ short/short, or a 1:2:1 ratio.

This was the ratio that Mendel recognized and used to develop his model of transmission genetics. Mendel used pure parents (selected to breed true for the one characteristic), so he knew when he had a generation in which all of the individuals had one of each gene.

As in the previous example, if it had been known that each of the parents had one of each gene, the ratio could have been predicted; conversely, by using the observed ratio, the information content of the parents could be deduced. Using a blood-type example, if one parent has blood-type genes AO and the other parent has the genes BO, the possible combinations observed in their offspring would be AB, AO, BO, and OO, each with the same probability of occurrence (½ A gene-bearing and ½ O gene-bearing cells in one parent × ½ B gene-bearing and ½ O gene-bearing cells in the other parent).

Reductional Division

Transmission genetics allows researchers to make predictions about specific crosses and explains the occurrence of characteristic expressions in the offspring. In genetic counseling, probabilities of the appearance of a genetic disease can be made when there is an affected child in the family or a family history of the condition. This is possible because, for most inherited characteristics, the pattern is established by the reduction of chromosome numbers that occurs when the reproductive cells are produced and by the random union of reproductive cells from the two parents. The recognition that the genes are located on the chromosomes and the description of the reductional division in which the like chromosomes separate, carrying the two copies of each gene into different cells during the reductional division of meiosis, provide the basis of the regularity of the transmission pattern. It is this regularity that allows the application of mathematical treatments to genetics. Two genes are present for each character in each individual, but only one is passed to each offspring by each parent; therefore, the 50 percent (or ½) probability becomes the basis for making predictions about the outcome of a cross for any single character.

The classical pattern of transmission genetics occurs because specialized reproductive cells, eggs and sperm, are produced by a special cell reproduction process (meiosis) in which the chromosome number is reduced from two complete sets to one set in each of the cells that result from the process. This reduction results because each member of a pair of chromosomes recognizes its partner, and the chromo-

somes come together. This joining (pairing) appears to specify that each chromosome in the pair will become attached to a "motor" unit from an opposite side of the cell that will move the chromosomes to opposite sides of the cell during cell division. The result is two new cells, each with only one of the chromosomes of the original pair. This process is repeated for each pair of chromosomes in the set.

Independent Genes

Humans have practiced selective breeding of plants and animals for centuries, but it was only during the nineteenth and twentieth centuries that the patterns of transmission of inherited characters were understood. This change occurred because the experimenters focused on a single characteristic and could understand the pattern for that characteristic. Previous attempts had been unsuccessful because the observers attempted to explain a large number of character patterns at the same time. Mendel expanded his model of transmission to show how observations become more complex as the number of characteristics examined is expanded. Consider a plant with three chromosomes and one simple character gene located on each chromosome. In the first parent, chromosome 1 contains the gene for tall expression, chromosome 2 contains the gene for expression of yellow seed color, and chromosome 3 contains the gene for purple flower color. In the other parent, chromosome 1 contains a gene for short height, chromosome 2 contains a gene for green seed color, and chromosome 3 contains a gene for white flower color. Each parent will transmit these genes to their offspring, who will have the genes tall/short, yellow/green, and purple/white. In the production of reproductive cells, the reductional division of meiosis will pass on one of the character expression genes for each of the three characters. (It is important to remember that the products of the reductional cell division have one of each chromosome. If this did not occur, information would be lost, and the offspring would not develop normally.) The characteristics are located on different chromosomes, and during the division process, these chromosome pairs act independently.

This means that the genes that came from any one parent (for example, the tall height, yellow seed, and purple flower expression genes from the one parent) do not have to go together during the division process. Since chromosome pairs act independently, different segregation patterns occur in different cells. The results from one meiosis may be a cell with the tall, green, and purple genes and one with the short, yellow, and white genes. In the same plant, another meiosis might produce a cell with the short, yellow, and purple genes, and the second cell would have the tall, green, and white genes.

Since these genes are independent, height does not influence seed color or flower color, nor does flower color influence seed color or height. The determining gene for each characteristic is located on a different chromosome, so the basic transmission model can be applied to each gene independently, and then the independent patterns can be combined. The tall/short height genes will segregate so that ½ of the cells will contain the tall gene and ½ will contain the short gene. Likewise, the yellow/green seed color genes will separate so that ½ of the cells will contain the yellow gene and ½ will contain a green gene. Finally, ½ of the cells will contain a purple flower gene and ½ will contain a white gene. These independent probabilities can be combined because the probability of any combination is the product of the independent probabilities. For example, the combination tall, purple, white will occur with a probability of $\frac{1}{2} \times \frac{1}{2} \times \frac{1}{2} = \frac{1}{8}$. This means that one should expect eight different combinations of these characters. The possible number of combinations for n chromosome pairs is 2^n. For humans, this means that any individual may produce 2^{23} different chromosome combinations. This is the same idea as tossing three coins simultaneously. Each coin may land with a head or a tail up, but how each coin lands is independent of how the other coins land. Knowledge of transmission patterns based on chromosome separation during meiosis allows researchers to explain the basic pattern for a single genetic character, but it also allows researchers to explain the great variation that is observed among individuals within a population in which genes for thousands of different characters are being transmitted.

Continuous Variation

The principles of transmission genetics were established by studying characters with discrete expressions—plants were tall or dwarf, seeds were yellow or green. In 1903, Danish geneticist Wilhelm Johannsen observed that characteristics that showed continuous variation, such as weight of plant seeds, fell into recognizable groups that formed a normal distribution. These patterns could also be explained by applying the principles of transmission genetics.

Assume a plant has two genes that influence its height and that these genes are on two different chromosomes (for example, 1 and 3). Each gene has two versions. A tall gene stimulates growth (increases the height), but a short gene makes no contribution to growth. A plant with the composition tall-1/tall-1, tall-3/tall-3 would have a maximum height because four genes would be adding to the plant's height. A short-1/short-1, short-3/short-3 plant would have minimum height because there would be no contribution to its height by these genes. Plants could have two contributing genes (tall-1/short-1, tall-3/short-3) or three contributing genes (tall-1/short-1, tall-3/tall-3). The number of offspring with each pattern would be determined by the composition of the parents and would be the result of gene segregation and transmission patterns. Many genes contributing to a single character expression apply to many interesting human characteristics, such as height, intelligence, amount of skin pigmentation, hair color, and eye color.

Linkage Groups

Mendel's model of the transmission of genes was supported by the observations of chromosome pair separation during the reductional division, but early in the twentieth century, it was recognized that some genes did not separate independently. Work in American geneticist Thomas Hunt Morgan's laboratory, especially by an undergraduate student, Alfred Sturtevant, showed that each chromosome contained determining genes for more than one characteristic and established that genes located close together on the same chromosome stayed together during the separation of the paired chromosomes during meiosis. If a pea plant had a chromosome with the tall height gene and, immediately adjacent to it, a gene for high sugar production, and if the other version of this chromosome had a gene for short height and a gene that limited the sugar production, the most likely products from meiosis would be two kinds of cells: one with the genes for tall height and high sugar production and one with the genes for short height and limited sugar production. These genes are said to be "linked," or closely associated on the same chromosome, because they go together as the chromosomes in the pair separate. It is generally accepted that humans contain approximately 21,000 genes, but there are only twenty-three kinds of chromosomes. This means that each chromosome contains many different genes. Each chromosome is considered a linkage group, and one of the goals of genetic study is to locate the gene responsible for each known characteristic to its proper chromosome.

A common problem in medical genetics is locating the gene for a specific genetic disease. Family studies may show that the disease is transmitted in a pattern consistent with the gene being on one of the chromosomes, but there is no way of knowing its location. Variations in DNA structure are also inherited in the classic pattern, and these DNA pattern modifications can be determined using modern molecular procedures. DNA variation patterns are analyzed for linkage to the disease condition. If a specific DNA pattern always occurs in individuals with the disease condition, it indicates that the DNA variation is on the same chromosome and close to the gene of interest because it is transmitted along with the disease-producing gene. This information locates the chromosome position of the gene, allowing further work to be done to study its structure. With the completion of the Human Genome Project, it is predicted that tracking down the genes responsible for genetic defects will be a much faster process than before. Many more genetic markers have now been identified, which, in theory, should greatly enhance the techniques used to locate a faulty gene.

—*D. B. Benner*

See also: Cell Division; Chromosome Mutation; Chromosome Structure; Chromosome Theory of Heredity; Dihybrid Inheritance; Epistasis; Extrachromosomal Inheritance; Genetic Code; Genetic Code: Cracking of; Hybridization and Introgression; Incomplete Dominance; Lamarckianism; Linkage Maps; Mendelian Genetics; Mitochondrial Genes; Mitosis and Meiosis; Monohybrid Inheritance; Multiple Alleles; Nondisjunction and Aneuploidy; Parthenogenesis; Penetrance; Polygenic Inheritance.

Further Reading

Cummings, Michael J. *Human Heredity: Principles and Issues.* 5th ed. Pacific Grove, Calif.: Brooks/Cole, 2000. College text that surveys topics such as genetics as a human endeavor; cells, chromosomes, and cell division; transmission of genes from generation to generation; cytogenics; the source of genetic variation; cloning and recombinant DNA; genes and cancer; genetics of behavior; and genes in populations.

Gonick, Larry, and Mark Wheelis. *The Cartoon Guide to Genetics.* New York: Harper Perennial, 1991. An easy-to-read presentation of the basic concepts of transmission genetics.

Lewis, Ricki. *Human Genetics: Concepts and Applications.* 5th ed. New York: McGraw-Hill, 2003. An introductory text for undergraduates with sections on fundamentals, transmission genetics, DNA and chromosomes, population genetics, immunity and cancer, and the latest genetic technology.

Moore, John A. *Science as a Way of Knowing.* Reprint. Cambridge, Mass.: Harvard University Press, 1999. Traces the development of scientific thinking with an emphasis on understanding hereditary mechanisms.

Stansfield, William D. *Schaum's Outline of Theory and Problems of Genetics.* 3d ed. New York: McGraw-Hill, 1991. Provides explanations of basic genetics concepts and an introduction to problem solving.

Cloning

Field of study: Genetic engineering and biotechnology

Significance: *Cloning includes both gene cloning and organismal cloning. Gene cloning, an important technique for understanding how cells work, has produced a multitude of useful products, including human medicines. Organismal cloning includes reproductive cloning and therapeutic cloning. Ethical and safety concerns have led to a consensus that human cloning should be banned, but therapeutic cloning is more controversial, since it could lead to treatments for many human diseases.*

Key terms

CLONING VECTOR: a plasmid or virus into which foreign DNA can be inserted to amplify the number of copies of the foreign DNA in the host cell or organism

DNA: dexoyribonucleic acid, a long-chain macromolecule, made of units called nucleotides and structured as a double helix joined by weak hydrogen bonds, that forms genetic material for most organisms

DNA HYBRIDIZATION: formation of a double-stranded nucleic acid molecule from single-stranded nucleic acid molecules that have complementary base sequences

LIGASE: an enzyme that joins recombinant DNA molecules together

PLASMID: a DNA molecule that replicates independently of chromosomes

RECOMBINANT DNA TECHNOLOGY: methods used to splice a DNA fragment from one organism into DNA from another organism and then clone the new recombinant DNA molecule

REPRODUCTIVE CLONING: cloning to produce individual organisms

RESTRICTION ENZYME: a protein (an enzyme) that recognizes a specific nucleotide sequence in a piece of DNA and causes a sequence-specific cleavage of the DNA

STEM CELLS: cells that are able to divide indefinitely in culture and to give rise to specialized cells

THERAPUTIC CLONING: cloning to produce a treatment for a disease

Types of Cloning

There are three different definitions of a clone. One is a group of genetically identical cells descended from a single common ancestor. This type of clone is often made by plant cell tissue culture in which a whole line of cells is made from a single cell ancestor. A second type of clone is a gene clone, or recombinant DNA clone, in which copies of a DNA sequence are made by genetic engineering. A third type of clone is an organism that is descended asexually from a single ancestor. A much-celebrated example of an organismal clone is the sheep Dolly (1997-2003), produced by placing the nucleus of a cell from an ewe's udder, with its genetic material (DNA), into an unfertilized egg from which the nucleus had been removed.

DNA Cloning

DNA is cloned to obtain specific pieces of DNA that are free from other DNA fragments. Clones of specific pieces of DNA are important for basic research. Once a piece of DNA is cloned, it can be sequenced (to determine the order of the four repeating nucleotides) to learn the details of genes within that DNA. Where does the gene begin and end? What type of control regions does the gene have? Cloned DNAs can be used as hybridization probes, where sequences that are complementary to the cloned DNA can be detected. Such DNA hybridization is useful to detect similarities between genes from different organisms, to detect the presence of specific disease genes, and to determine in what tissues that gene is expressed. The gene is expressed when a messenger RNA (mRNA) is made from the gene and the mRNA is translated into a protein product. A DNA clone is also used to produce the protein product for which that gene codes. When a clone is expressed, the protein made by that gene can be studied or an antibody against that protein can be made. An antibody is used to show in which tissues of an organism that protein is found. Also, a DNA clone may be expressed because the gene codes for a useful product. This is a way to obtain large amounts of the specific protein.

Products of Recombinant DNA Technology

Recombinant DNA technology has produced clones put to use for a wide variety of human purposes. For example, rennin and chymosin are used in cheese making. One of the most important applications, however, is in medicine. Numerous recombinant DNA products are useful in treating human diseases, including the production of human insulin (Humalin) for diabetics. Other human pharmaceuticals produced by gene cloning include clotting factor VIII to treat hemophilia A, clotting factor IX to treat hemophilia B, human growth hormone, erythropoietin to treat certain anemias, interferon to treat certain cancers and hepatitis, tissue plasminogen activator to dissolve blood clots after a heart attack or stroke, prolastin to treat genetic emphysemas, thrombate III to correct a genetic antithrombin III deficiency, and parathyroid hormone.

The advantages of the cloned products are their high purity, greater consistency from batch to batch, and the steady supply they offer.

How to Clone DNA

DNA is cloned by first isolating it from its organism. Vector DNA must also be isolated from bacteria. (A vector is a plasmid or virus into which DNA is inserted.) Both the DNA to be cloned and the vector DNA are cut with a restriction enzyme that makes sequence-specific cuts in the DNAs. The ends of DNA molecules cut with restriction enzymes are then joined together with an enzyme called ligase. In this way the DNA to be cloned is inserted into the vector. These recombinant DNA molecules (vector plus random pieces of the DNA to be cloned) are then introduced into a host, such as bacteria or yeast, where the vector can replicate. The recombinant molecules are analyzed to find the ones that contain the cloned DNA of interest.

These five cows on an Iowa farm in April, 2000, were cloned by Robert Lanza and colleagues of Advanced Cell Technologies in Worcester, Massachusetts. The cows' cells—unlike those of the first cloned vertebrate, Dolly the sheep—appeared to have a prolonged youth. (AP/Wide World Photos)

Regulation of DNA Cloning

In the 1970's the tools to permit cloning of specific pieces of DNA were developed. There was great concern among scientists about the potential hazards of some combinations of DNA from different sources. Concerns included creating new bacterial plasmids with new drug resistances and putting DNA from cancer-causing viruses into plasmids. In February, 1975, scientists met at a conference center in Asilomar, California, to discuss the need to regulate recombinant DNA research. The result of this conference was the formation of the Recombinant DNA Molecule Program Advisory Committee at the National Institutes of Health, and guidelines for recombinant DNA work were established.

Genetically Modified Organisms

Numerous cloned genes have been introduced into different organisms to produce genetically modified organisms (GMOs). Genes for resistance to herbicides and insects have been introduced into soybean, corn, cotton, and canola, and these genetically engineered plants are in cultivation in fields in the United States and other countries. Fish and fruit and nut trees that mature more rapidly have been created by genetic engineering. Edible vaccines have been made—for example, a vaccine for hepatitis B in bananas. A tomato called the Flavr Savr is genetically engineered to delay softening. Plants that aid in bioremediation by taking up heavy metals such as cadmium and lead are possible.

Concerns about genetically modified organisms include safety issues—for example, concerns that foreign genes introduced into food plants may contain allergens and that the antibiotic resistance markers used in creating the GMOs might be transferred to other organisms. There are concerns about the environmental impact of GMOs; for example, if these foreign genes are transferred to other plants by unintended crossing of a GMO with a weed plant, weeds may become difficult or impossible to eradicate and jeopardize crop growth. There is a concern about the use of genetically modified organisms as food. There is a concern about loss of biodiversity if only one, genetically modified, variety of a crop plant is cultivated. There are also ethical concerns surrounding whether certain GMOs might be made available only in rich countries, and there are concerns about careful labeling of GMOs so that consumers will be aware when they are using products from GMOs. All of these questions remain in flux as the marketing of GMOs proceeds.

Organismal Cloning

A goal of organismal cloning is to develop ways of efficiently altering animals genetically in order to reproduce certain animals that are economically valuable. Animals have been altered by the introduction of specific genes, such as human proteins that will create drug-producing animals. Some genes have been inactivated in organisms to create animal models of human diseases. For example, "knockout mice" are used as models for diabetes research. Another goal is to conduct research that might lead to the development of human organs for transplant produced from single cells. Similarly, animals might be genetically engineered to make their organs better suited for transplantation to humans. Finally, the cloning of a human might be a solution to human infertility.

Are Organismal Clones Normal?

There is, however, a concern about the health of cloned animals. First of all, when inserting a new nucleus into an egg from which the nucleus has been removed, and then implanting such eggs into surrogate mothers, only very few of the eggs develop properly. There are suggestions of other abnormalities in cloned animals that might be due to the cloning process. The first vertebrate to be successfully cloned, Dolly (1997-2003), developed first arthritis and then a lung disease when six years old; although neither condition was unusual in sheep, both appeared years earlier than normal, and Dolly was euthanized. Was she genetically older than her chronological age?

Stem Cells

Stem cells are unspecialized cells that are able to divide continuously and with the proper conditions be induced to give rise to specialized cell types. In the developing embryo they give rise to the hundreds of types of specialized cells that comprise the adult body. Embryonic stem cells can be isolated from three- to five-day-old embryos. Some tissues in the adult, such as bone marrow, brain, and muscle, contain adult stem cells that can give rise to cell types of the tissue in which they reside.

A goal of research on stem cells is to learn how stem cells become specialized cells. Human stem cells could be used to generate tissues or organs for transplantation and to generate specific cells to replace those damaged as a result of spinal cord injury, stroke, burns, heart disease, diabetes, osteoarthritis, rheumatoid arthritis, and other conditions.

Regulation of Organismal Cloning

Until the cloning of the sheep Dolly in 1997, it was thought that adult specialized cells could not be made to revert to nonspecialized cells that can give rise to any type of cell. However, Dolly was created from a specialized adult cell from a ewe's udder. After the publicity about Dolly, U.S. president Bill Clinton asked the National Bioethics Advisory Commission to form recommendations about the ethical, religious, and legal implications of human cloning. In June, 1997, that commission concluded that attempts to clone humans are "morally unacceptable" for safety and ethical reasons. There was a moratorium on using federal funds for human cloning. In January, 1998, the U.S. Food and Drug Administration (FDA) declared that it had the authority to regulate human cloning and that any human cloning must have FDA approval.

While there is general agreement in the United States and in many other countries that reproductive human cloning should be banned because of ethical and safety concerns, there is ongoing debate about whether or not to allow therapeutic cloning to treat human disease or research cloning to study how stem cells develop. The Human Cloning Prohibition Act of 2001 to ban both reproductive and therapeutic cloning passed in the U.S. House of Representatives, but the Senate did not support the ban. The ban was again considered by the lawmakers in 2002. In the meantime, individual states such as California and New Jersey have passed bills that approve of embryonic stem cell research. Such research might lead to treatments for diseases such as Parkinson's, diabetes, and Alzheimer's. The research is controversial because embryos must be destroyed to obtain the stem cells, and some groups believe that constitutes taking a human life. The embryos used are generally extra embryos left over from in vitro fertilizations. In December, 2002, and January, 2003, a company called Clonaid announced the births of several babies they claimed were the result of human cloning but then failed to to produce any scientific evidence that the babies were clones. February, 2003, the U.S. Congress considered a ban on both reproductive and therapeutic cloning. In late February, the House passed the Human Prohibition Cloning Act of 2003, banning the cloning of human beings but allowing limited research on some existing stem cell lines.

The tension between scientific possibility, public policy, and societal values continues in the arena of cloning. Through therapeutic cloning there is great potential for the treatment of human diseases, but the ethical concerns about such procedures must be carefully considered as well.

—*Susan J. Karcher*

See also: Animal Cloning; Biopharmaceuticals; Cloning: Ethical Issues; Cloning Vectors; DNA Replication; DNA Sequencing Technology; Gene Therapy; Gene Therapy: Ethical and Economic Issues; Genetic Engineering; Genetic Engineering: Agricultural Applications; Genetic Engineering: Historical Development; Genetic Engineering: Industrial Applications; Genetic Engineering: Medical Applications; Genetic Engineering: Risks; Genetic Engineering: Social and Ethical Issues; Genetically Modified (GM) Foods; Knockout Genetics and Knockout Mice; Plasmids; Polymerase Chain Reaction; Restriction Enzymes; Reverse Transcriptase; Shotgun Cloning; Stem Cells; Synthetic Genes; Transgenic Organisms; Xenotransplants.

Further Reading

Boylan, Michael. "Genetic Engineering." Chapter 6 in *Medical Ethics*, edited by Michael Boylan. Upper Saddle River, N.J.: Prentice Hall, 2000. Considers the ethical concerns of gene therapy and organismal cloning. Tables, list for further reading.

Cibelli, Jose B., Robert P. Lanza, Michael D. West, and Carol Ezzell. "The First Human Cloned Embryo." *Scientific American* 286, no. 1 (2002): 44-48. Describes the production of cloned early-stage human embryos and embryos generated only from eggs, not embryos.

Espejo, Roman, ed. *Biomedical Ethics: Opposing Viewpoints.* San Diego: Greenhaven Press, 2003. Presents debates about many aspects of organismal cloning. Illustrations, bibliography, index.

Fredrickson, Donald S. *The Recombinant DNA Controversy, a Memoir: Science, Politics, and the Public Interest, 1974-1981.* Washington, D.C.: ASM Press, 2001. An overview of the initial concerns about potential hazards of recombinant DNA cloning.

Klotzko, Arlene Judith, ed. *The Cloning Sourcebook.* New York: Oxford University Press, 2001.

Kreuzer, Helen, and Adrianne Massey. *Recombinant DNA and Biotechnology: A Guide for Teachers.* Washington, D.C.: ASM Press, 2001. Descriptions of recombinant DNA cloning methods and applications. Illustrations, index.

Lauritzen, Paul, ed. *Cloning and the Future of Human Embryo Research.* New York: Oxford University Press, 2001. Discusses cloning from the perspective of human embryo research and reproductive technology, seeing it as an extension of work that began with in vitro fertilization.

Schatten, G., R. Prather, and I. Wilmut. "Cloning Claim Is Science Fiction, Not Science." *Science* 299 (2003): 344. Letter to the editor written by prominent scientists expressing concern that evidence to support the claims of cloned humans has not been produced.

Web Site of Interest

Human Genome Project Information, Cloning Fact Sheet. http://www. doegenomes.org. Site links to information on cloning types—including DNA, reproductive, and therapeutic cloning—cloning organs for transplants, cloning risks, and more.

Cloning: Ethical Issues

Fields of study: Bioethics; Genetic engineering and biotechnology; Human genetics

Significance: *Although cloning of plants has been performed for hundreds of years and cloning from embryonic mammalian cells became commonplace in the early 1990's, the cloning of the sheep Dolly from adult cells in 1996 raised concerns that cloning might be used in a dangerous or unethical manner.*

Key terms

BIOETHICS: the study of human actions and goals in a framework of moral standards relating to use and abuse of biological systems

CLONE: an identical genetic twin of any organism or DNA sequence; clones can occur naturally or experimentally

CLONING: the process of producing a genetic twin in the laboratory by experimental means

Bioethics and Cloning

Bioethics was founded as a discipline by ethicist Van Rensselaer Potter (1911-2001) in the early 1970's as the formal study and application of ethics to biology and biotechnology. The discipline was initially created as an ethical values system to help guide scientists and others in making decisions that could affect the environment. The world has become even more complex since Potter's original vision of a planet challenged by ecological catastrophe. Humans have developed the ability to take genes from one organism and transfer them to another, creating something entirely new to nature, with unknown consequences. Moreover, humans have the ability to make endless genetic copies of these organisms by cloning. Bioethics now includes asking hard ethical questions about biotechnology, and, as Potter suggests, "promot[ing] the evolution of a better world for future generations."

Cloning involves making a genetic twin of an organism or of a DNA sequence. The focus of this article is on the cloning of whole organisms. The process of cloning has actually been performed with plants for centuries.

Cuttings can be removed from many species and induced to make roots. These cuttings are then grown into full-size, genetically identical copies of the parent plant. The emergence of crops that cannot be propagated in the standard fashion, such as seedless navel oranges, has led to whole groves of cloned siblings. Few would suggest that such cloning is inherently wrong or unethical. Animal cloning has been quietly occurring since the early 1990's. Eggs fertilized in vitro are allowed to develop to the eight-cell stage, at which point the cells are separated. Each individual cell then develops into an embryo that is implanted in a female. Thus, a single zygote can be used to make eight identical individuals. This type of cloning has been used routinely in animal husbandry to propagate desirable genetic traits.

In 1996, a team of scientists in Scotland headed by Ian Wilmut cloned a mammal—a sheep named Dolly—from adult cells for the first time. While bioethicists had seen no wrong in cloning orange trees and embryonic mammals, they were troubled by the cloning of a sheep. It is important to realize that the cloning of Dolly was not the key bioethical issue. Rather, the issue that worried the ethicists was the implication of the clone's existence: that scientists were only a small step away from cloning a human. If bioethics is concerned with protecting the evolution of future generations of humans, did the cloning of Dolly represent a potential threat? Could the natural progress of humans toward an unknown evolutionary future be sidetracked or derailed by the intervention and effects of cloning? What would be the social ramifications of human cloning? Would it have the potential to change humanity as it is now known forever? Was cloning simply wrong? Christian bioethicists, for example, were troubled by the implications of humans being able to manipulate themselves in this way, many considering it morally wrong.

Many scientists, including Wilmut, were quick to point out that they would never support human cloning but did not believe that cloning itself was unethical. Most ethicists agreed that cloning animals could help human society in many ways. Genetically engineered animals had the potential to be used to create vast quantities of protein-based therapeutic drugs. Commercial animals that are top producers, such as cows with high milk yields, could also be cloned. Human replacement organs could be grown in precisely controlled environments.

However, cloning, if misapplied, has frightening possibilities in the minds of many. Although only science fiction now, it is possible to envision a future world of human clones designed to fill certain roles, as genetically programmed soldiers, workers, or even an elite society of "perfect" cloned individuals. Others have envisioned the possibility of cloning an extra copy of themselves as donors of perfectly matched organs during old age. Even the possibility that individuals might be cloned without their knowledge or permission has been anticipated.

Human Cloning

Something about human cloning chafes at the human conscience. Bioethicist Karen Rothenberg, in statements delivered to the U.S. Senate's Public Health and Safety Subcommittee of the Labor and Human Resources Committee in the 1990's, suggested why society is made uneasy by the potential implications of human cloning. She broke her argument down into three Is. The first I is "interdependence." Cloning makes humans uneasy because it requires only one parent. People are humbled because it takes two humans to produce a baby. If part of the definition of humanity is the interdependence upon one another to reproduce, then a cloned human begs the question of just what is human. Rothenberg's second I is "indeterminateness." Cloning removes all randomness from human reproduction. With cloning, people predetermine whether they want to reproduce any physical or mental type available. They can control all possible genetic variables in cloning with a predicted outcome. However, does the same genetic variability that decides one's hereditary fate at conception also define some part of humanity? The last I is "individuality." It is disconcerting for people to imagine ten or one hundred copies of themselves walking around. Twins and triplets are common now, but what would such a vast change mean

Chief executive of Clonaid Brigitte Boisselier (left) and the founder of the Raelian movement, Claude Vorilhon, announced in January, 2003, the birth of a human clone, as well as imminent births of other cloned children. Physical evidence and independent confirmation of the cloning were never offered and the announcement was concluded to be a hoax, but in the wake of media attention the issue of human cloning became the focus of renewed public debate. (AP/Wide World Photos)

to individuality and the concept of the human soul? In closing, Rothenberg asked whether "the potential benefits of any scientific innovation [are] outweighed by its potential injury to our very concept of what it means to be human."

Andrew Scott of the Urban Institute takes a different view. He believes that bioethics does not apply to cloning but only to what happens after cloning. Cloning does not present a moral dilemma to Scott, assuming that the process does not purposely create "abnormalities." Scott states that "the clone [would] simply be another, autonomous human being . . . carrying the same genes as the donor, and [living] life in a normal, functional way." He suggests that as long as clones are not programmed to be "human drones" and are not used in an unethical way, cloning should not be a bioethical

worry. Many nonscientists miss the point that a clone is simply a genetically identical copy, not a copy in every aspect. If someone were to have cloned Albert Einstein, the cloned Einstein would not be identical behaviorally or in other ways to the original. What made Einstein who he was involved not merely his genes but also his many life experiences, which are impossible to duplicate in a clone. The same would be true of a cloned child brought to life by grieving parents who have lost their original child in an accident. The clone would be like a twin, not the same child.

Perhaps the right questions are not being asked. Better questions may be: Can humans be trusted not to abuse the technology of cloning? Can those in positions of power be trusted not to use cloning to their advantage and the endangerment of humanity? Probably the most basic question is, What compelling reason is there to clone a human in the first place? Carl B. Feldbaum, the president of the Biotechnology Industry Organization, believes that people should be wary of anyone who asks them to allow human cloning and states:

> In the future, society may determine that there are sound reasons to clone certain animals to improve the food supply, produce biopharmaceuticals, provide organs for transplantation and aid in research. I can think of no ethical reason to apply this technique to human beings, if in fact it can be applied.

The ethical issues are even more complicated than they first appear. Is the actual process of cloning, as performed by Wilmut, ethical if applied to humans? Wilmut's cloning process produced many failures before Dolly was conceived; only she survived of her 277 cloned sisters. Her early death at the age of six was also potentially precipitated by the cloning process. Bioethicists question whether manipulating human embryos to produce clones with only a 0.4 percent success rate is moral; to someone who believes that human life begins at conception, the cloning procedure as performed by Wilmut would almost certainly be unacceptable.

Of course, these questions remain irrelevant in most of the world. As of 2003, many developed nations had banned human cloning, including the United States. There is also some question as to whether the technology has progressed enough to make human cloning possible. Some believe the technology has reached this point and that cloning has been attempted secretly, at least somewhere in the world. A few even speculate that somewhere it might have already succeeded. Geneticists in several laboratories have carried out human cloning through the very early stages of embryogenesis, but there is no official case where it has been accomplished to the point of a healthy child being born.

Cloning offers a new and perhaps frightening view of life and the biological universe and brings with it a renewed respect for life. If almost any cell in the body can be used as the basis to clone an entirely new organism, this makes each cell the potential equivalent of a fertilized egg. While respect for life is renewed from this insight, life is simultaneously cheapened. If each cell contains all the genetic information needed to create a new individual, then what is a single cell worth among millions of copies? The answer may be "very little." When one million or one hundred million potential copies exist, then one copy is worth almost nothing. Therefore, the two contrary feelings of reverence and irreverence linger side by side. The question one must ask is, Which will win out in the end?

—*James J. Campanella, updated by Bryan Ness*

See also: Animal Cloning; Bioethics; Biological Weapons; Cloning; Cloning Vectors; Eugenics; Eugenics: Nazi Germany; Gene Therapy; Gene Therapy: Ethical and Economic Issues; Genetic Engineering: Medical Applications; Genetic Engineering: Risks; Genetic Engineering: Social and Ethical Issues; Genetics in Television and Films; Knockout Genetics and Knockout Mice; Polymerase Chain Reaction; Restriction Enzymes; Reverse Transcriptase; Shotgun Cloning; Stem Cells; Synthetic Genes; Transgenic Organisms; Xenotransplants.

Further Reading

Andrews, Lori B. *The Clone Age: Adventures in the New World of Reproductive Technology.* New York: Henry Holt, 1999. A lawyer specializing in reproductive technology, Andrews examines the legal ramifications of human cloning, from privacy to property rights.

Baudrillard, Jean. *The Vital Illusion.* Edited by Julia Witwer. New York: Columbia University Press, 2000. A sociological perspective on what human cloning means to the idea of what it means to be human.

Bonnicksen, Andrea L. *Crafting a Cloning Policy: From Dolly to Stem Cells.* Washington, D.C.: Georgetown University Press, 2002. Political and policy issues surrounding human cloning.

Brannigan, Michael C., ed. *Ethical Issues in Human Cloning: Cross-Disciplinary Perspectives.* New York: Seven Bridges Press, 2001. A collection of writings from a broad variety of Western and non-Western traditions and perspectives—philosophical, religious, scientific, and legal—good for sparking debate.

Klotzko, Arlene Judith, ed. *The Cloning Sourcebook.* New York: Oxford University Press, 2001. The editor, from the Center for Bioethics, University of Pennsylvania, and the Institute of Medical Ethics, Universtiy of Edinburgh, collects twenty-seven essays on the science, context, ethics, and policy issues surrounding cloning.

Lauritzen, Paul, ed. *Cloning and the Future of Human Embryo Research.* New York: Oxford Universiy Press, 2001. Places the ethical debate on human cloning in the larger context of reproductive technology.

MacKinnon, Barbara, ed. *Human Cloning: Science, Ethics, and Public Policy.* Urbana: University of Illinois Press, 2000. Experts from a variety of perspectives argue both for and against human cloning.

Rantala, M. L., and Arthur J. Milgram, eds. *Cloning: For and Against.* Chicago: Open Court, 1999. Scientists, journalists, ethicists, religious leaders, and legal experts represent all viewpoints, presenting all sides of the human cloning debate.

Shostak, Stanley. *Becoming Immortal: Combining Cloning and Stem-Cell Therapy.* Albany: State University of New York Press, 2002. Examines the question of whether human beings are equipped for potential immortality.

Yount, Lisa, ed. *The Ethics of Genetic Engineering.* San Diego: Greenhaven Press, 2002. Essays written by scientists, science writers, ethicists, and consumer advocates present the growing controversy over genetically modifying plants and animals, altering human genes, and cloning humans.

Web Sites of Interest

ActionBioScience.org. http://www.actionbio science.org/biotech/mcgee.html. A site that includes a "primer" on the ethics of cloning and useful links.

The President's Council on Bioethics. http://bioethics.gov. Government council that advises on ethical issues surrounding biomedical science and technology, including cloning. Includes links to bioethics literature.

Cloning Vectors

Field of study: Genetic engineering and biotechnology

Significance: *Cloning vectors are one of the key tools required for propagating (cloning) foreign DNA sequences in cells. Cloning vectors are vehicles for the replication of DNA sequences that cannot otherwise replicate. Expression vectors are cloning vectors that provide not only the means for replication but also the regulatory signals for protein synthesis.*

Key terms

BACTERIOPHAGE: a virus that infects bacterial cells, often simply called a phage

FOREIGN DNA: DNA taken from a source other than the host cell that is joined to the DNA of the cloning vector; also known as insert DNA

PLASMID: a small, circular DNA molecule that replicates independently of the host cell chromosome

RECOMBINANT DNA MOLECULE: a molecule of DNA created by joining DNA molecules from different sources, most often vector DNA joined to insert DNA

RESTRICTION ENZYME: an enzyme capable of cutting DNA at specific base pair sequences, produced by a variety of bacteria as a protection against bacteriophage infection

The Basic Properties of a Cloning Vector

Cloning vectors were developed in the early 1970's from naturally occurring DNA molecules found in some cells of the bacteria *Escherichia coli* (*E. coli*). These replicating molecules, called plasmids, were first used by the American scientists Stanley Cohen and Herbert Boyer as vehicles, or vectors, to replicate other pieces of DNA (insert DNA) that were joined to them. Thus the first two essential features of cloning vectors are their ability to replicate in an appropriate host cell and their ability to join to foreign DNA sequences to make recombinant molecules. Plasmid replication requires host-cell-specified enzymes, such as DNA polymerases that act at a plasmid sequence called the "origin of replication." Insert DNA is joined (ligated) to plasmid DNA through the use of two kinds of enzymes: restriction enzymes and DNA ligases. The plasmid DNA sequence must have unique sites for restriction enzymes to cut. Cutting the double-stranded circular DNA at more than one site would cut the plasmid into pieces and would separate important functional parts from one another. However, when a restriction enzyme cuts the circular plasmid at one unique site, it converts it to a linear molecule. Linear, insert DNA molecules, produced by cutting DNA with the same restriction enzyme as was used to cut the plasmid vector, can be joined to cut plasmid molecules using the enzyme DNA ligase. This catalyzes the covalent joining of the insert DNA and plasmid DNA ends to create a circular, recombinant plasmid molecule. Most cloning vectors have been designed to have many unique restriction enzyme cutting sites all in one stretch of the vector sequence. This part of the vector is referred to as the multiple cloning site.

In addition to an origin of replication and a multiple cloning site, most vectors have a third element: a selective marker. In order for the vector to replicate, it must be present inside an appropriate host cell. Introducing the vector into cells is often a very inefficient process. Therefore, it is very useful to be able to select, from a large population of host cells, those rare cells that have taken up a vector. This is the role of the selectable marker. The selectable marker is usually a gene that encodes resistance to an antibiotic to which the host is normally sensitive. For example, if a plasmid vector has a gene that encodes resistance to the antibiotic ampicillin, only those *E. coli* cells that harbor a plasmid will be able to grow on media containing ampicillin.

Many vectors have an additional selective marker that is rendered inactive when a plasmid is recombinant. A commonly used marker gene of this

kind is the *lacZ* gene, which encodes the enzyme beta-galactosidase. This enzyme breaks the disaccharide lactose into two monosaccharides. The pUC plasmid vector has a copy of the *lacZ* gene which has been carefully engineered to contain a multiple cloning site within it,

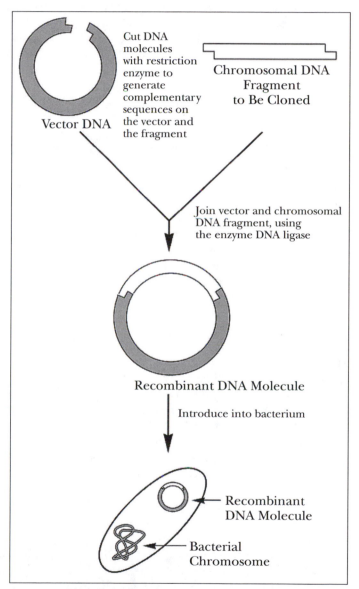

Segments of DNA from any organism can be cloned by inserting the DNA segment into a plasmid—a small, self-replicating circular molecule of DNA separate from chromosomal DNA. The plasmid can then act as a "cloning vector" when it is introduced into bacterial cells, which replicate the plasmid and its foreign DNA. This diagram from the Deparment of Energy's Human Genome Program site illustrates the process. (U.S. Department of Energy Human Genome Program, http://www.ornl.gov/hgmis.)

while maintaining the functionality of the expressed enzyme. When a DNA fragment is inserted into the multiple cloning site, the *lacZ* gene is no longer capable of making functional beta-galactosidase. This loss of function can be detected by putting X-gal into the growth media. X-gal has a structure similar to lactose but cannot be broken down by beta-galactosidase. Rather, beta-galactosidase modifies X-gal and produces a blue color. Thus, colonies of the bacterium *E. coli* containing recombinant plasmids will be normal colored, whereas those that have normal, nonrecombinant plasmids will be blue. Typical selection media then contain ampicillin and X-gal. The ampicillin only allows *E. coli* which contain a plasmid to grow, and the X-gal identifies which colonies have recombinant plasmids.

There are a number of procedures for introducing the plasmid vector into the host cell. Transformation is a procedure in which the host cells are chemically treated so that they will allow small DNA molecules to pass through the cell membrane. Electroporation is a procedure that uses an electric field to create pores in the host cell membrane to let small DNA molecules pass through.

Viruses and Cloning Vectors

In addition to plasmid cloning vectors, some bacteriophages (or phages) have been modified to serve as cloning vectors. Bacteriophages, like other viruses, are infectious agents that are made of a genome, either DNA or RNA, that is surrounded by a protective protein coat. Phage vectors are used similarly to the way plasmid vectors are used. The vector and insert DNAs are cut by restriction enzymes so that they subsequently can be joined by DNA ligase. The newly formed recombinant DNA molecules must enter an appropriate host cell to replicate. In order to introduce the phage DNA into cells, a whole phage particle must be built. This is referred to as "packaging" the DNA. The protein elements of the phage are mixed with the recombinant phage DNA and packaging enzymes to create an infectious phage particle. Appropriate host cells are then infected with it. The infected cells then make many copies of each recombinant molecule, along with the

proteins needed to make a completed phage particle. In many cases, the final step of viral infection is the lysis of the host cell. This releases the mature phage particles to infect nearby host cells. Phage vectors have two advantages relative to plasmid vectors: First, viral delivery of recombinant DNA to host cells is much more efficient than the transformation or electroporation procedures used to introduce plasmid DNA into host cells, and second, phage vectors can be used to clone larger fragments of insert DNA.

Viruses that infect cells other than bacteria have been modified to serve as cloning vectors. This permits cloning experiments using many different kinds of host cells, including human cells. Viral vectors, just like the natural viruses from which they are derived, have specific host and tissue ranges. A particular viral vector will be limited for use in specific species and cell types. The fundamental practice of all virally based cloning vectors involves the covalent joining of the insert DNA to the viral DNA to make a recombinant DNA molecule, introduction of the recombinant DNA into the appropriate host cell, and then propagation of the vector through the natural mechanism of viral replication. There are two fundamentally different ways that viruses propagate in cells. Many viruses, such as the phages already described, enter the host cell and subvert the cell's biosynthetic machinery to its own reproduction, which ultimately leads to lysis and thereby kills the host cell as the progeny viruses are released. The second viral life strategy is to enter the host cell and integrate the viral DNA into the host cell chromosome so that the virus replicates along with the host DNA. Such integrating viruses can be stably maintained in the host cell for long periods. The retroviruses, of which the human immunodeficiency virus (HIV) is an example, are a group of integrating viruses that are potentially useful vectors for certain gene therapy applications. Using cloning vectors and host cells other than bacteria allows scientists to produce some proteins that bacteria cannot properly make, permits experiments to determine the function of cloned genes, and is important for the development of gene therapy.

The Ti Plasmid of *Agrobacterium*

Two species of naturally occurring plant pathogenic bacteria, *Agrobacterium tumefaciens* and *Agrobacterium rhizogenes,* infect many plant species and have been harnessed through biotechnology to effect permanent genetic transformation of plants. Virulent (disease-causing) *Agrobacterium* species can infect plants and transfer a small portion of their own bacterial DNA, called T-DNA (transferred DNA), into the plant. The T-DNA is actually a small fragment of a large (approximately 200-kilobase-pair) plasmid called the Ti (tumor-inducing) plasmid in *A. tumefaciens* and the Ri (root-inducing) plasmid in *A. rhizogenes.*

The T-DNA fragment of the Ti plasmid is defined on both ends by 24-base-pair direct repeat sequences called the left-hand and right-hand border sequences. The T-DNA fragment is released from the plasmid by the action of endonucleases, which cut the DNA at specific points within the right-hand and left-hand border sequences. The endonucleases are two of the *Vir* (virulence) genes encoded on the Ti plasmid adjacent to the T-DNA. Several other *Vir* genes are produced when *Agrobacterium* cells are introduced into plant tissue, usually through a wound. Following infection of a plant, *Agrobacterium* cells sense the presence of phenolic wound compounds and the acidic environment within wounded plant tissues. These conditions trigger a series of several *Vir* genes to produce Vir proteins that direct excision of the T-DNA and facilitate transport and incorporation of the T-DNA segment into the plant's genome. Once the T-DNA is incorporated into the plant genome, expression of the T-DNA-encoded genes causes the plant to produce unusual quantities of plant hormones and other compounds that cause the plant cells to grow abnormally near the infection site, producing characteristic tumors.

Purposeful genetic transformation of plants requires a tool that can be used to insert new genes into a plant. This tool, regardless of its derivation, is called a vector. To date, the most common means for stable genetic transformation of plants involves the use of vectors derived from bacteria of the genus *Agrobacterium.* Biotechnologists have harnessed *Agrobacterium* to insert new genes of interest into plants by modifying the T-DNA segment of the bacterial DNA using standard recombinant methods. By deleting the genes on the T-DNA that cause tumors and then inserting desirable genes in their place, a wide variety of vectors can be produced to transfer desirable genes into plants. The genes transferred by way of *Agrobacterium* vectors become a permanent part of the plant's genome. DNA from plants, animals, bacteria, and viruses can be introduced into plants in this way.

One major drawback of *Agrobacterium* transformation is that insertion of T-DNA into the plant genome is essentially random. The genes on the T-DNA segment may not be efficiently transcribed at their location or the insertion of T-DNA may knock out an important plant gene by inserting in the middle of it. Therefore, a plant genetically transformed using an *Agrobacterium* vector is not necessarily guaranteed to perform as desired. A final drawback is that the vector works only with dicots, while many of the world's most important crops are monocots, such as wheat, rice, corn, and many other grain crops.

—*Robert A. Sinnott*

Expression Vectors

Expression vectors are cloning vectors designed to express the gene contained in the recombinant vector. In order to accomplish this, they must also provide the appropriate regulatory signals for the transcription and translation of the foreign gene. Regulatory sequences, which direct the cellular transcription machinery, are very different in bacteria and higher organisms. Thus, unless the vector provides the appropriate host regulatory sequences, foreign genes will not normally be expressed.

Expression vectors make it possible to produce proteins encoded by eukaryotic genes (that is, genes from higher organisms) in bacterial cells. Furthermore, producing proteins in this way often results in higher production rates than in the cells from which the gene was obtained. This technology not only is of immense benefit to scientists who study proteins but also is used by industry (particularly the pharmaceutical industry) to make valuable proteins. Proteins such as human insulin, growth hormone, and clotting factors that are difficult and extremely expensive to isolate from their natural sources are readily available because they can be produced much more cheaply in bacteria. An added benefit of expression vectors is that actual human proteins are produced by bacteria and therefore do not provoke allergic reactions as frequently as insulin that is isolated from other species.

Artificial Chromosomes

In 1987, a new type of cloning vector was developed by David Burke, Maynard Olson, and their colleagues. These new vectors, artificial chromosomes, filled the need created by the Human Genome Project (HGP) to clone very large insert DNAs (hundreds of thousands to millions of base pairs in length). One of the goals of the HGP—to map and ultimately sequence all the chromosomes of humans, as well as a number of other "model" organisms' genomic sequences—required a vector capable of propagating much larger DNA fragments than plasmid or phage vectors could propagate. The first artificial chromosome vector was developed in the yeast *Saccharomyces cerevisiae*. All the critical DNA sequence elements of a yeast chromosome were identified and isolated, and these were put together to create a yeast artificial chromosome (YAC). The elements of a YAC vector are an origin of replication, a centromere, telomeres, and a selectable marker suitable for yeast cells. A yeast origin of replication (similar to the origin of replication of bacterial plasmids) is a short DNA sequence that the host's replicative enzymes, such as DNA polymerase, recognize as a site to initiate DNA replication. In addition to replicating, the new copies of a chromosome must be faithfully partitioned into daughter cells during mitosis. The centromere sequence mediates the partitioning of the chromosomes during cell division because it serves as the site of attachment for the spindle fibers in mitosis. Telomeres are the DNA sequences at the ends of chromosomes. They are required to prevent degradation of the chromosome and for accurate replication of DNA at the ends of chromosomes.

YACs are used much as plasmid vectors are. Very large insert DNAs are joined to the YAC vector, and the recombinant molecules are introduced into host yeast cells in which the artificial chromosome is replicated just as the host's natural chromosomes are. YAC cloning technology allows very large chromosomes to be subdivided into a manageable number of pieces that can be organized (mapped) and studied. YACs also provide the opportunity to study DNA sequences that interact over very long distances. Since the development of YACs, artificial chromosome vectors for a number of different host cells have been created.

Impact and Applications

Cloning vectors are one of the key tools of recombinant DNA technology. Cloning vectors make it possible to isolate particular DNA sequences from an organism and make many identical copies of this one sequence in order to study the structure and function of that sequence apart from all other DNA sequences. Until the development of the polymerase chain reaction (PCR), cloning vectors and their host cells were the only means to collect many copies of one particular DNA sequence. For long DNA sequences (those over approximately ten thousand base pairs), cloning vectors are still the only means to do this.

Gene therapy is a new approach to treating and perhaps curing genetic disease. Many common diseases are the result of defective genes. Gene therapy aims to replace or supplement the defective gene with a normal, therapeutic gene. One of the difficulties faced in gene therapy is the delivery of the therapeutic gene to the appropriate cells. Viruses have evolved to enter cells, sometimes only a very specific subset of cells, and deliver their DNA or RNA genome into the cell for expression. Thus viruses make attractive vectors for gene therapy. An ideal vector for gene therapy would replace viral genes associated with pathogenesis with therapeutic genes; the viral vector would then target the therapeutic genes to just the right cells. One of the concerns related to the use of viral vectors for gene therapy is the random nature of the viral insertion into the target cell's chromosomes. Insertion of the vector DNA into or near certain genes associated with increased risk of cancer could theoretically alter their normal expression and induce tumor formation.

Plasmid DNA vectors encoding immunogenic proteins from pathogenic organisms are being tested for use as vaccines. DNA immunization offers several potential advantages over traditional vaccine strategies in terms of safety, stability, and effectiveness. Genes from disease-causing organisms are cloned into plasmid expression vectors that provide the regulatory signals for efficient protein production in humans. The plasmid DNA is inoculated intramuscularly or intradermally, and the muscle or skin cells take up some of the plasmid DNA and express the immunogenic proteins. The immune system then generates a protective immune response. There are two traditional vaccination strategies: One uses live, attenuated pathogenic organisms, and the other uses killed organisms. The disadvantage of the former is that, in rare cases, the live vaccine can cause disease. The disadvantage of the latter strategy is that the killed organism does not enter the patient's cells and make proteins like the normal pathogen. Therefore, one part of the immune response, the cell-mediated response, is usually not activated, and the protection is not as good. In DNA immunization, the plasmids enter the patient's cells, and the immunogenic proteins produced there result in a complete immune response. At the same time, there is no chance that DNA immunization will cause disease, because the plasmid vector does not carry all of the disease-causing organism's genes.

—*Craig S. Laufer, updated by Bryan Ness*

See also: Animal Cloning; Biopharmaceuticals; Cloning; Cloning: Ethical Issues; DNA Replication; DNA Sequencing Technology; Gene Therapy; Genetic Engineering; Genetic Engineering: Medical Applications; Genetic Engineering: Risks; Genomic Libraries; Knockout Genetics and Knockout Mice; Plasmids; Polymerase Chain Reaction; Protein Synthesis; Restriction Enzymes; Reverse Transcriptase; Shotgun Cloning; Stem Cells; Synthetic Genes; Telomeres; Transgenic Organisms; Xenotransplants.

Further Reading

Anderson, W. French. "Gene Therapy." *Scientific American* 124 (September, 1995). Provides a good review of the promises and problems of gene therapy.

Cohen, Philip. "Creators of the Forty-seventh Chromosome." *New Scientist* 34 (November 11, 1995). Describes the efforts to develop human artificial chromosomes.

Friedmann, Theodore. "Overcoming the Obstacles." *Scientific American* 96 (June, 1997). Elaborates on the relative merits of different delivery systems for gene therapy.

Hassett, Daniel E., and J. Lindsay Whitton. "DNA Immunization." *Trends in Microbiology* 307 (August, 1996). Reviews the process of DNA immunization and compares it to traditional vaccination strategies.

Jones, P., and D. Ramji. *Vectors: Cloning Applications and Essential Techniques*. New York: J. Wiley, 1998. A laboratory manual that allows quick and easy access to the key protocols required by those working with vectors.

Lu, Quinn, and Michael P. Weiner, eds. *Cloning and Expression Vectors for Gene Function Analysis*. Natick, Mass.: Eaton, 2001. Reprints forty-three articles from the journal *BioTechnique* to provide an overview of topics such as cloning vectors and strategies, protein expression and purification, gene tagging and epitope tagging strategies, and special purpose vectors.

Watson, James, et al. *Recombinant DNA*. New York: W. H. Freeman, 1992. Nobel laureate Watson uses accessible language and diagrams to address the methods, underlying concepts, and far-reaching applications of recombinant DNA technology. An excellent reference for details on how the different cloning vectors work and to what purposes each is particularly suited.

Color Blindness

Field of study: Diseases and syndromes

Significance: *Color blindness is a condition in people whose eyes lack one or more of the three color receptors present in most human eyes. It is an important condition to understand because so many people experience it to some degree. It is also a window into the inner workings of the eye and a marvelous example of the workings of Mendelian genetics.*

Key terms

CONES: the light-sensitive structures in the retina that function as color receptors and are the basis for color vision

DEUTERANOPES: people who lack the second kind of color receptor, the M cone, and thus are more sensitive to green

DICHROMATS: people whose eyes have only two of the three cones

PROTANOPES: people who lack the first kind of color receptor, the L cone, and thus are more sensitive to red

TRICHROMATS: people whose eyes have all three color receptors, or cones

TRITANOPES: people who lack the third kind of color receptor, the S cone, and thus are more sensitive to blue

Color Vision

Light-sensitive structures in the retina called cones are the basis for color vision. A person with normal vision can distinguish seven pure hues (colors) in the rainbow: violet, blue, cyan, green, yellow, orange, and red. People with normal vision are trichromats, meaning that they have three types of cones: L, M, and S, named for particular sensitivities to light of long, medium, and short wavelengths. The human vision system detects color by comparing the relative rates at which the L, M, and S cones react to light. For example, yellow light causes the M and L cones to signal at about the same rate, and the person "sees" yellow. Strangely, the right amounts of green and red stimulate these cones in the same fashion, and the person will again see the color yellow even though there is no yellow light present. Since people have only three types of color receptors, it takes the proper mix of intensities of only three primary colors to cause a person to "see" all the colors of the rainbow. A tiny droplet of water on the screen of a color television or computer monitor will act like a magnifying lens and reveal that the myriad colors that are displayed are formed from tiny dots of only blue, green, and red.

Dichromats

People are referred to as "color blind" if they are dichromats, that is, if they have only two of the three types of cones. Approximately 1.2 percent of males and 0.02 percent of females are protanopes (lack L cones); 1.5 percent of males and 0.01 percent of females are deuteranopes (lack M cones); but only 0.001 percent of males and females are tritanopes (lack S cones).

Dichromats can match all of the colors they see in the rainbow by mixing only two primary colors of light, but they see fewer (and different) hues in the rainbow than a person with normal vision. Protanopes and deuteranopes cannot distinguish between red and green. More exactly, protanopes tend to confuse reds, grays, and bluish blue-greens, while deuteranopes tend to confuse purples, grays, and greenish blue-greens.

S Pigment Genes

Tritanopes cannot distinguish between blue (especially greenish shades) and yellow. The genetic code for the S pigment lies on chromosome 7. The fact that the S pigment gene lies on an autosome explains why yellow-blue color blindness is manifested equally in males and females. The inheritance pattern is that of an autosomal dominant trait: Only one arm of the two arms of chromosome 7 has the defective allele in the affected parent, and since there is a 50 percent chance a child will receive the defective arm, 50 percent of the children will inherit the defect. In fact, the trait is often incompletely expressed, so that the majority of affected individuals retain some reduced S-cone function.

L and M Pigment Genes

Anomalous trichromats are more common than dichromats. They need three primary colors to match the hues of the rainbow, but they match them with different intensities than normal trichromats do because the peak sensitivities of their cones occur at wavelengths slightly different from normal. Their color confusion is similar to that of the dichromats, but less severe. About 1 percent of males and 0.03 percent of females have anomalous L cones, while 4.5 percent of males and 0.4 percent of females have anomalous M cones. The fact that far more males than females have some degree of red-green color blindness implies that the genetic information for the pigments in L and M cones lies on the X chromosome.

The gene structures for M-cone and L-cone pigments are 96 percent the same, so it is likely that one began as a mutation of the other. Small mutations in either gene can slightly shift the color of peak absorption in the cones and produce an anomalous trichromat. Generally these mutations make M and L cones more alike. The similarity between the genes and the fact that they are adjacent to each other on the X chromosome can lead to a variety of copying errors during meiosis. People with normal color vision have one L-cone gene and one to three M-cone genes. The complete omission of either type of gene will result in severe red-green color blindness: protanopia or deuteranopia. Hybrid genes that are a combination of L-cone and M-cone genes lead to less severe types of red-green color blindness, especially if there is also a normal copy of the gene present.

Red-green color blindness follows an X-gene recessive inheritance pattern. Suppose that Grandfather has a defective X gene (and is therefore color blind) and Grandmother is normal. Their male children are normal because they inherited their X genes from their mother, but their female children will be carriers because they had to inherit one X gene from their father. If the daughters married normal men, 50 percent of the grandsons got the defective gene from their mothers and were color blind, and 50 percent of the grandsons were normal. Likewise, 50 percent of the granddaughters were normal and 50 percent inherited the defective gene from their mothers and became carriers.

—*Charles W. Rogers*

See also: Classical Transmission Genetics; Congenital Defects; Dihybrid Inheritance; Hereditary Diseases; Monohybrid Inheritance.

Further Reading

Hsia, Yun, and C. H. Graham. "Color Blindness." In *The Science of Color.* Vol. 2 in *Readings on Color,* edited by Alex Byrne and David R. Hilbert. Cambridge, Mass.: MIT Press, 1997. A description of the genetics of color blindness for students with a good science background. Includes a series of color plates.

Nathans, Jeremy. "The Genes for Color Vision." In *The Science of Color.* Vol. 2 in *Readings on Color,* edited by Alex Byrne and David R. Hilbert. Cambridge, Mass.: MIT Press, 1997. An account of how the genes for color blindness were discovered, for students with a good science background. Includes a series of color plates.

Rosenthal, Odeda, and Robert H. Phillips. *Coping with Color Blindness.* Garden City Park, N.Y.: Avery, 1997. A description of color blindness aimed at nonspecialists and covering causes, testing, and coping strategies.

Wagner, Robert P. "Understanding Inheritance: An Introduction to Classical and Molecular Genetics." In *The Human Genome Project: Deciphering the Blueprint of Heredity,* edited by Necia Grant Cooper. Mill Valley, Calif.: University Science Books, 1994. A superb, well-illustrated discussion of Mendelian genetics and disorders.

Web Sites of Interest

Causes of Color. http://webexhibits.org/causesofcolor. A good introduction to light and color, including the genetics of color blindness. Includes demonstrations of how a scene looks to people with different types of color blindness and how color blindness tests are constructed.

Howard Hughes Medical Institute, Seeing, Hearing, and Smelling the World. http://www.hhmi.org/senses. Site that includes the articles "Color Blindness: More Prevalent Among Males" and "How Do We See Colors?"

Complementation Testing

Field of study: Techniques and methodologies

Significance: *Complementation testing is used to determine whether or not two mutations occur within the same gene.*

Key terms

ALLELE: a form of a gene; each gene (locus) in most organisms occurs as two copies called alleles

CISTRON: a unit of DNA that is equivalent to a gene; it encodes a single polypeptide

INBORN ERRORS OF METABOLISM: conditions that result from defective activity of an enzyme or enzymes involved in the synthesis, conversion, or breakdown of important molecules within cells

LOCUS (*pl.* LOCI): the location of a gene, often used as a more precise way to refer to a gene; each locus occurs as two copies called alleles

Finding Mutations

Most traits are the result of products from several genes. Mutations at any one of these genes may produce the same mutant phenotype. If the same mutant phenotype is observed in two different strains of an organism, there is no way, using simple observation, to determine whether this shared mutant phenotype represents a mutation in the same or different genes, or loci. One way of solving this problem is through complementation testing. If the mutations are alleles of the same locus, then a cross between mutant individuals from the two strains will only produce offspring with the mutant phenotype. In genetic terms, they fail to com-

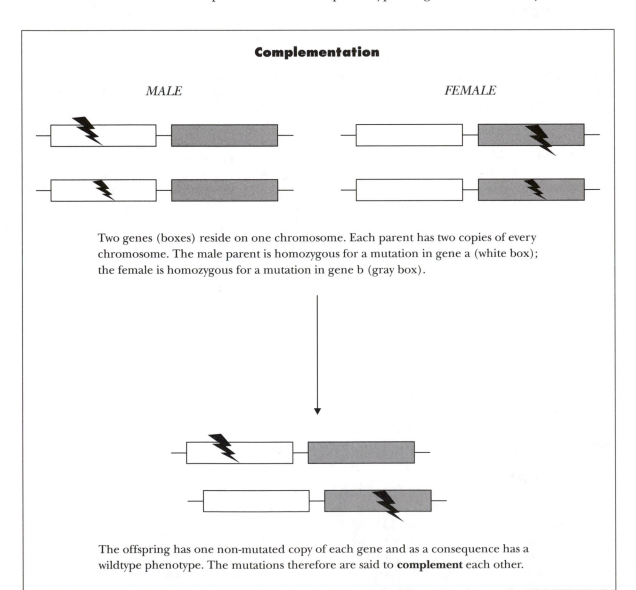

Complementation

MALE   *FEMALE*

Two genes (boxes) reside on one chromosome. Each parent has two copies of every chromosome. The male parent is homozygous for a mutation in gene a (white box); the female is homozygous for a mutation in gene b (gray box).

The offspring has one non-mutated copy of each gene and as a consequence has a wildtype phenotype. The mutations therefore are said to **complement** each other.

plement each other and are therefore members of the same complementation group. If from the same cross, all the offspring are normal; the two mutations are at the same locus and they are said to complement each other. Researchers often want to define multiple alleles of a single gene in order to understand the gene's function better.

Often a researcher is interested in the genetic control of a particular biological process, such as the biochemistry of eye color in fruit flies. As a first step, researchers often screen large numbers of individuals to find abnormal phenotypes involving the process in which they are interested. For instance, researchers studying eye color in fruit flies may screen hundreds of thousands of fruit flies for abnormal eye colors. Complementation testing is then used to organize the mutations into complementation groups.

Complementation Testing and Inborn Errors of Metabolism

Human genetic diseases that affect the function of cellular enzymes are known as inborn errors of metabolism and were defined by Sir Archibald Garrod long before DNA was determined to be the hereditary material. Garrod studied families with alkaptonuria, a disease that causes urine to turn dark upon exposure to air. He determined that this biochemical defect was inherited in a simple Mendelian fashion.

George Beadle and Edward Tatum studied mutant strains of *Neurospora* and expanded on Garrod's work. They used radiation to generate random mutations that resulted in strains of *Neurospora* that could not grow without specific nutritional supplements (essentially creating yeast with inborn errors of metabolism). Some of the mutant strains required the addition of a specific amino acid to the media. Each mutant strain had its own specific requirements for growth, and each strain was shown to have a single defective step in a metabolic pathway. When strains that had different defects were grown together, they were able to correct each other's metabolic defect. This correction was termed metabolic complementation. Using complementation tests, Beadle and Tatum were able to establish the number of genes required for a particular pathway. These studies formed the basis for the "one gene-one enzyme hypothesis": Each gene encodes a single enzyme required for a single step in a metabolic pathway. This hypothesis has since been renamed the "one gene-one polypeptide hypothesis" because some enzymes consist of multiple polypeptides, each of which is encoded by a single gene.

The Biochemical Basis for Complementation Testing

Complementation testing is useful for locating and identifying the genes affected by recessive or loss-of-function alleles. A researcher crosses two organisms that are each homozygous for a recessive mutation. If these two alleles affect the same gene, they will not complement each other, because the first-generation (F_1) offspring will inherit one mutant copy of the gene from one parent and a second mutant copy of the gene from the other parent, thus having no normal copies of the gene. If the mutations are alleles of two different genes, genes A and B, the F_1 offspring will receive a normal copy of A and a mutant copy of B from one parent and a mutant copy of A and a normal copy of B from the other, thus having one normal copy of each of the two genes and having a wild-type (normal) phenotype. The mutations are said to complement each other.

If a scientist is interested in a particular gene, obtaining as many alleles of that gene as possible will lead to a better understanding of how the gene works and what parts of the gene are essential for function. One way to identify new alleles of a gene is through an F_1 noncomplementation screen. In this type of screen, the researcher treats the model organism with radiation or chemicals to increase the rate of mutation. Any individuals from the screen that segregate the desired phenotype (white eyes, for example) in a Mendelian fashion are crossed with individuals carrying a known mutation in the gene of interest. If the progeny of this cross have white eyes (the mutant phenotype), then the two mutations have failed to complement each other and are most likely alleles of the same gene. Such noncomplementation screens have been used to identify genes involved in a wide variety of processes ranging from embryo development in fruit flies to spermatogeneis in *Caenorhabditis elegans*.

—*Michele Arduengo and Bryan Ness*

See also: Biochemical Mutations; Chemical Mutagens; Chromosome Mutation; Inborn Errors of Metabolism; Linkage Maps; Mutation and Mutagenesis; Model Organism: *Caenorhabditis elegans*; Model Organism: *Neurospora crassa*.

Further Reading

Hartl, D. L. *Genetics: Analysis of Genes and Genomes*. 5th ed. Boston: Jones and Bartlett, 2001. An excellent introductory genetics textbook.

Lewin, Benjamin. *Genes VII*. New York: Oxford University Press, 2001. Includes a summary of complementation and a discussion of complementation in bacterial systems as well.

Complete Dominance

Field of study: Classical transmission genetics

Significance: *Complete dominance represents one of the classic Mendelian forms of inheritance. In an individual that is heterozygous for a trait, the allele that displays complete dominance will determine the phenotype of the individual. Knowing whether the pattern of expression of a trait is dominant or recessive helps in making predictions concerning the inheritance of a particular genetic condition or disorder in a family's history.*

Key terms

ALLELES: different forms of a gene at a specific locus; for each genetic trait there are typically two alleles in most organisms, including humans

GENOTYPE: a description of the alleles at a gene locus

HETEROZYGOTE: an individual with two different alleles at a gene locus

HOMOZYGOTE: an individual with two like alleles at a gene locus

INCOMPLETE DOMINANCE: the expression of a trait that results when one allele can only partly dominate or mask the other

LOCUS (*pl.* LOCI): a gene, located at a specific location on a chromosome, which in humans and many other organisms occurs in the same location on homologous chromosomes

PHENOTYPE: the observed expression of a gene locus in an individual

The Discovery and Definition of Dominance

Early theories of inheritance were based on the idea that fluids carrying materials for the production of a new individual were transmitted to offspring from the parents. It was assumed that substances in these fluids from the two parents mixed and that the children would therefore show a blend of the parents' characteristics. For instance, individuals with dark hair mated to individuals with very light hair were expected to produce offspring with medium-colored hair. The carefully controlled breeding studies carried out in the 1700's and 1800's did not produce the expected blended phenotypes, but no other explanation was suggested until Gregor Mendel proposed his model of inheritance. In the 1860's, Mendel repeated studies using the garden pea and obtained the same results seen by other investigators, but he counted the numbers of each type produced from each mating and developed his theory based on those observations.

One of the first observations Mendel dealt with was the appearance of only one of the parental traits in the first generation of offspring (the first filial, or F_1, generation). For example, a cross of tall plants and dwarf plants resulted in offspring that were all tall. Mendel proposed that the character expression (in this case height) was controlled by a determining "factor," later called the "gene." He then proposed that there were different forms of this controlling factor corresponding to the different expressions of the characteristic and termed these "alleles." In the case of plant height, one allele produced tall individuals and the other produced dwarf individuals. He further proposed that in the cross of a tall (*D*) plant and a dwarf (*d*) plant, each parent contributed one factor for height, so the offspring were *Dd*. (Uppercase letters denote dominant alleles, while lowercase letters denote recessive alleles.) These plants contained a factor for both the tall expression and the dwarf expression, but the plants were all tall, so "tall" was designated the dominant phenotype for the height trait.

Mendel recognized from his studies that the determining factors occurred in pairs—each sexually reproducing individual contains two alleles for each inherited characteristic. When he made his crosses, he carefully selected pure breeding parents that would have two copies of the same allele. In Mendel's terminology, the parents would be homozygous: A pure tall parent would be designated *DD*, while a dwarf parent would be designated *dd*. His model also proposed that each parent would contribute one factor for each characteristic to each offspring, so the offspring of such a mating should be *Dd* (heterozygous). The tall appearance of the heterozygote defines the character expression (the phenotype) as dominant. Dominance of expression for any characteristic cannot be guessed but must be determined by observation. When variation is observed in the phenotype, heterozygous individuals must be examined to determine which expression is observed. For phenotypes that are not visible, such as blood types or enzyme activity variations, a test of some kind must be used to determine which phenotype expression is present in any individual.

Mendel's model and the appearance of the dominant phenotype also explains the classic 3:1 ratio observed in the second (F_2) generation. The crossing of two heterozygous individuals ($Dd \times Dd$) produces a progeny that is $\frac{1}{4}$ *DD*, $\frac{1}{2}$ *Dd*, and $\frac{1}{4}$ *dd*. Because there is a dominant phenotype expression, the $\frac{1}{4}$ *DD* and the $\frac{1}{2}$ *Dd* progeny all have the same phenotype, so $\frac{3}{4}$ of the individuals are tall. It was this numerical relation that Mendel used to establish his model of inheritance.

The Functional Basis of Dominance

The development of knowledge about the molecular activity of genes through the 1950's and 1960's provided information on the nature of the synthesis of proteins using the genetic code passed on in the DNA molecules. This knowledge has allowed researchers to explain variations in phenotype expression and to explain why a dominant allele behaves the way it does at the functional level. An enzyme's function is determined by its structure, and that structure is coded for in the genetic information. The simplest situation is one in which the gene product is an enzyme that acts on a specific chemical reaction that results in a specific chemical product, the phenotype. If that enzyme is not present or if its structure is modified so that it cannot properly perform its function, then the chemical action will not be carried out. The result will be an absence of the normal product and a phenotype expression that varies from the normal expression. For example, melanin is a brown pigment produced by most animals. It is the product of a number of chemical reactions, but one enzyme early in the process is known to be defective in albino animals. Lacking normal enzyme activity, these animals cannot produce melanin, so there is no color in the skin, eyes, or hair. When an animal has the genetic composition *cc* (*c* designates colorless, or albino), it has two alleles that are the same, and neither can produce a copy of the normal enzyme. Animals with the genetic composition *CC* (*C* designates colored, or normal) have two copies of the allele that produces normal enzymes and are therefore pigmented. When homozygous normal (*CC*) and albino (*cc*) animals are crossed, heterozygous (*Cc*) animals are produced. The *c* allele codes for production of an inactive enzyme, while the *C* allele codes for production of the normal, active enzyme. The presence of the normal enzyme promotes pigment production, and the animal displays the pigmented phenotype. The presence of pigment in the heterozygote leads to the designation that the pigmented phenotype is dominant to albinism or, conversely, that albinism is a recessive phenotype because it is seen only in the homozygous (*cc*) state.

The same absence or presence of an active copy of an enzyme explains why blood types A and B are both dominant to blood type O. When an *A* allele or a *B* allele is present, an active enzyme promotes the production of a substance that is identified in a blood test; the blood type A expression or the blood type B expression is seen. When neither of these alleles is present, the individual is homozygous *OO*. There is no detectable product present, and the blood test is negative; therefore, the individual has blood type O. When the *A* allele and the *B* allele are both present in a heterozygous individual, each produces an active enzyme, so both the *A* and the *B* product are detected in blood tests; such an individual has blood type AB. The two phenotypes are both expressed in the heterozygote, a mode of gene expression called codominance.

When there are a number of alleles present for the expression of a characteristic, a dominance relation among the phenotype expressions can be established. In some animal coats, very light colors result from enzymes produced by a specific allele that is capable of producing melanin but at a much less efficient rate than the normal version of the enzyme. In the rabbit, chinchilla (c^{ch}) is such an allele. In the Cc^{ch} heterozygote, the normal allele (C) produces a normal, rapidly acting enzyme, and the animal has normal levels of melanin. The normal pigment phenotype expression is observed because the animals are dark in color, so this expression is dominant to the chinchilla phenotype expression. In the heterozygote $c^{ch}c$, the slow-acting enzyme produced by the c^{ch} allele is present and produces pigment, in a reduced amount, so the chinchilla phenotype expression is observed and is dominant to the albino phenotype expression. The result is a dominance hierarchy in which the normal pigment phenotype is dominant to both the chinchilla and the albino phenotypes, and the chinchilla expression is dominant to the albino expression.

It is important to note that the dominant phenotype is the result of the protein produced by each allele. In the previous examples, both the albino allele and the chinchilla allele produce a product—a version of the encoded enzyme—but the normal allele produces a version of the enzyme that produces more pigment. The relative ability of the enzymes to carry out the function determines the observed phenotype expression and therefore the dominance association. The *C* allele does not inhibit the activity of either of the other two alleles or their enzyme products, and the allele does not,

therefore, show dominance; rather, its enzyme expression does.

Dominant Mutant Alleles

Dominance of a normal phenotype is fairly easy to explain at the level of the functioning protein because the action of the normal product is seen, but dominance of mutant phenotypes is more difficult. Polydactyly, the presence of extra fingers on one hand or extra toes on one foot, is a dominant phenotype. The mechanism that leads to this expression and numerous other developmental abnormalities is not yet understood. One insight comes from the genetic expression of enzymes that are composed of two identical polypeptide subunits. In this situation, the gene locus codes for one polypeptide, but it takes two polypeptide molecules joined together to form a functional enzyme molecule. In order to function normally, both of the polypeptide subunits must be normal. A heterozygote can have one allele coding for a normal polypeptide and the other allele coding for a mutant, nonfunctional polypeptide. These polypeptides will join together at random to form the enzyme. The possible combinations will be defective-defective, which results in a nonfunctional enzyme; defective-normal, which also results in a nonfunctional enzyme; and normal-normal, which is a normal, functional enzyme. The majority of the enzyme molecules will be nonfunctional, and their presence will interfere with the action of the few normal units. The normal function will be, at best, greatly reduced, and the overall phenotype will be abnormal. One form of hereditary blindness is dominant because the presence of abnormal proteins interferes with the transport of both protein types across a membrane to their proper location in the cells that react to light. The abnormal phenotype appears in the heterozygote, so the abnormal phenotype is dominant. A number of human disease conditions, including some forms of cancer, display a dominant mode of inheritance.

Sometimes a trait that appears to be dominant is actually more complex. The Manx trait in cats, which results in a very short, stubby tail, occurs only in heterozygous individuals. On the surface, this would appear to be a simple case of dominance, where the Manx allele, *T*, is dominant to the normal tail allele, *t*. Recall that when two heterozygotes are mated, the expected phenotype ratio in the offspring is 3:1, dominant:recessive. If two Manx cats are mated, the phenotype ratio in the offspring is 2:1, Manx:normal, because kittens that are homozygous for the Manx allele (*TT*) die very early in development and are reabsorbed by the mother cat. Therefore, the Manx allele does not display complete dominance, but rather incomplete dominance. The Manx allele is lethal in the homozygous state and causes a short, stubby tail in the heterozygous state. This occurs because the Manx allele causes a developmental defect that affects spinal development. If one normal allele is present, the spine develops enough for the cat to survive, although it will display the Manx trait. In mutant homozygotes (*TT*) the spine is unable to develop, which proves lethal to the developing fetus.

Impact and Applications

One of the aims of human genetic research is to find cures for inherited conditions. When a condition shows the recessive phenotype expression, treatment may be effective. The individual lacks a normal gene product, so supplying that product can have a beneficial effect. This is the reason for the successful treatment of diabetes using insulin. There are many technical issues to be considered in such treatments, but current successes give hope for the treatment of other recessive genetic conditions.

Dominant disorders, on the other hand, will be much more difficult to treat. An affected heterozygous individual has a normal allele that produces normal gene product. The nature of the interactions between the products results in the defective phenotype. Supplying more normal product may not improve the situation. A great deal more knowledge about the nature of the underlying mechanisms will be needed to make treatment effective.

—*D. B. Benner, updated by Bryan Ness*

See also: Biochemical Mutations; Dihybrid Inheritance; Epistasis; Incomplete Dominance; Mendelian Genetics; Monohybrid Inheritance; Multiple Alleles.

Further Reading

Campbell, Neil. *Biology.* 6th ed. San Francisco: Benjamin Cummings, 2001. A college-level biology text that provides an introduction to many topics relating to genetics.

Russell, Peter J. *Fundamentals of Genetics.* 2d ed. San Francisco: Benjamin Cummings, 2000. Introduces the three main areas of genetics: transmission genetics, molecular genetics, and population and quantitative genetics. Reflects advances in the field, such as the structure of eukaryotic chromosomes, alternative splicing in the production of mRNAs, and molecular screens for the isolation of mutants.

Strachan, Tom, and Andrew Read. *Human Molecular Genetics.* 2d ed. New York: Wiley, 1999. Provides introductory material on DNA and chromosomes and describes principles and applications of cloning and molecular hybridization. Surveys the structure, evolution, and mutational instability of the human genome and human genes, and examines mapping of the human genome, study of genetic diseases, and dissection and manipulation of genes.

Congenital Defects

Field of study: Diseases and syndromes
Significance: *Congenital defects are malformations caused by abnormalities in embryonic or fetal development that may interfere with normal life functions or cause a less severe health problem. The defect may be morphological or biochemical in nature. Understanding the causes of birth defects has led to improved means of detection and treatment.*

Key terms

SENSITIVE PERIOD: a critical time during development when organs are most susceptible to teratogens

TERATOGEN: any agent that is capable of causing an increase in the incidence of birth defects

TERATOLOGY: the science or study of birth defects

Normal Development

In order to understand the causes of birth defects, it is necessary to have some understanding of the stages of normal development. If the time and sequence of development of each organ are not correct, an abnormality may result. It has been useful to divide human pregnancy into three major periods: the preembryonic stage, the embryonic stage, and the fetal stage.

The preembryonic stage is the first two weeks after fertilization. During this stage, the fertilized egg undergoes cell division, passes down the Fallopian tube, and implants in the uterine wall, making a physical connection with the mother. It is of interest to note that perhaps as many as one-half of the fertilized eggs fail to implant, while the half, which do implant, do not survive the second week. The second stage, the embryonic stage, runs from the beginning of the third week through the end of the eighth week. There is tremendous growth and specialization of cells during this period, as all of the body's organs are formed. The embryonic stage is the time during which most birth defects are initiated.

The fetal stage runs from the beginning of the ninth week to birth. Most organs continue their rapid growth and development during this final period of gestation leading up to birth. By the end of the eighth week, the embryo, although it has features of a human being, is only about 1 inch (2.54 centimeters) long. Its growth is amazing during this period, reaching 12 inches (30 centimeters) by the end of the fifth month and somewhere around 20 inches (50 centimeters) by birth. It is evident from the description of normal development that the changes the embryo and fetus undergo are very rapid and complicated. It is not

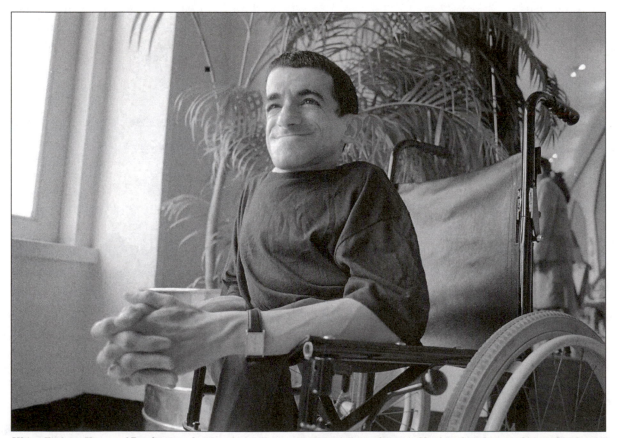

Writer Firdaus Kanga of Bombay was born with the disease osteogenesis imperfecta. A film based on his autobiographical novel Try to Grow *was produced in the mid-1990's.* (AP/Wide World Photos)

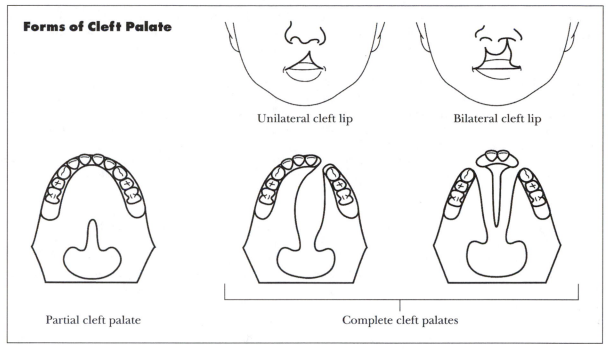

Forms of Cleft Palate

Unilateral cleft lip

Bilateral cleft lip

Partial cleft palate

Complete cleft palates

(Hans & Cassidy, Inc.)

unexpected that mistakes can happen, leading to congenital disorders.

Causes of Birth Defects

Throughout history, examples of birth defects have been described by all cultures and ethnic groups. Although the incidence of specific malformations may vary from group to group, the overall incidence of birth defects is probably similar in all people on earth. It is estimated that three out of every hundred newborns have some sort of major or minor disorder. An additional 2 to 3 percent have malformations that fully develop sometime after birth. When it is also realized that perhaps another 5 percent of all fertilized eggs have severe enough malformations to lead to an early, spontaneous abortion, the overall impact of birth defects is considerable.

Humans have long sought an explanation for why some couples have babies afflicted with serious birth defects. Such children were long regarded as "omens" or warnings of a bad event to come. The word "teratology" (Greek for "monster causing") was coined by scientists to reflect the connection of "monster" births with

warnings. Frequently, ancient people sacrificed such babies. It was thought that such pregnancies resulted from women mating with animals or evil spirits. Maternal impression has long been invoked as an explanation for birth defects, and from early Greek times until even relatively recent times, stories and superstitions abounded.

Of the birth defects in which a specific cause has been identified, it has been found that some are caused by genetic abnormalities, including gene mutations and chromosomal changes, while others are caused by exposure of the pregnant woman and her embryo or fetus to some sort of environmental toxin such as radiation, viruses, drugs, or chemicals.

Examples of Birth Defects

Many birth defects are caused by changes in the number or structure of chromosomes. The best-known chromosomal disorder is Down syndrome, which results from individuals having an extra chromosome 21, giving them forty-seven chromosomes rather than the normal forty-six. A person with Down syndrome characteristically has a flattened face, square-shaped ears, epicanthal folds of the eye, a short neck, poor muscle tone, slow development, and subnormal intelligence. Cystic fibrosis is an example of a defect caused by a single gene. Affected people inherit a recessive gene from each parent. The disorder is physiological in nature and results in a lack of digestive juices and the production of thick and sticky mucus that tends to clog the lungs, pancreas, and liver. Respiratory infections are common, and death typically occurs by the age of thirty. Cleft lip, or cleft palate, is multifactorial in inheritance (some cases are caused by chromosomal abnormalities or by single-gene mutations). Multifactorial traits are caused by many pairs of genes, each having a small effect, and are usually influenced by factors in the environment. The result is that such traits do not follow precise, predictable patterns in a family.

Genetic factors account for the great majority (perhaps 85 to 90 percent) of the birth defects in which there is a known cause. The remaining cases of known cause are attributed to maternal illness; congenital infections; exposure to chemicals, drugs, and medicines; and physical factors such as X rays, carbon dioxide, and low temperature. The "government warning" on liquor bottles informs pregnant women that if they drink alcohol during a sensitive period of prenatal development, they run the risk of having children with fetal alcohol syndrome. There is a wide variation in the effects of alcohol on a developing fetus. Alcohol exposure can lead to an increased frequency of spontaneous abortion, and it depresses growth rates, both before and after birth. Facial features of a child exposed to alcohol may include eye folds, a short nose, small mid-face, a thin upper lip, a flat face, and a small head. These characteristics are likely to be associated with mental retardation. Frequently, however, otherwise normal children have learning disorders and only a mild growth deficiency. Variation in the symptoms of prenatal alcohol exposure has made it difficult to estimate the true incidence of the fetal alcohol syndrome. Estimates for the United States range from 1 to 3 per 1,000 newborns.

In 50 to 60 percent of babies born with a major birth disorder, no specific cause can be identified. Because of this rather large gap in knowledge, nonscientific explanations about the causes of birth defects flourish. What is known is that most congenital defects, whether caused by a genetic factor or an environmental factor, are initiated during the embryonic period. It is also known that some disorders, such as learning disorders, frequently result from damage to the fetus during the last three months of pregnancy. Knowledge about what can be done by parents to avoid toxic exposure and activity that could cause birth defects is critical.

—*Donald J. Nash*

See also: Albinism; Color Blindness; Consanguinity and Genetic Disease; Cystic Fibrosis; Developmental Genetics; Down Syndrome; Dwarfism; Fragile X Syndrome; Hemophilia; Hereditary Diseases; Hermaphrodites; Huntington's Disease; Inborn Errors of Metabolism; Klinefelter Syndrome; Metafemales; Mitochondrial Diseases; Neural Tube Defects; Phenylketonuria (PKU); Prader-Willi and Angelman Syndromes; Prion Diseases: Kuru and Creutzfeldt-Jakob Syndrome; Pseudohermaphrodites; Sickle-Cell Disease; Tay-Sachs Disease; Testicular Feminization Syndrome; Thalidomide and Other Teratogens; Turner Syndrome; XYY Syndrome.

Further Reading

Berul, Charles I., and Jeffrey A. Towbin, eds. *Molecular Genetics of Cardiac Electrophysiology.* Boston: Kluwer Academic, 2000. Reviews research regarding single-cell electrophysiology, animal models, and hereditary diseases, including structural anomalies.

Edwards, Jesse E. *Jesse E. Edwards' Synopsis of Congenital Heart Disease.* Edited by Brooks S. Edwards. Armonk, N.Y.: Futura, 2000. A comprehensive review of more than thirty-five categories of congenital cardiac lesions. Aimed at students, clinicians, and health care providers.

Harvey, Richard P., and Nadia Rosenthal, eds. *Heart Development.* San Diego: Academic Press, 1999. A broad discussion of the molecular basis of cardiovascular development, including the lineage origins and morphogenesis of the developing cardiovascular system, the genetic dissection of cardiovascular development in a variety of model organisms, and the molecular basis of congenital heart defects.

Kramer, Gerri Freid, and Shari Maurer. *The Parent's Guide to Children's Congenital Heart Defects: What They Are, How to Treat Them, How to Cope with Them.* Foreword by Sylvester Stallone and Jennifer Flavin-Stallone. New York: Three Rivers Press, 2001. Collects the expertise of more than thirty leading experts in pediatric cardiology—cardiologists, surgeons, nurses, nutritionists, counselors, and social workers—to give detailed answers to parents' concerns of managing a child's heart defect.

Riccitiello, Robina, and Jerry Adler. "Your Baby Has a Problem." *Newsweek*, Spring/Summer, 1997. Discusses how advances in medicine have reduced the number of birth defects and how surgeries have been designed to correct some birth defects before babies are born.

Rossen, Anne E. "Understanding Congenital Disorders." *Current Health* 18 (May, 1992). A useful article describing some congenital disorders, including some that are not apparent early in life, such as Huntington's disease. Some of the environmental factors causing congenital defects are also covered.

"'Tis the Season." *Psychology Today* 28 (1995). Describes how certain birth defects may be related to different seasons of the year and also discusses some of the biological and environmental conditions that cause birth defects.

Tomanek, Robert J., and Raymond B. Runyan, eds. *Formation of the Heart and Its Regulation.* Foreword by Edward B. Clark. Boston: Birkhauser, 2001. Details the major events in heart development and their control via genes, cell-cell interactions, growth factors, and other contributing elements.

Wynbrandt, James, and Mark D. Ludman. *The Encyclopedia of Genetic Disorders and Birth Defects.* 2d ed. New York: Facts On File, 2000. Written for the general public, a guidebook to clinical and research information on hereditary conditions and birth defects. Includes a general essay on the basics of genetic science and its medical applications.

More than 600 cross-referenced entries are alphabetically arranged and cover genetic anomalies, diagnostic procedures, causes of mutations, and high risk groups.

Web Sites of Interest

March of Dimes Birth Defects Foundation. http://www.marchofdimes.com. Includes fact sheets and links to resources on birth defects.

Medline Plus. http://www.nlm.nih.gov/medlineplus. Medline, sponsored by the National Institudes of Health, is one of the first stops for any medical question, and it offers information and references on most genetic diseases, birth defects, and disorders.

National Birth Defects Network. http://www.nbdpn.org. National support group with information, resources, links.

National Institutes of Health, National Library of Medicine. http://www.nlm.nih.gov/medlineplus/birthdefects.html. Government site that includes dozens of links to resources on birth defects, with information on genetics, treatments, statistics, and more.

Consanguinity and Genetic Disease

Fields of study: Diseases and syndromes; Population genetics

Significance: *The late onset of sexual maturity and the random mating habits of most humans make studying rare mutations in human populations especially difficult. Small, isolated communities in which mates are chosen only from within the population lead to consanguineous populations that can serve as natural laboratories for the study of human genetics, especially in the area of human disease.*

Key terms

ALLELES: genetic variants of a particular gene

CONSANGUINEOUS: literally, "of the same blood," or sharing a common genetic ancestry; members of the same family are consanguineous to varying degrees

ISOLATE: a community in which mates are chosen from within the local population rather than from outside populations

The Importance of Isolates

When studying the genetics of the fruit fly or any other organism commonly used in the laboratory, a researcher can choose the genotypes of the flies that will be mated and can observe the next few generations in a reasonable amount of time. Experimenters can also choose to mate offspring flies with their siblings or with their parents. As one might expect, this is not possible when studying the inheritance of human characteristics. Thus, progress in human genetics most often relies on the observation of the phenotypes of progeny that already exist and matings that have already occurred. Many genetic diseases only appear when a person is homozygous for two recessive alleles; thus a person must inherit the same recessive allele from both parents. Since most recessive alleles are rare in the general population, the chance that both parents carry the same recessive allele is small. This makes the study of these diseases very difficult. The chance that both parents carry the same recessive allele is increased whenever mating occurs between individuals who share some of the same genetic background. These consanguineous matings produce measurably higher numbers of offspring with genetic diseases, especially when the degree of consanguinity is at the level of second cousin or closer.

In small religious communities in which marriage outside the religion is forbidden, and in small, geographically isolated populations in which migration into the population from the outside is at or near zero, marriages often occur between two people who share some common ancestry; therefore, the level of consanguinity can be quite high. These communities thus serve as natural laboratories in which to study genetic diseases. Geographically isolated mountain and island communities are found in many areas of the world, including the Caucasus Mountains of Eurasia, the Appalachian Mountains of North America, and many areas in the South Pacific. Culturally isolated communities are also of worldwide distribution.

Among the Druse, a small Islamic sect, first-cousin marriages approach 50 percent of all marriages. The Amish, Hutterites, and Dunkers in the United States are each descended from small groups of original settlers who immigrated in the eighteenth and nineteenth centuries and rarely mated with people from outside their religions.

The Amish

There are many reasons the Amish serve as a good example of an isolate. The original immigration of Amish to America consisted of approximately two hundred settlers. In subsequent generations, the available mates came from the descendants of the original settlers. With mate choice this limited, it is inevitable that some of the marriages will be consanguineous. Consanguinity increases as further marriages take place between the offspring of consanguineous marriages. Current estimates are that the average degree of consanguinity of Amish marriages in Lancaster County, Pennsylvania, is at the level of marriages between second cousins.

Other factors that make the Amish good subjects for genetic research are their high fertility and their high level of marital fidelity. Thus, if both parents happen to be heterozygous for a particular genetic disease, the chance that at least one of the offspring will show the disease is high. In families of two children, there is a 44 percent chance that at least one child will show the trait. This increases to 70 percent of the families with four children and to more than 91 percent of the families with eight children, a common number among the Amish. Because of the high marital fidelity, researchers do not have to worry about illegitimacy when making these estimates.

Many genetic diseases that are nearly nonexistent in the general population are found among the Amish. The allele for a type of dwarfism known as the Ellis-van Creveld syndrome is found in less than 0.1 percent of the general population; among the Lancaster Amish, however, the allele exists in approximately 7 percent of the population. Other genetic diseases at higher levels among the Amish include cystic fibrosis, limb-girdle muscular dystrophy, pyruvate kinase-deficient hemolytic anemia, and several inherited psychological disorders. Having more families and individuals with these diseases to study helps geneticists and physicians discover ways to treat the problems and even prevent them from occurring.

—*Richard W. Cheney, Jr.*

See also: Cystic Fibrosis; Dwarfism; Genetic Load; Hardy-Weinberg Law; Hereditary Diseases; Heredity and Environment; Inbreeding and Assortative Mating; Lateral Gene Transfer; Mendelian Genetics; Natural Selection; Polyploidy; Population Genetics; Punctuated Equilibrium; Quantitative Inheritance; Sociobiology; Tay-Sachs Disease.

Further Reading

Cross, Harold. "Population Studies of the Old Order Amish." *Nature* 262 (July, 1976). Describes the advantages of isolates and some of the genetic characteristics seen in Amish populations.

McKusick, Victor, et al. "Medical Genetic Studies of the Amish with Comparison to Other Populations." In *Population Structure and Genetic Disorders*, edited by A. W. Eriksson et al. New York: Academic Press, 1981. Describes many of the inherited conditions seen among the Amish.

Criminality

Field of study: Human genetics and social issues

Significance: *The pursuit of genetic links to criminality is a controversial field of study that has produced several intriguing examples of the apparent contribution of genetic defects to criminal behavior. However, most of these defects involve major alterations in metabolic pathways that, in turn, affect numerous characteristics. Experts disagree on the validity and significance of these data. These research efforts have also come under strong criticism by opponents who fear that such discoveries will be used to charge that certain ethnic or racial groups are genetically predisposed to deviant behaviors such as criminality.*

Key terms

METABOLIC PATHWAY: a biochemical process that converts certain chemicals in the body to other, often more useful, chemicals with the help of proteins called enzymes

NEUROTRANSMITTER: a chemical that carries messages between nerve cells

Biochemical Abnormalities

Early attempts to identify the root of the tendency for criminal behavior fell under the auspices of biological determinism, which sought to explain and justify human society as a reflection of inborn human traits. For example, Italian physician Cesare Lombroso reported in *L'uomo delinquente* (1876; criminal man) that certain "inferior" groups, by virtue of their apish appearance, were evolutionary throwbacks with criminal tendencies. Since that time, more sophisticated scientific methods have been employed to seek the "root causes" of criminality. Among the most prominent findings are those that indicate that certain biochemical imbalances, particularly in neurotransmitters, may lead to a range of "abnormal" behaviors. For example, levels of the neurotransmitter serotonin have been found to be low in many people who have attempted suicide and in people with poor impulse control, such as children who torture animals and impulsive arsonists. However, the environment itself may lower or raise serotonin levels, calling into question the importance of genetic influence. The psychological effects of serotonin are also far-reaching, with antidepressant drugs such as Prozac functioning by increasing the amount of time serotonin remains in the system after its release (latency).

Abnormalities in dopamine levels (the primary neurotransmitter in the brain's "pleasure center") have also been implicated in aggressive, antisocial behavior. In 1995, researchers reported that increased latency of dopamine might be associated with a tendency toward violence among alcoholics. A genetic abnormality on the X chromosome that causes a defect in the enzyme monoamine oxidase A was reported by researchers in 1993. This enzyme is responsible for degrading certain neurotransmitters, including dopamine and epinephrine. This defect was linked to a heritable history of low intelligence quotient (IQ) and violent acts in one Dutch family. Males who possess an extra Y chromosome (XYY syndrome males) also demonstrate a variety of behavioral difficulties and are overrepresented in prisons and mental institutions. However, no link to criminal behavior has been established.

In all cases, such genetic abnormalities affect numerous characteristics (often including mental capabilities) and manifest themselves as any number of unassociated antisocial behaviors ranging from exhibitionism to arson. Since criminality simply refers to a violation of the law and since there are numerous types of crimes and motivations for them (such as anger, revenge, and financial gain), it is difficult to make claims of definitive, nonenvironmental links between biochemical disorders and criminal behavior. Poorly defined, multifaceted social descriptors (for example, violence, aggression, and intelligence) are usually used to represent such behaviors and, as such, cannot be considered true "characters." As child psychiatrist Michael Rutter has said, to claim that there is a gene for crime is "like saying there's a gene for Roman Catholicism."

Impact and Applications

Genetic links to criminality entered the public spotlight in the early 1990's as part of the U.S. government's Violence Initiative, championed by Secretary of Health and Human Services Louis Sullivan. The uproar began in 1992 when Frederick Goodwin, then director of the Alcohol, Drug Abuse, and Mental Health Administration, made comments comparing urban youth to aggressive jungle primates. The public feared that research on genetic links to criminality would be used to justify the disproportionate numbers of African Americans and Hispanics in the penal system. Psychiatrist Peter Breggin also warned that unproved genetic links would be used as an excuse to screen minority children and give them sedating drugs to intervene in their impending aggression and criminality. After all, forced sterilization laws had been enacted in thirty U.S. states in the 1920's to prevent reproduction by the "feebleminded" and "moral degenerate." In 1993, public protest led to the temporary cancella-

tion of Genetic Factors in Crime, a federally funded conference organized by philosopher David Wasserman. A similar symposium, Genetics of Criminal and Antisocial Behaviour, was held in London in 1995. However, the public remains highly suspicious of the motivation for such research.

In an era in which genes have been implicated in everything from manic depression to the propensity to change jobs, the belief that genes are responsible for criminal behavior is very enticing. However, such a belief may have severe ramifications. To the extent that society accepts the view that crime is the result of pathological and biologically deviant behavior, it is possible to ignore the necessity to change social conditions such as poverty and oppression that are also linked to criminal behavior. Moreover, this view may promote the claim by criminals themselves that their "genes" made them do it. While biochemical diagnosis and treatment with drugs may be simpler and therefore more appealing than social intervention, it is reminiscent of the days when frontal lobotomy (surgery of the brain) was the preferred method of biological intervention for aggressive mental patients. In the future, pharmacological solutions to social problems may be viewed as similarly inhumane.

—Lee Anne Martínez

See also: Aggression; Alcoholism; Altruism; Behavior; Biological Determinism; Developmental Genetics; DNA Fingerprinting; Eugenics; Eugenics: Nazi Germany; Forensic Genetics; Hardy-Weinberg Law; Heredity and Environment; RFLP Analysis; Sociobiology; Sterilization Laws; XYY Syndrome.

Further Reading

Andreasen, Nancy C. *Brave New Brain: Conquering Mental Illness in the Era of the Genome.* New York: Oxford University Press, 2001. Surveys the way in which advances in the understanding of the human brain and the human genome are coming together in an ambitious effort to conquer mental illness.

Faraone, Stephen V., Ming T. Tsuang, and Debby W. Tsuang. *Genetics of Mental Disorders: A Guide for Students, Clinicians, and Researchers.* New York: Guilford Press, 1999. Introduces the investigative methods of human genetics as applied to mental disorders, their clinical applications, and some of the biological, ethical, and legal implications of the investigative processes and conclusions.

Fukuyama, Francis. "Is It All in the Genes?" *Commentary* 104 (September, 1997). Compares the philosophies of cultural versus biological effects on behavior.

Gilbert, Paul, and Kent G. Bailey Hove, eds. *Genes on the Couch: Explorations in Evolutionary Psychotherapy.* Philadelphia: Brunner-Routledge, 2000. Examines models and interventions in psychotherapy based on evolutionary findings and includes topics such as psychotherapy in the context of Darwinian psychiatry, Jungian analysis, gender, and the syndrome of rejection sensitivity.

Rose, Steven. "The Rise of Neurogenetic Determinism." *Nature* 373 (February, 1995). Comments on how technological advances have revived genetic explanations for behavior.

Sapolsky, Robert. "A Gene for Nothing." *Discover* 18 (October, 1997). An entertaining account of the complex interaction between genes and the environment.

Wasserman, David, and Robert Wachbroit, eds. *Genetics and Criminal Behavior.* New York: Cambridge University Press, 2001. Explores issues surrounding causation and responsibility in the debate over genetic research into criminal behavior. Chapters include "Understanding the Genetics of Violence Controversy," "Separating Nature and Nurture," "Genetic Explanations of Behavior," "On the Explanatory Limits of Behavioral Genetics," "Degeneracy, Criminal Behavior, and Looping," "Genetic Plans, Genetic Differences, and Violence," "Crime, Genes, and Responsibility," "Genes, Statistics, and Desert," "Genes, Electrotransmitters, and Free Will," "Moral Responsibility Without Free Will," "Strong Genetic Influence and the New 'Optimism,'" and "Genetic Predispositions to Violent and Antisocial Behavior."

Williams, Juan. "Violence, Genes, and Prejudice." *Discover* 15 (November, 1994). Gives an excellent account of the controversy and debate that accompanied the U.S. government's funding of research on genetic links to violence and crime.

Wright, Robert. "The Biology of Violence." *The New Yorker* 71 (March 15, 1995). Discusses evolutionary psychology's view that violent responses to oppressive environments may be adaptive rather than genetically inflexible.

Web Sites of Interest

Bioethics and Genetics at Bioethics.net. http://bioethics.net/genetics. Site links to an article exploring the debate about genetics and criminal behavior, entitled "Not Guilty, by Reason of Genetic Determinism," which includes a bibliography for further study.

Cystic Fibrosis

Field of study: Diseases and syndromes
Significance: *Cystic fibrosis, although a rare disease, is the most common lethal inherited disease among Caucasians in the United States and the United Kingdom. Advances in genetic screening and treatment may someday result in the prevention or elimination of this disease.*

Key terms

EPITHELIAL CELLS: cells responsible for transporting salt and water
RECESSIVE GENE: a gene that in diploid organisms gets expressed only when it represents both copies at a gene locus

Causes and Symptoms

Cystic fibrosis is caused by an abnormal recessive gene that must be inherited from both parents. If both parents carry the gene, their child has a 25 percent chance of inheriting the abnormal gene from both parents and thus having the disease. The child has a 50 percent chance of having one normal and one abnormal gene, thus becoming a carrier of the disease.

Cystic fibrosis is chronic and has no known cure. Generally, symptoms are apparent shortly after birth and become progressively more serious. Abnormally thick mucus blocks the ducts

At the National Heart, Lung, and Blood Institute in Bethesda, Maryland, Dr. Ronald Crystal works on a nasal spray of proteins that have been genetically engineered to break down mucus in the lungs of individuals with cystic fibrosis. (AP/Wide World Photos)

in sweat glands and glands in the lungs and pancreas. Diagnosis usually involves a simple test that measures excessive sodium and chloride (salt) in the person's sweat. Ten percent of newborns with cystic fibrosis cannot excrete undigested material in their intestines and so develop a blocked intestine. Symptoms of this condition include a swollen abdomen, vomiting, constipation, and dehydration. The child may become severely undernourished because the digestive system is not functioning properly. In turn, the malnutrition causes poor weight gain, impaired blood clotting, slow bone growth, and poor overall growth. Digestive problems usually increase as the child ages.

Although cystic fibrosis can damage other organs, the most serious complications from the disease involve the lungs. Respiratory problems develop as the affected person ages. The lungs may appear normal in early life but will later malfunction. Normal lungs fight off infection by secreting mucin, the primary component of mucus. The mucin helps trap germs and foreign particles in the lungs. The mucus is swept toward the throat by hairlike projec-

tions known as cilia and is then expelled from the respiratory tract. Lungs also produce a natural antibiotic called defensin, which destroys germs.

For people with cystic fibrosis, the defense mechanisms are crippled. The lungs produce mucin that is too thick and sticky to be flushed away by the cilia. The mucus stays in the lungs and provides an ideal breeding ground for bacteria. The person suffers from repeated respiratory infections such as bronchitis and pneumonia. In addition, cystic fibrosis prevents the body from properly absorbing salt into the epithelial cells. The lungs cannot absorb the salt, causing a buildup of salt outside the cells. This buildup disables the natural antibiotic defensin, which in turn leads to bacterial infections that cause the increased production of mucus. The damage from infections and the buildup of mucus in the airways eventually makes breathing impossible.

Treatment and Outlook

Approximately thirty thousand Americans have cystic fibrosis, and one in twenty-five Caucasians are carriers of the disease. In the 1950's, an infant with cystic fibrosis seldom survived past the first year or two. With aggressive medical treatment and therapy to relieve the severe symptoms, the average life span of a person with cystic fibrosis had increased to twenty-nine years by the early 1990's, and a person with the disorder may live as long as forty years.

Treatment depends on the organs involved. People with cystic fibrosis follow special diets and take dietary supplements so that they have proper nutrition and receive the necessary digestive enzymes, salts, and vitamins that cannot pass through the blocked ducts. The disease makes the digestion and absorption of fats and proteins difficult because certain enzymes are depleted as a result of blocked ducts in the pancreas.

The affected person also undergoes daily backslapping designed to break up the mucus in the lungs. Antibiotics can reduce infections of the lungs. In the 1990's, therapy involving the inhalation of a special enzyme began. This enzyme helps break down the thick mucus so that it is easier to cough out. In late 1997, a potent antibiotic was recommended to the Food and Drug Administration. This antibiotic, called Tobi (tobramycin for inhalation), was the first inhaled antibiotic and appeared to increase the lung function of some cystic fibrosis patients.

Research continues in an effort to determine the nature of the genetic defect that causes cystic fibrosis so that the normal function of the defective gene can be replaced by specially designed therapies. Recent and ongoing attempts at gene therapy have not yet been successful. A corrected form of the gene has been engineered and recombinant adenoviruses have been used to try to deliver the corrected gene to epithelial cells in the lungs. The recombinant viruses can easily be inhaled by atomizing a solution containing the viruses. Unfortunately, patients in clinical trials have often developed immune responses to the adenoviruses and the transfer efficiency of the adenoviruses was too low to cause much lasting im-

Gene Therapy for Cystic Fibrosis

Once scientists discovered the cystic fibrosis gene, *CF*, and its protein product, cystic fibrosis transmembrane regulator (CFTR), attempts at gene therapy were quickly initiated. Since most of the life-threatening complications of this disease are seen in the respiratory system, that system became the main target for gene replacement therapy.

Early attempts at gene therapy involved the attachment of a functional *CF* gene to a virus that acts as a vector and the subsequent introduction of this virus to the respiratory epithelium in an aerosol. Several problems arose. Although a cystic fibrosis patient's immune system does not function properly, especially in the respiratory system, the immune system is active enough to prevent many of the viruses from entering the target cells. Those that did penetrate and inserted the normal *CF* gene induced only a transient benefit. This most likely occurred because of the high turnover rate of surface epithelial cells. The epithelial cells could incorporate the gene that codes for normal CFTR, but cells that had not been repaired would soon replace the repaired cells. Continued aerosol applications were also not helpful, because the body began producing antibodies to the viral vector, which further reduced the virus's ability to enter cells and introduce an active *CF* gene. Another problem was the inflammation caused by the virus itself.

To surmount some of these difficulties, other approaches have been tried. A team of Australian researchers has looked at preconditioning the respiratory epithelium with a detergent-like substance found in normal lungs as a way of increasing virus uptake by the epithelium. This system has had success in mice and has led to longer-term improvements of lung function. These researchers speculate that long-tem improvement occurs when some epithelial stem cells have had defective DNA replaced by the DNA for functional CFTR. In Cleveland, researchers have tried to insert the *CF* gene directly into cells without a viral vector. They have accomplished this by compacting the DNA into a particle small enough to enter the cell.

Another novel gene therapy has been labeled SMaRT by its proponents. This therapy takes advantage of the need to remove introns (noncoding intervening sequences) from pre-messenger RNA (pre-mRNA) in eukaryotes and then to splice the exons (coding sequences) together to form functional mRNA. In this procedure, multiple copies of a "minigene" that contain a good copy of the exon that normally contains the defect in the *CF* gene are introduced to the epithelial cells. When the pre-mRNA is processed, there are so many more copies of the corrected exon that it is usually spliced into the CFTR mRNA. This technique has the advantage of not disrupting the cells' normal regulation of the CFTR protein. However, the viruses involved in the transfer of the minigenes face the same barriers that all viral vectors face in cystic fibrosis gene therapy.

—*Richard W. Cheney, Jr.*

provement. Researchers still hope to improve gene therapy methods and eventually find ways either to cure cystic fibrosis fully or at least to reduce some of the more serious symptoms.

Genetic testing for cystic fibrosis can locate the defect responsible for the disease in about 75 percent of afflicted people. Since cystic fibrosis can cause more than one hundred other genetic mutations, however, a simple test to detect the variations may be very difficult to develop. In addition, the symptoms of the disease can vary from severe to extremely mild. Research focuses on the development of inexpensive and accurate diagnosis as well as sound genetic counseling in order to reduce the occurrence of the disease. In early 1997, the National Institutes of Health (NIH) recommended that all couples planning to have children should be offered the option of testing for the cystic fibrosis gene mutations.

—*Virginia L. Salmon, updated by Bryan Ness*

See also: Amniocentesis and Chorionic Villus Sampling; Biochemical Mutations; Chromosome Mutation; Chromosome Walking and Jumping; Congenital Defects; Gene Therapy; Gene Therapy: Ethical and Economic Issues; Genetic Counseling; Genetic Engineering; Genetic Screening; Genetic Testing: Ethical and Economic Issues; Hereditary Diseases; Human Genetics; Inborn Errors of Metabolism; Multiple Alleles.

Further Reading

Hodson, Margaret E., and Duncan M. Geddes, eds. *Cystic Fibrosis.* 2d ed. New York: Oxford University Press, 2000. Focuses on the diagnosis and management of cystic fibrosis and its complications, also giving an extensive review of cystic fibrosis, including chapters on both clinical and basic science.

Oliwenstien, Lori. "How Salt Can Kill." *Discover* 18 (January, 1997). Discusses the link between salt and the lungs.

Orenstein, David M., Beryl J. Rosenstein, and Robert C. Stern. *Cystic Fibrosis: Medical Care.* Philadelphia: Lippincott Williams & Wilkins, 2000. Examines the principles and practices of cystic fibrosis care, providing chapters on the molecular and cellular bases of the disease and its diagnosis, the major organ systems affected, transplantation (lung and liver), hospitalization, terminal care, special populations, and laboratory testing.

Shale, Dennis. *Cystic Fibrosis.* London: British Medical Association, 2002. Topics include clinical management, genetic disorders and developments, lung injury, transplantation, and nutrition.

Sternberg, S. "Cystic Fibrosis Puzzle Coming Together." *Science News* 151 (February 8, 1997). An informative overview of the lung symptoms.

Travis, John. "Cystic Fibrosis Controversy." *Science News* 151 (May 10, 1997). Discussion of a controversial treatment for cystic fibrosis.

Yankaskas, James R., and Michael R. Knowles, eds. *Cystic Fibrosis in Adults.* Philadelphia: Lippincott-Raven, 1999. Advances in medicine have meant longer survival for cystic fibrosis patients and thus a gradual evolution of the clinical manifestations of the disorder. Details the typical pulmonary and gastrointestinal/nutritional problems of patients, as well as the challenges of adult patients, like diabetes mellitus, osteoporosis, and pregnancy, that were once relatively uncommon.

Web Sites of Interest

Cystic Fibrosis Foundation. http://www.cff.org. This national organization's site includes information on the genetics of cystic fibrosis.

Dolan DNA Learning Center, Your Genes Your Health. http://www.ygyh.org. Sponsored by the Cold Spring Harbor Laboratory, this site, a component of the DNA Interactive Web site, offers information on more than a dozen inherited diseases and syndromes, including cystic fibrosis.

Cytokinesis

Field of study: Cellular biology

Significance: *Cytokinesis is a process, usually occurring concurrent with mitosis, in which the cytoplasm and organelles are divided into two new cells. In eukaryotes, mitosis and meiosis involve division of the nucleus, while cytokinesis is the division of the cytoplasm.*

Key terms

BINARY FISSION: cell division in prokaryotes in which the plasma membrane and cell wall grow inward and divide the cell in two

CELL CYCLE: a regular and repeated sequence of events during the life of a cell; it ends when a cell completes dividing

DAUGHTER CELLS: cells that result from cell division

INTERPHASE: the phase that precedes mitosis in the cell cycle, a period of intense cellular activities that include DNA replication

MEIOSIS: a type of cell division that leads to production of gametes (sperm and egg) during sexual reproduction

MITOSIS: nuclear division, a process of allotting a complete set of chromosomes to two daughter nuclei

Events Leading to Cytokinesis

Cytokinesis is the division or partitioning of the cytoplasm during the equal division of genetic material into the daughter cells. Before a cell can divide, its genetic material, DNA, has to be duplicated through DNA replication. The identical copies of DNA are then separated into one of the two daughter cells through a multistep process, which varies among prokaryotes, plants, and animals. With a single chromosome and no nucleus, prokaryotes (such as bacteria) utilize a simple method of cell division called binary fission (meaning "splitting in two"). The single circular DNA molecule is replicated rapidly and split into two. Each of the two circular DNAs then migrates to the opposite pole of bacterial cell. Eventually, one bacterial cell splits into two through binary fission. On average, a bacterial cell can go through the whole process of cell division within twenty minutes.

In eukaryotes, cell division is a more complex process given the presence of a nucleus and multiple DNA molecules (chromosomes). Each chromosome needs to be replicated in preparation for the division. The replication process is completed during the interphase. Once replicated, the copies of each chromosome, called sister chromatids, are connected together in a region called the centromere. The chromosomes then go through a process of shortening, condensing, and packing with proteins that make them visible using a light microscope. Chromosomes then migrate and line up at the equator of the parent cell. Then the sister chromatids are separated and pulled to opposite poles. These multiple steps include interphase (cell growth and DNA replication), prophase (disintegration of nuclear envelope, formation of spindle fibers, condensation of chromosomes), metaphase (lining up of chromosomes at equator plate), anaphase (split of two sister chromatids), and telophase (completion of migration of chromatids to opposite poles). Although animal and plant cells share many common features in DNA replication and mitosis, some noticeable differences in interphase and cytokinesis exist. Even within the animal kingdom, cytokinesis may vary with the type of cell division. Particularly during oogenesis (the process of forming egg), both meiosis I and meiosis II engage in unequal partitioning of cytoplasm that is distinct from normal mitosis of animal and plant cells. In some cases, a cell will complete mitosis without cytokinesis, resulting in a multinucleate cell.

Cytokinesis in Animals

In animal cells, cytokinesis normally begins during anaphase or telophase and is completed following the completion of chromosome segregation. First, microfilaments attached to the plasma membrane and form a ring around the equator of the cell. This ring then contracts and constricts the cell's equator, forming a cleavage furrow, much like pulling the drawstring around the waist of a pair of sweatpants. Eventually the "waist" is pinched through and contracts down to nothing, partitioning the cytoplasm equally into two daughter cells. Partitioning the cytoplasm includes distributing cellular organelles so each daughter cell has what is needed for cellular processes.

Cytokinesis in Plants

Cytokinesis in plant cells is different from that in animal cells. The presence of a tough cell wall (made up of cellulose and other materials) makes it nearly impossible to divide plant cells in the same manner as animals cells. Instead, it begins with formation of a cell plate. In early telophase, an initially barrel-shaped system of microtubules called a phragmoplast forms between the two daughter nuclei. The cell plate is then initiated as a disk suspended in the phragmoplast.

The cell plate is formed by fusion of secretory vesicles derived from the Golgi apparatus. Apparently, the carbohydrate-filled vesicles are directed to the division plane by the phragmoplast microtubules, possibly with the help of motor proteins. The vesicles contain matrix molecules, hemicelluloses, and/or pectins. As the vesicles fuse, their membranes contribute to the formation of the plasma membrane on either side of the cell plate. When enough vesicles have fused, the edges of the cell plate merge with the original plasma membrane around the circumference of the cell, completing the separation of the two daughter cells. In between the two plasma membranes is the middle lamella. Each of the two daughter cells then deposits a primary wall next to the middle lamella and a new layer of primary wall around the entire protoplast. This new wall is continuous with the wall at the cell plate. The original wall of the parent cell stretches and ruptures as the daughter cells grow and expand.

Cytokinesis in Sexual Reproduction

In animal oogenesis, the formation of ova, or eggs, occurs in the ovaries. Although the daughter cells resulting from the two meiotic divisions receive equal amounts of genetic material, they do not receive equal amounts of cytoplasm. Instead, during each division, almost all the cytoplasm is concentrated in one of the two daughter cells. In meiosis I, unequal partitioning of cytoplasm during cytokinesis produces the first polar body almost void of cytoplasm, and the secondary oocyte with almost all cytoplasm from the mother cell. During meiosis II, cytokinesis again partitions almost all cytoplasm to one of the two daughter cells, which

will eventually grow and differentiate into a mature ovum, or egg. Another daughter cell, the secondary polar body, receives almost no cytoplasm. This concentration of cytoplasm is necessary for the success of sexual reproduction because a major function of the mature ovum is to nourish the developing embryo following fertilization.

—Ming Y. Zheng

See also: Cell Cycle, The; Cell Division; Mitosis and Meiosis; Polyploidy; Totipotency.

Further Reading

Grant, M. C. "The Trembling Giant." *Discover,* October, 1993. Excellent illustrations on asexual reproduction (by reference to the aspen tree) through mitosis of plant cells and tissues.

Murray, A. W., and Tim Hunt. *The Cell Cycle: An Introduction.* New York: W. H. Freeman, 1993. An informative overview for both students and general readers, without too much scientific jargon. Bibliographical references, index.

Murray, A. W., and M. W. Kirschner. "What Controls the Cell Cycle." *Scientific American,* March, 1991. An illuminating description of a group of proteins that are involved in cell cycle control. The synthesis, processing, and degradation of these proteins seems to regulate the progression of a cell through various stages of the cell cycle.

Shaul, Orit, Marc van Montagu, and Dirk Inze. "Regulation of Cell Divisions in *Arabidopsis.*" *Critical Reviews in Plant Sciences* 15 (1996): 97-112. A review of what is known about plant cell cycle regulation and cell divisions. For serious students.

Staiger, Chris, and John Doonan. "Cell Divisions in Plants." *Current Opinion in Cell Biology* 5 (1993): 226-231. A condensed version on plant cell divisions. Provides a quick overview.

Developmental Genetics

Field of study: Developmental genetics
Significance: *The discovery of the genes responsible for the conversion of a single egg cell into a fully formed organism has greatly increased our understanding of development. Common developmental mechanisms exist for diverse organisms and experimental manipulation of particular genes could potentially lead to treatments or cures for cancers and developmental abnormalities in humans.*

Key terms

DIFFERENTIATION: the process by which a cell changes its phenotype, or outward appearance, and becomes different from its parent cell, usually by altering its gene expression

EPIGENESIS: the formation of differentiated cell types and specialized organs from a single, homogeneous fertilized egg cell without any preexisting structural elements

GENE EXPRESSION: the combined biochemical processes, called transcription and translation, that convert the linearly encoded information in the bases of DNA into the three-dimensional structures of proteins

INDUCTION: an easily observed event in which a cell or group of cells signals an adjacent cell to pursue a different developmental pathway and so become differentiated from its neighboring cells

MORPHOGEN: a chemical compound or protein made by cells in an egg that creates a concentration gradient affecting the developmental fate of surrounding cells by altering their gene expression or their ability to respond to other morphogens

Early Hypotheses of Development in Diverse Organisms

From the earliest times, people noted that a particular organism produced offspring very much like itself in structure and function, and the fully formed adult consisted of numerous cell types and other highly specialized organs and structures, yet it came from one simple egg cell. How could such simplicity, observed in the egg cell, give rise to such complexity in the adult and always reproduce the same structures?

In the seventeenth century, the "preformationism" hypothesis was advanced to answer these questions by asserting that a miniature organism existed in the sperm or eggs. After fertilization, this miniature creature simply grew into the fully formed adult. Some microscopists of the time claimed to see a "homunculus," or little man, inside each sperm cell. That the preformationism hypothesis was ill-conceived became apparent when others noted that developmental abnormalities could not be explained satisfactorily, and it became clear that another, more explanatory hypothesis was needed to account for these inconsistencies.

In 1767, Kaspar Friedrich Wolff published his "epigenesis" hypothesis, in which he stated that the complex structures of chickens developed from initially homogeneous, structureless areas of the embryo. Many questions remained before this new hypothesis could be validated, and it became clear that the chick embryo was not the best experimental system for answering them. Other investigators focused their efforts on the sea squirt, a simpler organism with fewer differentiated tissues.

Work with the sea squirt, a tiny sessile marine animal often seen stuck to submerged rocks, led to the notion that development followed a mosaic pattern. The key property of mosaic development was that any cell of the early embryo, once removed from its surroundings, grew only into the structure for which it was destined or determined. Thus the early embryo consisted of a mosaic of cell types, each determined to become a particular body part. The determinants for each embryonic cell were found in the cell's cytoplasm, the membrane-bound fluid surrounding the nucleus. Other scientists, most notably Hans Driesch in 1892 and Theodor Boveri (working with sea urchin embryos) in 1907, noted that a two-cell-stage embryo could be teased apart into separate cells, each of which grew into a fully formed sea urchin. These results appeared to disagree with the mosaic developmental mechanism. Working from an earlier theory, the "germ-plasm" theory of August Weismann (1883), Driesch and Boveri proposed a new mechanism called regulative development.

The key property of regulative development was that any cell separated from its embryo could regulate its own development into a complete organism. In contrast to mosaic development, the determinants for regulative development were found in the nuclei of embryonic cells, and Boveri hypothesized that gradients of these determinants, or morphogens, controlled the expression of certain genes. Chromosomes were assumed to play a major role in controlling development; however, how they accomplished this was not known, and Weismann mistakenly implied that genes were lost from differentiated cells as more and more specific structures formed.

In spite of the inconsistencies among the several hypotheses, a grand synthesis was soon formed. Working with roundworm, mollusc, sea urchin, and frog embryos, investigators realized that both mosaic and regulative mechanisms operate during development, with some organisms favoring one mechanism over the other. The most important conclusion coming from these early experiments suggested that certain genes on the chromosomes interacted with both the cytoplasmic and nuclear morphogenetic determinants to control the proliferation and differentiation of embryonic cells. What exactly were these morphogens, where did they originate, and how did they form gradients in the embryo? How did they interact with genes?

The Morphology of Development

Before the "how and why" mechanistic questions of morphogens could be answered, more answers to the "what happens when" questions were needed. Using new, powerful microscopes in conjunction with cell-specific stains, many biologists were able to precisely map the movements of cells during embryogenesis and to create "fate maps" of such cell migrations. Fate maps were constructed for sea squirt, roundworm, mollusc, sea urchin, and frog embryos, which showed that specific, undifferentiated cells in the early embryo gave rise to complex body structures in the adult.

In addition, biologists observed an entire stepwise progression of intervening cell types and structures that could be grouped into various stages and that were more or less consistent from one organism to another. Soon after fertilization, during the very start of embryogenesis, specific zones with defining, yet structureless, characteristics were observed. These zones consisted of gradients of different biochemical compounds, some of which were morphogens, and they seemed to function by an induction process. Some of these morphogen gradients existed in the egg before fertilization; thus it became evident that the egg was not an entirely amorphous, homogeneous cell but one with some amount of preformation. This preformation took the form of specific morphogen gradients.

After these early embryonic events and more cell divisions, in which loosely structured patterns of morphogen gradients were established to form the embryo's polar axes, the cells aggregated into a structure called a "blastula," a hollow sphere of cells. The next stage involved the migration of cells from the surface of the blastula to its interior, a process called gastrulation. This stage is important because it forms three tissue types: the ectoderm (for skin and nerves), the mesoderm (for muscle and heart), and the endoderm (for other internal organs). Continued morphogenesis generates a "neurula," an embryo with a developing nervous system and backbone. During axis formation and cell migrations, the embryonic cells are continually dividing to form more cells that are undergoing differentiation into specialized tissue types such as skin or muscle. Eventually, processes referred to as "organogenesis" transform a highly differentiated embryo into one with distinct body structures that will grow into a fully formed adult.

Experimental Systems for Studying Developmental Genes

In order to understand the details of development, biologists normally study organisms with the simplest developmental program, ones with the fewest differentiated cell types that will still allow them to answer fundamental questions about the underlying processes. Sea squirts and roundworms have been valuable, but they exhibit a predominantly mosaic form of development and are not the best systems for studying morphogen-dependent induction.

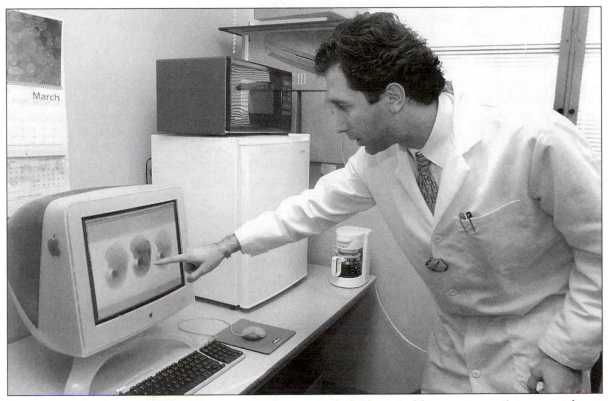

Bahri Karacay, a postdoctoral fellow at the University of Iowa, studies the development of the nervous system in a mouse embryo as part of a project that seeks to treat brain tumors in humans. (AP/Wide World Photos)

Frog embryogenesis, with both mosaic and regulative processes, was well described and contributed greatly to answering the "what and when" questions of sequential events, but no effective genetic system existed for examining the role of genes in differentiation necessary for answering the "why and how" questions.

The issue was finally resolved by focusing once again on the morphogens. These mediators of cellular differentiation were found only in trace amounts in developing embryos and thus were difficult if not impossible to isolate in pure form for experimental investigation. An alternative to direct isolation of morphogens was to isolate the genes that make the morphogens. The organism deemed most suitable for such an approach was the fruit fly *Drosophila melanogaster*, even though its development was more complex than that of the roundworm. Fruit flies could be easily grown in large numbers in the laboratory, and many mutants could be generated quickly; most important, an effec-

tive genetic system already existed in *Drosophila*, making it easier to create and analyze mutants. The person who best used the fruit fly system and greatly contributed to the understanding of developmental genetics was Christiane Nüsslein-Volhard, who shared a 1995 Nobel Prize in Physiology or Medicine with Edward B. Lewis and Eric Wieschaus.

The Genes of Development

The first important developmental genes discovered in *Drosophila* were the latest acting in morphogenesis, which led to the isolation of the gene for one of the morphogens controlling the anterior-posterior axis of the embryo, the *bicoid* gene. The study of mutants, such as those with legs in place of antennae, allowed the discovery of many other developmental genes, referred to generally as "homeotic" genes.

The *bicoid* gene's discovery validated the gradient hypothesis originally proposed by Boveri because its gene product functioned as a "typical" morphogen. It was a protein that existed in the highest concentration at the egg's anterior pole and diffused to lower concentrations toward the posterior pole, thus forming a gradient. Through the use of more fruit fly mutants, geneticists showed that the BICOID protein stimulated the gene expression of another early gene, called *hunchback*, which in turn affected the expression of other genes: *Krüppel* and *knirps*. The BICOID protein controls the *hunchback* gene by binding to the gene's control region.

Since these initial discoveries, a plethora of new developmental genes have been discovered. It is now clear that some fifty genes are involved in development of a fruit fly larva from an egg, with yet more genes responsible for development of the larva into an adult fly. These genes are grouped into three major categories: maternal effect genes, segmentation genes, and homeotic genes. Maternal effect genes include the *bicoid* gene. These genes, located in special "nurse" cells of the mother, make proteins that contribute to the initial morphogen gradients along the egg's axes before fertilization. Segmentation genes comprise three subgroups: gap, pair-rule, and segment polarity genes. Each of these types of segmentation genes determines a different aspect of the segments that make up a developing fruit fly. The *hunchback*, *Krüppel*, and *knirps* genes are all gap genes. Homeotic genes ultimately determine the segment identity of previously differentiated cell groups.

Pattern Formation

Through the use of highly specific stains to track the morphogens in normal and mutant embryos, a fascinating picture of the interactions among developmental genes has emerged. Even before fertilization, shallow, poorly defined gradients are established by genes of the mother, such as the *bicoid* gene and related genes. These morphogen gradients establish the anterior-to-posterior and dorsal-to-ventral axes. After fertilization, these morphogens bind to the control regions of gap genes, whose protein products direct the formation of broadly defined zones which will later develop into several specific segments. The gap proteins then bind to the control regions of pair-rule genes, whose protein products direct further refinements in the segmentation process. The last group of segmentation genes, the segment polarity genes, direct the completion of the segmentation patterns observable in the embryo and adult fly, including definition of the anterior-posterior orientation of each segment. Homeotic genes then define the specific functions of the segments, including what appendages will develop from each one. Mutations in any of these developmental genes cause distinct and easily observed changes in the developing segment patterns. Genes such as *hunchback*, *giant*, *gooseberry*, and

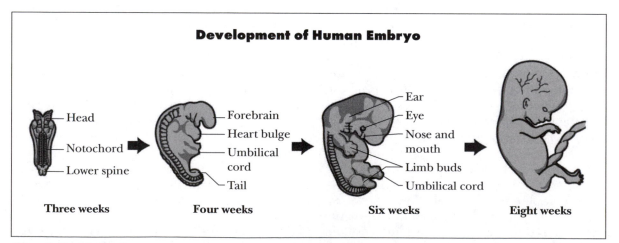

Development of Human Embryo

Head — Notochord — Lower spine

Three weeks

Forebrain — Heart bulge — Umbilical cord — Tail

Four weeks

Ear — Eye — Nose and mouth — Limb buds — Umbilical cord

Six weeks

Eight weeks

(Hans & Cassidy, Inc.)

Caenorhabditis Studies Tracing Cell Fates

Caenorhabditis elegans, a free-living soil nematode (a type of worm) 1 millimeter in length, has proved invaluable as a model organism for studying development. In addition to its small size, it has a rapid life cycle, going from egg to sexual maturity in three and a half days and living only two to three weeks. The presence of rudimentary physiological systems, including digestive, nervous, muscular, and reproductive systems, enables comparative studies between *Caenorhabditis* and "higher" organisms, such as mice and humans. Because the animal is transparent, the formation of every cell in the 959-celled adult can be observed microscopically and manipulated to illuminate its developmental program.

In 1963 Sydney Brenner set out to learn everything there was to know about *Caenorhabditis elegans*. In a 1974 publication he demonstrated how specific mutations could be induced in the *C. elegans* genome through chemical mutagenesis and showed how these mutations could be linked to specific genes and specific effects on organ development. Proving the utility of the organism as a genetic model encouraged a cadre of researchers to pursue research with *C. elegans*.

One of Brenner's students, John Sulston, developed techniques to track and study cell divisions in the nematode, from fertilized egg through adult. Microscopic examination of individual cell nuclei of the animal as it developed, along with electron microscopy of serial sections of the animal, enabled scientists to trace each of the adult worm's 959 cells back to a single fertilized egg. This "lineage map" was then used to track the fates of cells in animals that had been experimentally manipulated. Using a fine laser beam, scientists could kill a single cell at some point in development of the animal, then determine what changes, if any, awaited the remaining cells. These studies proved that the *C. elegans* cell lineage is invariant; that is, every worm underwent exactly the same sequence of cell divisions and differentiation.

Studies on cell fate and lineage mapping also led to the discovery that specific cells in the lineage always die through programmed cell death. Robert Horvitz, another of Brenner's students, discovered two "death genes" in *C. elegans* as well as genes that protect against cell death and direct the elimination of the dead cell. He also identified the first counterparts of the death gene in humans.

The characterization of the invariant cell lineage of *C. elegans* and the genetic linkages have been of great value to understanding basic principles of development, including signaling pathways in multicellular organisms and pathways controlling cell death. This knowledge has been invaluable to medicine, where it has helped researchers to understand mechanisms by which bacteria and viruses invade cells and has provided insights into the cellular mechanisms involved in neurodegenerative diseases, autoimmune disorders, and cancer. For their pioneering work in the "genetic regulation of organ development and programmed cell death" Brenner, Sulston, and Horvitz were awarded the 2002 Nobel Prize in Physiology or Medicine.

—Karen E. Kalumuck

---

hedgehog were all named with reference to the specific phenotypic changes that result from improper control of segmentation.

Homeotic genes are often called the "master" genes because they control large numbers of other genes required to make a whole wing or leg. Several clusters of homeotic genes have been discovered in *Drosophila*. Mutations in a certain group of genes of the *bithorax* complex result in adult fruit flies with two sets of wings. Similarly, mutations in some of the genes in the *antennapedia* complex can result in adult fruit flies with legs, rather than antennae, on the head.

A general principle applying to developmental processes in all organisms has emerged from the elegant work with *Drosophila* mutants: Finer and finer patterns of differentiated cells are progressively formed in the embryo along its major axes by morphogens acting on genes in a cascading manner, in which one gene set controls the next in the sequence until a highly complex pattern of differentiated cells results. Each cell within its own patterned zone then responds to the homeotic gene products and contributes to the formation of distinct, identifiable body parts.

Another important corollary principle was substantiated by the genetic analysis of development in *Drosophila* and other organisms.

In direct contrast to Weismann's implication about gene loss during differentiation, convincing evidence showed that genes were not systematically lost as egg cells divided and acquired distinguishing features. Even though a muscle cell was highly differentiated from a skin cell or a blood cell, each cell type retained the same numbers of chromosomes and genes as the original, undifferentiated, but fertilized egg cell. What changed in each cell was the pattern of gene expression, so that different proteins were made by specific genes while other genes were turned off. The morphogens, working in complex combinatorial patterns during the course of development, determined which genes would stay "on" and which would be turned "off."

Impact and Applications

The discovery and identification of the developmental genes in *Drosophila* and other lower organisms led to the discovery of similarly functioning genes in higher organisms, including humans. The base-pair sequences of many of the developmental genes, especially shorter subregions coding for sections of the morphogen that bind to the control regions of target genes, are conserved, or remain the same, across diverse organisms. This conservation of gene sequences has allowed researchers to find similar genes in humans. For example, some forty homeobox genes have been found in mice and humans, even though only eight were initially discovered in *Drosophila*. Some of the late-acting human homeobox genes are responsible for such developmental abnormalities as fused fingers and extra digits on the hands and feet. One of the most interesting abnormalities is craniosynostosis, a premature fusion of an infant's skull bones that can cause mental retardation. In 1993, developmental biologist Robert Maxson and his research group at the University of Southern California's Norris Cancer Center were the first to demonstrate that a mutation in a human homeobox gene *MSX2* was directly responsible for craniosynostosis and other bone/limb abnormalities requiring corrective surgeries. Maxson made extensive use of "knockout" mice, genetically engineered mice lacking particular genes, to test his human gene isolates. He and his research group made great progress in understanding the role of the *MSX2* gene as inducer of surrounding cells in the developing embryo. When this induction process fails because of defective *MSX2* genes, the fate of cells destined to participate in skull and bone formation and fusion changes, and craniosynostosis occurs.

A clear indication of the powerful cloning methods developed in the late 1980's was the discovery and isolation in 1990 of an important mouse developmental gene called *brachyury* ("short tails"). The gene's existence in mutant mice had been inferred from classical genetic studies sixty years prior to its isolation. In 1997, Craig Basson, Quan Yi Li, and a team of coworkers isolated a similar gene from humans and named it T-box brachyury (*TBX5*). Discovered first in mice, the "T-box" is one of those highly conserved subregions of a gene, and it allowed Basson and Li to find the human gene. When mutated or defective in humans, *TBX5* causes a variety of heart and upper limb malformations referred to as Holt-Oram syndrome. *TBX5* codes for an important morphogen affecting the differentiation of embryonic cells into mesoderm, beginning in the gastrulation phase of embryonic development. These differentiated mesodermal cells are destined to form the heart and upper limbs.

One of the important realizations emerging from the explosive research into developmental genetics in the 1990's was the connection between genes that function normally in the developing embryo but abnormally in an adult, causing cancer. Cancer cells often display properties of embryonic cells, suggesting that cancer cells are reverting to a state of uncontrolled division. Some evidence indicates that mutated developmental genes participate in causing cancer. Taken together, the collected data from many isolated human developmental genes, along with powerful reproductive and cloning technologies, promise to lead to cures and preventions for a variety of human developmental abnormalities and cancers.

—Chet S. Fornari, updated by Bryan Ness
See also: Aging; Animal cloning; Cell Cycle, The; Cell Division; Congenital Defects; Cytoki-

nesis; DNA Structure and Function; Evolutionary Biology; Genetic Engineering; Hereditary Diseases; Homeotic Genes; In Vitro Fertilization and Embryo Transfer; Model Organism: *Caenorhabditis elegans*; Model Organism: *Drosophila melanogaster*; Model Organism: *Mus musculus*; RNA Structure and Function; Stem Cells; Telomeres; Totipotency; X Chromosome Inactivation.

Further Reading

Beurton, Peter, Raphael Falk, and Hans-Jorg Rheinberger, eds. *The Concept of the Gene in Development and Evolution: Historical and Epistemological Perspectives*. New York: Cambridge University Press, 2000. A collection of essays that examines the question of what genes actually are, for philosophers and historians of science.

Bier, Ethan. *The Coiled Spring: How Life Begins*. Cold Spring Harbor, N.Y.: Cold Spring Harbor Laboratory Press, 2000. A basic overview of the development of embryos in both plants and animals.

Cronk, Quentin C. B., Richard M. Bateman, and Julie A. Hawkins, eds. *Developmental Genetics and Plant Evolution*. New York: Taylor & Francis, 2002. Developmental genetics for botanists.

DePamphilis, Melvin L., ed. *Gene Expression at the Beginning of Animal Development*. New York: Elsevier, 2002. Developmental genetics for zoologists.

Gilbert, Scott F. *Developmental Biology*. Sunderland, Mass.: Sinauer Associates, 2003. Presents a detailed description of all aspects of development.

Lewin, Benjamin. *Genes VII*. New York: Oxford University Press, 2001. Includes a comprehensive, clear discussion of genes and development, with excellent illustrations.

Nüsslein-Volhard, Christiane. "Gradients That Organize Embryo Development." *Scientific American* (August, 1996). The Nobel laureate reports on findings from the *Drosophila* studies.

Nüsslein-Volhard, Christiane, and J. Kratzschmar, eds. *Of Fish, Fly, Worm, and Man: Lessons from Developmental Biology for Human Gene Function and Disease*. New York: Springer, 2000. Designed for researchers, a consideration of the next phase of biology following the sequencing of several large genomes (accomplished at the turn of the millennium): determining the functions of genes and the interplay between them and their protein products.

Web Sites of Interest

Institute for Developmental Genetics. http://www.gsf.de/idg. Based in Germany, maintains data on mouse and zebra fish genetics "to unravel the molecular genetic networks controlling neuronal pattern formation, neuronal specification and differentiation during development."

Society for Developmental Biology. http://sdb.bio.purdue.edu. Professional society for biologists and geneticists interested in problems of development and growth of organisms. Site links to developmental biology sites and features the "Developmental Biology Cinema," which links to video sequences of developing organisms.

Virtual Library of Developmental Biology, Society for Developmental Biology. http://sdb.bio.purdue.edu. Primarily a collection of annotated links to laboratories by subject or organism, useful for its lists of departments, programs, and organizations.

Diabetes

Field of study: Diseases and syndromes

Significance: *Diabetes mellitus is a syndrome in which the body cannot metabolized glucose appropriately. The subsequent elevated levels cause significant damage to the eyes, heart, kidneys, and other organs. Diabetes is a significant public health problem with more than 17 million persons affected in the United States alone, of whom more than 90 percent have adult-onset (Type II) diabetes.*

Key terms

AUTOIMMUNE RESPONSE: an immune response of an organism against its own cells

LOCUS (*pl.* LOCI): the physical location of a gene, which in most organisms occurs as two copies called alleles, one copy on each of the chromosomes in a homologous pair

Types of Diabetes

Diabetes mellitus actually comprises a number of different diseases, broadly categorized into Type I and Type II diabetes. In both forms, the body's ability to process sugars is impaired, with consequences that can lead to death if untreated. Genetics plays a role in both types of diabetes, although both are thought to be the result of the interaction between genetics and the environment.

Glucose is a simple sugar that is required by all cells for normal functioning. Most of the body's glucose initially comes from carbohydrates broken down during digestion. Normally, the glucose level in a person's blood rises when carbohydrates are ingested. When the blood glucose reaches a certain level, it triggers the pancreas to release insulin, which causes the glucose level in the blood to drop by increasing its uptake in muscle, fat, the liver, and the gut. One theory is that as the glucose level drops, the person becomes hungry. Eating begins a new cycle of blood sugar elevation and insulin response.

Patients with either type of diabetes have difficulty metabolizing glucose, with a subsequent rise in fasting and postprandial blood sugar levels. In Type I diabetes, also called juvenile-onset or insulin-dependent diabetes, this is due to destruction of the insulin-secreting cells in the pancreas. In Type II, also called adult-onset, maturity-onset, or non-insulin-dependent diabetes, cells become resistant to the effects of insulin even though the pancreas is still producing some insulin. Both types lead to increased risk of heart and vascular disease, kidney problems, blindness, neurological problems, and other serious medical consequences.

Type I Diabetes

Type I diabetes mellitus is a chronic autoimmune disease that results from a combination of genetic and environmental factors. Certain persons are born with a genetic susceptibility to the disease. Before the disease can develop, however, some kind of trigger in the environment must be present. The environmental trigger, possibly a viral infection like measles or mumps, triggers an autoimmune response; that is, the person's own immune system aids in the destruction of the beta cells, those responsible for the secretion of insulin, in the pancreas. Typically, autoantibodies appear a few years ahead of the actual disease. The first recognizable symptom is a condition called pre-diabetes in which the usual insulin release in response to elevated blood sugar levels in the blood is diminished. At a certain point, most commonly between the ages of ten and fourteen, the person develops full-blown diabetes, with excessive thirst and urination, as well as weight loss despite adequate or increased caloric intake. If untreated, the person can become comatose or even die. Regular treatment with insulin (by injection or by an insulin pump) is required for the rest of the person's life. There is also evidence that weight reduction and exercise can alleviate symptoms to some extent.

The earliest evidence for a genetic basis for Type I diabetes was the observation by epidemiologists that it often occurs in families. In white Americans, the risk of diabetes is 0.12 percent overall, but in children of persons with diabetes the risk is much greater, 1-15 percent. Worldwide, the incidence of Type I diabetes is highest in Scandinavia, Northern Europe, and areas of the United States where Scandinavian immigrants settled. The lowest incidence of this type of diabetes is in China and parts of South America.

The genetic basis for developing Type I diabetes appears not so much to involve mutant genes per se, but rather a bad combination of particular alleles. Most of the genes implicated so far are found in the major histocompatibility complex, known as the HLA (human leukocyte antigen) complex. Certain combinations of alleles at these loci seem to confer a much higher susceptibility than normal. Some of the autoantigens themselves have also been identified, insulin being one of them, which should be no surprise. In addition, a rare type of autoimmune diabetes, resembling Type I, occurs as part of a syndrome called autoimmune polyendocrinopathy-candidiasis-ectodermal dystrophy (APECED), which is caused by mutation in *Aire*, an autoimmune regulator gene. Although the function of *Aire* is not known, expression of the gene has been detected in the thymus, pancreas, and adrenal cortex, and developmental studies suggest that mutations in *Aire* might cause the thymus (which is integral to proper immune system function) to develop incorrectly.

Type II Diabetes

Diabetes mellitus Type II is by far the more common type of diabetes. For example, about 90 percent of diabetics in the United States have Type II. It is a disease that occurs primarily in older adults, although the incidence in younger people is increasing as the incidence of obesity increases and as more children lead sedentary lives. The children at greatest risk in the United States are those from ethnic minority backgrounds.

Type II diabetes appears to be a group of diseases, rather than a single disease, in which there are two defects: (1) beta-cell dysfunction, leading to somewhat decreased production of insulin (although elevated levels of insulin also occur), and (2) tissue resistance to insulin. As with Type I, it appears that people who develop Type II are born with a genetic susceptibility but the development of actual disease is dependent upon an environmental trigger. Some possible triggers include aging, sedentary lifestyle, and abdominal obesity. Obesity plays a particularly significant role in the development of Type II diabetes. Among North Americans, Europeans, and Africans with Type II diabetes, between 60 and 70 percent are obese.

As with Type I, epidemiologic evidence suggests a strong genetic component to Type II diabetes. In identical twins over forty years of age, for example, the likelihood is about 70 percent that the second twin will develop Type II diabetes once the first twin has developed the disease. Diabetes Type II is also found in nearly 100 percent of obese Pima Indians and some Pacific Islanders.

Mutant alleles for a number of genes have been implicated in susceptibility and development of Type II diabetes. The first genes to be implicated were the insulin gene, genes encoding important components of the insulin secretion pathways, and other genes involved in glucose homeostasis. Mutations are very diverse and can include mutation not only in the genes themselves but also in transcription factors and control sequences. As more genes and their mutant alleles are discovered, better treatment options should become available, possibly even some that are tailored to specific types of mutations.

Treatments for Type II diabetes available at the beginning of the twenty-first century include such lifestyle changes as increased activity and weight loss, as well as oral drugs that increase tissue sensitivity to circulating insulin, stimulate increased insulin secretion, or alter insulin action. Some patients may ultimately have to add insulin to their treatment regimen. Once the genetic factors have been completely elucidated for both types of diabetes, treatments to modify the genes may become a reality.

—Rebecca Lovell Scott and Bryan Ness

See also: Autoimmune Disorders; Bacterial Genetics and Cell Structure; Biopharmaceuticals; Cloning; Gene Therapy: Ethical and Economic Issues; Genetic Engineering; Heart Disease; Hereditary Diseases.

Further Reading

American Diabetes Association. *American Diabetes Association Complete Guide to Diabetes: The Ultimate Home Reference from the Diabetes Experts.* New York: McGraw-Hill, 2002. Written for the consumer, this book includes everything a person with diabetes or a caregiver needs to know, including information on symptoms, complications, exercise and nutrition, blood sugar control, sexual issues, drug therapies, insulin regimes, and daycare.

_____. *Type II Diabetes.* 3d ed. New York: McGraw-Hill, 2000. Includes a broad range of information, from who gets Type II diabetes and warning signs to diet plans and self-care guides.

Becker, Gretchen. *The First Year: Type 2 Diabetes, An Essential Guide for the Newly Diagnosed.* New York: Marlowe, 2001. The author suffers from Type II diabetes and provides firsthand advice for coping with the disease, beginning with the day of diagnosis, day by day through the first week, and beyond.

Froguel, Philippe, and Gilberto Velho. "Genetic Determinants of Type 2 Diabetes." *Recent Progress in Hormone Research* 56 (2001): 91-106. An in-depth review article discussing many of the genes involved in the development of Type II diabetes.

Kahn, C. R., Gordon C. Weir, George L. King, Alan C. Moses, Robert J. Smith, and Alan M. Jacobson. *Joslin's Diabetes Mellitus.* 14th ed. Philadelphia: Lippincott Williams and Wilkins, 2003. A classic medical book, this bible of diabetes was first published in 1916 and is periodically updated and reissued. This volume reviews the dramatic advances in diabetes in addition to sections on origin, treatment, complications, and other aspects of the disease.

Lowe, William L., Jr., ed. *Genetics of Diabetes Mellitus.* Boston: Kluwer Academic, 2001. An in-depth, scientifically based book written by multiple experts in the field of diabetes research.

Milchovich, Sue K., and Barbara Dunn-Long. *Diabetes Mellitus: A Practical Handbook.* 8th ed. Boulder, Colo.: Bull, 2003. A basic reference book that contains comprehensive information on living with diabetes.

Notkins, Abner Louis. "Immunologic and Genetic Factors in Type I Diabetes." *The Journal of Biological Chemistry* 277, no. 46 (2002): 43,545-43,548. An overview of the major lines of evidence used to consider Type I diabetes primarily an autoimmune disease. Also provides specifics about the gene defects involved in Type I diabetes.

Web Sites of Interest

American Diabetes Association. http://www .diabetes.org. Site includes information on genetics and diabetes.

National Institute of Diabetes & Digestive & Kidney Diseases. http://www.niddk.nih.gov. This arm of the National Institues of Health offers resources and links to research on lactost intolerance.

Dihybrid Inheritance

Field of study: Classical transmission genetics

Significance: *The simultaneous analysis of two different hereditary traits may produce more information than the analysis of each trait separately. In addition, many important hereditary traits are controlled by more than one gene. Traits controlled by two genes serve as an introduction to the more complex topic of traits controlled by many genes.*

Key terms

ALLELES: different forms of the same gene; any gene may exist in several forms having very similar but not identical DNA sequences

DIHYBRID: an organism that is heterozygous for both of two different genes

HETEROZYGOUS: a condition in which the two copies of a gene in an individual (one inherited from each parent) are different alleles

HOMOZYGOUS: a condition in which the two copies of a gene in an individual are the same allele; synonymous with "purebred"

Mendel's Discovery of Dihybrid Inheritance

Austrian botanist Johann Gregor Mendel was the first person to describe both monohybrid and dihybrid inheritance. When he crossed purebred round-seed garden peas with purebred wrinkled-seed plants, they produced only monohybrid round seeds. He planted the monohybrid round seeds and allowed them to fertilize themselves; they subsequently produced ¾ round and ¼ wrinkled seeds. He concluded correctly that the monohybrid generation was heterozygous for an allele that produces round seeds and another allele that produces wrinkled seeds. Since the monohybrid seeds were round, the round allele must be dominant to the wrinkled allele. He was able to explain the 3:1 ratio in the second generation by assuming that each parent contributes only one copy of a gene to its progeny. If W represents the round allele and w the wrinkled allele, then the original true-breeding parents are WW and ww. When eggs and pollen are produced, they each contain only one copy of the gene. Therefore the monohybrid seeds are heterozygous *Ww.* Since these two alleles will separate during meiosis when pollen and eggs are produced, ½ of the eggs and pollen will be *W* and ½ will be *w.* Mendel called this "segregation." When the eggs and pollen combine randomly during fertilization, ¼ will produce *WW* seeds, ½ will produce *Ww* seeds, and ¼ will produce *ww* seeds. Since *W* is dominant to *w,* both the *WW* and *Ww* seeds will be round, producing ¾ round and ¼ wrinkled seeds. When Mendel crossed a purebred yellow-seed plant with a purebred green-seed plant, he observed an entirely analogous result in which the yellow allele (*G*) was dominant to the green allele (*g*).

Once Mendel was certain about the nature of monohybrid inheritance, he began to experiment with two traits at a time. He crossed purebred round, yellow pea plants with purebred wrinkled, green plants. As expected, the dihybrid seeds that were produced were all round and yellow, the dominant form of each trait. He planted the dihybrid seeds and allowed them to fertilize themselves. They produced ⁹⁄₁₆ round, yellow seeds; ³⁄₁₆ round, green seeds; ³⁄₁₆ wrinkled, yellow seeds; and ¹⁄₁₆ wrinkled, green seeds. Mendel was able to explain this dihybrid ratio by assuming that in the dihybrid flowers,

the segregation of *W* and *w* was independent of the segregation of *G* and *g.* Mendel called this "independent assortment." Thus, of the ¾ of the seeds that are round, ¾ should be yellow and ¼ should be green, so that ¾ × ¾ = ⁹⁄₁₆ should be round and yellow, and ¾ × ¼ = ³⁄₁₆ should be round and green. Of the ¼ of the seeds that are wrinkled, ¾ should be yellow and ¼ green, so that ¼ × ¾ = ³⁄₁₆ should be wrinkled and yellow, and ¼ × ¼ = ¹⁄₁₆ should be wrinkled and green. This relationship can be seen in the table headed "Dihybrid Inheritance and Sex Linkage."

Sex Chromosomes

Humans and many other species have sex chromosomes. In humans, normal females have two X chromosomes and normal males have one X and one Y chromosome. Therefore, sex-linked traits, which are controlled by genes on the X or Y chromosome, are inherited in a different pattern than the genes that have already been described. Since there are few genes on the Y chromosome, most sex-linked traits are controlled by genes on the X chromosome.

Every daughter gets an X chromosome from each parent, and every son gets an X from his mother and a Y from his father. Human red-

Dihybrid Inheritance and Sex Linkage

		Pollen			
		W;G	*W;g*	*w;G*	*w;g*
Eggs	*W;G*	*W W;G G* round, yellow	*W W;G g* round, yellow	*W w;G G* round, yellow	*W w;G g* round, yellow
	W;g	*W W;G g* round, yellow	*W W;g g* round, green	*W w;G g* round, yellow	*W w;g g* round, green
	w;G	*W w;G G* round, yellow	*W w;G g* round, yellow	*w w;G G* wrinkled, yellow	*w w;G g* wrinkled, yellow
	w;g	*W w;G g* round, yellow	*W w;g g* round, green	*w w;G g* wrinkled, yellow	*w w;g g* wrinkled, green

Note: Semicolons indicate that the two genes are on different chromosomes.

green color blindness is controlled by the recessive allele (*r*) of an X-linked gene. A red-green color-blind woman (*rr*) and a normal man (*R*Y) will have normal daughters (all heterozygous *Rr*) and red-green color-blind sons (*r* Y). Conversely, a homozygous normal woman (*RR*) and a red-green color-blind man (*r*Y) will have only normal children, since their sons will get a normal X from the mother (*R*Y) and the daughters will all be heterozygous (*Rr*). A heterozygous woman (*Rr*) and a red-green color-blind man (*r*Y) will have red-green color-blind sons (*r*Y) and daughters (*rr*), and normal sons (*R*Y) and daughters (*Rr*) in equal numbers.

A dihybrid woman who is heterozygous for red-green color blindness and albinism (a recessive trait that is not sex linked) can make four kinds of eggs with equal probability: *R;A*, *R;a*, *r;A*, and *r;a*. A normal, monohybrid man who is heterozygous for albinism can make four kinds of sperm with equal probability: *R;A*, *R;a*, Y;*A*, and Y;*a*. By looking at the table headed "Mixed Sex-Linked and Autosomal Traits," it is easy to predict the probability of each possible kind of child from this mating.

The probabilities are $\frac{6}{16}$ normal female, $\frac{2}{16}$ albino female, $\frac{3}{16}$ normal male, $\frac{3}{16}$ red-green color-blind male, $\frac{1}{16}$ albino male, and $\frac{1}{16}$ albino, red-green color-blind male. Note that the probability of normal coloring is $\frac{1}{4}$ and the probability of albinism is $\frac{1}{4}$ in both sexes. There is no change in the inheritance pattern for the gene that is not sex linked.

Other Examples of Dihybrid Inheritance

A hereditary trait may be controlled by more than one gene. To one degree or another, almost every hereditary trait is controlled by many different genes, but often one or two genes have a major effect compared with all the others, so they are called single-gene or two-gene traits. Dihybrid inheritance can produce traits in various ratios, depending on what the gene products do. A number of examples will be presented, but they do not exhaust all of the possibilities.

The comb of a chicken is the fleshy protuberance that lies on top of the head. There are four forms of the comb, each controlled by a different combination of the two genes that control this trait. The first gene exists in two forms (*R* and *r*), as does the second (*P* and *p*). In each case, the form represented by the uppercase letter is dominant to the other form. Since there are two copies of each gene (with

Mixed Sex-Linked and Autosomal Traits

		Sperm			
		A;R	*a;R*	*A;*Y	*a;*Y
Eggs	*A;R*	*A A;R R* normal female	*A a;R R* normal female	*A A;R*Y normal male	*A a;R*Y normal male
	A;r	*A A;R r* normal female	*A a;R r* normal female	*A A;r*Y red-green color-blind male	*A a;r*Y red-green color-blind male
	a;R	*A a;R R* normal female	*a a;R R* albino female	*A a;R*Y normal male	*a a;R*Y albino male
	a;r	*A a;R r* normal female	*a a;r r* albino female	*A a;r*Y red-green color-blind male	*a a;r*Y albino, red-green color-blind male

Note: Semicolons indicate that the two genes are on different chromosomes.

Partial Dominance

	Pollen			
	A;B	A;b	a;B	a;b
A;B	A A;B B red	A A;B b medium red	A a;B B medium red	A a;B b light red
A;b	A A;B b medium red	A A;b b light red	A a;B b light red	A a;b b very light red
a;B	A a;B B medium red	A a;B b light red	a a;B B light red	a a;B b very light red
a;b	A a;B b light red	A a;b b very light red	a a;B b very light red	a a;b b white

Eggs (row label at left)

Note: Semicolons indicate that the two genes are on different chromosomes. Dihybrid ratios may change if both genes are on the same chromosome.

the exception of genes on sex chromosomes), the first gene can be present in three possible combinations: RR, Rr, and rr. Since R is dominant, the first two combinations produce the same trait, so the symbols $R\_$ and $P\_$ can be used to represent either of the two combinations. Chickens with $R\_;P\_$ genes have what is called a walnut comb, which looks very much like the meat of a walnut. The gene combinations $R\_;pp$, $rr;P\_$, and $rr;pp$ produce combs that are called rose, pea, and single, respectively. If two chickens that both have the gene combination $Rr;Pp$ mate, they will produce progeny that are 9/16 walnut, 3/16 rose, 3/16 pea, and 1/16 single (see Table 1 for an explanation of these numbers).

White clover synthesizes small amounts of cyanide, which gives clover a bitter taste. There are some varieties that produce very little cyanide (sweet clover). When purebred bitter clover is crossed with some varieties of purebred sweet clover, the progeny are all bitter. However, when the hybrid progeny is allowed to fertilize itself, the next generation is 9/16 bitter and 7/16 sweet. This is easy to explain if it is assumed that bitter/sweet is a dihybrid trait. The bitter parent would have the gene combination $AA;BB$ and the sweet parent $aa;bb$, where A and

B are dominant to a and b, respectively. The bitter dihybrid would have the gene combination $Aa;Bb$. When it fertilizes itself, it would produce 9/16 $A\_;B\_$, which would be bitter, and 3/16 $A\_;bb$, 3/16 $aa;B\_$, and 1/16 $aa;bb$, all of which would be sweet. Clearly, both the A allele and the B allele are needed in order to synthesize cyanide. If either is missing, the clover will be sweet.

Absence of Dominance

In all of the previous examples, there was one dominant allele and one recessive allele. Not all genes have dominant and recessive alleles. If a purebred snapdragon with red flowers (RR) is crossed with a purebred snapdragon with white flowers (rr), all the monohybrid progeny plants will have pink flowers (Rr). The color depends on the number of R alleles present: two Rs will produce a red flower, one R will produce a pink flower, and no Rs will produce a white flower. This is an example of partial dominance or additive inheritance.

Consider a purebred red wheat kernel ($AA;BB$) and a purebred white wheat kernel ($aa;bb$) (see the table headed "Partial Dominance"). If the two kernels are planted and the resulting plants are crossed with each other, the progeny dihybrid kernels will be light red (*Aa;Bb*). If the dihybrid plants grown from the dihybrid kernels are allowed to self-fertilize, they will produce $\frac{1}{16}$ red (*AA;BB*), $\frac{4}{16}$ medium red (*AA;Bb* and *Aa;BB*), $\frac{6}{16}$ light red (*AA;bb*, *Aa;Bb*, and *aa;BB*), $\frac{4}{16}$ very light red (*Aa;bb* and *aa;Bb*), and $\frac{1}{16}$ white (*aa;bb*). The amount of red pigment depends on the number of alleles (*A* and *B*) that control pigment production. Although it may appear that this is very different than the example in the first table, they are in fact very similar.

All of the inheritance patterns that have been discussed are examples of "independent assortment," in which the segregation of the alleles of one gene is independent of the segregation of the alleles of the other gene. That is exactly what would be expected from meiosis if the two genes are not on the same chromosome. If two genes are on the same chromosome and sufficiently close together, they will not assort independently and the progeny ratios will not be like any of those described. In that case, the genes are referred to as "linked" genes.

—*James L. Farmer*

See also: Chromosome Theory of Heredity; Classical Transmission Genetics; Complete Dominance; Dihybrid Inheritance; Epistasis; Incomplete Dominance; Linkage Maps; Mendelian Genetics; Monohybrid Inheritance; Multiple Alleles.

Further Reading

Madigan, Michael M., et al., eds. *Brock Biology of Microorganisms*. 10th ed. Englewood Cliffs, N.J.: Prentice Hall, 2002. A college-level text organized into six units on the principles of microbiology, evolutionary microbiology and microbial diversity, immunology and pathogenicity, microbial diseases, and microorganisms as tools for industry and research.

Tortora, Gerard. *Microbiology: An Introduction*. 7th ed. San Francisco: Benjamin Cummings, 2001. An accessible introduction to the basic principles of microbiology, the interaction between microbe and host, and human diseases caused by microorganisms. Gives a general overview of antibiotics and how bacterial resistances to antibiotics occur.

Wolf, Jason B., et al. *Epistasis and the Evolutionary Process*. New York: Oxford University Press, 2000. Primary focus is on the role of gene interactions (epistasis) in evolution. Leading researchers examine how epistasis impacts the evolutionary processes in overview, theoretical, and empirical chapters.

Diphtheria

Fields of study: Bacterial genetics; Diseases and syndromes

Significance: *Diphtheria is an acute bacterial disease known best for damaging the respiratory system. Afflicted individuals die from this as well as from damage to the heart, nerves, and kidney. Genetic research has led to better understanding of diphtheria's cause, action, and treatment.*

Key terms

ANAPHYLAXIS: a severe, sometimes fatal allergic reaction

ANTIBODIES: proteins that help identify and destroy foreign pathogens and other molecules in the body

ANTITOXIN: a vaccine containing antibodies against a specific toxin

CUTANEOUS: related to or affecting skin

Diphtheria Symptoms and Cure

The acute bacterial disease diphtheria is caused by rod-shaped *Corynebacterium diphtheriae* (*C. diphtheriae*), discovered in 1883 by Edwin Klebs and Friedrich Löffler. Diphtheria involves the respiratory tract, nerves, and heart in ways that can be lethal. After 1950, the disease became uncommon in industrialized nations because of immunization by vaccination with antitoxin originally isolated from horses by Emil Adolf von Behring in the 1880's. In such nations, diphtheria is contracted by contact with travelers coming from developing nations, where it is much more common, who may be asymptomatic carriers or have active diphtheria.

C. diphtheriae usually enters the body through mucous membranes of the mouth or nose, though it can also enter via breaks in the skin (cutaneous route). After infection and a two- to five-day incubation period, diphtheria's first symptoms are localized inflammation that kills cells in the respiratory tract or skin. Respiratory diphtheria initially appears as a sore throat in which a dirty gray membrane (diphtheria pseudomembrane) forms and spreads through the respiratory system. The pseudomembrane (made mostly of dead cells, bacteria, and white blood cells) causes a husky voice and is accompanied by swollen lymph glands. In severe cases, diphtheria kills by heart failure or throat paralysis as little as one day after the initial symptoms appear. Fortunately, such lethality occurs mostly in unimmunized individuals. Cutaneous *C. diphtheriae* infections most often produce only skin lesions, though they can cause death if the bacteria spreads widely through the blood and damages the heart, nerves, and kidneys. Damage depends upon the bacterial entry site, individual immunization status, and the amount of toxin made.

Although most people in industrialized nations are immunized, the consequences of diphtheria can be so severe that therapy by diphtheria antitoxin should begin as soon as symptoms suggest the disease. Cure of diphtheria requires, in addition to the antitoxin, destruction of all *C. diphtheriae* in afflicted individuals. Immunization is the first line of defense, so it is crucial to ensure that the suspected diphtheria sufferer is not sensitive to antitoxin because incautious antitoxin administration may cause lethal anaphylaxis in sensitive people. Individual sensitivity is identified by scratch tests with diluted antitoxin. In sensitive people, desensitization is achieved through the sequential administration of increasing doses of antitoxin in an intensive care unit until effective doses are safely reached.

Diphtheria is so dangerous that all patient contacts are tested for *C. diphtheriae*. Afflicted individuals are given penicillin, erythromycin, and/or antitoxin, depending on the presence, absence, and severity of diphtheria symptoms. Though adequate universal immunization is a sure diphtheria control, booster shots—like those for tetanus prophylaxis—should be given every ten years in addition to childhood shots. Because of the extremely infectious and fatal nature of diphtheria, all people positive for *C. diphtheriae* must be kept in bed, isolated, and treated until symptoms and bacteria are absent after antibiotic therapy stops. This may require four to six weeks.

Genetics and Diphtheria

Diphtheria symptoms are caused by diphtheria toxin, a protein so lethal that 6 micrograms will kill a 150-pound human. Most often, the toxin first localizes in respiratory mucosa cells or cutaneous sites, where it causes diphtheria pseudomembrane or skin lesions by interacting with the protein translocase. Translocase is essential to synthesis of proteins needed for body cell growth, survival, and reproduction. Diphtheria toxin and translocase interact through a process called adenine ribosylation, similar to that in cholera. Diphtherial adenine ribosylation inactivates translocase, preventing its action and killing affected cells. Dead respiratory cells form diphtheria pseudomembrane, which closes off the throat. In skin, toxin-killed cells cause skin lesions. Destruction of nerve, heart, and kidney cells leads to damage in those tissues.

The diphtheria pseudomembrane may cut off breathing. In such cases, suffocation is prevented by a tracheotomy (a surgical incision in the neck that creates an airway). Major causes of quick diphtheria fatality are damage to nerves and the heart. The toxin is a protein made by genes that are present only in certain strains, and *C. diphtheriae* strains that do not produce the toxin are harmless. In addition, genetic studies have identified interaction of the toxin with respiratory mucosa cell translocase as well as similar action in many other tissues. Use of bacterial genetics has also enabled more scientific production of diphtheria antitoxin. The antitoxin is useful to visitors of regions where the disease is common. Its universal use has led to a worldwide decrease in diphtheria fatalities to fewer than five deaths per million people. The immunization is effective for ten years.

Impact and Applications

Diphtheria has long been a serious, worldwide threat. During the twentieth century, its danger greatly diminished in industrialized nations with the advent of antitoxin and the wide use of antibiotics to kill *C. diphtheriae.* In poorer nations, diphtheria still flourishes and is a severe threat, partly because of less advanced medical practices and the public's fear of immunization.

Prevention of diphtheria relies mostly on immunization via antitoxin. The isolation and identification of diphtheria toxin and the development of antitoxin have depended on genetic methods that now protect most people from the disease. Wherever it afflicts people, diphtheria treatment also requires the use of antibiotics. Hence, advanced diphtheria prevention and treatment will be best effected by using genetic, immunologic, and biochemical methods to produce vaccines effective for more than ten years and to produce more potent antibiotics. Efforts toward these ends will most likely utilize molecular genetics to clearly define why diphtheria is intractable to lifelong vaccination. Especially valuable will be DNA sequence analysis, when a genome sequence becomes available.

—*Sanford S. Singer*

See also: Cholera; Emerging Diseases; Hereditary Diseases; Smallpox.

Further Reading

Beers, Mark H., and Robert Berkow, eds. *The Merck Manual of Diagnosis and Therapy.* 17th ed. Whitehouse Station, N.J.: Merck Research Laboratories, 1999. Contains medical details on epidemiology, symptoms, diagnosis, treatment, outbreak management, and antitoxin sensitivity.

Parker, James N., ed. *Diphtheria: The Official Patient's Sourcebook: A Revised and Updated Directory for the Internet Age.* San Diego: ICON Health, 2002. Tells patients how to look for information covering virtually all topics related to diphtheria, from the essentials to the most advanced areas of research.

Rakel, Robert E., et al., eds. *Conn's Current Therapy.* Philadelphia: W. B. Saunders, 2003. Describes the latest advances in therapeutics, including a succinct overview of diphtheria and its treatment for general readers. More than three hundred leading practitioners from fifteen countries describe their preferred techniques for managing hundreds of common disorders affecting every organ system.

Stratton, Kathleen R., Cynthia J. Howe, and Richard B. Johnston, Jr., eds. *DPT Vaccine and Chronic Nervous System Dysfunction: A New Analysis.* Washington, D.C.: National Academy Press, 1994. Examines nervous system diseases in infancy and childhood, their etiology, and the adverse effects of the vaccines.

Web Site of Interest

Diphtheria Hub. http://www.healthubs.com/diptheria. Provides dozens of links to information on diphtheria, including overviews, diagnosis, and treatment.

DNA Fingerprinting

Field of study: Human genetics and social issues

Significance: *DNA fingerprinting includes a variety of techniques in which individuals are uniquely identified through examination of specific DNA sequences that are expected to vary widely among individuals. Uses for these technologies include not only practical applications in forensic analysis and paternity tests but also basic research in paternity, breeding systems, and ecological genetics for many nonhuman species.*

Key terms

MICROSATELLITE: a type of VNTR in which the repeated motif is 1 to 6 base pairs; synonyms include simple sequence repeat (SSR) and short tandem repeat (STR)

MINISATELLITE: a type of VNTR in which the repeated motif is 12 to 500 base pairs in length

POLYMERASE CHAIN REACTION (PCR): a laboratory procedure for making millions of identical copies of a short DNA sequence

VARIABLE NUMBER TANDEM REPEAT (VNTR): a type of DNA sequence in which a short sequence is repeated over and over; chromosomes from different individuals frequently have different numbers of the basic repeat, and if many of these variants are known, the sequence is termed a hypervariable

Genetic Differences Among Individuals

All individuals, with the exception of twins and other clones, are genetically unique. Theoretically it is therefore possible to use these genetic differences, in the form of DNA sequences, to identify individuals or link samples of blood, hair, and other features to a single individual. In practice, individuals of the same species typically share the vast majority of their DNA sequences; in humans, for example, well over 99 percent of all of the DNA is identical. For individual identification, this poses a problem: Most of the sequences that might be examined are identical (or nearly so) among randomly selected individuals. The solution to this problem is to focus only on the small regions of the DNA that are known to vary widely among individuals. These regions, termed hypervariable, are typically based on repeat sequences in the DNA.

Imagine a simple DNA base sequence, such AAC (adenine-adenine-cytosine), which is repeated at a particular place (or locus) on a human chromosome. One chromosome may have eleven of these AAC repeats, while another might have twelve or thirteen, and so on. If one could count the number of repeats on each chromosome, it would be possible to specify a diploid genotype for this chromosomal locus: An individual might have one chromosome with twelve repeats, and the other with fifteen. If there are many different chromosomal variants in the population, most individuals will have different genotypes. This is the conceptual basis for most DNA fingerprinting.

DNA fingerprint data allow researchers or investigators to exclude certain individuals: If, for instance, a blood sample does not match an individual, that individual is excluded from further consideration. However, if a sample and

A criminalist at the Phoenix Police Department prepares samples of DNA taken from a crime scene for comparison to the DNA fingerprints of suspects. (AP/Wide World Photos)

an individual match, this is not proof that the sample came from that individual; other individuals might have the same genoytpe. If a second locus is examined, it becomes less likely that two individuals will share the same genotype. In practice, investigators use enough independent loci that it is extremely unlikely that two individuals will have the same genotypes over all of the loci, making it possible to identify individuals within a degree of probability expressed as a percentage, and very high percentages are possible.

The First DNA Fingerprints

Alec Jeffreys, at the University of Leicester in England, produced the first DNA fingerprints in the mid-1980's. His method examined a twelve-base sequence that was repeated one right after another, at many different loci in the human genome. Once collected from an individual, the DNA was cut using restriction enzymes to create DNA fragments that contained the repeat sequences. If the twelve-base sequence was represented by more repeats, the fragment containing it was that much longer. Jeffreys used agarose gel electrophoresis to separate his fragments by size, and he then used a specialized staining technique to view only the fragments containing the twelve-base repeat. For two samples from the same individual, each fragment, appearing as a band on the gel, should match. This method was used successfully in a highly publicized rape and murder case in England, both to exonerate one suspect and to incriminate the perpetrator.

While very successful, this method had certain drawbacks. First, a relatively large quantity of DNA was required for each sample, and results were most reliable when each sample compared was run on the same gel. This meant that small samples, such as individual hairs or tiny blood stains, could not be used, and also that it was difficult to store DNA fingerprints for use in future investigations.

Variable Number Tandem Repeat Loci

The type of sequence Jeffreys exploited is now included in the category of variable number tandem repeats (VNTRs). This type of DNA sequence is characterized, as the name implies, by a DNA sequence which is repeated, one copy right after another, at a particular locus on a chromosome. Chromosomes vary in the number of repeats present.

VNTRs are often subcategorized based on the length of the repeated sequence. Minisatellites, like the Jeffreys repeat, include repeat units ranging from about twelve to several hundred bases in length. The total length of the tandemly repeated sequences may be several hundred to several thousand bases. Many different examples have since been discovered, and they occur in virtually all eukaryotes. In fact, the Jeffreys repeat first discovered in humans was found to occur in a wide variety of other species.

Shorter repeat sequences, typically 1 to 6 bases in length, were subsequently termed microsatellites. In humans, AC (adenine-cytosine) and AT (adenine-thymine) repeats are most common; an estimate for the number of AC repeat loci derived from the Human Genome Project suggests between eighty thousand and ninety thousand different AC repeat loci spread across the genome. Every eukaryote studied to date has had large numbers of microsatellite loci, but they are much less common in prokaryotes.

The Polymerase Chain Reaction

The development of the polymerase chain reaction (PCR) in the mid-1980's, and its widespread use and optimization in DNA labs a few years later offered an alternative approach to DNA fingerprinting. The PCR technique makes millions of copies of short segments of DNA, with the chromosomal location of the fragments produced under the precise control of the investigator. PCR is extremely powerful and can be used with extremely small amounts of DNA. Because the fragments amplified are small, PCR can also be used on partially degraded samples. The size and chromosomal location of the fragments produced depends on the DNA primers used in the reaction. These are short, single-stranded DNA molecules that are complementary to sequences that flank the region to be amplified.

With this approach, an investigator must find and determine the DNA sequence of a re-

DNA TESTING RESULTS - SDPD

Description	D3S1358	vWA	FGA	AMEL	D8S1179	D21S11	D18S51	D5S818	D13S317	D7S820	D16S539	TH01	TPOX	CSF1PO	POSSIBLE SOURCE	FREQUENCY
#3 Underpants	17,19	14,16	22,23	X	12,13	28,31.2	14,16	11,12	8,12	14	9,12	7,9.3	8,9	11,12		
#84 Blood- RV Carpet	17,19	14,16	22,23	X	12,13	28,31.2	14,16	11,12	8,12	14	9,12	7,9.3	8,9	ND	DANIELLE VAN DAM	1 in 130,000,000,000,000,000
#94D-2 Blood- Jacket - Shoulder	17,19	14,16	22,23	X	12,13	28,31.2	14,16	11,12	8,12	14	9,12	7,9.3	8,9	11,12	DANIELLE VAN DAM	1 in
#95 David Westerfield	16	15,17	23,24	X,Y	13,15	28,32.2	15,16	11,14	11,12	8,10	10,13	6,9.3	8	10,13		
#94D-1 Blood- Jacket - Lapel	16	15,17	ND	X,Y	13,15	(tr 28)	ND	11 (tr14)	ND	ND	NT	NT	NT	NT	DAVID WESTERFIELD	1 in

Legend: (tr) = trace results   ND = Not Detected   NT = NOT TESTED

Co-founder of the San Diego DNA laboratory Annette Peer testifies about DNA evidence at the David Westerfield murder trial on June 20, 2002. (AP/Wide World Photos)

gion containing a VNTR. Primers are designed to amplify the VTNR region, together with some flanking DNA sequences on both ends. The fragments produced in the reaction are then separated by length using gel electrophoresis so that differences in length, attributable to different numbers of the repeat, become apparent. For a dinucleotide repeat like AC, fragments representing different numbers of repeats, and hence different alleles, differ by a multiple of two. For instance, a researcher might survey a number of individuals and find fragments of 120, 122, 124, 128, and 130 base pairs in length.

Current Approaches

Most current approaches to DNA fingerprinting use data collected simultaneously from a number of different VNTR loci, most commonly microsatellites. Preferably, the loci are PCR amplified using primers with fluorescent dyes attached, so that fragments from different loci are uniquely tagged with different colors. The fragments are then loaded in polyacrylamide DNA gels of the type used for DNA sequencing and separated by size. The fluorescent colors and sizes of the fragments are determined automatically, using the same automated machines typically used for DNA sequencing.

DNA fingerprint data generated in this way are easily stored and saved for future comparisons. Since each allelic variant is represented by a specific DNA fragment length, and because these are measured very precisely, the initial constraint of running samples for comparison on the same gel is avoided.

Human Forensic and Paternity Testing

Although several different systems have been developed and used, a widely employed current standard comprises the Federal Bureau of Investigation's Combined DNA Index System (CODIS), with thirteen core loci. These thirteen are tetranucleotide (TCTA) microsatellite repeat loci, located on autosomes. Each locus has many known alleles, in some cases more than forty; the genetic variation is well characterized, and databases of variation within a variety of ethnic groups are available.

In addition to its role in criminal cases, this technique has seen widespread use to establish or exclude paternity, in immigration law to prove relatedness, and to identify the remains of casualties resulting from military combat and large disasters.

Other Uses for VNTR Genotyping

Soon after VNTRs were discovered in humans and used for DNA fingerprinting, researchers demonstrated that the same or similar types of sequences were found in all animals, plants, and other eukaryotes. The method pioneered by Jeffreys was, only a few years later, used for studies of paternity in wild bird populations. Since then, microsatellite analysis has come to dominate studies of relatedness, paternity, breeding systems, and other questions of individual identification in wild species of all kinds, including plants, insects, fungi, and vertebrates. Researchers now know, for example, that among the majority of birds which appear monogamous, between 10 and 15 percent of all progeny are fathered by males other than the recognized mate.

—*Paul R. Cabe*

See also: Criminality; Forensic Genetics; Genetic Testing; Genetics, Historical Development of; Human Genetics; Paternity Tests; Repetitive DNA.

Further Reading

Burke, Terry, R., Wolf, G. Dolf, and A. Jeffreys, eds. *DNA Fingerprinting: Approaches and Applications.* Boston: Birkhauser, 2001. Describes repetitive DNA and the broad variety of practical applications to law, medicine, politics, policy, and more. Aimed at the layperson.

Fridell, Ron. *DNA Fingerprinting: The Ultimate Identity.* New York: Scholastic, 2001. The history of the technique, from its discovery to early uses. Aimed at younger readers and nonspecialists.

Herrmann, Bernd, and Susanne Hummel, eds. *Ancient DNA: Recovery and Analysis of Genetic Material from Paleographic, Archaeological, Museum, Medical, and Forensic Speciments.* New York: Springer-Verlag, 1994. Written when DNA fingerprinting was just coming to the fore and films such as *Jurrasic Park* were in theaters, this collection of papers by first-generation researchers reflects the broad applications of the technology, including paleontological investigations.

Hummel, Susanne. *Fingerprinting the Past: Research on Highly Degraded DNA and Its Applications.* New York: Springer-Verlag, 2002. Manual about typing ancient DNA.

Rudin, Norah, and Keith Inman. *An Introduction to Forensic DNA Analysis.* Boca Raton, Fla.: CRC Press, 2002. An overview of many DNA typing techniques, along with numerous examples and a discussion of legal implications.

Wambaugh, Joseph. *The Blooding.* New York: Bantam Books, 1989. The policeman-turned-writer offers a fascinating account of the British rape and murder case in which DNA fingerprinting was first used.

Web Sites of Interest

Earl's Forensic Page. http://members.aol.com/EarlNMeyer/DNA.html. Summarizes how DNA fingerprinting works and its use in crime investigations and in determining paternity.

Iowa State University Extension and Office of Biotechnology, DNA Fingerprinting in Agricultural Genetics Programs. http://www.biotech.iastate.edu/biotech_info_series. Site links to a comprehensive and illustrative article on the role of DNA fingerprinting in agriculture.

DNA Isolation

Fields of study: Genetic engineering and biotechnology; Molecular genetics

Significance: *Before it can be manipulated and studied, DNA must be isolated from other substances such as complex carbohydrates, proteins, and RNA. The isolation process is central to biotechnology and genetic engineering.*

Key terms

CHLOROFORM/ISOAMYL ALCOHOL (CIA): a mixture of two chemicals used in DNA isolation to rid the extract of the contaminating compound phenol

LYSIS: the breaking open of a cell

OSMOTIC SHOCK: the lysing of cells by moving them from a liquid environment with a high solute concentration to an environment with a very low solute concentration

PHENOL: a simple chemical used in DNA extraction to precipitate proteins and aid in their removal

DNA Discovery and Extraction

Deoxyribonucleic acid (DNA) was discovered in 1869 by the Swiss physician Friedrich Miescher, who studied white blood cells in pus obtained from a surgical clinic. Miescher found that when bandages that had been removed from the postoperative wounds of injured soldiers were washed in a saline solution, the cells on the bandages swelled into a gelatinous mass that consisted largely of DNA. Miescher had isolated a denatured form of DNA—that is, DNA not in the normal double-stranded conformation. After a series of experiments, Miescher concluded that the substance he had isolated originated in the nuclei of the blood cells; he first called the substance nuclein and later nucleic acid.

The first problem when extracting DNA is lysing, or breaking open, the cell. Bacteria, yeast, and plant cells usually have a thick cell wall protecting their plasma membrane, which makes lysis more difficult. Bacteria, such as *Escherichia coli*, are the easiest of these cells to open by a process called alkaline lysis, in which cells are treated with a solution of sodium hydroxide and detergent that degrades both the cell wall and the cell membrane. Yeast cells are often broken open with enzymes such as lysozyme that degrade cell walls or by using a "French press," a piston in an enclosed chamber that forces cells open under high pressure. Plant tissue is usually mechanically broken into a fine cell suspension before extraction by grinding frozen tissue in a mortar and pestle. Once the suspension of cells is obtained, the tissue may be treated with a variety of enzymes to break down cell walls or with strong detergents, such as sodium lauryl sarcosine, that disrupt and dissolve both cell walls and cell membranes. Animal cells, such as white blood cells, do not have cell walls and can generally be opened by osmotic shock, the lysing of cells by moving them from a liquid environment with a high solute concentration to an environment with a very low solute concentration.

Isolation and Purification

Although lysis methods differ according to cell type, the process of DNA isolation and purification is more standardized. The isolation process may be imagined as a series of steps designed to remove either naturally occurring biological contaminants from the DNA or contaminants added by the scientist during the extraction process. The biological contaminants already present in cells are proteins and ribonucleic acid (RNA); additionally, plant cells have high levels of complex carbohydrates. Contaminants intentionally added by scientists may include salts and various chemicals.

After cells are lysed, a high-speed centrifugation is performed to form large-scale, insoluble cellular debris, such as membranes and organelles, into a pellet. The liquid extract remaining still contains dissolved proteins, RNA, and DNA. If salts are not already present in the extract, they are added; salt must be present later for the DNA to precipitate efficiently. Proteins must be removed from the extract since some not only degrade DNA but also inhibit enzymatic reactions with DNA. Proteins are precipitated by mixing the extract with a chemical called phenol. When phenol and the extract are mixed in a test tube, they separate into two parts like oil and water. If these fluids are centrifuged, precipitated proteins will actually collect between the two liquids at a spot called the interphase. The liquid layer containing the dissolved DNA is then drawn up and away from the precipitated protein.

The protein-free solution still contains DNA, RNA, salts, and traces of phenol dissolved into the extract. To remove the contaminating phenol, the extract is mixed with a chloroform/isoamyl alcohol solution (CIA). Again like oil and water, the DNA extract and CIA separate

Differential Isolation of Organelle DNA

Discussions of DNA isolation usually concern isolation of DNA from the nucleus. While the nucleus is the location of most of the genetic information in the cell, DNA molecules also exist in other organelles, such as mitochondria and chloroplasts. Chromosomes of these organelles, referred to as nonnuclear or cytoplasmic DNA, contain a small subset of genes, mostly encoding proteins needed by these organelles.

Most standard DNA isolation techniques isolate both nuclear and nonnuclear DNA together. For a person working with nuclear DNA, this is usually not a concern because the amount of nuclear DNA is much greater than the amount of nonnuclear DNA. In working with nonnuclear DNA however, the presence of nuclear DNA can often cause problems. Some techniques used to examine nonnuclear DNA, such as the polymerase chain reaction (PCR), are not affected by the presence of nuclear DNA, but for other techniques, pure nonnuclear DNA is required.

Isolation strategies for nonnuclear DNA usually involve two steps. The first step is the isolation of intact mitochondria or chloroplasts from the cells, followed by the lysing of the mitochondria or chloroplasts to release the DNA so it can be purified. The process is the same for isolation of both mitochondrial and chloroplast DNA. Isolation of intact mitochondria (for example) requires that the membranes of the cells be lysed in a way that does not rupture the mitochondria. To achieve this goal, gentle mechanical, chemical, or enzymatic methods (de-pending on the nature of the cell membrane and whether there is a cell wall) are employed to break open the cells and release the cytoplasmic contents. The lysis of the cells is usually done in an osmotically stabilized buffer. The solutes in this buffer match the concentration of the solutes inside the mitochondria, which prevents the mitochondria from bursting when the cells are lysed.

Once the cells are lysed, the lysate is centrifuged at low speed (usually between one thousand and three thousand times the force of gravity) to remove nuclei, membrane fragments, and other debris. The resulting supernatant contains the mitochondria in suspension. To concentrate the mitochondria, the supernatant is centrifuged at high speed (twelve thousand times the force of gravity). The pellet formed by this centrifugation will contain mitochondria and can be suspended in a small volume of liquid to create a concentrated suspension of mitochondria. This suspension may be treated with the enzyme DNase, which will degrade any nuclear DNA that remains without crossing the intact mitochondrial membrane. The enzyme will then be deactivated, and the mitochondria will be lysed. Lysis of the mitochondria is achieved by adding a strong detergent to the suspension of the mitochondria. Once the mitochondria have been lysed, the free mitochondrial DNA can be purified just as nuclear DNA would be, using phenol extraction and ethanol precipitation.

—*Douglas H. Brown*

into two layers. If the two layers are mixed vigorously and separated by centrifugation, the phenol will move from the DNA extract into the CIA layer. At this point the extract—removed to a new test tube—contains RNA, DNA, and salt.

The extract is next mixed with 100 percent ethanol, inducing the DNA to precipitate out in long strands. The DNA strands may be isolated by either spooling the sticky DNA around a glass rod or by centrifugation. If spooled, the DNA is placed in a new test tube; if centrifuged, the liquid is decanted from the pellet of DNA. The precipitated DNA, with salt and RNA present, is still not pure. It is washed for a final time with 70 percent ethanol, which does not dis-solve the DNA but forces salts present to go into solution. The DNA is then reisolated by spooling or centrifugation and dried to remove all traces of ethanol. At this point, only DNA and RNA are left; this mixture can be dissolved in a low-salt buffer containing the enzyme RNase, which degrades any RNA present, leaving pure DNA.

Technological advances have allowed deproteinization by the use of "spin columns" without the employment of toxic phenol. The raw DNA extract is placed on top of a column containing a chemical matrix that binds proteins but not DNA; the column is then centrifuged in a test tube. The raw extract passes through the chemical matrix and exits protein-free into the collection tube. These newer methods not only increase safety and reduce the production of toxic waste; they are also much faster.

—*James J. Campanella*

See also: Ancient DNA; DNA Replication; DNA Sequencing Technology; DNA Structure and Function; RFLP Analysis; RNA Isolation; RNA Structure and Function.

Further Reading

Gjerde, Douglas T., Christopher P. Hanna, and David Hornby. *DNA Chromatography.* Weinheim, Germany: Wiley-VCH, 2002. In chapters about instrumentation and operation, chromatographic principles, size-based separations, purification of nucleic acids, RNA chromatography, and special techniques, among others, this book bridges the chasm between the work of analytic chemists and molecular biologists. Illustrated.

Mirsky, Alfred. "The Discovery of DNA." *Scientific American*, June, 1968. The fascinating story of Friedrich Miescher's work.

Roe, Bruce A., Judy S. Crabtree, and Akbar S. Khan, eds. *DNA Isolation and Sequencing.* New York: John Wiley & Sons, 1996. Focus is on protocol, describing the most commonly used methods for DNA isolation, DNA sequencing, sequence analysis, and allied molecular biology techniques. Illustrated.

Sambrook, Joseph, and David W. Russell, eds. *Molecular Cloning: A Laboratory Manual.* 3d ed. 3 vols. Cold Spring Harbor, N.Y.: Cold Spring Harbor Laboratory Press, 2001. A standard for researchers, covering plasmids, bacteriophage, high-capacity vectors, gel electrophoresis, eukaryotic genomic DNA preparation and analysis, eukaryotic mRNA, polymerase chain reaction techniques, and more. Bibliographical references and index.

Trevors, J. T., and J. D. van Elsas, eds. *Nucleic Acids in the Environment.* New York: Springer, 1995. A laboratory manual that details molecular biological techniques such as DNA/RNA extraction and purification, and polymerase chain reaction methods. Illustrated.

Watson, James, et al. *Recombinant DNA.* New York: W. H. Freeman, 1992. Uses accessible language and exceptional diagrams to give a concise background on the methods, underlying concepts, and far-reaching applications of recombinant DNA technology.

Weissman, Sherman M., ed. *cDNA Preparation and Characterization.* San Diego: Academic Press, 1999. Examines the analysis and mapping of messenger RNA, gene mapping DNA, complementary isolation and purification DNA, and chromosome-mapping methods. Includes six pages of plates, illustrations.

DNA Repair

Field of study: Molecular genetics
Significance: *To protect the integrity of their genetic material, cells are able to correct damage to DNA. Many of these mechanisms are found in organisms ranging from bacteria to humans, indicating that they evolved early in the history of life. Disruption of DNA repair mechanisms in humans has been associated with the development of cancers.*

Key terms

BASE: the component of a nucleotide that gives it its identity and special properties

NUCLEOTIDE: the basic unit of DNA, consisting of a five-carbon sugar, a nitrogen-containing base, and a phosphate group

DNA Structure and DNA Damage

All living things are continually exposed to agents such as radiation or chemicals that can damage their genetic material. In addition, damage to DNA can occur spontaneously, or as a by-product of other cellular processes. Because DNA is the blueprint for directing the functions of the cell; it must be accurately maintained. The integrity of DNA also assures that all daughter cells receive the same genetic information. DNA damage can include a change in the meaning of a gene (a mutation), a break in a DNA molecule, or the abnormal joining of two DNA molecules. To a bacterial cell, DNA damage may mean death. To a multicellular organism, damaged DNA in some of its cells may mean loss of function of organs or tissues or it may lead to cancer.

A brief overview of the structure of DNA will clarify how it can be damaged. DNA is assembled from nucleotides, of which there are four types, each defined by the base it contains. If the DNA double helix is pictured as a twisted ladder, the outside supports (sometimes referred to as the "backbone" of the DNA) are alternating units of sugar and phosphate, and the "rungs" of the ladder are bases. There are four types of bases found in DNA: the double-ring purines, adenine and guanine, and the single-ring pyrimidines, cytosine and thymine. The structure of each base allows it to pair with only one other base: adenine pairs with thymine, and cytosine pairs with guanine. Base pairing holds the two strands of the double helix together and is essential for the synthesis of new DNA molecules (DNA replication) and for the transfer of information from DNA to RNA in the process of transcription. DNA replication is carried out by an enzyme called DNA polymerase, which reads the information (the sequence of bases) on a single strand of DNA, brings the appropriate nucleotide to pair with the existing strand one nucleotide at a time, and joins it to the end of a growing chain. In addition to copying entire long strands of DNA every time a cell divides, DNA polymerases are also responsible for repairing short, damaged regions of DNA. Transcription occurs through a process similar to DNA replication, except that an RNA polymerase copies only a portion of one of the strands of DNA (a gene), making an RNA copy. The RNA can then direct the production of a particular protein, which is the ultimate product of the most genes.

One of the most frequent forms of DNA damage is loss of a base. Purines are particularly unstable, and many are lost each day in human cells. If a base is absent, the DNA cannot be copied correctly during DNA replication. Another common type of DNA damage is a pyrimidine dimer, an abnormal linkage between two cytosines, two thymines, or a cytosine and a thymine next to each other in a DNA strand. These are caused by the action of ultraviolet light on DNA. A pyrimidine dimer creates a distortion in the double helix that interferes with the processes of DNA replication and transcription. Another form of DNA damage is a break

in the backbone of one or both strands of the double helix. Breaks can block DNA replication, create problems during cell division, or cause rearrangements of the chromosomes. DNA replication itself can cause problems by inserting an incorrect base or an additional or too few bases in a new strand. While DNA replication errors are not DNA damage as such, they can also lead to mutations and are subject to repair.

DNA Repair Systems

DNA repair systems are found in most organisms. Even some viruses, such as bacteriophages (which infect bacteria) and herpesviruses (which infect animals), are capable of repairing some damage to their genetic material. The DNA repair systems of single-celled organisms, including bacteria and yeasts, have been extensively studied for many years. In the 1980's and 1990's, techniques including the use of recombinant DNA methods revealed that DNA repair systems of multicellular organisms such as humans, animals, and plants are quite similar to those of microorganisms.

Scientists generally classify DNA repair systems into three categories on the basis of complexity, mechanism, and the fate of the damaged DNA. "Damage reversal" systems are the simplest: They usually require only a single enzyme to directly act on the damage and restore it to normal, usually in a single step. "Damage removal" systems are somewhat more complicated: These involve cutting out and replacing a damaged or inappropriate base or section of nucleotides and require several proteins to act together in a series of steps. "Damage tolerance" systems are those that respond to and act on damaged DNA but do not actually repair the original damage. Instead, they are ways for cells to cope with DNA damage in order to continue growth and division.

Damage Reversal Systems

Photoreactivation is one of the simplest and perhaps oldest known repair systems: It consists of a single enzyme that can split pyrimidine dimers in the presence of light. An enzyme called photolyase catalyzes this reaction; it is found in many bacteria, lower eukaryotes, insects, and plants but seems to be absent in mammals (including humans). A similar gene is present in mammals but may code for a protein that functions in another type of repair.

X rays and some chemicals such as peroxides can cause breaks in the backbone of DNA. Simple breaks in one strand are rapidly repaired by the enzyme DNA ligase. Mutant strains of microorganisms with reduced DNA ligase activity tend to have high levels of recombination since DNA ends are very "sticky" and readily join with any other fragment of DNA. While recombination is important in generating genetic diversity during sexual reproduction, it can also be dangerous if DNA molecules are joined inappropriately. The result can be aberrant chromosomes that do not function properly.

Damage Removal Systems

Damage removal systems are accurate and efficient but require the action of several enzymes and are more energetically "expensive" to the cell. There are three types of damage removal systems that work in the same general way but act on different forms of DNA damage. In "base excision" repair, an enzyme called a DNA glycosylase recognizes a specific damaged or inappropriate base and cuts the base-sugar linkage to remove the base. Next, the backbone is cut by another protein that removes the baseless sugar; then a new nucleotide is inserted to replace the damaged one by a DNA polymerase enzyme. Finally, the break in the backbone is sealed by DNA ligase. There are a number of specific glycosylases for particular types of DNA damage caused by radiation and chemicals.

The "nucleotide excision" repair system works on DNA damage that is "bulky" and that creates a block to DNA replication and transcription, such as ultraviolet-induced pyrimidine dimers and some kinds of DNA damage created by chemicals. It probably does not recognize a specific abnormal structure but sees a distortion in the double helix. Several proteins joined in a complex scan the DNA for helix distortions. When one is found, the complex binds to the damage and creates two cuts in the DNA strand containing the damaged bases on either side of the damage. The short segment

with the damaged bases (around twenty-seven nucleotides in humans) is removed from the double helix, leaving a short gap that can be filled by DNA polymerase using the intact nucleotides in the other DNA strand as a guide. In the last step, DNA ligase rejoins the strand. Mutants that are defective in nucleotide excision repair have been isolated in many organisms and are extremely sensitive to mutation induction by ultraviolet light and similar-acting chemical mutagens. Humans with the hereditary disease xeroderma pigmentosum are sunlight-sensitive and have a very high risk of skin cancers on sun-exposed areas of their bodies. These individuals have defective copies of genes that code for proteins involved in nucleotide excision repair. A comparison of the genes defective in xeroderma pigmentosum patients and those involved in nucleotide excision repair in simpler organisms reveals a great deal of similarity, indicating that this repair system evolved early in the history of life.

"Mismatch" repair occurs during DNA replication as a last "spell check" on its accuracy. By comparing mutation rates in *Escherichia coli* bacteria that either have or lack mismatch repair systems, scientists have estimated that this process adds between one hundred and one thousand times more accuracy to the replication process. It is carried out by a group of proteins that can scan DNA and look for incorrectly paired bases (or unpaired bases). The incorrect nucleotide is removed as part of a short stretch, and then the DNA polymerase gets a second try to insert the correct sequence. In 1993, Richard Fishel, Bert Vogelstein, and their colleagues isolated the first genes for human mismatch repair proteins and showed that they are very similar to those of the bacterium *Escherichia coli* and the simple eukaryote baker's yeast. Further studies in the 1990's revealed that mismatch repair genes are defective in people with hereditary forms of colon cancer.

Damage Tolerance Systems

Not all DNA damage is or can be removed immediately; some of it may persist for a while. If a DNA replication complex encounters DNA damage such as a pyrimidine dimer, it will normally act as a block to further replication of that DNA molecule. In eukaryotes, however, DNA replication initiates at multiple sites and may be able to resume downstream of a damage site, leaving a "gap" of single-stranded, unreplicated DNA in one of the two daughter molecules. The daughter-strand gap is potentially just as dangerous as the original damage site, if not more so. The reason for this is that if the cell divides with a gap in a DNA molecule, there will be no way accurately to repair that gap or the damage in one of its two daughter cells. To avoid this problem, cells have developed a way to repair daughter-strand gaps by recombination with an intact molecule of identical or similar sequence. The "recombinational" repair process, which requires a number of proteins, yields two intact daughter molecules, one of which still contains the original DNA damage. In addition to dealing with daughter-strand gaps, recombinational repair systems can also repair single- and double-strand breaks caused by the action of X rays and certain chemicals on DNA. Many of the proteins required for recombinational repair are also involved in the genetic recombination that occurs in meiosis, the sexual division process of higher cells. In 1997, it was shown that the products of the breast cancer susceptibility genes *BRCA1* and *BRCA2* participate in both recombinational repair and meiotic recombination.

An alternative choice for a DNA polymerase blocked at a DNA damage site is to change its specificity so that it can insert any nucleotide opposite the normally nonreadable damage and continue DNA replication. This type of "damage bypass" is very likely to cause a mutation, but if the cell cannot replicate its DNA, it will not be able to divide. In *Escherichia coli* bacteria, there is a set of genes that are turned on when the bacteria have received a large amount of DNA damage. Some of these gene products alter the DNA polymerase and allow damage bypass. This system has been termed the "SOS response" to indicate that it is a system of last resort. Other organisms, including humans, seem to have similar damage bypass mechanisms that allow a cell to continue growth despite DNA damage at the price of mutations. For this reason, damage bypass systems are sometimes referred to as "error-prone" or mutagenic repair systems.

Impact and Applications

DNA repair systems are an important component of the metabolism of cells. Studies in microorganisms have shown that as little as one unrepaired site of DNA damage per cell can be lethal or lead to permanent changes in the genetic material. The integrity of DNA is normally maintained by an elaborate series of interrelated checks and surveillance systems. The greatly increased risk of cancer suffered by humans with hereditary defects in DNA repair shows how important these systems are in avoiding genetic changes. As the relationship between mutations in DNA repair genes and cancer susceptibility becomes clearer, this information may be used in directing the course of cancer therapy and possibly in providing gene therapy to individuals with cancer.

—Beth A. Montelone

See also: Aging; Biochemical Mutations; Breast Cancer; Cancer; Chemical Mutagens; DNA Structure and Function; Human Genetics; Immunogenetics; Model Organism: *Escherichia coli*; Mutation and Mutagenesis; Oncogenes; Protein Structure; Protein Synthesis; RNA Structure and Function; RNA Transcription and mRNA Processing; Telomeres; Tumor-Suppressor Genes.

Further Reading

Dizdaroglu, Miral, and Ali Esat Karakaya, eds. *Advances in DNA Damage and Repair: Oxygen Radical Effects, Cellular Protection, and Biological Consequences.* New York: Plenum Press, 1999. Covers advances in research and contains contributions from scientists working in the fields of biochemistry, molecular biology, enzymology, biomedical science, and radiation biology.

Gilchrest, Barbara A., and Vilhelm A. Bohr, eds. *The Role of DNA Damage and Repair in Cell Aging.* New York: Elsevier, 2001. Topics include aging in mitotic and post-mitotic cells, age-associated changes in DNA repair and mutation rates, human premature aging syndromes as model systems, and gene action at the Werner helicase locus. Illustrated.

Henderson, Daryl S., ed. *DNA Repair Protocols: Eukaryotic Systems.* Totowa, N.J.: Humana Press, 1999. Fifty-two chapters cover all aspects of the cellular response to DNA damage, including nucleotide and base excision repair, DNA strand break repair, mismatch repair, and tolerance mechanisms.

Mills, Kevin D. *Silencing, Heterochromatin, and DNA Double Strand Break Repair.* Boston: Kluwer Academic, 2001. Presents new directions in research regarding the involvement of chromatin in the repair of broken DNA, concentrating on the study of the budding yeast *Saccharomyces cerevisiae* conducted in the laboratory of Leonard Guarente at the Massachusetts Institute of Technology.

Science. December 23, 1994. The magazine declared the DNA repair enzyme "Molecule of the Year" in 1994 and published three short reviews in this special issue that discuss three repair processes: "Mechanisms of DNA Excision Repair," by Aziz Sancar; "Transcription-Coupled Repair and Human Disease," by Philip C. Hanawalt; and "Mismatch Repair, Genetic Stability, and Cancer," by Paul Modrich.

Smith, Paul J., and Christopher J. Jones, eds. *DNA Recombination and Repair.* New York: Oxford University Press, 1999. Explores the cellular processes involved in DNA recombination and repair by highlighting current research, including strategies for dealing with DNA mismatches or lesions, the enzymology of excision repair, and the integration of DNA repair into cellular pathways.

Vaughan, Pat, ed. *DNA Repair Protocols: Prokaryotic Systems.* Totowa, N.J.: Humana Press, 2000. Divided into two sections, the book examines the classic identification, purification, and characterization of DNA repair enzymes and provides several protocols for the applied use of DNA repair proteins in the latest molecular biology techniques, including mutation detection, cloning, and genome diversification.

Weinberg, Robert A. "How Cancer Arises." *Scientific American* 197 (September, 1996). Discusses cancer and the roles of DNA repair genes.

DNA Replication

Field of study: Genetic engineering and biotechnology

Significance: *Cells and organisms pass hereditary information from generation to generation. To assure that offspring contain the same genetic information as their parents, the genetic material must be accurately reproduced. DNA replication is the molecular basis of heredity and is one of the most fundamental processes of all living cells.*

Key terms

REPLICATION: the process by which one DNA molecule is converted to two DNA molecules identical to the first

TRANSCRIPTION: the process of forming an RNA according to instructions contained in DNA

TRANSLATION: the process of forming proteins according to instructions contained in an RNA molecule

X-RAY DIFFRACTION: a method for determining the structure of molecules which infers structure by the way crystals of molecules scatter X rays as they pass through

DNA Structure and Function

The importance of chromosomes in heredity has been known since early in the twentieth century. Chromosomes consist of both DNA and protein, and in the early twentieth century there was considerable controversy concerning which component was the hereditary molecule. Early evidence favored the proteins. In 1944, however, a series of classic experiments by Oswald Avery, Maclyn McCarty, and Colin MacLeod lent strong support to the proponents favoring DNA as the genetic material. They showed that a genetic transforming agent of bacteria was DNA and not protein. In experiments reported in 1952, Alfred Hershey and Martha Chase provided evidence that DNA was the genetic material of bacteriophages (viruses that infect bacteria). Combined with additional circumstantial evidence from many sources, DNA became favored as the hereditary molecule, and a heated race began to determine its molecular structure.

In 1953, James Watson and Francis Crick published a model for the atomic structure of DNA. Their model was based on known chemical properties of DNA and X-ray diffraction data obtained from Rosalind Franklin and Maurice Wilkins. The structure itself made it clear that DNA was indeed the molecule of heredity and provided evidence for how it might be copied. The molecule resembles a ladder. The "rails" are composed of repeating units of sugar and phosphate, forming a backbone for the molecule. Each "rung" consists of a pair of nitrogenous bases, one attached to each of the two rails and held together in the middle through weak bonds called hydrogen bonds. Since there are thousands to hundreds of millions of units on a DNA molecule, the hydrogen bonds between each pair of bases add up to a strong force that holds the two strands together. DNA, then, consists of two strands, each consisting of a repeating sugar-phosphate backbone and nitrogenous bases with the two strands held together by base-pair interactions. The two strands are oriented in opposite directions. The ends of a linear DNA molecule can be distinguished by which part of the backbone sugar is exposed and are referred to as the 5' (five prime) end and the 3' end, named for a particular carbon atom on the ribose sugar. If one DNA strand is oriented 5' to 3', its complementary partner is oriented 3' to 5'. This organization has important implications for the mechanism of DNA replication.

There are four different bases: adenine (A), guanine (G), cytidine (C), and thymine (T). They can be arranged in any order on a DNA strand, allowing the enormous diversity necessary to encode the blueprint of every organism. A key feature of the double-stranded DNA molecule is that bases have strict pairing restrictions: A can only pair with T; G can only pair with C. Thus if a particular base is known on one strand, the corresponding base is automatically known on the other. Each strand can serve as a template, or mold, dictating the precise sequence of bases on the other. This feature is fundamental to the process of DNA replication.

The genome (the complete DNA content of an organism) stores all the genetic information that determines the features of that organism. The features are expressed when the DNA is transcribed to a messenger molecule, mRNA, which is used to construct a protein. The proteins encoded by the organism's genes in its DNA carry out all of the activities of the cell.

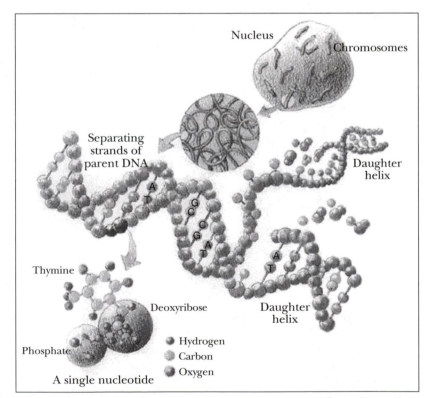

This illustration from the Human Genome Program of the Department of Energy shows the basic context of DNA replication from the cellular nucleus, which contains the chromosomes, to the separation of DNA strands and their replication at the molecular and atomic levels into daughter helixes. (U.S. Department of Energy Human Genome Program, http://www.ornl.gov/hgmis.)

The Cell Cycle

In eukaryotic organisms (most organisms other than bacteria), cells progress through a series of four stages between cell divisions. The stages begin with a period of growth (G_1 phase), followed by replication of the DNA (S phase). A second period of growth (G_2 phase) is followed by division of the cell (M phase). The two cells resulting from the cell division each go through their own cell cycle or may enter a dormant stage (G_0 phase). The passage from one stage to the next is tightly regulated and directed by internal and external signals to the cell.

The transition from G_1 into S phase marks the beginning of DNA replication. In order to enter S phase, the cell must pass through a checkpoint or restriction point in which the cell determines the quality of its DNA: If there is any damage to the DNA, entry into S phase will be delayed. This prevents the potentially lethal process of beginning replication of a DNA molecule that has damage that would prevent completion of replication. If conditions are determined to be acceptable, a "molecular switch" is thrown, triggering the initiation of DNA replication. What is the nature of this molecular switch? There are many different proteins that participate in the process of DNA replication, and they can have their activity turned off and on by other proteins. Addition or removal of a chemical group called a phosphate is a common mechanism of chemical switching. This reaction is catalyzed by a class of enzymes called kinases. Certain key proteins are phosphorylated at the boundary of the G_1 and S phases of the cell cycle by kinases, switching on DNA replication.

Origins and Initiation

If the human genome were replicated from one end to the other, it would take several years to complete the process. The DNA molecule is simply too large to be copied end to end. Instead, replication is initiated at many different sites called origins of replication, and DNA synthesis proceeds from each site in both directions until regions of copied DNA merge. The region of DNA copied from a particular origin is called a replicon. Using hundreds to tens of thousands of initiation sites and replicons, the genome can be copied in a matter of hours. The structure of replication origins has been difficult to identify in all but a few organisms, most notably yeast. Origins consist of several hundred base pairs of DNA comprising sequences that attract and bind a set of proteins called the origin recognition complex (ORC). The exact mechanism by which the origin is activated is still under investigation, but a favored model is supported by all of the available evidence.

The ORC proteins are believed to be bound to the origin DNA throughout the cell cycle but become activated at the G_1/S boundary through the action of kinases. Kinases add phosphate groups to one or more of the six ORC proteins, activating them to initiate DNA replication. Different replicons are initiated at different times throughout S phase. It is unclear how the proposed regulatory system distinguishes between replicons that have been replicated in a particular S phase and those that have not, since each must be used once and only once during each cell division cycle.

A number of different enzymatic activities are required for the initiation process. The two strands of DNA must be unwound or separated, exposing each of the parent strands so they can be used as templates for the synthesis of new, complementary strands. This unwinding is mediated by an enzyme called a helicase. Once unwound, the single strands are stabilized by the binding of proteins called single-strand binding proteins (SSBs). The resulting structure resembles a "bubble" or "eye" in the DNA strand. This structure is recognized by the DNA replication machinery that is recruited to the site, and DNA replication begins. As replication proceeds, the DNA continues to unwind through the action of helicase. The site at which unwinding and DNA synthesis are occurring is at either end of the expanding eye or bubble, called a replication fork.

DNA Synthesis

The DNA synthesis machinery is not able to synthesize a strand of DNA from scratch; rather, a short stretch of RNA is used to begin the new strands. The synthesis of the RNA is catalyzed by an enzyme called primase. This

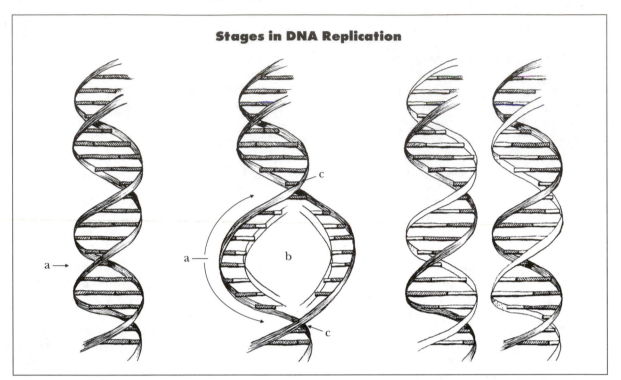

Stages in DNA Replication

At left, a double-stranded DNA molecule, with its sides formed by sugar-phosphate molecules and its "rungs" formed by base pairs. Replication begins at point (a), with the separation of a base pair as a result of the action of a special initiator protein (b). The molecule splits, or "unzips," in opposite directions (c) as each parental strand is used as a template for the daughter strand, which is formed when bases form hydrogen bonds with their appropriate "mate" bases to form new ladder "rungs." Finally (right), one parental strand and its newly synthesized daughter strand form a new double helix, while the other parental strand and its daughter strand form the second double helix. (Kimberly L. Dawson Kurnizki)

short piece of RNA, or primer, is extended using DNA nucleotides by the enzymes of DNA synthesis, called DNA polymerases. The RNA primer is later removed and replaced by DNA. Nucleotide monomers align with the exposed template DNA strand one at a time and are joined by the DNA polymerase. The joining of nucleotides into a growing DNA chain requires energy. This energy is supplied by the nucleotide monomers themselves. A high-energy phosphate bond in the nucleotide is split, and the breakage of this high-energy bond provides the energy to drive the polymerase reaction.

The two strands of DNA are not synthesized in the same way. The two strands are oriented opposite one another, but DNA synthesis only occurs in one direction: 5' to 3'. Therefore, one strand, called the leading strand, is synthesized continuously in the same direction that the replication fork is moving, while the lagging strand is synthesized away from the direction of fork movement. Since the lagging-strand DNA synthesis and fork movement are in opposite directions, this strand of DNA must be made in short pieces that are later joined. Lagging-strand synthesis is therefore said to be discontinuous. These short intermediates are called Okazaki fragments, named for their discoverer, Reiji Okazaki. Overall, DNA replication is said to be semidiscontinuous.

The DNA synthesis machine operating at the replication fork is a complex assembly of proteins. Many different activities are necessary to carry out the process of DNA replication efficiently. Several proteins are necessary to recognize the unwound origin and assemble the rest of the complex. Primase must function to begin both new strands and is then required periodically throughout synthesis of the lagging strand. A doughnut-shaped clamp called

PCNA functions as a "processivity factor" to keep the entire complex attached to the DNA until the job is completed. Helicase is continuously required to unwind the template DNA and move the fork along the parent molecule. As the DNA is unwound, strain is created on the DNA ahead of the replication fork. This strain is alleviated through the action of topoisomerase enzymes. Single-strand binding proteins are needed to stabilize the regions of unwound DNA that exist before the DNA is actually copied. Finally, an enzyme called ligase is

The Replication Process

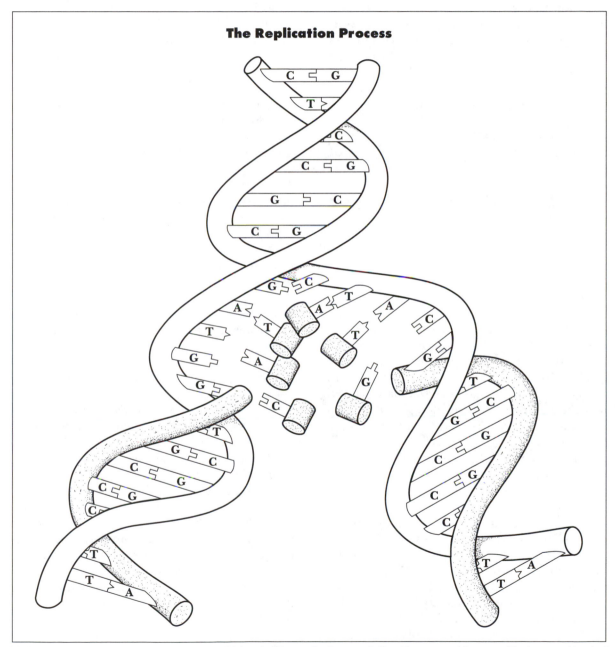

A detailed schematic of DNA replication, in which a double-standard parent helix splits apart and reassembles into two identical daughter helixes. The amino acid base pairs are reproduced exactly, because cytosine (C) pairs only with guanine (G), and adenine (A) pairs only with thymine (T). (Electronic Illustrators Group)

necessary to join the regions replicated from different origins and to attach all of the Okazaki fragments of the lagging strand. All of these proteins are part of a well-orchestrated, efficient machine ideally suited to its task of copying the genetic material.

DNA polymerases are not perfect. At a relatively low frequency, they can add an incorrect nucleotide to a growing chain, one that does not match the template strand as dictated by the base-pairing rules. However, because the DNA molecules are so extremely large, novel mechanisms for proofreading have evolved to ensure that the genetic material is copied accurately. DNA polymerases can detect the miscorporation of a nucleotide and use an additional enzymatic activity to correct the mistake. Specifically, the polymerase can "back up" and cut out the last nucleotide added, then try again. With this and other mechanisms to correct errors, the observed error rate for DNA synthesis is a remarkable one error in every billion nucleotides added.

Impact and Applications

DNA replication is a fundamental cellular process: Proper cell growth cannot occur without it. It must be carefully regulated and tightly controlled. Despite its basic importance, the details of the mechanisms that regulate DNA replication are poorly understood. Even with all of the checks and balances that have evolved to ensure a properly replicated genome, occasional mistakes do occur. Attempting to replicate a genome damaged by chemical or other means may simply lead to death of a single cell. Far more ominous are genetic errors that lead to loss of regulating mechanisms. Without regulation, cell growth and division can proceed without normal limits, resulting in cancer. Much of the focus for the study of cell growth and regulation is to set a foundation for the understanding of how cancer cells develop. This knowledge may lead to new techniques for selective inhibition or destruction of cancer cells.

Manipulation of DNA replication and cell cycle control are the newest tools for progress in genetic engineering. In early 1997, the first successful cloning of an adult mammal, Dolly the sheep, raised important new issues about the biology and ethics of manipulating mammalian genomes. The technology now exists to clone human beings, although such experiments are not likely to be carried out. More relevant is the potential impact on agriculture. It is now possible to select for animals that have the most desirable traits, such as lower fat content or disease resistance, and create herds of genetically identical animals. Of direct relevance to humans is the potential impact on the understanding of fertility and possible new treatments for infertility.

A new class of genetic diseases was discovered in the 1980's called triplet repeat diseases. Regions of DNA consist of copies of three nucleotides (such as CGG) that are repeated up to fifty times. Through unknown mechanisms related to DNA replication, the number of repeats may increase from generation to generation, at some point reaching a threshold level at which disease symptoms appear. Diseases found to conform to this pattern include fragile X syndrome, Huntington's disease (Huntington's chorea), and Duchenne muscular dystrophy.

The process of aging is closely related to DNA replication. Unlike bacteria, eukaryotic organisms have linear chromosomes. This poses problems for the cell, both in maintaining intact chromosomes (ends are unstable) and in replicating the DNA. The replication machinery cannot copy the extreme ends of a linear DNA molecule, so organisms have evolved alternate mechanisms. The ends of linear chromosomes consist of telomeres (short, repeated DNA sequences that are bound and stabilized by specific proteins), which are replicated by a separate mechanism using an enzyme called telomerase. Telomerase is inactivated in mature cells, and there may be a slow, progressive loss of the telomeres that ultimately leads to the loss of important genes, resulting in symptoms of aging. Cancer cells appear to have reactivated their telomerase, so potential anticancer therapies are being developed based on this information.

—Michael R. Lentz

See also: Animal Cloning; Cancer; Cell Cycle, The; Cell Division; Cloning; DNA Sequencing Technology; DNA Structure and Function;

Genetic Code; Genetic Engineering; Molecular Genetics; Mutation and Mutagenesis; Protein Structure; Protein Synthesis; Restriction Enzymes; RNA Structure and Function; RNA Transcription and mRNA Processing; Telomeres.

Further Reading

Abstracts of Papers Presented at the 2001 Meeting on Eukaryotic DNA Replication: September 5-September 9, 2001. Arranged by Thomas Kelly and Bruce Stillman. Cold Spring Harbor, N.Y.: Cold Spring Harbor Laboratory, 2001. This annual meeting results in similar publications yearly.

Cann, Alan J. *DNA Virus Replication.* New York: Oxford University Press, 2000. Gives an analysis of protein-protein interactions in DNA virus replication, covering all major DNA virus groups: hepatitis B virus, papillomavirus, herpes simplex virus, Epstein-Barr virus, Kaposi's sarcoma herpesvirus (KSHV), human cytomegalovirus, and adenoviruses. Illustrated.

Cotterill, Sue, ed. *Eukaryotic DNA Replication: A Practical Approach.* New York: Oxford University Press, 1999. Serves as a comprehensive lab manual that describes key aspects of current techniques for investigating DNA replication in eukaryotes. More than one hundred reliable protocols, including methods for studying the origin of replication, replication proteins, and the synthesis of telomeres.

DePamphilis, Melvin L., ed. *Concepts in Eukaryotic DNA Replication.* Cold Spring Harbor, N.Y.: Cold Spring Harbor Laboratory Press, 1999. A broad account of the basic principles of DNA replication and related functions such as DNA repair and protein phosphorylation. One chapter surveys the most recent advances in the field.

Drlica, Karl. *Understanding DNA and Gene Cloning: A Guide for the Curious.* Rev. ed. New York: Wiley, 2003. An excellent introduction to the basic properties of DNA and its modern applications. Consists of four sections: basic molecular biology, manipulation of DNA, insights gained through the use of gene cloning (including a chapter on retroviruses), and human genetics.

Holmes, Frederic Lawrence. *Meselson, Stahl, and the Replication of DNA: A History of "the Most Beautiful Experiment in Biology."* New Haven, Conn.: Yale University Press, 2001. Traces the evolution of Matthew Meselson and Frank Stahl's 1957 landmark experiment, which confirmed that DNA replicates as predicted by the double helix structure Watson and Crick had recently proposed. Illustrations.

Kornberg, Arthur. *For the Love of Enzymes: The Odyssey of a Biochemist.* Reprint. Cambridge, Mass.: Harvard University Press, 1991. Kornberg discovered the enzymes that replicate DNA and was awarded the Nobel Prize for his work. This autobiography is a rich history of the process of science and discovery.

Watson, James. *The Double Helix.* New York: Simon and Schuster, 2001. Watson's account of the race to solve the structure of the DNA molecule.

DNA Sequencing Technology

Field of study: Genetic engineering and biotechnology

Significance: *The genetic code is contained in the ordered, linear arrangement of the four DNA nucleotides: adenine, cytosine, guanine, and thymine. Determining this sequence is called DNA sequencing. Advances in DNA sequencing technology have increased speed and accuracy of sequencing by several orders of magnitude.*

Key terms

AUTOMATED FLUORESCENT SEQUENCING: a modification of dideoxy termination sequencing which uses fluorescent markers to identify the terminal nucleotides, allowing the automation of sequencing in which robots can carry out large scale projects

BASE PAIR (BP): often used as a measure of the size of a DNA fragment or the distance along a DNA molecule between markers; both the singular and plural are abbreviated bp

MAXAM-GILBERT SEQUENCING: A method of base-specific chemical degradation to determine DNA sequence

PRIMER: A short piece of single-stranded DNA that can hybridize to denatured DNA and provide a start point for extension by a DNA polymerase

SANGER SEQUENCING: Also known as dideoxy termination sequencing, a method using nucleotides that are missing the 3′ hydroxyl group in order to terminate the polymerization of new DNA at a specific nucleotide

The Need for Sequencing

DNA was first discovered in the 1869 as a viscous material in pus, and its basic chemical composition was well established by the 1930's. By 1950, the critical role of DNA as the hereditary material was clearly determined. In the 1950's, the classic papers by James Watson and Francis Crick and Matthew Meselson and Frank Stahl gave scientists a clear picture of the structure and mode of replication of DNA. Crick demonstrated that the genetic code con-

Frederick Sanger developed one of the first methods for sequencing DNA and published the first genome sequence. (© The Nobel Foundation)

sisted of triplet codons in 1961. However, there was no system to read the sequence and uncover the actual words that spelled out the code of life.

The discovery of rapid sequencing methods in the 1970's created a flood of new discoveries in biology. The coding region and control elements could be identified and compared. The sequence changes in different alleles of the same gene could be identified, homologous genes could be identified in divergent species, and evolutionary changes could be studied. Today, an entire genome can be sequenced, meaning the identification of every nucleotide in the correct order along every chromosome, in a matter of months. This ability to sequence the genomes of entire organisms has created a whole new field called genomics, the study and comparison of whole genomes of different organisms. Sequencing is now at the core of many of the new discoveries in biology.

Principles of DNA Sequencing

Molecular biologists cannot observe DNA molecules directly, even through a microscope, so they must devise controlled chemical reactions whose outcomes are indicative of what occurs at the submicroscopic level. In DNA sequencing, the key is to use a chemical method that allows analysis of the base sequence one base at a time. Such a method needs to produce a collection of DNA fragments whose lengths can be used to detect the identity of the base located at the end of each different-sized fragment. For example, if fragments of the short DNA sequence ACGTCCGATCG can be predictably produced, then the size of each fragment could be used to determine the locations of each base. If the fragment is cut to the right of each of the thymine bases, two fragments of 4 and 9 base pairs (bp) will be produced. Doing the same for the other three nucleotides could identifiy their positions. Reading from smallest to largest fragment and seeing which reaction generates each piece provides the DNA sequence. Although this is a very simple example, the principles apply to all current sequencing methods. Electrophoresis in denaturing polyacrylamide gels (to keep the DNA single-stranded) is used to separate fragments that are

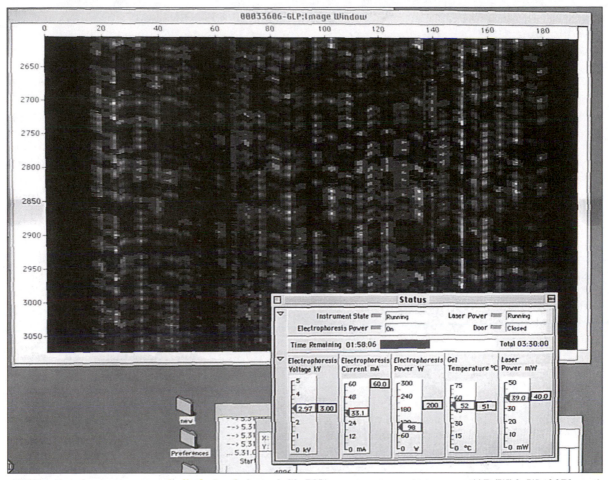

A DNA sequencing program at work, displaying the bar-code-like DNA sequence on a computer screen. (AP/Wide World Photos)

hundreds of base pairs in length but differ by only a single nucleotide. The DNA is labeled with either radioactive or fluorescent markers so that the bands can be detected.

Maxam-Gilbert Sequencing

Organic chemists working with DNA have identified many chemicals that react with specific bases and cleave the backbone of the strand at that location. To sequence DNA with this method, the DNA fragment to be sequenced is isolated and the 5′ end of only one of the strands is labeled by the placement in the terminal phosphate of the radioactive atom phosphorus 32. This creates the endpoint. In separate tubes, the DNA is reacted with chemicals that will cleave at one of the four nucleotides. Since only one strand is broken, it is nec-

essary to denature the DNA to separate the strands before separating them by size so that the fragments will correspond in length to the distance from the terminal phosphate label to the cleavage point. The method requires dangerous chemicals and does not easily lend itself to automation, so it is rarely used today.

Sanger Sequencing

This method requires that the sequence of a short stretch of DNA adjacent to the region to be sequenced is known so that a short synthetic oligonucleotide can be made which can hybridize onto the region to act as a primer for DNA synthesis in the direction of the DNA to be sequenced. This is usually not a problem, since the DNA to be sequenced is often cloned into a plasmid vector whose sequence is known.

The DNA is denatured and the primer is allowed to anneal. A DNA polymerase is added and extends the DNA for a short distance in the presence of radioactive nucleotides, which labels the new DNA. The reaction is then divided into four equal parts and added into four separate reaction tubes. Each tube contains all four DNA nucleotides, but a small percentage of one nucleotide is missing the 3′ hydroxyl group. Without the hydroxyl group, no more bases can be added and the reaction terminates. Since the dideoxy nucleotide constitutes only a small percentage of the available nucleotides, the reactions will terminate at random places along the DNA strand. Since the terminated fragment is attached to the larger template strand, the DNA must be denatured before gel electrophoresis so that the size will correspond accurately to the position of the terminated base.

Automated Sequencing

Automated sequencing is a variant of Sanger sequencing. Each of the four dideoxy bases has a different fluorescent dye attached. When the elongation is blocked, the fragment will also be labeled with a specific color indicating which nucleotide is in the terminal position. As a result, four separate reactions (one for each nucleotide) are not needed. Also, the polymerase chain reaction (PCR), which requires much smaller amounts of DNA, is often used in automated sequencers. The reaction products are electrophoresed through a narrow tube of polyacrylamide with a laser and fluorescence detector at the bottom. As the different-sized fragments reach the bottom, the detector will register the colors as they pass. The data are logged on a computer, which outputs the DNA sequence. This system has been automated with robot arms moving the samples into reaction tubes and loading them into the tube gels and computers compiling and comparing the sequence data. An automated sequencer may cost more than $250,000 but can generate 10,000 bp of new sequence data per day.

Future Directions

Scientists have developed silicon chips that can bind to DNA and change their electrical properties when they bind. Currently, the technology allows interaction only with specific sequences. However, the goal is to develop new systems that will generate electrical outputs indicative of the sequence of the DNA with which the chip interacts. This would allow the generation of new data at phenomenal rates, would allow field scientists to sequence the genomes of new organisms instantly, and would allow police investigators to generate sequence evidence right at the crime scene. While such developments would be very exciting from a scientific perspective, they would also open many new questions about personal privacy as prospective employers (and mates) could easily scan a person's genome for undesirable sequences.

—*J. Aaron Cassill*

See also: Cloning; Cloning Vectors; DNA Replication; Genetic Code; Genetic Engineering; Genome Libraries; Genome Size; Genomics; Human Genome Project; Knockout Genetics and Knockout Mice; Model Organism: *Escherichia coli*; Molecular Clock Hypothesis; Polymerase Chain Reaction; Population Genetics; Pseudogenes; Repetitive DNA; Restriction Enzymes; Reverse Transcriptase; RFLP Analysis; Shotgun Cloning; Synthetic Genes; Transposable Elements.

Further Reading

Maxam, Allan M., and Walter Gilbert. "A New Method for Sequencing DNA." *Proceedings of the National Academy of Sciences* 74 (1977): 560. The original description of sequencing by chemical cleavages.

Reilly, Philip R. *Abraham Lincoln's DNA and Other Adventures in Genetics.* Cold Spring Harbor, N.Y.: Cold Spring Harbor Laboratory Press, 2000. A series of brief articles about the social and moral implications of uncovering DNA information in humans.

Sanger, F., S. Nicklen, and A. R. Coulson. "DNA Sequencing with Chain-Terminating Inhibitors." *Proceedings of the National Academy of Sciences* 74 (1977): 5463. The original description of dideoxy termination sequencing.

Smith, Lloyd M., et al. "Fluorescence Detection in Automated DNA Sequence Analysis." *Nature* 321 (1986): 674. The original description of automated sequencing techniques.

DNA Structure and Function

Field of study: Molecular genetics
Significance: *Structurally, DNA is a relatively simple molecule; functionally, however, it has wide-ranging roles in the cell. It functions primarily as a stable repository of genetic information in the cell and as a source of genetic information for the production of proteins. Greater knowledge of the characteristics of DNA has led to advances in the fields of genetic engineering, gene therapy, and molecular biology in general.*

Key terms

DOUBLE HELIX: the molecular shape of DNA molecules, which resembles a ladder that twists, or spirals

GENE EXPRESSION: the processes (transcription and translation) by which the genetic information in DNA is converted into protein

TRANSCRIPTION: the process by which genetic information in DNA is converted into messenger RNA (mRNA)

TRANSLATION: the process by which the genetic information in an mRNA molecule is converted into protein

Chemical and Physical Structure of DNA

Deoxyribonucleic acid (DNA) is the genetic material found in all cells. Chemically, it is classified as a nucleic acid, a relatively simple molecule composed of nucleotides. A nucleotide consists of a sugar (deoxyribose), a phosphate group, and one of the nitrogenous bases: adenine (A), cytosine (C), guanine (G), or thymine (T). In fact, nucleotides differ only in the particular nitrogenous base that they contain. Ribonucleic acid (RNA) is the other type of nucleic acid found in the cell; however, it contains ribose as its sugar instead of deoxyribose and has the nitrogenous base uracil (U) instead of thymine. Nucleotides can be assembled into long chains of nucleic acid via connections between the sugar on one nucleotide and the phosphate group on the next, thereby creating a sugar-phosphate "backbone" in the molecule. The nitrogenous base on each nucleotide is positioned such that it is perpendicular to the backbone, as shown in the following diagram:

```
sugar — phosphate — sugar — phosphate — sugar — phosphate — sugar
  |                   |                   |                   |
 base                base                base                base
```

Any one of the four DNA nucleotides (A, C, G, or T) can be used at any position in the molecule; it is therefore the specific sequence of nucleotides in a DNA molecule that makes it unique and able to carry genetic information. The genetic information is the sequence itself.

In the cell, DNA exists as a double-stranded molecule; this means that it consists of two chains of nucleotides side by side. The double-stranded form of DNA can most easily be visualized as a ladder, with the sugar-phosphate backbones being the sides of the ladder and the nitrogenous bases being the rungs of the ladder, as shown in the following diagram:

```
sugar — phosphate — sugar — phosphate — sugar — phosphate — sugar
  |                   |                   |                   |
 base                base                base                base
  |                   |                   |                   |
 base                base                base                base
  |                   |                   |                   |
sugar — phosphate — sugar — phosphate — sugar — phosphate — sugar
```

This ladder is then twisted into a spiral shape. Any spiral-shaped molecule is called a "helix," and since each strand of DNA is wound into a spiral, the complete DNA molecule is often called a "double helix." This molecule is extremely flexible and can be compacted to a great degree, thus allowing the cell to contain large amounts of genetic material.

The Discovery of DNA as the Genetic Material

Nucleic acids were discovered in 1869 by the physician Friedrich Miescher. He isolated these molecules, which he called nuclein, from the nuclei of white blood cells. This was the first association of nucleic acids with the nucleus of the cell. In the 1920's, experiments performed by other scientists showed that DNA could be located on the chromosomes within the nucleus. This was strong evidence for the role of DNA in heredity, since at that time there was already a link between the activities of chromosomes during cell division and the inheritance of particular traits, largely because of the work

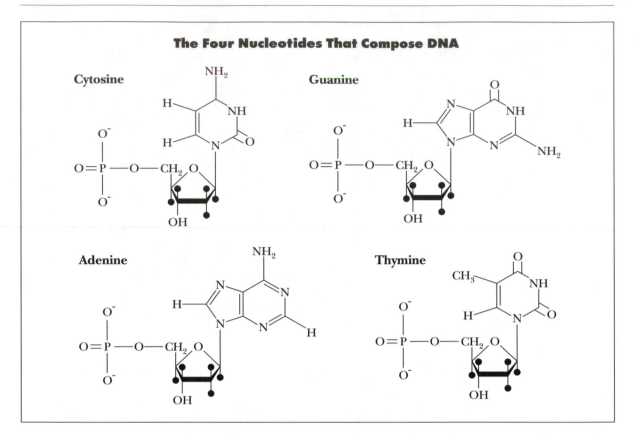

The Four Nucleotides That Compose DNA

Cytosine

Guanine

Adenine

Thymine

of the geneticist Thomas Hunt Morgan about ten years earlier.

However, it was not immediately apparent, based on this evidence alone, that DNA was the genetic material. In addition to DNA, proteins are present in the nucleus of the cell and are an integral part of chromosomes as well. Proteins are also much more complex molecules than nucleic acids, having a greater number of building blocks; there are twenty amino acids that can be used to build proteins, as opposed to only four nucleotides for DNA. Moreover, proteins tend to be much more complex than DNA in terms of their three-dimensional structure as well. Therefore, it was not at all clear in the minds of many scientists of the time that DNA had to be the genetic material, since proteins could not specifically be ruled out.

In 1928, the microbiologist Frederick Griffith supplied some of the first evidence that eventually led to the identification of DNA as the genetic material. Griffith's research involved the bacterium *Streptococcus pneumoniae,* a common cause of lung infections. He was working primarily with two different strains of this bacterium: a strain that was highly virulent (able to cause disease) and a strain that was nonvirulent (not able to cause disease). Griffith noticed that if he heat-killed the virulent strain and then mixed its cellular debris with the living, nonvirulent strain, the nonvirulent strain would be "transformed" into a virulent strain. He did not know what part of the heat-killed virulent cells was responsible for the transformation, so he simply called it the "transforming factor" to denote its activity in his experiment. Unfortunately, Griffith never took the next step necessary to reveal the molecular identity of this transforming factor.

That critical step was taken by another microbiologist, Oswald Avery, and his colleagues in 1944. Avery essentially repeated Griffith's experiments with two important differences: Avery partially purified the heat-killed virulent strain preparation and selectively treated this preparation with a variety of enzymes to see if the transforming factor could be eliminated, thereby eliminating the transformation itself. Avery showed that transformation was prevented only when the preparation was treated with deoxyribonuclease, an enzyme that specifically attacks and destroys DNA. Other enzymes that specifically destroy RNA or proteins could not prevent transformation from occurring. This was extremely strong evidence that the genetic material was DNA.

Experiments performed in 1952 by molecular biologists Alfred Hershey and Martha Chase using the bacterial virus *T2* finally demonstrated conclusively that DNA was indeed the genetic material. Hershey and Chase studied how *T2* infects bacterial cells to determine what part of the virus, DNA or protein, was responsible for causing the infection, thinking that whatever molecule directed the infection would have to be the genetic material of the virus. They found that DNA did directly participate in infection of the cells by entering them, while the protein molecules of the viruses stayed outside the cells. Most strikingly, they found that the original DNA of the "parent" viruses showed up in the "offspring" viruses produced by the infection, directly demonstrating inheritance of DNA from one generation to another. This was an important element of the argument for DNA as the genetic material.

The Watson-Crick Double-Helix Model of DNA

With DNA conclusively identified as the genetic material, the next step was to determine the structure of the molecule. This was finally accomplished when the double-helix model of DNA was proposed by molecular biologists James Watson and Francis Crick in 1953. This model has a number of well-defined and experimentally determined characteristics. For example, the diameter of the molecule, from one sugar-phosphate backbone to the other, is 20 angstroms. (There are 10 million angstroms in one

millimeter, which is one-thousandth of a meter.) There are 3.4 angstroms from one nucleotide to the next, and the entire double helix makes one turn for every ten nucleotides, a distance of about 34 angstroms. These measurements were determined by the physicists Maurice Wilkins and Rosalind Franklin around 1951 using a process called X-ray diffraction, in which crystals of DNA are bombarded with X rays; the resulting patterns captured on film

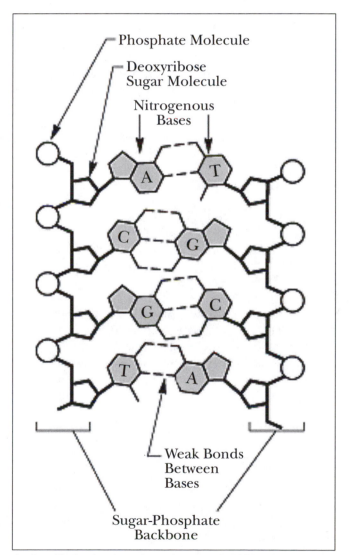

A schematic showing the major components of a DNA molecule, including the four bases that compose DNA—adenine (A) and thymine (T), cytosine (C) and guanine (G)—and how they form the "rungs" of the DNA "ladder" by forming hydrogen bonds. (U.S. Department of Energy Human Genome Program, http://www.ornl.gov/hgmis)

gave Wilkins and Franklin, and later Watson and Crick, important clues about the physical structure of DNA.

Another important aspect of Watson and Crick's double-helix model is the interaction between the nitrogenous bases in the interior of the molecule. Important information about the nature of this interaction was provided by molecular biologist Erwin Chargaff in 1950. Chargaff studied the amounts of each nitrogenous base present in double-stranded DNA from organisms as diverse as bacteria and humans. He found that no matter what the source of the DNA, the amount of adenine it contains is always roughly equal to the amount of thymine; there are also equal amounts of guanine and cytosine in DNA. This information led Watson and Crick to propose an interaction, or "base pairing," between these sets of bases such that A always base pairs with T (and vice versa) and G always base pairs with C. Another name for this phenomenon is "complementary base pairing": A is said to be the "complement" of T, and so on.

The force that holds complementary bases, and therefore the two strands of DNA, together is a weak chemical interaction called a "hydrogen bond," which is created whenever a hydrogen atom in one molecule has an affinity for nitrogen or oxygen atoms in another molecule. The affinity of the atoms for each other draws the molecules together in the hydrogen bond. A-T pairs have two hydrogen bonds between them because of the chemical structure of the bases, whereas G-C pairs are connected by three hydrogen bonds, making them slightly stronger and more stable than A-T pairs. The entire DNA double helix, although it is founded upon the hydrogen bond, one of the weakest bonds in nature, is nonetheless an extraordinarily stable structure because of the combined force of the millions of hydrogen bonds holding most DNA molecules together. However, these hydrogen bonds can be broken under certain conditions in the cell. This usually occurs as part of the process of the replication of the double helix, in which the two strands of DNA must come apart in order to be duplicated. In the cell, the hydrogen bonds are broken with the help of enzymes. Under artificial conditions in the laboratory, hydrogen bonds in the double helix can easily be broken just by heating a solution of DNA to high temperatures (close to the boiling point).

Other Features of the Watson-Crick Model

Watson and Crick were careful to point out that their double-helix model of DNA was the first model to immediately suggest a mechanism by which the molecule could be replicated. They knew that this replication, which must occur before the cell can divide, would be a necessary characteristic of the genetic material of the cell and that an adequate model of DNA must help explain how this duplication could occur. Watson and Crick realized that the mechanism of complementary base pairing that was an integral part of their model was a potential answer to this problem. If the double helix is separated into its component single-strand molecules, each strand will be able to direct the replacement of the opposite, or complementary, strand by base pairing properly with only the correct nucleotides. For example, if a single-strand DNA molecule has the sequence TTAGTCA, the opposite complementary strand will always be AATCAGT; it is as if the correct double-stranded structure is "built in" to each single strand. Additionally, as each of the single strands in a double-strand DNA molecule goes through this addition of complementary nucleotides, two new DNA double helices are produced where there was only one before. Further, these new DNA molecules are completely identical to each other, barring any mistakes that might have been made in the replication process.

A strand of DNA also has a certain direction built into it; the DNA double helix is often called "antiparallel" in reference to this aspect of its structure. "Antiparallel" means that although the two strands of the DNA molecule are essentially side by side, they are oriented in different directions relative to the position of the deoxyribose molecules on the backbone of the molecule. To help keep track of the orientation of the DNA molecule, scientists often refer to a 5′ to 3′ direction. This designation comes from numbering the carbon atoms on the deoxyribose molecule (from 1′ to 5′) and takes

The Structure of DNA

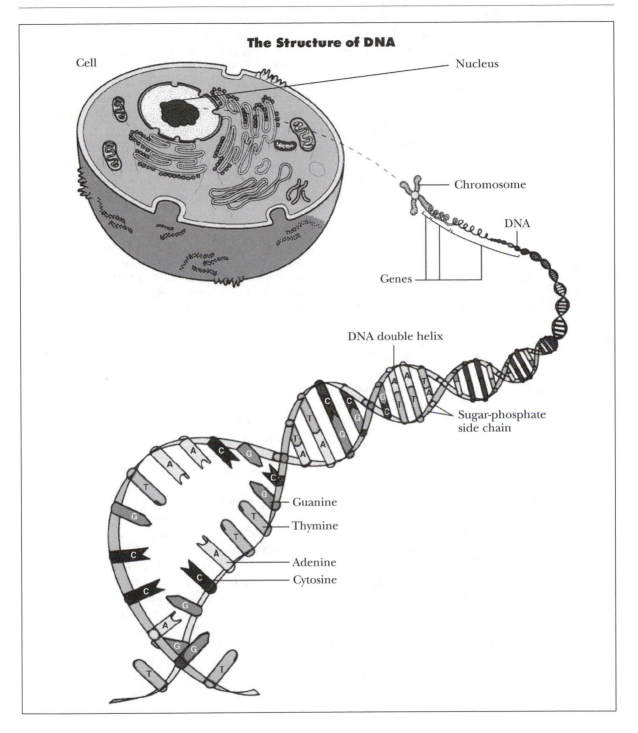

Cell

Nucleus

Chromosome

DNA

Genes

DNA double helix

Sugar-phosphate side chain

Guanine

Thymine

Adenine

Cytosine

note of the fact that the deoxyribose molecules on the DNA strand are all oriented in the same direction in a head-to-tail fashion. If it were possible to stand on a DNA molecule and walk down one of the sugar-phosphate backbones, one would encounter a 5′ carbon atom on a sugar, then the 3′ carbon, and so on all the way down the backbone. If one were walking on the other strand, the 3′ carbon atom would always be encountered before the 5′ carbon. The concept of an antiparallel double helix has important implications for the ways that DNA is produced and used in the cell. Generally, the cellular enzymes that are involved in processes concerning DNA are restricted to recognizing just one direction. For example, DNA polymerase, the enzyme that is responsible for making DNA in the cell, can only make DNA in a 5′ to 3′ direction, never the reverse.

Watson and Crick postulated a right-handed helix as part of their double-helix model; this means that the strands of DNA turn to the right, or in a counterclockwise fashion. This is now regarded as the "biological" (B) form of DNA because it is the form present inside the nucleus of the cell and in solutions of DNA. However, it is not the only possible form of DNA. In 1979, an additional form of DNA was discovered by molecular biologist Alexander Rich that exhibited a zigzag, left-handed double helix; he called this form of DNA Z-DNA. Stretches of alternating G and C nucleotides most commonly give rise to this conformation of DNA, and scientists think that this alternative form of the double helix is important for certain processes in the cell in which various molecules bind to the double helix and affect its function.

The Function of DNA in the Cell

DNA plays two major roles in the cell. The first is to serve as a storehouse of the cell's genetic information. Normally, cells have only one complete copy of their DNA molecules, and this copy is, accordingly, highly protected. DNA is a chemically stable molecule; it resists damage or destruction under normal conditions; and, if it is damaged, the cell has a variety of mechanisms to ensure the molecule is rapidly repaired. Furthermore, when the DNA in the cell is duplicated in a process called DNA replication, this duplication occurs in a regulated and precise fashion so that a perfect copy of DNA is produced. Once the genetic material of the cell has been completely duplicated, the cell is ready to divide in two in a process called mitosis. After cell division, each new cell of the pair will have a perfect copy of the genetic material; thus these cells will be genetically identical to each other. DNA thus provides a

mechanism by which genetic information can be transferred easily from one generation of cells (or organisms) to another.

The second role of DNA is to serve as a blueprint for the ultimate production of proteins in the cell. This process occurs in two steps. The first step is the conversion of the genetic information in a small portion of the DNA molecule, called a gene, into messenger RNA (mRNA). This process is called transcription, and here the primary role of the DNA molecule is to serve as a template for synthesis of the mRNA molecule. The second step, translation, does not involve DNA directly; rather, the mRNA produced during transcription is in turn used as genetic information to produce a molecule of protein. However, it is important to note that genetic information originally present in the DNA molecule indirectly guides the synthesis and final amino acid sequence of the finished protein. Both of these steps, transcription and translation, are often called gene expression. A single DNA molecule in the form of a chromosome may contain thousands of different genes, each providing the information necessary to produce a particular protein. Each one of these proteins will then fulfill a particular function inside or outside the cell.

Impact and Applications

Knowledge of the physical and chemical structure of DNA and its function in the cell has undoubtedly had far-reaching effects on the science of biology. However, one of the biggest effects has been the creation of a new scientific discipline: molecular biology. With the advent of Watson and Crick's double-helix model of DNA, it became clear to many scientists that, perhaps for the first time, many of the important molecules in the cell could be studied in detail and that the structure and function of these molecules could also be elucidated. Within fifteen years of Watson and Crick's discovery, a number of basic genetic processes in the cell had been either partially or completely detailed, including DNA replication, transcription, and translation. Certainly the seeds of this revolution in biology were being planted in the decades before Watson and Crick's 1953 model, but it was the double helix that allowed scientists to investigate the important issues of genetics on the cellular and molecular levels.

An increased understanding of the role DNA plays in the cell has also provided scientists with tools and techniques for changing some of the genetic characteristics of cells. This is demonstrated by the rapidly expanding field of genetic engineering, in which scientists can precisely manipulate DNA and cells on the molecular level to achieve a desired result. Additionally, more complete knowledge of how the cell uses DNA has opened windows of understanding into abnormal cellular processes such as cancer, which is fundamentally a defect involving the cell's genetic information or the expression of that information.

Through the tools of molecular genetics, many scientists hope to be able to correct almost any genetic defect that a cell or an organism might have, including cancer or inherited genetic defects. The area of molecular biology that is concerned with using DNA as a way to correct cellular defects is called gene therapy. This is commonly done by inserting a normal copy of a gene into cells that have a defective copy of the same gene in the hope that the normal copy will take over and eliminate the effects of the defective gene. It is hoped that this sort of technology will eventually be used to overcome even complex problems such as Alzheimer's disease and acquired immunodeficiency syndrome (AIDS).

One of the most unusual and potentially rewarding applications of DNA structure was introduced by computer scientist Leonard Adleman in 1994. Adleman devised a way to use short pieces of single-stranded DNA in solution as a rudimentary "computer" to solve a relatively complicated mathematical problem. By devising a code in which each piece of DNA stood for a specific variable in his problem and then allowing these single-stranded DNA pieces to base pair with each other randomly in solution, Adleman obtained an answer to his problem in a short amount of time. Soon thereafter, other computer scientists and molecular biologists began to experiment with other applications of this fledgling technology, which represents an exciting synthesis of two formerly separate disciplines. It may be that this research will prove to be the seed of another biological revolution with DNA at its center.

—*Randall K. Harris*

See also: Ancient DNA; Antisense RNA; Chromosome Structure; DNA Isolation; DNA Repair; DNA Replication; Genetic Code; Genetic Code, Cracking of; Molecular Genetics; Noncoding RNA Molecules; One Gene-One Enzyme Hypothesis; Protein Structure; Protein Synthesis; Repetitive DNA; RNA Isolation; RNA Structure and Function; RNA Transcription and mRNA Processing.

Further Reading

Banaszak, Leonard. *Foundations of Structural Biology.* San Diego: Academic Press, 2000. Provides visualization skills with three-dimensional imaging to assist students in understanding the implications of the three-dimensional coordinates for a molecule with several thousand atoms.

Bradbury, E. Morton, and Sandor Pongor, eds. *Structural Biology and Functional Genomics.* Boston: Kluwer Academic, 1999. Topics include DNA repeats in the human genome, modeling DNA stretching for physics and biology, chromatin control of HIV-1 gene expression, and exploring structure space.

Frank-Kamenetskii, Maxim D. *Unraveling DNA.* Reading, Mass.: Addison-Wesley, 1997. Melds history, biographical details, and science to provide a general discussion of DNA and its basic structure and function.

McCarty, Maclyn. *The Transforming Principle: Discovering That Genes Are Made of DNA.* New York: W. W. Norton, 1994. Gives an insider's view of the circumstances surrounding Oswald Avery's pivotal experiments.

Maddox, Brenda. *Rosalind Franklin: The Dark Lady of DNA.* New York: HarperCollins, 2002. Tells the other side of the story in the discovery and structure of DNA, focusing on the often neglected role of Franklin. Illustrations, bibliography, index.

Rosenfield, Israel, et al. *DNA for Beginners.* New York: W. W. Norton, 1983. Provides an entertaining, yet factual, cartoon account of basic DNA structure and function as well as more advanced topics in molecular biology.

Smith, Paul J., and Christopher J. Jones, eds. *DNA Recombination and Repair.* New York: Oxford University Press, 1999. Explores the cellular processes involved in DNA recombination and repair by highlighting current research, including strategies for dealing with DNA mismatches or lesions, the enzymology of excision repair, and the integration of DNA repair into other cellular pathways.

Watson, James. *The Double Helix.* New York: Simon and Schuster, 2001. Watson's account of the race to solve the structure of the DNA molecule.

Watson, James, et al. *Recombinant DNA.* New York: W. H. Freeman, 1992. Uses accessible language and exceptional diagrams to give a concise background on the methods, underlying concepts, and far-reaching applications of recombinant DNA technology.

Web Sites of Interest

Deakin University, Australia. http://agrippina .bcs.deakin.edu.au/bcs_courses/forensic/ chemical%20detective/dna_type.htm. A richly illustrated guide to the molecular structure and function of DNA.

Left-Handed DNA Hall of Fame. http://www .lecb.ncifcrf.gov/~toms/lefthanded.DNA .html. Molecular information theorist Tom Schneider created this site to document media and book illustrations in which DNA is shown incorrectly twisting to the left.

University of Massachusetts. DNA Structure. http://molvis.sdsc.edu/dna/index.htm. An interactive, animated, downloadable tutorial on the molecular composition and structure of DNA for high school students and college freshmen. Available in Spanish, German, and Portuguese.

Down Syndrome

Field of study: Diseases and syndromes
Significance: *Down syndrome is one of the most thoroughly studied genetic diseases. The discovery that the syndrome is usually caused by the presence of an extra chromosome was a landmark in the understanding of the causes of genetic defects.*

Key terms

ANEUPLOID: a cell or individual with one a few missing or extra chromosomes
MEIOSIS: a process of nuclear division that produces cells containing half the number of chromosomes as the original cell
NONDISJUNCTION: the failure of homologous chromosomes to separate correctly during cell division
TRISOMY: the condition of having an extra chromosome in a set present, such as having three copies of chromosome 21, as in Down syndrome

Discovery and Cause

Down syndrome is one of the most common chromosomal defects in human beings. According to some studies, it occurs in one in seven hundred live births; other studies place the number at one in nine hundred. Further, it occurs in about one in every two hundred conceptions. This syndrome (a pattern of characteristic abnormalities) was first described in 1866 by the English physician John Langdon Down. While in charge of an institution housing the profoundly mentally retarded, he noticed that almost one in ten of his patients had a flat face and slanted eyes causing Down to use the term "mongolism" to describe the syndrome; this term, however, is misleading. Males and females of every race and ethnicity can and do have this syndrome. To eliminate the unintentionally racist implications of the term "mongolism," Lionel Penrose and his colleagues changed the name to Down syndrome.

Although Down syndrome was observed and reported in the 1860's, it was almost one hundred years before the cause was discovered. In 1959, the French physician Jérôme Lejeune and his associates realized that the presence of an extra chromosome 21 was the apparent cause. This fact places Down syndrome in the broader category of aneuploid conditions. All human cells have forty-six chromosomes or strands made up of the chemical called deoxyribonucleic acid (DNA). The sections or subdivisions along these forty-six strands, called genes, are responsible for producing all the proteins that determine our specific human characteristics. An aneuploid is a cell with forty-

The Cause of Down Syndrome

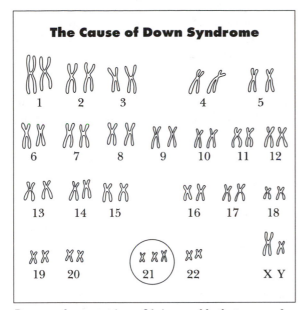

Down syndrome, or trisomy 21, is caused by the presence of an abnormal third chromosome in pair 21. (Electronic Illustrators Group)

five or forty-seven or more chromosomes, with the missing or extra strands of DNA leaving the individual with too few or too many genes. This aneuploid condition then results in significant alterations in one's traits and a great number of potential abnormalities.

In a normal individual, the forty-six strands are actually twenty-three pairs of chromosomes that are referred to as homologous because each pair is the same size and contains the same genes. In most cases of Down syndrome, there are three copies of chromosome 21. An aneuploid with three of a particular chromosome is called trisomic; thus Down syndrome is often called trisomy 21. The extra chromosome is gained because either the egg or sperm that came together at fertilization contained an extra one. This error in gamete (egg or sperm) production is called nondisjunction and occurs during the process of meiosis. When meiosis proceeds normally, the homologous chromosome pairs are separated from each other, forming gametes with twenty-three chromosomes, one from each pair. If nondisjunction occurs, a pair fails to separate, producing a gamete type with twenty-two chromosomes and a second gamete type with twenty-four chromo-

somes. If the pair that has failed to separate is chromosome 21, then the potential exists for twenty-three chromosomes in a normal gamete to combine with a gamete containing twenty-four, creating a trisomic individual with forty-seven chromosomes.

Symptoms of Down Syndrome

The slanted appearance of the eyes first reported by Down is caused by a prominent fold of skin called an epicanthic fold (a fold in the upper eyelid near the corner of the eye). This fold of skin is accompanied by excess skin on the back of the neck and abnormal creases in the skin of the palm. In addition, the skull is wide, with a flat back and a flat face. The hair on the skull is sparse and straight. The rather benign physical abnormalities are minor compared to the defects in internal organ systems. Almost 40 percent of Down syndrome patients suffer from serious heart defects. They are very prone to cancer of the white blood cells (acute leukemia), the formation of cataracts, and serious recurring respiratory infections. Short of stature with poorly formed joints, they often have poor reflexes, weak muscle tone, and an unstable gait. The furrowed, protruding tongue that often holds the mouth partially open is an external sign of the serious internal digestive blockages frequently present. These blockages must often be surgically repaired before the individual's first birthday. Many suffer from major kidney defects that are often irreparable. Furthermore, a suppressed immune system can easily lead to death from an infectious disease such as influenza or pneumonia.

With all these potential physical problems, it is not surprising that nearly 50 percent of Down syndrome patients die before the age of one. For those who live, there are enormous physical, behavioral, and mental challenges. The mental retardation that always accompanies Down syndrome ranges from quite mild to profound. This mental retardation makes all learning difficult and speech acquisition in particular very slow. Yet most Down syndrome individuals have warm, loving personalities and enjoy art and music very much.

Modern Understanding of Down Syndrome

Although this syndrome was recognized by Down in 1866, true understanding of it dates from the work that Lejeune began in 1953. The seemingly innocuous characteristic of abnormal palm prints and fingerprints fostered an important insight for him. Since those prints are laid down very early in the child's prenatal development, they suggest a profoundly altered embryological course of events. His intuition told him that not one or two altered genes but rather a whole chromosome's genes must be at fault. In 1957, he discovered, by the culturing of cells from Down syndrome children in dishes in the laboratory, that those cells contained forty-seven chromosomes. This work eventually resulted in his 1959 publication, which was soon followed by the discovery that the extra chromosome present was a third copy of chromosome 21.

The modern development of more sophisticated methods of identifying individual parts of chromosomes has shed much light on the possible mechanisms by which the symptoms are caused. Some affected individuals do not have a whole extra chromosome 21; rather, they possess a third copy of some part of that chromosome. A very tiny strand of DNA, chromosome 21 contains only about fifteen hundred genes. Of these fifteen hundred, only a few hundred are consistently present in those who suffer from Down syndrome, namely the genes in the bottom one-third of the chromosome. Among those genes are several that could very likely cause certain symptoms associated with Down syndrome. A leukemia-causing gene and a gene for a protein in the lens of the eye that could trigger cataract formation have both been identified. A gene for the production of the chemicals called purines has been located. The overabundance of purines produced when three copies of this gene are present has been linked to the mental retardation usually seen. Even the fact that Down syndrome individuals have a greatly reduced life expectancy is validated by the presence of an extra gene for the enzyme superoxide dismutase, which seems involved in the normal aging process. Like Alzheimer's disease patients, Down syndrome patients who live past forty years of age have gummy tangles of protein strands called amyloid fibers in their brains. Since one form of inherited Alzheimer's is caused by a gene on chromosome 21, scientists continue to search for links between the impaired mental functioning characteristic of both diseases.

Other modern research has shed light on the long-recognized relationship between the age of the mother and an increased risk of having a Down syndrome child. Using more and more elaborate methods of chromosome banding, geneticists can determine whether the extra

Down syndrome is characterized by impaired mental ability and a complex of physical traits that may include short stature, stubby fingers and toes, protruding tongue, a single transverse palm crease, slanting of the eyes, small nose and ears, abnormal finger orientation, congenital heart defects, and other traits that vary from individual to individual. Despite some or all of these traits, many Down syndrome children, supported by parents and organizations such as the Special Olympics and Boy Scouts, engage fully in life. (AP/Wide World Photos)

Familial Down Syndrome

Down syndrome always involves either an extra portion of or a complete extra copy of chromosome 21. There are three mechanisms by which this can occur. Between 92 and 95 percent of cases result from nondisjunction during meiosis, in which two copies of chromosome 21 migrate to the same pole and end up in the same daughter cell. This most often happens in women, and if an egg with two copies of chromosome 21 is fertilized, the zygote will have three copies and all cells throughout the developing fetus will have an extra chromosome 21.

The second mechanism, mosaic trisomy 21, involves an error in cell division shortly after conception. This error produces two populations or lines of cells, some with 46 chromosomes and some with 47—the ones that have the additional chromosome 21. This mechanism occurs in 2 to 4 percent of Down syndrome live births. Covert mosaicism in parents used to be suspected as causing familial Down syndrome but is no longer indicated.

Between 1 and 4 percent of children with Down syndrome have translocation trisomy 21, which occurs when extra genetic material from chromosome 21 has been translocated to another chromosome. A family history of Down syndrome is an indication that this may be the cause of the defect. The occurrence of more than one case of Down syndrome in a family is relatively rare, but when it does occur, translocation trisomy is often suspected. Carrier parents usually do not display any genetic abnormalities. Not until there is miscarriage of a fetus with Down syndrome or birth of a child with Down syndrome do couples discover that one of them is a translocation carrier. Carriers can produce (1) noncarrier, chromosomally normal, children (which usually happens); (2) carrier children, just like the carrier parent, who are translocation heterozygotes; or (3) children with Down syndrome. Carrier mothers produce children with Down syndrome about 12 percent of the time. Carrier fathers produce children with Down syndrome about 3 percent of the time. Why greater risk exists for mothers is not clear.

Though maternal age is the most frequent predisposing factor for Down syndrome, it is uncorrelated with familial Down syndrome; translocation trisomy 21 occurs with equal frequency in younger and older women. Rarely, a carrier parent will have a translocation between both twenty-first chromosomes, a translocation carrier homozygote. This parent has a 100 percent chance for producing children with Down syndrome.

Ever since the genes on chromosome 21 were fully mapped, pedigree research (family recurrence studies) and epidemiological research (studies of chance occurrence among populations) have supported that these chromosomal abnormalities and uneven distributions of genetic material are inherited, and most often through mothers. Though cryptic parental mosaicism is no longer suspected and there is promising research investigating mitochondrial DNA in the form of cytoplasmic inheritance, the specific genetic mechanism of familial Down syndrome remains elusive.

—Paul Moglia

chromosome 21 came from the mother or the father. In 94 percent of children, the egg brings the extra chromosome. Since the first steps of meiosis to produce her future eggs occur before the mother's own birth, the older the mother, the longer these egg cells have been exposed to potentially harmful chemicals or radiation. On the other hand, paternal age is not a factor because all the steps of meiosis in males occur in cells produced in the few weeks before conception. The continued study of the age factor as well as new insights from genomics are leading to a greater understanding for all those affected by Down syndrome.

—Grace D. Matzen

See also: Aging; Amniocentesis and Chorionic Villus Sampling; Congenital Defects; Fragile X Syndrome; Genetic Counseling; Genetic Testing; Genetics, Historical Development of; Intelligence; Mutation and Mutagenesis; Nondisjunction and Aneuploidy; Prenatal Diagnosis; Proteomics.

Further Reading

Beighton, Peter. *The Person Behind the Syndrome.* Rev. ed. New York: Springer-Verlag, 1997. Biographical details about John Langdon Down.

Cohen, William I., Lynn Nadel, and Myra E. Madnick, eds. *Down Syndrome: Visions for the*

Twenty-first Century. New York: Wiley-Liss, 2002. Reviews findings from a 2000 conference, providing a comprehensive treatment of the current issues of self-determination, education, and advocacy, as well as the most recent research developments.

Hogenboom, Marga. *Living with Genetic Syndromes Associated with Intellectual Disability*. Philadelphia: Jessica Kingsley, 2001. Addresses not only Down syndrome but also Williams, Angelman, and and Prader-Willi syndromes from both a psychological and a practical standpoint. Valuable to the genetics student for its introductory material on the genetics of these syndromes.

Lubec, G. *The Molecular Biology of Down Syndrome*. New York: Springer, 1999. Twenty-five chapters examine different aspects of Down syndrome, including neuropathology, molecular pathology, mechanisms of neuronal death, oxidative stress, and apoptosis.

_____. *Protein Expression in Down Syndrome Brain*. New York: Springer, 2001. Both original research and current opinions on Down syndrome, with an emphasis on the molecular biology at the protein (rather than the nucleic acid) level, from studies using fetal brains with Down syndrome.

Newton, Richard. *The Down's Syndrome Handbook: A Practical Guide for Parents and Caregivers*. New York: Arrow Books, 1997. Helpful advice about the capabilities of affected people. Combines medical knowledge with sympathetic common sense, to provide help and advice to caregivers of young Down syndrome patients.

Selikowitz, Mark. *Down Syndrome: The Facts*. 2d ed. New York: Oxford University Press, 1997. Covers the entire life span of patients, from infancy to adulthood, and deals with the developmental, educational, and social-sexual needs of individuals with Down syndrome as well as their medical needs.

Shannon, Joyce Brennfleck. *Mental Retardation Sourcebook: Basic Consumer Health Information About Mental Retardation and Its Causes, Including Down Syndrome, Fetal Alcohol Syndrome, Fragile X Syndrome, Genetic Conditions, Injury*. Detroit, Mich.: Omnigraphics, 2000. Provides basic consumer health information about mental retardation, its causes, and prevention strategies. Topics include parenting issues, educational implications, health care needs, employment and economic matters, and legal issues. Glossary.

Web Sites of Interest

Dolan DNA Learning Center, Your Genes Your Health. http://www.ygyh.org. Sponsored by the Cold Spring Harbor Laboratory, this site, a component of the DNA Interactive Web site, offers information on more than a dozen inherited diseases and syndromes, including Down syndrome.

National Down Syndrome Society. http://www.ndss.org. A comprehensive site that includes information on research into the genetics of the syndrome and links to related resources.

Dwarfism

Field of study: Diseases and syndromes

Significance: *Dwarfism in humans is a term used to describe adults who are less than 50 inches (127 centimeters) in height. Dwarfism can be caused by genetic factors, endocrine malfunction, acquired conditions, or growth hormone deficiency. Individuals with dwarfism usually have normal intelligence and have an average life span. Dwarfism may result in multiple medical problems that can lead to death. Dwarfs who have normal body proportions are referred to as midgets. Disproportioned human dwarfs are referred to as dwarfs. Both scientific terms are socially unacceptable; human dwarfs prefer to be referred to as little people.*

Key terms

ALLELE: one of the different forms of a particular gene (locus)

AUTOSOMAL DOMINANT ALLELE: an allele of a gene (locus) on one of the nonsex chromosomes that is always expressed, even if the other allele is normal

AUTOSOMAL RECESSIVE ALLELE: an allele of a gene which will be expressed only if there are two identical copies

Types and Symptoms of Dwarfism

Dwarfism, of which there are several hundred forms, occurs in approximately one in every ten thousand births. Most dwarfs are born to parents of average height. The most common type of dwarfism, achondroplasia, is an autosomal dominant trait, but in 80 percent of cases it appears in children born to normal parents as a result of mutations in the sperm or egg.

Dwarfisms in which body proportions are normal usually result from metabolic or hormonal disorders in infancy or childhood. Chromosomal abnormalities, pituitary gland disorders, problems with absorption, malnutrition, kidney disease, and extreme emotional distress can also interfere with normal growth. When body parts are disproportioned, the dwarfism is usually due to a genetic defect.

Skeletal dysplasias are the most common causes of dwarfism and are the major cause of disproportionate types of dwarfism. More than five hundred skeletal dysplasias have been identified. Chondrodystrophic dwarfism occurs when cartilage cells do not grow and divide as they should and cause defective cartilage cells. Most chondrodystrophic dwarfs have abnormal body proportions. The defective cells occur only in the spine or only in the arms and legs.

Short-limb dwarfism includes individuals with achondroplasia, diastropic dysplasia, and Hunter-Thompson chondrodysplasia. Achondroplasia is the most common skeletal dysplasia and affects more than 50 percent of all dwarfs. Achondroplasia is caused by an autosomal dominant allele and is identified by a disproportionate short stature consisting of long trunk and short upper arms and legs. Eighty percent of all cases of achondroplasia result from a mutation on chromosome 4 in a gene that codes for a fibroblast growth factor receptor. Achondroplasia is seen in both males and females, occurs in all races, and affects approximately one in every twenty thousand births. If one parent has achondroplasia and the other does not, a child born to them would have a 50 percent chance of inheriting achondroplasia. On the other hand, if both parents have achondroplasia, their offspring have a 50 percent chance of inheriting achondroplasia, a 25 percent chance of being normal, and a 25 percent chance of inheriting the abnormal allele from each parent and suffering often fatal skeletal abnormalities. Children who do not inherit the defective gene will never have achondroplasia, and cannot pass it on their offspring, unless a mutation occurs in the sperm or egg of the parents. Geneticists have observed that fathers who are 40 years of age or older are more likely to have children with achondroplasia due to mutations in their sperm.

Diastrophic dysplasia is a relatively common form of short-limb dwarfism that occurs in approximately one in 100,000 births and is identified by the presence of short arms and calves, clubfeet, and short, broad fingers with a thumb that has a hitchhiker type appearance. Infant mortality can be high as a result of respiratory complications, but if they survive infancy, short-limbed dwarfs have a normal life span. Orthopedic dislocations of joints are common. Scoliosis is seen especially in the early teens, and progressive cervical kyphosis and partial dislocation of the cervical spine eventually cause compression of the spinal cord. Diastrophic dysplasia is an inherited autosomal recessive condition linked to chromosome 5. Parents have a 25 percent chance that each additional child will get diastrophic dysplasia.

Short-trunk dwarfism includes individuals with spondyloepiphyseal dysplasia, which results from abnormal growth in the spine and long bones that leads to a shortened trunk. In spondyloepiphyseal dysplasia tarda, the lack of growth may not be recognized until five to ten years of age. Those affected have progressive joint and back pain and eventually develop osteoarthritis. Spondyloepiphyseal dysplasia congenita is caused by autosomal dominant gene mutations and is evidenced by a short neck and trunk, and barrel chest at birth. It is not uncommon for cleft palate, hearing loss, myopia, and retinal detachment to be present. Morquio syndrome, which was first described in 1929, is classified as a mucopolysaccharidosis (MPS) disease caused by the body's inability to produce enzymes that help to breakdown and recycle dead cells. Consequently, wastes are stored in the body's cells.

Fred (left) and Toby Gill, in a photograph taken September 6, 1996. The twins were born with a form of short-limb dwarfism known as diastropic dysplasia, which occurs in approximately one in 100,000 births and is identified by the presence of short arms and calves, clubfeet, and short, broad fingers with a thumb that has a hitchhiker-type appearance. (AP/Wide World Photos)

chromosomes in most of the cells in the body. Short stature and the failure to develop sexually are hallmarks of Turner syndrome. Learning difficulties, skeletal abnormalities, heart and kidney problems, infertility, and thyroid dysfunction may also occur. Turner syndrome can be treated with human growth hormones and by replacing sex hormones.

Treatments for Dwarfism

Some forms of dwarfism can be treated through surgical and medical interventions such as bone-lengthening procedures, reconstructive surgery, and growth and sex hormone replacement. The Human Genome Project continues to investigate genetic links to dwarfism. Prenatal counseling and screening for traits of dwarfism, along with genetic counseling and support groups, are avenues to pursue for family and individual physical, psychological, and social well-being and to make informed choices. Family and public education regarding dwarfism and growth problems offers a means to bring greater awareness of dwarfism to communities.

—*Sharon Wallace Stark*

See also: Congenital Defects; Consanguinity and Genetic Disease; Hereditary Diseases; Human Growth Hormone; Pedigree Analysis.

Hunter-Thompson chondrodysplasia is a form of dwarfism caused by a mutation in growth factor genes. Affected individuals have shortened and misshapened bones in the lower arms, the legs, and the joints of the hands and feet. Fingers are shortened and toes are ball-shaped.

Growth hormone, a protein that is produced by the pituitary ("master") gland, is vital for normal growth. Hypopituitarism results in a deficiency of growth hormone and afflicts between ten thousand and fifteen thousand children in the United States.

Turner syndrome affects one in every two thousand female infants and is characterized by the absence of or damage to one of the X

Further Reading

Ablon, J. *Living with Difference: Families with Dwarf Children.* Westport, Conn.: Greenwood, 1988. Exploration of developmental and medical problems, school experiences, social world of the dwarf children, and how dwarf children fit into family systems.

Apajasalo, M., et al. "Health-Related Quality of Life of Patients with Genetic Skeletal Dysplasias." *European Journal of Pediatrics* 157 (1998): 114-121. Presents tools for assessing the well-being of individuals with dwarfism and discusses results of a survey conducted by the authors.

Krakow, D., et al. "Use of Three-Dimensional Ultrasound Imaging in the Diagnosis of Prenatal-Onset Skeletal Dysplasias." *Ultrasound in Obstetrics and Gynecology* 21 (2003): 4676-4678. Describes in detail the newest approach to prenatal detection of dysplasias.

Page, Nick. *Lord Minimus: The Extraordinary Life of Britain's Smallest Man.* New York: St. Martin's Press, 2002. Relates the exciting life led by Jeffrey Hudson, a dwarf, in the court of King Charles I and Queen Henrietta Maria.

Ranke, M., and G. Gilli. *Growth Standards, Bone Maturation, and Idiopathic Short Stature.* Farmington, Conn.: S. Karger, 1996. KABI International Growth Study to establish global guidelines and standards for diagnosis and treatment of growth disorders and definition of idiopathic short stature (ISS).

Richardson, John H. *In the Little World: A True Story of Dwarfs, Love, and Trouble.* San Francisco: HarperCollins, 2001. A wide-ranging look at the world of the "little people," as many people with dwarfism prefer to be called.

Thorner, M., and R. Smith. *Human Growth Hormone: Research and Clinical Practice.* Vol. 19. Totowa, N.J.: Humana Press, 1999. Examines the use of human growth hormone therapies in the treatment of short stature and various diseases.

Ulijaszek, J. S., Francis E. Johnston, and Michael A. Preece. *Cambridge Encyclopedia of Human Growth and Development.* New York: Cambridge University Press, 1998. Broadly discusses genetic growth anomalies in relation to environmental, physiological, social, economic, and nutritional influences on human growth.

Vajo, Zoltan, Clair A. Francomano, and Douglas J. Wilkin. "The Molecular and Genetic Basis of Fibroblast Growth Factor Receptor 3 Disorders: The Achondroplasia Family of Skeletal Dysplasias, Muenke Craniosynostosis, and Crouzon Syndrome with Acanthosis Nigricans." *Endocrine Reviews* 21, no. 1 (2000): 23-39. Aimed at researchers.

Zelzer, Elazar, and Bjorn R. Olsen. "The Genetic Basis for Skeletal Diseases." *Nature* 423 (2003): 343-348. Aimed at researchers but understandable by a wider audience.

Web Sites of Interest

Centralized Dwarfism Resources. http://www.dwarfism.org. Offers information on types of dwarfism and links to other informative sites.

National Center for Biotechnology Information. Online Mendelian Inheritance in Man. http://www.ncbi.nlm.nih.gov/Omim. A catalog on genes and genetic disorders, including dwarfism, searchable by keyword.

Emerging Diseases

Fields of study: Diseases and syndromes; Viral genetics

Significance: *Pandemics caused by emerging diseases have the potential to sweep the globe and kill millions of persons. These diseases typically arise by evolutionary processes—natural selection or artificial selection via the misuse of antibiotics—by swapping of plasmids or other genetic elements, or else they represent pathogens normally found in animals that make the jump into humans.*

Key term

DRUG RESISTANCE: a phenomenon in which pathogens no longer respond to drug therapies that used to control them; resistance can arise by recombination, by mutation or by several methods of gene transfer, or by misuse of existing drugs

EMERGING DISEASE: a disease whose incidence in humans or other target organisms has increased

INTERSPECIFIC TRANSMISSION: when a pathogen infecting one host infects another host of a different species

A Global Scare

Early in 2003, a new virus appeared. The illness it caused, severe acute respiratory syndrome (SARS), began with flu-like symptoms—fever, chills, headache followed by a dry cough—but appeared to be quite deadly. By mid-April of the year, more than 3,500 persons around the globe had been diagnosed with SARS. Of those, more than 150 had died. As the illness spread from its apparent source in Guangdong Province, China, individuals infected with or exposed to it were quarantined. International travel was restricted, damping an already sluggish global economy. Also by mid-April of that year, the World Heath Organization announced that an international team of researchers had determined that a coronavirus—a type of virus related to those that cause the common cold—was the cause of the syndrome. However, it was of a type never before seen. SARS is a prime example of the potential threat posed by emerging infectious diseases.

Epidemiological Background

Emerging diseases are new illnesses or old ailments whose incidence in humans or other target organisms, such as economically important plants or animals, have increased. The key to an emergent or reemergent disease is that the change in its status is sudden and unexpected. For example, acquired immunodeficiency syndrome (AIDS)—caused by the human immunodeficiency virus (HIV)—is an example of a new disease that emerged from an isolated, mysterious wasting syndrome to a pandemic that has killed millions. Tuberculosis, caused by *Mycobacterium tuberculosis*, is an example of an old, more or less controlled condition that has reemerged in an age when many people have compromised immune systems and when antibiotic misuse has promoted the development of drug-resistant strains.

In many cases, emerging diseases are not the result of genetic changes. They instead gain a foothold in a new species by transmission from another host. The probability of interspecific transmission rises as contacts increase between species that formerly had few opportunities to interact. For example, the Ebola virus may have made the leap into humans as humans encroached upon the forests of central Africa and came into greater contact with the natural hosts of the disease. HIV may have made a similar leap. Influenza viruses infect humans, swine, horses, poultry, and waterfowl. New strains of the virus often emerge in regions, such as southern China, where an agricultural mode of life prevails and humans live in close proximity to farm animals. The close proximity of several susceptible hosts makes it easy for mutant strains to develop in one species and spread to others.

The Mutants

In 1997, a new strain of influenza emerged in Hong Kong. The strain (H5N1) leapt from chickens to humans and was armed with a mutation that allowed it to disable a part of the immune system. H5N1 killed only six people—but that small number was one-third of the persons infected, a very large proportion. Normally, the flu virus would be contained within the lungs or intestines of the chicken, but this particular strain spread to other tissues and had the potential to kill victims quickly. Subsequent research traced the change to one gene, *PB2*. The mutation in *PB2* gave H5N1 the ability to infect people. Because of the dangers posed by the 1997 strain, Hong Kong authorities ordered the destruction of more than a million fowl. The mass killing was repeated during another outbreak in 2001 and 2002, during which 4 million fowl were destroyed.

A mutation can affect as little as a single base pair in a strand of DNA or RNA, or it can involve much longer segments. The classes of mutations include substitutions, deletions, and insertions. Substitutions occur when a nitrogenous base or a sequence of bases in a nucleic acid is replaced by another sequence of identical length. Substitutions can be silent—that is, they can have no effect because the changed sequence calls for the same amino acid as the original. However, substitutions can cause missense mutations, in which a different amino acid is called for, or nonsense mutations, which signals the end of a polypeptide chain. Deletions occur when a portion of a gene is cut out and lost. Insertions occur when more nucleotides are inserted into the sequence of an existing gene.

Reshuffling the Genome

Some of the world's most lethal pandemics arise through recombination in the genome of the pathogen. Influenza is a prime example. The flu viruses are notorious for their ability to change—one reason flu vaccinations are given annually. The 1918 Spanish flu pandemic, caused by a strain that actually originated in North America, spread quickly, infecting about half of the human population at the time and killed tens of millions before running its course.

Genetic analysis of the Spanish flu virus revealed changes in the gene that codes for a protein, hemagglutinin, that helps the virus attach to host cells. Parts of the Spanish flu hemagglutinin gene resembled that of strains that typically infected

humans. Other parts resembled those of strains that normally infected swine. Such swapping of similar genetic sequences is called recombination. The new combinations may have arisen as the swine strain infected humans, or as the human strain infected swine—or they may have arisen through many infections back and forth. While no one is sure how the combination of human and swine characteristics proved so deadly—or how it arose—no one can dispute its lethal effectiveness.

In the spring and early summer of 2003, a new pandemic, severe acute respiratory syndrome or SARS, emerged from China's Guangdong Province and was quickly spread across the globe by world travelers. After it was recognized as a new and highly infectious coronavirus, the World Health Organization issued guidelines for its containment, which seemed to occur as the summer drew to a close. Here a woman stands in front of a poster in Taipei, Taiwan, that urges anti-SARS practices on the part of the public. (AP/Wide World Photos)

Mix and Match and Mix

One of the types of recombination seen in influenza viruses is a called reassortment. It occurs·in viruses, like influenza viruses, with segmented genomes—their genetic material does not come in one piece, like a chromosome does, but in separate units. If two or more strains of a virus infect a host, they can exchange those units via reassortment and give rise to a new strain with characteristics of the parental strains. The rotaviruses are another group of viruses that recombine their genomes via reassortment. A human rotavirus is one of the most common causes of gastroenteritis in small children around the world and kills hundreds of thousands of children each year.

Viruses are capable of classical recombination, in which homologous segments of DNA or RNA break away and reattach to related strands of genetic material. Another type of recombination requires DNA or double-stranded RNA. It is called copy-choice recombination and occurs when the polymerase enzyme used to make a copy of a nucleic acid strand switches strands in the middle of the process.

Bacterial Gene Swapping

Bacteria swap via plasmids—tiny rings of DNA that can be transferred directly from bacterial cell to bacterial cell via conjugation. Plasmids can also be transferred indirectly as some cells release them and others take them up. Plasmids can also be transferred via bacteriophages—viruses that infect bacteria. Viruses also stash their own genetic material inside the bacterial genome.

Such gene swapping is the primary mechanism by which bacteria acquire new characteristics such as antibiotic resistance—which is now a serious threat in some circumstances. The resistance itself may arise through a mutation but spreads among other microbes via gene exchange. Jim Henson, the renowned puppeteer who created the Muppets, died in 1990 at a young and typically healthy age from an infection of drug-resistant *Streptococcus*, a bacterium that normally gives humans only a sore throat and that normally responds quickly to antibiotic treatment.

New Threats

Influenza is an ancient scourge of humankind, but because of its ability to change, remains a significant pandemic threat. HIV and severe acute respiratory syndrome (SARS) are much more recent emergent viral diseases. Others include Ebola, described in 1976; hepatitis C, identified in 1989; Sin nombre virus, isolated in 1993; the H5N1 strain of the flu virus; and West Nile virus, which emerged in 2002 in the United States.

Several bacterial diseases have likewise emerged in recent decades. They include Legionnaires' disease, caused by *Legionella pneumophilia*; *Escherichia coli* strain O157, a virulent form of a bacterium that normally lives in the human gut; and Lyme disease, caused by *Borrelia burgdorferi*. Antibiotic resistance itself is on the rise and poses an increasing danger to humans. Bacteria for which resistance has been documented include *Mycobacterium tuberculosis*, *Neisseria gonorrhoea*, *Staphylococcus aureus*, and *Streptococcus*.

Monitoring

Public health agencies have recognized the importance of surveillance in any attempt to contain emerging or reemerging diseases. Disease monitoring has been going on for some time: The U.S. Congress in 1878 authorized the U.S. Marine Hospital Service to collect data on four diseases—cholera, smallpox, plague, and yellow fever—from U.S. consuls around the world. The Centers for Disease Control and Prevention now administer disease surveillance activities within the United States. The World Health Organization (WHO) in 2001 established a Department of Communicable Disease Surveillance and Response in order to improve monitoring activities globally. Reports of new diseases, such as SARS, are forwarded to the WHO headquarters in Geneva, Switzerland, which coordinates a global response.

—David M. Lawrence

See also: Bacterial Resistance and Super Bacteria; Biological Weapons; Diphtheria; Mutation and Mutagenesis; Smallpox; Swine Flu; Viral Genetics; Viroids and Virusoids.

Further Reading

DeSalle, Rob, ed. *Epidemic! The World of Infectious Disease.* New York: The New Press, 1999. An accessible discussion of infectious diseases caused by viruses, bacteria, and parasites. Topics include ecology and evolution, modes of transmission, infectious processes, outbreaks, and public health policies.

Drexler, Madeline. *Secret Agents: The Menace of Emerging Infections.* Washington, D.C.: Joseph Henry Press, 2002. Drexler discusses the ever-changing rules of engagement in the war between humans and microbes, recounts the history of past microbial killers, and makes a case for increasing the ability of the public health system to respond to emerging diseases.

Garrett, Laurie. *The Coming Plague: Newly Emerging Diseases in a World Out of Balance.* New York: Penguin, 1995. The author, a reporter for *Newsday*, explains in plain language the reasons infectious diseases remain a threat to humanity. Well researched and documented, the book provides a readable introduction to the topic of infectious disease.

Kolata, Gina. *Flu: The Story of the Great Influenza Pandemic of 1918 and the Search for the Virus That Caused It.* New York: Simon and Schuster, 2001. The discovery of pieces of this flu virus in Arctic remains, as well as government laboratories, prompted this history of the pandemic. Describes the science surrounding the possible recurrence of one of the deadliest plagues to attack humankind.

Levy, Stuart B. *The Antibiotic Paradox: How Miracle Drugs Are Destroying the Miracle.* New York: Plenum Press, 1992. Levy presents a frightening and authoritative indictment of how misuse of antibiotics is leading to the emergence of drug-resistant microbes against which humanity has no defense. This classic and accessible account is still timely.

McNeill, William H. *Plagues and Peoples.* New York: Anchor Books, 1998. An updated edition of a classic work on the effect of disease outbreaks on humans throughout history.

Web Sites of Interest

National Center for Infectious Diseases, Emerging Infectious Diseases Resources. http://www.cdc.gov/ncidod. This searchable government site includes information on emerging diseases such as Severe Acute Respiratory Syndrome (SARS) and others.

The DNA Files, Genetics of Infectious Disease. http://www.dnafiles.org/resources. Site includes articles on emerging diseases, a link to the site of the CDC's 2002 international conference on emerging diseases, and a useful bibliography for further study.

World Health Organization. http://www.who.int/en. This U.N. agency monitors disease outbreaks and emerging diseases worldwide, working with government health agencies to coordinate scientific and clinical efforts.

Epistasis

Field of study: Classical transmission genetics

Significance: *Epistasis is the interaction of genes such that the alleles at one locus can modify or mask the expression of alleles at another locus. Dihybrid crosses involving epistasis produce progeny ratios that are non-Mendelian, that is, different from the kinds of ratios discovered by Gregor Mendel.*

Key terms

ALLELE: an alternate form of a gene at a particular locus; a single locus can possess two alleles

DIHYBRID CROSS: a cross between parents that involve two specified genes, or loci

F_1: first filial generation, or the progeny resulting from the first cross in a series

F_2: second filial generation, or the progeny resulting from cross of the F_1 generation

LOCUS (*pl.* LOCI): a more precise word for gene; in diploid organisms, each locus has two alleles

Definition and History

The term "epistasis" is of Greek and Latin origin, meaning "to stand upon" or "stoppage." The term was originally used by geneticist William Bateson at the beginning of the twentieth century to define genes that mask the expression of other genes. The gene at the initial location (locus) is termed the epistatic gene. The genes at the other loci are "hypostatic" to the initial gene. In its strictest sense, it describes a nonreciprocal interaction between two or more genes, such that one gene modifies, suppresses, or otherwise influences the expression of another gene affecting the same phenotypic (physical) character or process. By this definition, simple additive effects of genes affecting a single phenotypic character or process would not be considered an epistatic interaction. Similarly, interactions between alternative forms (alleles) of a single gene are governed by dominance effects and are not epistatic. Epistatic effects are interlocus interactions. Therefore, in terms of the total genetic contribution to phenotype, three factors are involved: dominance effects, additive effects, and epistatic effects. The analysis of epistatic effects can suggest ways in which the action of genes can control a phenotype and thus supply a more complete un-

derstanding of the influence of genotype on phenotype.

A gene can influence the expression of other genes in many different ways. One result of multiple genes is that more phenotypic classes can result than can be explained by the action of a single pair of alleles. The initial evidence for this phenomenon came out of the work of Bateson and British geneticist Reginald C. Punnett during their investigations on the inheritance of comb shape in domesticated chickens. The leghorn breed has a "single" comb, brahmas have "pea" combs, and wyandottes have "rose" combs. Crosses between brahmas and wyandottes have "walnut" combs. Intercrosses among walnut types show four different types of F_2 (second-generation) progeny, in the ratio 9 walnut: 3 rose: 3 pea: 1 single. This ratio of phenotypes is consistent with the classical F_2 ratio for dihybrid inheritance. The corresponding ratio of genotypes, therefore, would be 9 $A\_ B\_$:3 $A\_ bb$:3 $aa B\_$:1 $aa bb$, respectively. (The underscore is used to indicate that the

A Punnett Square Showing Flower Pigmentation

		White CCpp	×		White ccPP

F_1 Purple CcPp

F_2		CP	Cp	cP	cp
	CP	CCPP purple	CCPp purple	CcPP purple	CcPp purple
	Cp	CCPp purple	CCpp white	CcPp purple	Ccpp white
	cP	CcPP purple	CcPp p22urple	ccPP white	ccPp white
	cp	CcPp purple	Ccpp white	ccPp white	ccpp white

When white-flowered sweet pea plants were crossed, the first-generation progeny (F_1) all had purple flowers. When these plants were self-fertilized, the second-generation progeny (F_2) revealed a ratio of nine purple to seven white. This result can be explained by the presence of two genes for flower pigmentation, P (dominant) and p (recessive) and C or c. Both dominant forms, P and C, must be present in order to produce purple flowers.

second gene can be either dominant or recessive; for example, $A\_$ means that both AA and Aa will result in the same phenotype.) In this example, one can recognize that two independently assorting genes can affect a single trait. If two gene pairs are acting epistatically, however, the expected 9:3:3:1 ratio of phenotypes is altered in some fashion. Thus, although the preceding example involves interactions between two loci, it is not considered a case of epistasis, because the phenotype ratio is a classic Mendelian ratio for a dihybrid cross. Five basic examples of two-gene epistatic interactions can be described: complementary, modifying, inhibiting, masking, and duplicate gene action.

Complementary Gene Action

For complementary gene action, a dominant allele of two genes is required to produce a single effect. An example of this form of epistasis again comes from the observations of Bateson and Punnett of flower color in crosses between two white-flowered varieties of sweet peas. In their investigation, crosses between these two varieties produced an unexpected result: All of the F_1 (first-generation) progeny had purple flowers. When the F_1 individuals were allowed to self-fertilize and produce the F_2 generation, a phenotypic ratio of nine purple-flowered to seven white-flowered individuals resulted. Their hypothesis for this ratio was that a homozygous recessive genotype for either gene (or both) resulted in the lack of flower pigmentation. A simple model to explain the biochemical basis for this type of flower pigmentation is a two-step process, each step controlled by a separate gene and each gene having a recessive allele that eliminates pigment formation. Given this explanation, each parent must have had complementary genotypes (AA bb and aa BB), and thus both had white flowers. Crosses between these two parents would produce double heterozygotes (Aa Bb) with purple flowers. In the F_2 generation, $9/16$ would have the genotype $A\_$ $B\_$ and would have purple flowers. The remaining $7/16$ would be homozygous recessive for at least one of the two genes and, therefore, would have white flowers. In summary, the phenotypic ratio of the F_2 generation would be 9:7.

Modifying Gene Action

The term "modifying gene action" is used to describe a situation whereby one gene produces an effect only in the presence of a dominant allele of a second gene at another locus. An example of this type of epistasis is aleurone color in corn. The aleurone is the outer cell layer of the endosperm (food storage tissue) of the grain. In this system, a dominant gene ($P\_$) produces a purple aleurone layer only in the presence of a gene for red aleurone ($R\_$) but expresses no effect in the absence of the second gene in its dominant form. Thus, the corresponding F_2 phenotypic ratio is 9 purple:3 red:4 colorless. The individuals without aleurone pigmentation would, therefore, be of the genotype $P\_$ rr ($3/16$) or pp rr ($1/16$). Again, a two-step biochemical pathway for pigmentation can be used to explain this ratio; however, in this example, the product of the second gene (R) acts first in the biochemical pathway and allows for the production of red pigmentation and any further modifications to that pigmentation. Thus, the phenotypic ratio of the F_2 generation would be 9:3:4.

Inhibiting Gene Action

Inhibiting action occurs when one gene acts as an inhibitor of the expression of another gene. In this example, the first gene allows the phenotypic expression of a gene, while the other gene inhibits it. Using a previous example (the gene R for red aleurone color in corn seeds), the dominant form of the first gene R does not produce its effect in the presence of the dominant form of the inhibitor gene I. In other words, the genotype $R\_$ $i\_$ results in a phenotype of red aleurone ($3/16$), while all other genotypes result in the colorless phenotype ($13/16$). Thus gene R is inhibited in its expression by the expression of gene I. The F_2 phenotypic ratio would be 13:3. This ratio, unlike the previous two examples, includes only two phenotypic classes and highlights a complicating factor in determining whether one or two genes may be influencing a given trait. A 13:3 ratio is close to a 3:1 ratio (the ratio expected for the F_2 generation of a monohybrid cross). Thus it emphasizes the need to look at an F_2 population of sufficient size to discount the possibility of a single gene phenomenon over an inhibiting epistatic gene interaction.

Masking Gene Action

Masking gene action, a form of modifying gene action, results when one gene is the primary determinant of the phenotype of the offspring. An example of this phenomenon is fruit color in summer squash. In this example, the F_2 ratio is 12:3:1, indicating that the first gene in its dominant form results in the first phenotype (white fruit); thus this gene is the primary determinant of the phenotype. If the first gene is in its recessive form and the second gene is in its dominant form, the fruit will be yellow. The fruit will be green at maturity only when both genes are in their recessive form ($\frac{1}{16}$ of the F_2 population).

Duplicate Gene Interaction

Duplicate gene interaction occurs when two different genes have the same final result in terms of their observable influence on phenotype. This situation is different from additive gene action in that either gene may substitute for the other in the expression of the final phenotype of the individual. It may be argued that duplicate gene action is not a form of epistasis, since there may be no interaction between genes (if the two genes code for the same protein product), but this situation may be an example of gene interaction when two genes code for similar protein products involved in the same biochemical pathway and their combined interaction determines the final phenotype of the individual. An example of this type of epistasis is illustrated by seed capsule shape in the herb shepherd's purse. In this example, either gene in its dominant form will contribute to the final phenotype of the individual (triangular shape). If both genes are in their recessive form, the seed capsule has an ovoid shape. Thus, the phenotypic ratio of the F_2 generation is 15:1.

Impact and Applications

Nonallelic gene interactions have considerable influence on the overall functioning of an individual. In other words, the genome (the entire genetic makeup of an organism) determines the final fitness of an individual, not only as a sum total of individual genes (additive effects) or by the interaction between different forms of a gene (dominance effects) but also by the interaction between different genes (intragenomic or epistatic effects). This situation is something akin to a chorus: Great choruses not only have singularly fine voices, but they also perform magnificently as finely tuned and coordinated units. Knowledge of what contributes to a superior genome would, therefore, lead to a fuller understanding of the inheritance of quantitative characters and more directed approaches to genetic improvement. For example, most economically important characteristics of agricultural species (such as yield, pest and disease resistance, and stress tolerance) are quantitatively inherited, the net result of many genes and their interactions. Thus an understanding of the combining ability of genes and their influence on the final appearance of domesticated breeds and crop varieties should lead to more efficient genetic improvement schemes. In addition, it is thought that many important human diseases are inherited as a complex interplay among many genes. Similarly, an understanding of genomic functioning should lead to improved screening or therapies.

—Henry R. Owen

See also: Chromosome Structure; Chromosome Theory of Heredity; Classical Transmission Genetics; Complete Dominance; Dihybrid Inheritance; Extrachromosomal Inheritance; Hybridization and Introgression; Incomplete Dominance; Lamarckianism; Mendelian Genetics; Monohybrid Inheritance; Multiple Alleles; Nondisjunction and Aneuploidy; Parthenogenesis; Penetrance; Polygenic Inheritance; Quantitative Inheritance.

Further Reading

Frankel, Wayne N., and Nicholas J. Schork. "Who's Afraid of Epistasis?" *Nature Genetics* 14 (December, 1996). A reexamination of the whole concept of epistasis, with statistical implications.

Russell, Peter J. *Fundamentals of Genetics.* 2d ed. San Francisco: Benjamin Cummings, 2000. Introduces the three main areas of genetics:

transmission genetics, molecular genetics, and population and quantitative genetics.

Wolf, Jason B., Edmund D. Brodie III, and Michael J. Wade. *Epistasis and the Evolutionary Process.* New York: Oxford University Press, 2000. Primary focus is on the role of gene interactions (epistasis) in evolution. Leading researchers examine how epistasis impacts the evolutionary processes in overview, theoretical, and empirical chapters. Illustrations, index.

Eugenics

Field of study: Human genetics and social issues

Significance: *The eugenics movement sought to speed up the process of natural selection through the use of selective breeding and led to the enactment of numerous laws requiring the sterilization of "genetically inferior" individuals and limiting the immigration of supposedly defective groups. Such flawed policies were based on an inadequate understanding of the complexity of human genetics, an underestimation of the role of the environment in gene expression, and the desire of certain groups to claim genetic superiority and the right to control the reproduction of others.*

Key terms

BIOMETRY: the measurement of biological and psychological variables

NEGATIVE EUGENICS: improving human stocks through the restriction of reproduction

POSITIVE EUGENICS: improving human stocks by encouraging the "naturally superior" to breed extensively with other superior humans

The Founding of the Eugenics Movement

With the publication of Charles Darwin's *On the Origin of Species by Means of Natural Selection* (1859), the concept of evolution began to revolutionize the way people thought about the human condition. Herbert Spencer and other proponents of what came to be known as social Darwinism adhered to the belief that social class structure arose through natural selection, seeing class stratification in industrial societies, including the existence of a permanently poor underclass, as a reflection of the underlying, innate differences between classes.

During this era there was also a rush to legitimize all sciences by using careful measurement and quantification. There was a blind belief that attaching numbers to a study would ensure its objectivity.

Francis Galton, an aristocratic inventor, statistician, and cousin of Darwin, became one of the primary promoters of such quantification. Obsessed with mathematical analysis, Galton measured everything from physiology and reaction times to boredom, the efficacy of prayer, and the beauty of women. He was particularly interested in the differences between human races. Galton eventually founded the field of biometry by applying statistics to biological problems.

A hereditarian, Galton assumed that talent in humans was subject to the laws of heredity. Although Galton did not coin the term "eugenics" until 1883, he published the first discussion of his ideas in 1865, in which he recognized the apparent evolutionary paradox that those of talent often have few, if any, children and that civilization itself diminishes the effects of natural selection on human populations. Fearing that medicine and social aid would lead to the propagation of weak individuals, Galton advocated increased breeding by "better elements" in the population (positive eugenics), while at the same time discouraging breeding of the "poorer elements" (negative eugenics).

Like most in his time, Galton believed in "blending inheritance," whereby hereditary material would mix together like different colors of paint. Trying to reconcile how superior traits would avoid being swamped by such blending, he came up with the statistical concept of the correlation coefficient, and in the process connected Darwinian evolution to the "probability revolution." His work focused on the bell-shaped curve or "normal distribution" demonstrated by many traits and the possibility of shifting the mean by selection pressure at either extreme. His statistical framework deepened the theory of natural selection. Unfortunately, the mathematical predictability he studied has often been misinterpreted as inevitability. In 1907, Galton founded the Eugenics Education Society of London. He also carefully cataloged eminent families in his *Hereditary Genius* (1869), wherein the Victorian world was assumed to be the ultimate level that society could attain and the cultural transmission of status, knowledge, and social connections were discounted.

Early Eugenics in Britain

Statistician and social theorist Karl Pearson was Galton's disciple and first Galton Professor of Eugenics at the Galton Laboratory at the University of London. His *Grammar of Science* (1892) outlined his belief that eugenic management of society could prevent genetic deterioration and ensure the existence of intelligent rulers, in part by transferring resources from inferior races back into the society. According to philosopher David J. Depew and biochemist Bruce H. Weber, even attorney Thomas Henry Huxley, champion of Darwinism, balked at this "pruning" of the human garden by the administrators of eugenics. For the most part, though, British eugenicists focused on improving the superior rather than eliminating the inferior.

Another of Galton's followers, comparative anatomist Walter Frank Weldon, like Galton before him, set out to measure all manner of things, showing that the distribution of many human traits formed a bell-shaped curve. In a study on crabs, he showed that natural selection can cause the mean of such a curve to shift, adding fuel to the eugenicists' conviction that they could better the human race through artificial selection.

Population geneticist Ronald A. Fisher was Pearson's successor as the Galton Professor of Eugenics. Fisher cofounded the Cambridge Eugenics Society and became close to Charles Darwin's sons, Leonard and Horace Darwin. In a speech made to the Eugenics Education Society, Fisher called eugenicists the "agents of a new phase of evolution" and the "new natural nobility," with the view that humans were becoming responsible for their own evolution. The second half of his book *The Genetical Theory of Natural Selection* (1930) deals expressly with eugenics and the power of "good-making traits" to shape society. Like Galton, he believed that those in the higher social strata should be provided with financial subsidies to counteract the "resultant sterility" caused when upper class individuals opt to have fewer children for their own social advantage.

British embryologist William Bateson, who coined the terms "genes" and "genetics," championed the Mendelian genetics that finally unseated the popularity of Galton's ideas in England. In a debate that lasted thirty years, those that believed in Austrian monk Gregor Mendel's particulate inheritance argued against the selection touted by the biometricians, and vice versa. Bateson, who had a deep distrust of eugenics, successfully replicated Mendel's experiments. Not recognizing that the two arguments were not mutually exclusive, Pearson and Weldon rejected genetics, thus setting up the standoff between the two camps.

Fisher, on the other hand, tried to model the trajectory of genes in a population as if they were gas molecules governed by the laws of thermodynamics, with the aim of converting natural selection into a universal law. He used such "genetic atomism" to propose that continuous variation, natural selection, and Mendelian genetics could all coexist. Fisher also mathematically derived Galton's bell-shaped curves based on Mendelian principles. Unfortunately, by emulating physics, Fisher underestimated the degree to which environment dictates which traits are adaptive.

Early Eugenics in the United States

While Mendelians and statisticians were debating in Britain, in the United States, Harvard embryologist Charles Davenport and others embarked on a mission of meshing early genetics with the eugenics movement. In his effort, Davenport created the Laboratory for Experimental Evolution at Cold Springs Harbor, New York. The laboratory was closely linked to his Eugenics Record Office (ERO), which he established in 1910. Davenport raised much of the money for these facilities by appealing to wealthy American families who feared unrestricted immigration and race degeneration. Though their wealth depended on the availabil-

In the first half of the twentieth century, thousands of people in the United States, many of them teenagers thought to be weak or abnormal, were sterilized to prevent their genes from passing on to the next generation. Here Sarah Jane Wiley revisits the Virginia Colony of the Epileptic and Feebleminded in Lynchburg, Virginia, where she and her brother were both sterilized in 1959. (AP/ Wide World Photos)

ity of cheap labor guaranteed by immigration, these American aristocrats feared the cultural impact of a flood of "inferior immigrants."

Unlike the British, U.S. eugenicists thought of selection as a purifying force and thus focused on how to stop the defective from reproducing. Davenport wrongly felt that Mendelian genetics supported eugenics by reinforcing the effects of inheritance over the environment. He launched a hunt to identify human defects and link specific genes (as yet poorly understood entities) to specific traits. His primary tool was the family pedigree chart. Unfortunately, these charts were usually based on highly subjective data, such as questionnaires given to schoolchildren to determine the comparative social traits of various races.

The Eugenics Research Association was founded in 1913 to report the latest findings. In 1918, the Galton Society began meeting regularly at the American Museum of Natural History in New York, and in 1923 the American Eugenics Society was formed. These efforts paid

off. By the late 1920's and early 1930's, eugenics was a topic in high school biology texts and college courses across the United States.

Among eugenics supporters was psychologist Lewis M. Terman, developer of the Stanford-Binet intelligence quotient (IQ) test, and Harvard psychologist Robert M. Yerkes, developer of the Army IQ test, who both believed that IQ test performance (and hence intelligence) was hereditary. The administration of such tests to immigrants by eugenicist Henry Goddard represented a supposedly "objective and quantitative tool" for screening immigrants for entry into the United States. Biologist Garland Allen reports that Goddard, in fact, determined that more than 80 percent of the Jewish, Hungarian, Polish, Italian, and Russian immigrants were mentally defective.

Fear that immigrants would take jobs away from hardworking Americans, supported by testimony from ERO's superintendent, Harry Laughlin, and the findings of Goddard's IQ tests, resulted in the Johnson Act of 1924, which severely restricted immigration. In the end, legal sterilization and immigration restrictions became more widespread in the United States than in any country other than Nazi Germany. By 1940, more than thirty states in the United States had enacted compulsory sterilization laws. Most were not repealed until after the 1960's.

Eugenics and the Progressive Era

During the Progressive Era, the eugenics movement became a common ground for such diverse groups as biologists, sociologists, psychologists, militarists, pacifists, socialists, communists, liberals, and conservatives. The progressive ideology, exemplified by Theodore Roosevelt's Progressive Party, sought the scientific management of all parts of society. Eugenics attracted the same crowd as preventative medicine, since both were seen as methods of harnessing science to reduce suffering and misfortune. For example, cereal entrepreneur John Harvey Kellogg founded the Race Betterment Foundation, mixing eugenics with hygiene, diet, and exercise. During this period, intellectuals of all stripes were attracted by the promise of "the improvement of the human race by better breeding." The genetics research of this time focused on improving agriculture, and eugenics was seen as the logical counterpart to plant and animal husbandry.

Davenport did not hesitate to play on their sympathies by making wild claims about the inheritance of "nomadism," "shiftlessness," "love of the sea," and other "traits" as if they were single Mendelian characteristics. Alcoholism, pauperism, prostitution, rebelliousness, criminality, feeblemindedness, chess expertise, and industrial sabotage were all claimed to be determined by one or two pairs of Mendelian genes. In particular, the progressives were lured by the idea of sterilizing the "weak minded," especially after the publication of articles about families in Appalachia and New Jersey that supposedly documented genetic lines cursed by a preponderance of habitual criminal behavior and mental weakness.

Having the allure of a "social vaccination," the enthusiasm to sterilize the "defective" spread rapidly among intellectuals, without regard to political or ideological lines. Sweden's Social Democrats forcibly sterilized some sixty thousand Swedes under a program that lasted from 1935 to 1976 organized by the state-financed Institute for Racial Biology. Grounds for sterilization included not only "feeblemindedness" but also "gypsy features," criminality, and "poor racial quality." The low class or mentally slow were institutionalized in the Institutes for Misled and Morally Neglected Children and released only if they would agree to be sterilized. Involuntary sterilization policies were also adopted in countries ranging from Switzerland and Austria to Belgium and Canada, not to be repealed until the 1970's.

Hermann Müller, a eugenicist who emigrated to the Soviet Union (and later returned to the United States), attacked Davenport's style of eugenics at the International Eugenics Congress in 1932. Müller, a geneticist who won the 1946 Nobel Prize for Physiology or Medicine for his discovery of the mutagenic power of X rays, instead favored the style of eugenics envisioned by English novelist Aldous Huxley's *Brave New World* (1932), with state nurseries, artificial insemination, and the use of other scientific techniques to produce a genetically engineered socialist society.

According to journalist Jonathan Freedland, the British left, including a large number of socialist intellectuals such as playwright George Bernard Shaw and philosopher Bertrand Russell, was convinced that it knew what was best for society. Concerned with the preservation of their higher intellectual capacities, they joined the fashionable and elitist Eugenics Society in the 1930's, where they advocated the control of reproduction, particularly favoring the idea of impregnating working-class women with sperm of men with high IQs.

The American Movement Spreads to Nazi Germany

The eugenics movement eventually led to grave consequences in Nazi Germany. Negative eugenics reached its peak there, with forced sterilization, euthanasia or "mercy killing," experimentation, and ultimately genocide being used in the name of "racial hygiene." Eugenicists in the United States and Germany formed close and direct alliances, especially after the Nazis came to power in 1933. The ERO's Laughlin gave permission for his article "Eugenical Sterilization" to be reprinted in German in 1928. It soon became the basis of Nazi sterilization policy. Davenport even arranged for a group of German eugenicists to participate in the three hundredth anniversary of Harvard's founding in 1936.

Inspired by the U.S. eugenics movement and spurred by economic hardship that followed World War I, the Nazi Physician's League took a stand that those suffering from incurable disease caused useless waste of medications and, along with the crippled, the feeble-minded, the elderly, and the chronic poor, posed an economic drain on society. Hereditary defects were considered to be the cause of such maladies, and these people were dubbed "lives not worth living." In 1933, the German Law on Preventing Hereditarily Diseased Progeny made involuntary sterilization of such people, including the blind, deaf, epileptic, and poor, legal. The Nazis set up "eugenics courts" to decide cases of involuntary sterilization. Frederick Osborn, secretary of the American Eugenics Society, wrote a 1937 report summarizing the German sterilization programs, indicative of the fascination American eugenicists had for the Nazi agenda and the Nazi's ability to move this experiment to a scale never possible in the United States.

The Demise of Eugenics

With the Great Depression in 1929, the U.S. eugenics movement lost much of its momentum. Geneticist and evolutionary biologist Sewall Wright, although himself a member of the American Eugenics Society, found fault with the genetics and the ideology of the movement: "Positive eugenics seems to require . . . the setting up of an ideal of society to aim at, and this is just what people do not agree on." He also wrote several articles in the 1930's challenging the assumptions of Fisher's genetic atomism model. In a speech to the Eugenics Society in New York in 1932, Müller pointed out the economic disincentive for middle and upper classes to reproduce, epitomized by the failure of many eugenicists to have children. Galton himself died childless. This inverse relationship between fertility and social status, coupled with the apparent predatory nature of the upper class, seemed to doom eugenics to failure.

Evolutionary biologist Stephen Jay Gould claimed that the demise of the eugenics movement in the United States was more a matter of Adolf Hitler's use of eugenic arguments for sterilization and racial purification than it was of advances in genetic knowledge. Once the Holocaust and other Nazi atrocities became known, eugenicists distanced themselves from the movement. Depew and Weber have written that Catholic conservatives opposed to human intervention in reproduction and progressives, who began to abandon eugenics in favor of behaviorism (nurture rather than nature), were political forces that began to close down the eugenics movement, while Allen points out that the movement had outlived its political usefulness. Russian geneticist Theodosius Dobzhansky had by this time recognized the prime importance of context in genetics and consequently rejected the premise of eugenics, helping to push it into the realm of phony genetics.

Implications

The term "euphenics" is used to describe human genetic research that is aimed at improving the human condition, replacing the tainted term eugenics. Euphenics deals primarily with medical or genetic intervention that is designed to reduce the impact of defective genotypes on individuals (such as gene therapy for those with cystic fibrosis). However, in this age of increasing information about human genetics, it is necessary to keep in mind the important role played by environment and the malleability of human traits.

Allen argues that the eugenics movement may reappear (although probably under a different name) if economic problems again make it attractive to eliminate "unproductive" people. His hope is that a better understanding of genetics, combined with the lessons of Nazi Germany, will deter humans from ever again going down that path that journalist Jonathan Freedland calls "the foulest idea of the 20th century."

—*Lee Anne Martínez*

See also: Artificial Selection; Bioethics; Bioinformatics; Biological Determinism; Cloning: Ethical Issues; Eugenics: Nazi Germany; Evolutionary Biology; Gene Therapy: Ethical and Economic Issues; Genetic Counseling; Genetic Engineering: Social and Ethical Issues; Genetic Screening; Genetic Testing: Ethical and Economic Issues; Heredity and Environment; Human Genetics; Insurance; Intelligence; Miscegenation and Antimiscegenation Laws; Natural Selection; Patents on Life-Forms; Paternity Tests; Race; Sociobiology; Stem Cells; Sterilization Laws.

Further Reading

Allen, Garland E. "Eugenics and Public Health in American History." *Technology Review* 99 (August/September, 1996). Discusses the connection between the eugenics movement and periods of economic or social hardship.

Depew, David, and Bruce Weber. *Darwinism Evolving: Systems Dynamics and the Genealogy of Natural Selection.* Boston: MIT Press, 1995. Discusses the relationship between eugenics and Darwinian evolution and the role played by statistics in the origin of this movement.

Gillham, Nicholas Wright. *A Life of Sir Francis Galton: From African Exploration to the Birth of Eugenics.* New York: Oxford University Press, 2001. A biography of the founder of the eugenics movement.

Herbert, Wray. "The Politics of Biology." *U.S. News and World Report* (April, 1997). Considers the fluctuating viewpoint on the validity and social remedy for the genetics of human behavior.

Kevles, Daniel J. *In the Name of Eugenics: Genetics and the Uses of Human Heredity.* Cambridge, Mass.: Harvard University Press, 1995. Traces the history of eugenics, mainly in the United States and Britain, from the nineteenth century to the late twentieth century. Individuals such as Karl Pearson, C. B. Davenport, R. A. Fisher, and J. B. S. Haldane, who have been associated with eugenics in various ways, are discussed.

Kühl, Stefan. *The Nazi Connection: Eugenics, American Racism, and German National Socialism.* New York: Oxford University Press, 2002. Exposes the ties between the American eugenics movement and the Nazi program of "racial hygiene."

Mazumdar, Pauline Margaret. *Eugenics, Human Genetics, and Human Failings.* London: Routledge, 1991. A thorough historical approach that examines the eugenics movement from its origin to its heyday as the source of a science of human genetics.

Pernick, Martin S. "Science Misapplied: The Eugenics Age Revisited." *American Journal of Public Health* 87 (November, 1997). A fascinating exploration of the overlap between the goals, values, and concepts of public health and the eugenics movements in the early twentieth century.

Web Sites of Interest

Cold Spring Harbor Laboratory, Image Archive on the American Eugenics Movement. http://www.eugenicsarchive.org/eugenics. Comprehensive and extensively illustrated site that covers the eugenics movement in the United States, including its scientific history and origins, research methods and flaws, sterilization laws, and more.

National Reference Center for Bioethics Literature. http://www.georgetown.edu/research/nrcbl/scopenotes/sn28.html. An introduction to eugenics and a comprehensive annotated bibliography of sources for further study.

Eugenics: Nazi Germany

Field of study: History of genetics; Human genetics and social issues;

Significance: *Fueled by economic hardship and racial prejudice, the largest-scale application of eugenics occurred in Nazi Germany, where numerous atrocities, including genocide, were committed in the name of the genetic improvement of the human species. The German example raised worldwide awareness of the dangers of eugenics and did much to discredit eugenic theory.*

Key terms

ARYAN: a "race" believed by Nazis to have established the civilizations of Europe and India

EUTHANASIA: the killing of suffering people, sometimes referred to as "mercy" killing

NORDIC: the northernmost of the Aryan groups of Europe, believed by the Nazis to be the highest and purest racial group

Origins of Nazi Eugenic Thought

Nazi eugenic theory and practice grew out of two traditions: the eugenics movement, founded by British scientist Francis Galton, and racial theories of human nature. Most historians trace the origin of modern racial theories to French diplomat and writer Joseph-Arthur de Gobineau, who maintained that all great civilizations had been products of the Aryan, or Indo-Germanic, race. Through the late nineteenth and early twentieth centuries, German thinkers applied Galton's ideas to the problem of German national progress. The progress of the nation, argued scientists and social thinkers, could be best promoted by improving the German people through government-directed control of human reproduction. This type of eugenic thinking became known as "racial hygiene"; in 1904, eugenicists and biologists formed the Racial Hygiene Society in Berlin.

The Aryan mythology of Gobineau also grew in popularity. In 1899, an English admirer of Germany, Houston Stewart Chamberlain, published a widely read book entitled *The Foundations of the Nineteenth Century*. Chamberlain, heavily influenced by Gobineau, maintained that Europe's accomplishments had been the work of ethnic Germans, members of a healthy and imaginative race. Opposed to the Germans were the Jews, who were, according to Chamberlain, impure products of crossbreeding among the peoples of the Middle East.

Basics of Nazi Eugenics

The Law to Prevent Hereditarily Sick Offspring, requiring sterilization of people with hereditary diseases and disabilities, was drafted and decreed in Germany in 1933. Before the Nazis came to power, many segments of German society had supported sterilization as a way to improve future generations, and Adolf Hitler's emergence as a national leader provided the pressure to ensure the passage of the law. Between 1934 and 1945, an estimated 360,000 people (about 1 percent of the German population) who were believed to have hereditary ailments were sterilized. Despite this law, the Nazis did not see eugenics primarily as a matter of discouraging the reproduction of unhealthy individuals and encouraging the reproduction of healthy individuals. Following the theories of Chamberlain, Adolf Hitler and his followers saw race, not individual health or abilities, as the distinguishing characteristic of human beings.

The Schutzstaffel (SS) organization was a key part of Nazi eugenic activities. In January, 1929, Heinrich Himmler was put in charge of the SS, a police force aimed at establishing order among the street fighters who formed a large part of the early Nazi Party. In addition to disciplining rowdy Nazis, the SS quickly emerged as a racial elite, the spearhead of an intended German eugenic movement. Himmler recruited physicians and biologists to help ensure that only those of the purest Nordic heritage could serve in his organization. In 1931, the agriculturalist R. Walther Darre helped Himmler draw up a marriage code for SS men, and Himmler appointed Darre head of an SS Racial Office. Himmler hoped to create the seeds of a German super race by directing the marriages and reproduction of the "racially pure" members of the SS.

Since the Nazis saw Germans as a "master race," a race of inherently superior people, they attempted to improve the human stock by encouraging the birth of as many Germans as possible and by encouraging those seen as racially pure to reproduce. The Nazis declared that women should devote themselves to bearing and caring for children. Hitler's mother's birthday was declared the Day of the German Mother. On this day, public ceremonies awarded medals to women with large numbers of children. The SS set up and maintained an organization of maternity homes for unmarried mothers of acceptable racial background and orphanages for their children; these institutions were known as the Lebensborn ("fountain of life"). There is some evidence that young women with desired racial characteristics who were not pregnant were brought to the

On Wehrmacht Day, 1935, in Nazi Germany (from left): German chancellor Adolf Hitler, head of the air force Hermann Göring, army commander Wernher von Fritsch, minister of war Werner Eduard Fritz von Blomberg, commander of the navy Erich Raeder, and other Nazi officials. During the late 1930's and early 1940's, the Nazi government conducted the extreme and brutal form of eugenics that culminated in the Holocaust and the murder of millions of innocent Jews and other "undesirables." (Library of Congress)

Lebensborn to have children by the SS men to create "superior" Nordic children.

Impact

In addition to encouraging the reproduction of those seen as racially pure, the Nazis sought to eliminate the unhealthy and the racially undesirable. In August, 1939, a committee of physicians and government officials, operating under Hitler's authority, issued a secret decree under which all doctors and midwives would have to register births of malformed or handicapped children. By October of that year, Hitler had issued orders for the "mercy killing" of these children and all those with incurable diseases. This euthanasia movement expanded from sick and handicapped children to those believed to belong to "sick" races. The T4 euthanasia organization, designed for efficient and secret killing, experimented with lethal injections and killing by injection and became a pilot program for the mass murder of the Jews during the Holocaust.

German racial hygienists had long advocated controlling marriages of non-Jewish Germans with Jews in order to avoid "contaminating" the German race. In July, 1941, Nazi leader Hermann Göring appointed SS officer Reinhard Heydrich to carry out the "final solution" of the perceived Jewish problem. At the Wannsee Conference in January, 1942, Hitler and his close associates agreed on a program of extermination. According to conservative estimates, between four million and five million European Jews died in Nazi extermination camps. When the murderous activities of the Nazis were revealed to the world after the war, eugenics theory and practice fell into disrepute.

—*Carl L. Bankston III*

See also: Bioethics; Bioinformatics; Biological Determinism; Eugenics; Evolutionary Biology; Gene Therapy: Ethical and Economic Issues; Genetic Counseling; Genetic Engineering: Social and Ethical Issues; Genetic Screening;

Genetic Testing: Ethical and Economic Issues; Heredity and Environment; Human Genetics; Insurance; Intelligence; Miscegenation and Antimiscegenation Laws; Patents on Life-Forms; Paternity Tests; Race; Sociobiology; Stem Cells; Sterilization Laws.

Further Reading

Goldhagen, Daniel J. *Hitler's Willing Executioners: Ordinary Germans and the Holocaust.* New York: Random House, 1996. Argues that the German people participated in the mass murder of Jews because Germans had come to see Jews as a racial disease.

Henry, Clarissa, and Marc Hillel. *Of Pure Blood: An Investigation into the Creation of a Super Race.* Video, produced by Maryse Addison and Peter Bate. Maljack Productions, 1976. Oak Forest, Ill.: MPI Home Video, 1985. Investigates the Lebensborn organization, a Nazi plan to breed and distill the German children into a pure Aryan race.

Kühl, Stefan. *The Nazi Connection: Eugenics, American Racism, and German National Socialism.* New York: Oxford University Press, 2002. Exposes the ties between the American eugenics movement and the Nazi program of racial hygiene.

Laffin, John. *Hitler Warned Us: The Nazis' Master Plan for a Master Race.* Totowa, N.J.: Barnes and Noble, 1998. Using photographs and propagandist ephemera, Laffin, a military historian, questions why Hitler was allowed by other leaders and nations to engage in his destructive drive for power and domination.

Weindling, Paul. *Health, Race, and German Politics Between National Unification and Nazism, 1870-1945.* 1989. Reprint. Cambridge, Mass.: Cambridge University Press, 1993. Offers a definitive history of the origins, social composition, and impact of eugenics in the context of the social and political tension of the rapidly industrializing Nazi empire.

Web Site of Interest

National Information Resource on Ethics and Human Genetics. http://www.georgetown.edu/research/nrcbl/nirehg/quickbibs.htm. Searchable "quick bib" on resources about eugenics and Nazi Germany.

Evolutionary Biology

Fields of study: Evolutionary biology; Population genetics

Significance: *While the existence of evolutionary change is firmly established, many questions remain about its causes in particular groups of organisms. The science of evolutionary biology focuses on reconstructing the actual history of life and on understanding how evolutionary mechanisms operate in nature.*

Key terms

ADAPTATION: a genetically based characteristic that increases the ability of an organism to survive and reproduce under prevailing environmental conditions

EVOLUTION: the process of change in the genetic structure of a population over time; descent with modification

FITNESS: the relative reproductive contribution of one individual to the next generation as compared to that of others in the population

GENETIC DRIFT: chance fluctuations in allele frequencies within a population, resulting from random variation in the number and genotypes of offspring produced by different individuals

GENOTYPE: the genetic makeup of an individual or group

NATURAL SELECTION: the phenomenon of differing survival and reproduction rates among various genotypes; the frequency of the favored genotypes increases in succeeding generations

PHYLOGENY: the history of descent of a group of species from a common ancestor

An Evolutionary Context

Life is self-perpetuating, with each generation connected to previous ones by the thread of DNA passed from ancestors to descendants. Life on earth thus has a single history much like the genealogy of a family, the shape and characteristics of which have been determined by internal and external forces. The effort to uncover that history and describe the forces that shape it constitutes the field of evolutionary biology.

As an example of the need for this perspective, consider three vertebrates of different species, two aquatic (a whale and a fish) and one terrestrial (a deer). The two aquatic species share a torpedolike shape and oarlike appendages. These two species differ, however, in that one lays eggs and obtains oxygen from the water using gills, while the other produces live young and must breathe air at the surface. The terrestrial species has a different, less streamlined, shape and appendages for walking, but it too breathes air using lungs and produces live young. All three species are the same in having a bony skeleton typical of vertebrates. In order to understand why the various organisms display the features they do, it is necessary to consider what forces or historical constraints influence their genotypes and subsequent phenotypes.

It is logical to hypothesize that a streamlined shape is beneficial to swimming creatures, as is the structure of their appendages. This statement is itself an evolutionary hypothesis; it implies that streamlined individuals will be more successful than less streamlined ones and so will become prevalent. It may initially be difficult to reconcile the differences between the two aquatic forms swimming side-by-side with the similarities between one of them and the terrestrial species walking around on dry land. However, if it is understood that the whale is more closely related to the terrestrial deer than it is to the fish, much of the confusion disappears. Using this comparative approach, it is unnecessary, and scientifically unjustified, to construct an elaborate scenario whereby breathing air at the surface is more advantageous to a whale than gills would be; the simpler explanation is that the whale breathes air because it (like the deer) is a mammal, and both species inherited this trait from a common ancestor sometime in the past.

Organisms are thus a mixture of two kinds of traits. Ecological traits are those the particular form of which reflects long-term adaptation to the species' habitat. Two species living in the same habitat might then be expected to be similar in such features and different from species in other habitats. Evolutionary characteristics, on the other hand, indicate common ancestry rather than common ecology. Here, similarity between two species indicates that they are related to each other, just as familial similarity can be used to identify siblings in a crowd of people. In reality, all traits are somewhere along a continuum between these two extremes, but this distinction highlights the importance of understanding the evolutionary history of organisms and traits. The value of an evolutionary perspective comes from its comparative and historical basis, which allows biologists to place their snapshot-in-time observations within the broader context of the continuous history of life.

Early Evolutionary Thought

Underlying evolutionary theory is Mendelian genetics, which provides a mechanism whereby advantageous traits can be passed on to offspring. Both Mendelian genetics and the theory of evolution are at first glance (and in retrospect) remarkably simple. The theory of evolution, however, is paradoxical in that it leads to extremely complex predictions and thus is often misunderstood, misinterpreted, and misapplied.

It is important to distinguish between the phenomenon of evolution and the various processes or mechanisms that may lead to evolution. The idea that species might be mutable, or subject to change over generations, dates back to at least the mid-eighteenth century, when the French naturalist Georges-Louis Leclerc, comte de Buffon, the Swiss naturalist Charles Bonnet, and even the Swedish botanist Carolus Linnaeus suggested that species (or at least "varieties") might be modified over time by intrinsic biological or extrinsic environmental factors. Other biologists after that time also promoted the idea that populations and species could evolve. Nevertheless, with the publication of *On the Origin of Species by Means of Natural Selection* in 1859, Charles Darwin became the first to propose that all species had descended from a common ancestor and that there was a single "tree of life." These claims regarding the history of evolution, however, are distinct from the problem of how, or through what mechanisms, evolution occurs.

In the first decade of the nineteenth century, Jean-Baptiste Lamarck promoted the the-

Charles Darwin is credited as the father of the theory of natural selection, on which modern evolutionary biology is based. (Library of Congress)

ory of inheritance of acquired characteristics to explain how species could adapt over time to their environments. His famous giraffe example illustrates the Lamarckian view: Individual giraffes acquire longer necks as a result of reaching for leaves high on trees, then pass that modified characteristic to their offspring. According to Lamarck's theories, as a result of such adaptation, the species—and, in fact, each individual member of the species—is modified over time. While completely in line with early nineteenth century views of inheritance, this view of the mechanism of evolution has since been shown to be incorrect.

Darwinian Evolution: Natural Selection

In the mid-nineteenth century, Darwin and Alfred Russel Wallace independently developed the theory of evolution via natural selection, a theory that is consistent with the genetics of inheritance as first described by Gregor Mendel. Both Darwin's and Wallace's arguments center on four observations of nature and a logical conclusion derived from those observations

(presented here in standard genetics terminology, although Darwin and Wallace used different terms).

First, variation exists in the phenotypes of different individuals in a population. Second, some portion of that variation is heritable, or capable of being passed from parents to offspring. Third, more individuals are produced in a population than will survive. Fourth, some individuals are, because of their particular phenotype, better able to survive and reproduce than others. From this, Darwin and Wallace deduced that those individuals whose phenotypes conferred on them greater fitness for survival would produce more offspring (genetically and phenotypically similar to themselves) than would less fit individuals; therefore, the frequency of individuals with the favored genotype would increase in the next generation, though each individual would be unchanged throughout its lifetime. This process would continue as long as new genetic variants continued to arise and selection favored some over others. The theory of natural selection provided a workable and independently testable natural mechanism by which evolution of complex and sometimes very different adaptations could occur within and among species.

Evolutionary Biology After Darwin

Despite their theoretical insight, Darwin and Wallace had no knowledge of the genetic basis of inheritance. Mendel published his work describing the particulate theory of inheritance in 1866 (he had reported the results before the Natural Sciences Society earlier, in February and March of 1865), but Darwin and Wallace appear to have been unaware throughout their lives that this vexing problem had been solved. In fact, Mendel's work went unnoticed by the entire scientific community for nearly fifty years; it was rediscovered, and its significance appreciated, in the first decade of the twentieth century. Over the next three decades, theoreticians integrated Darwin's theory of natural selection with the principles of Mendelian genetics. Simultaneously, Ernst Mayr, G. Ledyard Stebbins, George Gaylord Simpson, and Julian Huxley demonstrated that evolution of species and the patterns in the fossil record could be readily explained using Darwinian principles. This effort culminated in the 1930's and 1940's in the "modern synthesis," a fusion of thought that resulted in the development of the field of population genetics, a discipline in which biologists seek to describe and predict, quantitatively, evolutionary changes in populations and higher groups of organisms.

Since the modern synthesis (also called the neo-Darwinian synthesis), biologists have concentrated their efforts on applying the theories of population genetics to understanding the evolutionary dynamics of particular groups of organisms. More recently, techniques of phylogenetic systematics have been developed to provide a means of reconstructing phylogenetic relationships among species. This effort has emphasized the need for a comparative and evolutionary approach to biology, which is essential to correct interpretation of data.

In the 1960's, Motoo Kimura proposed the neutral theory of evolution, which challenged the "selectionist" view that patterns of genetic and phenotypic variation in most traits are determined by natural selection. The "neutralist" view maintains that much genetic variation, especially that seen in the numerous alleles of enzyme-coding genes, has little effect on fitness and therefore must be controlled by mechanisms other than selection. The last remaining frontier in the quest for a unified model of evolution is the integration of evolutionary theory with the understanding of the processes of development.

Evolutionary Mechanisms

Natural selection as described by Darwin and Wallace leads to the evolution of adaptations. However, many traits (perhaps the majority) are not adaptations; that is, differences in the particular form of those traits from one member of the species to the next do not lead to differences in fitness among those individuals. Such traits cannot evolve through natural selection, yet they can and do evolve. Thus, there must be additional mechanisms that lead to changes in the genetic structure of biological systems over time.

Evolutionary mechanisms are usually envisioned to act on individual organisms within a population. For example, natural selection may eliminate some individuals while others survive and produce a large number of offspring similar to themselves. As a result, evolution occurs within those populations. A key tenet of Darwinian evolution (which distinguishes it from Lamarckian evolution) is that populations evolve, but the individual organisms that constitute that population do not. While evolution of populations is certainly the most familiar scenario, this is not the only level at which evolution occurs.

Richard Dawkins energized the scientific discussion of evolution with his book *The Selfish Gene*, first published in 1976. Dawkins argued that natural selection could operate on any type of "replicator," or unit of biological organization that displayed a faithful but imperfect mechanism of copying itself and that had differing rates of survival and reproduction among the variant copies. Under this definition, it is possible to view individual genes or strands of DNA as focal points for evolutionary mechanisms such as selection. Dawkins used this

Alfred Russel Wallace is now considered the co-author of modern evolutionary theory along with Darwin. (National Library of Medicine)

framework to consider how the existence of DNA selected to maximize its chances of replication (or "selfish DNA") would influence the evolution of social behavior, communication, and even multicellularity.

Recognizing that biological systems are arranged in a hierarchical fashion from genes to genomes (or cells) to individuals through populations, species, and communities, Elisabeth Vrba and Niles Eldredge in 1984 proposed that evolutionary changes could occur in any collection of entities (such as populations) as a result of mechanisms acting on the entities that make up that collection (individuals). Because each level in the biological hierarchy (at least above that of genes) has as its building blocks the elements of the preceding one, evolution may occur within any of them. Vrba and Eldredge further argued that evolution could be viewed as resulting from two general kinds of mechanisms: those that introduce genetic variation and those that sort whatever variation is available. At each level, there are processes that introduce and sort variation, though they may have different names depending on the level being discussed.

Natural selection is a sorting process. Other mechanisms that sort genetic variation include sexual selection, whereby certain variants are favored based on their ability to enhance reproductive success (though not necessarily survival), and genetic drift, which is especially important in small populations. While these forces are potentially strong engines for driving changes in genetic structure, their action—and therefore the direction and magnitude of evolutionary changes that they can cause—is constrained by the types of variation available and the extent to which that variation is genetically controlled.

Processes such as mutation, recombination, development, migration, and hybridization introduce variation at one or more levels in the biological hierarchy. Of these, mutation is ultimately the most important, as changes in DNA sequences constitute the raw material for evolution at all levels. Without mutation, there would be no variation and thus no evolution. Nevertheless, mutation alone is a relatively weak evolutionary force, only really significant in driv-

ing evolutionary changes when coupled with processes of selection or genetic drift that can quickly change allele frequencies. Recombination, development, migration, and hybridization introduce new patterns of genetic variation (initially derived from mutation of individual genes) at the genome, multicellular-organism, population, and species levels, respectively.

The Reality of Evolution

It is impossible to prove that descent with modification from a common ancestor is responsible for the diversity of life on earth. In fact, this dilemma of absolute proof exists for all scientific theories; as a result, science proceeds by constructing and testing potential explanations, gradually accepting those best supported by new observations until they are either clearly disproved or replaced by another theory even more consistent with the data.

Darwin's concept of a single tree of life is supported by vast amounts of scientific evidence. In fact, the theory of evolution is among the most thoroughly tested and best-supported theories in all of science. The view that evolution has and continues to occur is not debated by biologists; there is simply too much evidence to support its existence, across every biological discipline.

On a small scale, it is possible to demonstrate evolutionary changes experimentally or through observation. Spontaneous mutations that introduce genetic variation are well documented; the origination and spread of drug-resistant forms of viruses and other pathogens is clear evidence of this potential. Agricultural breeding programs and other types of artificial selection illustrate that the genetic structure of lineages containing heritable variation can be changed over time. For example, work by John Doebley begun in the late 1980's suggested that the evolution of corn from its wild ancestor teosinte may have involved changes in as few as five major genes and that this transition likely occurred as a result of domestication processes established in Mexico between seven thousand and ten thousand years ago. The effects of natural selection can likewise be observed in operation: Peter Grant and his colleagues demonstrated that during drought periods, when seed is limited, deep-billed individuals of the Galápagos Island finch *Geospiza fortis* increase in proportion to the general population of the species, as only the deep-billed birds can crack the large seeds remaining after the supply of smaller seeds is exhausted. These and similar examples demonstrate that the evolutionary mechanisms put forward by Darwin and others do occur and lead to microevolution, or evolutionary change within single species.

Attempts to account for larger-scale macroevolutionary patterns, such as speciation and the origin of major groups of organisms, rely on indirect tests using morphological and genetic comparisons among different species, observed geographic distributions of species, and the fossil record. Such comparative studies rely on the concept of homology, the presence of corresponding and similarly constructed features among species.

At the most basic level, organization of the genetic code is remarkably similar across species; only minor variations exist among organisms as diverse as archaea (bacteria found in extreme environments such as hot springs, salt lakes, and habitats lacking in oxygen), bacteria, and eukaryotes (organisms whose cells contain a true nucleus, including plants, animals, fungi, and their unicellular counterparts). This genetic homology extends as well to the presence of shared and similarly functioning gene sequences across biological taxa, such as homeotic genes, common among all eukaryotes. Morphological homologies are also widespread; the limbs of mammals, birds, amphibians, and reptiles, for example, are all built out of the same arrangement of bones (although the particular shapes of these bones can vary greatly among groups). The conclusion that emerges from this weight of independent evidence is that structural homologies reflect an underlying evolutionary homology, or descent from a common ancestor.

Punctuated Equilibrium

Although the order of appearance of organisms in the fossil record is consistent with evolutionary theory in general, it has been troubling in two major ways: (1) there are no unequivocal transition fossils for the major evolutionary splits, such as reptiles to birds or nonflowering plants to flowering plants; and (2) paleontologists have long emphasized that gradualism—that is, evolution by gradual changes, eventually producing major changes—is not supported by the fossil record. The fossil record more often shows a pattern of almost no change over millions of years, followed by much shorter periods of rapid change. Stephen Jay Gould and Niles Eldredge, both paleontologists, proposed a new theory called punctuated equilibrium to explain this discrepancy.

Gould and Eldredge's theory accepts the fact that the fossil record shows long periods of stasis (no change) followed by periods of rapid change and consider this the primary mode for evolution. Instead of the strict neo-Darwinian view of gradual changes leading to large changes over time, Gould and Eldredge suggest that large changes are the result of a series of larger steps over a much shorter period of time. Some of the discoveries in developmental biology—of genes that, when mutated, can cause fruit flies (*Drosophila melanogaster*) to grow legs on their heads instead of antennae, or that can cause every other body segment to be missing—have helped provide some plausible mechanisms for rapid change. If genes like those in the fruit fly, which are master control genes, undergo mutation, the result could be large changes in a very short time. When first proposed, the punctuated equilibrium theory was not readily accepted, but it has gained more acceptance over time.

The Practice of Evolutionary Biology

Contemporary evolutionary biology builds upon the theoretical foundations established by Darwin, Wallace, the framers of the modern synthesis, and now Gould and Eldredge. While the reality of evolution is no longer in doubt, considerable debate remains about the importance of the various mechanisms in the history of particular groups of organisms. Much effort continues to be directed at reconstructing the particular historical path that life on earth has taken and that has led to the enormous diversity of species. Likewise, scientists seek a fuller understanding of how new species arise, as the process of splitting lineages represents a watershed event separating microevolution and macroevolution.

Unlike many other fields of biology, evolutionary biology is not amenable to tests of simple cause-and-effect hypotheses. Much of what evolutionary biologists are interested in understanding occurred in the past and over vast periods of time. In addition, the evolutionary outcomes observed in nature depend on such a large number of environmental, biological, and random factors that re-creating and studying the circumstances that could have led to a particular outcome is virtually impossible. Finally, organisms are complex creatures exposed to conflicting evolutionary pressures, such as the need to attract mates while simultaneously attempting to remain hidden from predators; such compromise-type situations are hard to simulate under experimental conditions.

Many evolutionary studies rely on making predictions about the patterns one would expect to observe in nature if evolution in one form or another were to have occurred, and such studies often involve synthesis of data derived from fieldwork, theoretical modeling, and laboratory analysis. While such indirect tests of evolutionary hypotheses are not based on the sort of controlled data that are used in direct experiments, if employed appropriately the indirect tests can be equally valid and powerful. Their strength comes from the ability to formulate predictions based on one species or type of data that may then be supported or refuted by examining additional species or data from another area of biology. In this way, evolutionary biologists are able to use the history of life on earth as a natural experiment, and, like forensic scientists, to piece together clues to solve the greatest biological mystery of all.

—*Doug McElroy, updated by Bryan Ness*

See also: Ancient DNA; Artificial Selection; Classical Transmission Genetics; Genetic Code; Genetic Code, Cracking of; Genetics, Historical Development of; Hardy-Weinberg Law; Human Genetics; Lamarckianism; Mendelian Genetics; Molecular Clock Hypothesis; Mutation and Mutagenesis; Natural Selection; Population Genetics; Punctuated Equilibrium; Repetitive DNA; RNA World; Sociobiology; Speciation; Transposable Elements.

Further Reading

Darwin, Charles. *On the Origin of Species by Means of Natural Selection*. 1859. Reprint. New York: Modern Library, 1998. While difficult (partly as a result of its nineteenth century language and style), Darwin's seminal work is an enormously thorough and visionary treatise on evolution and natural selection.

Dawkins, Richard. *The Blind Watchmaker: Why the Evidence of Evolution Reveals a Universe Without Design*. New York: W. W. Norton, 1996. Argues the case for Darwinian evolution, criticizing the prominent punctuationist school and taking issue with the views of creationists and others who believe that life arose by design of a deity.

_____. *Climbing Mount Improbable*. New York: W. W. Norton, 1997. Using "Mount Improbable" as a metaphor, discusses genetics, natural selection, and embryology for hundreds of species spanning millions of years in a fascinating, instructive way.

_____. *The Selfish Gene*. 2d ed. New York: Oxford University Press, 1990. This pathbreaking book reformulated the notion of natural selection by positing the existence of true altruism in a genetically "selfish" world. This edition contains two new chapters.

Eldredge, Niles, and Stephen Jay Gould. "Punctuated Equilibria: An Alternative to Phyletic Gradualism." In *Models in Paleobiology*, edited by Thomas J. M. Schopf. San Francisco: Freeman, Cooper, 1972. The 1972 paper that introduced the theory of punctuated equilibrium to the scientific community. Illustrations, bibliography.

Freeman, Scott, and Jon C. Herron. *Evolutionary Analysis*. 2d ed. Upper Saddle River, N.J.: Prentice Hall, 2000. An excellent textbook that presents evolutionary biology as a dynamic field of scientific inquiry.

Gould, Stephen Jay. *Eight Little Piggies*. New York: W. W. Norton, 1994. In this collection of essays originally published by Gould in the *Natural History*, the author of the theory of punctuated equilibrium considers the potential for mass extinctions of species in the face of ongoing degradation of the environment.

Quammen, David. *Song of the Dodo*. New York: Simon & Schuster, 1997. Chronicles the rich experiences of the unsung theorist of evolution Alfred Russel Wallace, whose research paralleled that of Charles Darwin.

Singh, Rama S., and Costas B. Krimbas, eds. *Evolutionary Genetics: From Molecules to Morphology.* New York: Cambridge University Press, 2000. Focuses on the necessary role of evolutionary genetics in evolutionary biology. Published in recognition of Richard Lewontin's work in evolutionary biology. Illustrations, bibliography, tables, diagrams, and index.

Weiner, Jonathan. *The Beak of the Finch: A Story of Evolution in Our Time.* New York: Random House, 1995. Describes the work of Peter and Rosemary Grant on the evolution of Charles Darwin's finches in the Galápagos Islands.

Web Sites of Interest

Evolutionary Genetics Group, Gröningen University. http://www.rug.nl/biologie/onder zoek/onderzoekGroepen/evolutionary Genetics/index. Concerned with how genetic structure and variation affect evolutionary and ecological processes.

Harvard University, Department of Organismic and Evolutionay Biology. http://www.oeb .harvard.edu. Site has a time line of significant figures in the history of evolutionary biology and a survey of its foundations.

Society for the Study of Evolution. http://www .evolutionandsociety.org. This site offers an illustrative guide, "Evolution, Science, and Society and the National Research Agenda," which includes discussion of the foundations, social effects, and future of evolutionary biology.

Extrachromosomal Inheritance

Field of study: Cellular biology

Significance: *Extrachromosomal inheritance refers to the transmission of traits that are controlled by genes located in non-nuclear organelles such as chloroplasts and mitochondria. Nuclear or chromosomal traits are determined equally by both parents, but the site of non-nuclear DNA, the cytoplasm, is almost always contributed by the female parent. The understanding of this extrachromosomal inheritance is crucial, since many important traits in plants and animals—as well as mutations implicated in disease and aging—display this type of transmission.*

Key terms

GENOME: hereditary material in the nucleus or organelle of a cell

PLASMAGENE: a self-replicating gene in a cytoplasmic organelle

PLASMON: the entire complement of genetic factors in the cytoplasm of a cell (plasmagenes or cytogenes); a plastid plasmon is referred to as a "plastome"

PLASTID: organelles, including chloroplasts, located in the cytoplasm of plant cells which form the site for metabolic processes such as photosynthesis

MITOCHONDRIA: small structures enclosed by double membranes found in the cytoplasm of all higher cells, which produce chemical power for the cells and harbor their own DNA

Discovery of Extrachromosomal Inheritance

Carl Correns, one of the three geneticists who rediscovered Austrian botanist Gregor Mendel's laws of inheritance in 1900, and Erwin Baur first described, independently, extrachromosomal inheritance of plastid color in 1909. However, they did not know then that they were observing the transmission patterns of organelle genes. Correns studied the inheritance of plastid color in the albomaculata strain of four-o'clock plants (*Mirabilis jalapa*), whereas Baur investigated garden geraniums (*Pelargonium zonate*). Correns observed that seedlings resembled the maternal parent regardless of the color of the male parent (uniparental-maternal inheritance). Seeds obtained from plants with three types of branches—with green leaves, white leaves, and variegated (a mixture of green and white) leaves—provided interesting results. Seeds from green-leaved branches produced only green-leaved seedlings, and seeds from white-leaved branches produced only white-leaved seedlings. However, seeds from branches with variegated leaves resulted in varying ratios of green-leaved, white-leaved, and variegated-leaved offspring. The explanation is that plastids in egg cells of the green-leaved branches and white-leaved branches were only of one type (homoplasmic or homoplastidic)—that is, normal chloroplasts in the green-leaved cells and white plastids (leukoplasts) in the white-leaved cells. The cells of the variegated branches, on the other hand, contained both chloroplasts and leukoplasts (heteroplasmic or heteroplastidic) in varying proportions. Some descendants of the heteroplastidic cells received only chloroplasts, some received only leukoplasts, and some received a mixture of the two types of plastids in varying proportions in the next generation, hence variegation.

Baur observed similar progeny from reciprocal crosses between normal green and white *Pelargonium* plants. Progeny in both cases were of three types: green, white, and variegated, in varying ratios. This indicated that cytoplasm was inherited from the male as well as the female parent; however, the transmission of plastids was cytoplasmic. Male transmission of plastids has also been observed in oenothera, snapdragons, beans (*Phaseolus*), potatoes, and rye. Rye is the only member of the grass family that exhibits both maternal and paternal inheritance of plastids.

The investigations on plastid inheritance also clearly established that in plants exhibiting uniparental-maternal inheritance, a variegated maternal parent always produces green, white, and variegated progeny in varying proportions because of its heteroplastidic nature. Crosses between green and white plants always yield green or white progeny, depending upon the maternal parent, when the parental plants are homoplasmic for plastids.

Extrachromosomal Inheritance vs. Nuclear Inheritance

Extrachromosomal inheritance has been found in many plants, including barley, maize, and rice. Traits are inherited through chloroplasts, mitochondria, or plasmids (small, self-replicating structures). Inheritance of traits that are controlled by organelle genomes (plasmons) can be called nonnuclear or cytoplasmic. The cytoplasm contains, among other organelles, mitochondria in all higher organisms and mitochondria and chloroplasts in plants. Because cytoplasm is almost always totally contributed by the female parent, this type of transmission may also be called maternal or uniparental inheritance.

Most chromosomally inherited traits obey Mendel's law of segregation, which states that a pair of alleles or different forms of a gene separate from each other during meiosis (the process that halves the chromosome number in gamete formation). They also follow the law of independent assortment, in which two alleles of a gene assort and combine independently with two alleles of another gene. Such traits may be called Mendelian traits. Extrachromosomal inheritance is one of the exceptions to Mendelian inheritance. Thus, it can be called non-Mendelian inheritance. (Mendel only studied and reported on traits controlled by nuclear genes.) Mendelian heredity is characterized by regular ratios in segregating generations for qualitative trait differences and identical results from reciprocal crosses. On the contrary, non-Mendelian inheritance is characterized by a lack of regular segregation ratio and non-identical results from reciprocal crosses.

The mitochondria are the sites of aerobic respiration (the breaking down of organic substances to release energy in the presence of oxygen) in both plants and animals. They are, like plastids, self-replicating entities and exhibit genetic continuity. The mitochondrial genes do not exhibit the Mendelian segregation pattern either. Mitochondrial genetics began around 1950 with the discovery of "petite" mutations in baker's yeast (*Saccharomyces cerevisiae*). Researchers observed that one or two out of every one thousand colonies grown on culture medium were smaller than normal colonies. The petite colonies bred true (produced only petite colonies). The petite mutants were respiration deficient under aerobic conditions. The slow growth of the petite colonies was related to the loss of a number of respiratory (cytochrome) enzymes that occur in mitochondria. These mitochondrial mutants, termed "vegetative petites," can be induced with acriflavine and related dyes. Another type of mutation, called a "suppressive petite," was found to be caused by defective, rapidly replicating mitochondrial DNA (mtDNA). Petite mutants that are strictly under nuclear gene control have also been reported and are called segregational petite mutants. Most respiratory enzymes are under both nuclear and mitochondrial control, which is indicative of collaboration between the two genetic systems.

In the fungus *Neurospora*, mitochondrial inheritance has been demonstrated for mutants referred to as "poky" (a slow-growth characteristic). The mutation resulted from an impaired mitochondrial function related to cytochromes involved in electron transport. The mating between poky female and normal male yields only poky progeny, but when the cross is reversed, the progeny are all normal, confirming maternal inheritance for this mutation.

According to a 1970 study, cytoplasmic male sterility is found in about eighty plant species. The molecular basis of cytoplasmic male sterility in maize through electrophoretic separation of restriction-endonuclease-created fragments of DNA was traced to mitochondrial DNA. Cytoplasmic male sterility can be overcome by nuclear genes. The plasmids that reside in mitochondria are also important extrachromosomal DNA molecules that are especially important in antibiotic resistance. Plasmids have been found to be extremely useful in genetic engineering.

Mutator Genes

Plastome mutations can be induced by nuclear genes. A gene that increases the mutation rate of another gene is called a "mutator." One such gene is the recessive, nuclear iojap (*ij*) mutation in maize. In the homozygous (*ij ij*) condition, it induces a plastid mutation. The name "iojap" has been derived from "Iowa" (the maize strain in which the mutation is found) and "japonica" (a type of striped variety that the mutation resembles). Once the plastid gene mutation caused by the *ij* gene has been initiated, the inheritance is non-Mendelian, and it no longer depends on the nuclear *ij* gene. As long as the iojap plants are used as female parents, the inheritance of the trait is similar to that for plastids in the albomaculata variety of four o'clock plants.

The *chm* mutator gene causes plastid mutations in the plant *Arabidopsis*, and mutator "striata" in barley causes mutations in both plastids and mitochondria. Cases of mutator-induced mutations in the plastome have also been reported in rice and catnip.

Chloroplast and Mitochondrial DNA

Plastids contain DNA, have their own DNA polymerase (the enzyme responsible for DNA replication), and undergo mutation. The chloroplast DNA (cpDNA) is a circular, self-replicating system that carries genetic information that is transcribed (from DNA to RNA) and translated (from RNA to protein) in the plastid. It replicates in a semiconservative manner—that is, an original strand of DNA is conserved and serves as the template for a new strand in a manner similar to replication.

The soluble enzyme ribulose biphosphate carboxylase/oxygenase (Rubisco) is involved in photosynthetic carbon dioxide fixation. In land plants and green algae, its large subunit is a cpDNA product, while its small subunit is controlled by a nuclear gene family. Thus, the Rubisco protein is, as are chloroplast ribosomes, a product of the cooperation between the nuclear and chloroplast genes. In all other algae, both the large and small subunits of Rubisco are encoded in cpDNA.

Mitochondrial DNA (mtDNA) molecules are also circular and self-replicating. Human, yeast, and higher plant mtDNAs are the major systems that have been studied. The human mtDNA has a total of 16,569 base pairs. The yeast mtDNA is five times larger than that (84 kilobases), and maize mtDNA is much larger than the yeast mtDNA. Every base pair of human mtDNA may be involved in coding for a mitochondrial messenger RNA (mRNA) for a protein, a mitochondrial ribosomal RNA (rRNA), or a mitochondrial transfer RNA (tRNA). It is compact, showing no intervening, noncoding base sequences between genes. It has only one major promoter (a DNA region to which an RNA polymerase binds and initiates transcription) on each strand. Most codons—triplets of nucleotides (bases) in messenger

RNA carrying specific instructions from DNA—have the same meaning as in the universal genetic code, except the following differences: UGA represents a "stop" signal (universal), but represents tryptophan in yeast and human mtDNA; AUA represents isoleucine (universal), but methionine in human mtDNA; CUA represents leucine (universal), but threonine in yeast mtDNA; and CGG represents arginine (universal), but tryptophan in plant mtDNA.

The mtDNA carries the genetic code (plasmagene names in parentheses) for proteins, such as cytochrome oxidase subunits I (coxl), II (cox2), and III (cox3); cytochrome B (cytb); and ATPase subunits 6 (atp6), 8 (atp8), and 9 (atp9). It also contains the genetic codes for several ribosomal RNAs, such as mtrRNA 16s and 12s in the mouse; mtrRNA 9s, 15s, and 21s in yeast; and mtrRNA 5s, 18s, and 26s in maize. In addition, twenty-two transfer RNAs in mice, twenty-four in yeast, and three in maize are encoded in mtDNA.

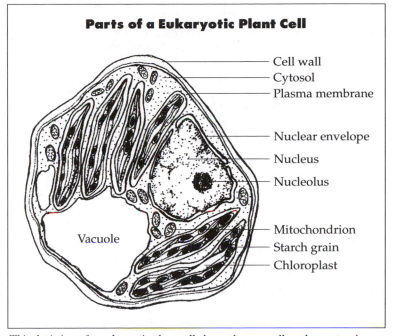

Parts of a Eukaryotic Plant Cell

Cell wall
Cytosol
Plasma membrane
Nuclear envelope
Nucleus
Nucleolus
Mitochondrion
Starch grain
Chloroplast
Vacuole

This depiction of a eukaryotic plant cell shows the organelles where extrachromosomal DNA is found: mitochondria, and in plants chloroplasts and other plastids. (Kimberly L. Dawson Kurnizki)

Chlamydomonas reinhardtii

Chlamydomonas reinhardtii is a unicellular green algaa in which chloroplast and mitochondrial genes show uniparental transmission. In 1954, Ruth Sager discovered the chloroplast genetic system. Resistance to high levels of streptomycin (a trait controlled by chloroplast genes) has been shown to be transmitted uniparentally by the mt⁺ mating type parent. The mt⁻ mating type transmits the mitochondrial genes uniparentally. Mutants in chloroplasts have been identified for antibiotic and herbicide resistance. Genetic recombination is common in *C. reinhardtii*, which occurs in zygotes (the fused gametes of opposite sexes) when biparental cytogenes are in a heterozygous (union of unlike genes) state. This is an ideal system among plants for recombination studies, since there is only one large plastid per cell. In higher plants, study of genetic recombination is difficult because of a large number of plastids in cells and a lack of genetic markers.

Mutations in mitochondria of *C. reinhardtii* can be induced with acriflavine or ethidium bromide dyes. Point mutations for myxothiazol resistance mapping in the *cytb* gene have been isolated. The mitochondrial genome of this species of algae has been completely sequenced. It encodes five of more than twenty-five subunits of the reduced nicotinamide-adenine dinucleotide (NADH) dehydrogenase of complex I (nad1, nad2, nad4, nad5, and nad6), the COX I subunit of cytochrome oxidase (cox1), and the apocytochrome b (cob) subunit of complex III. All of these proteins have a respiratory function.

Origin of Plastid and Mitochondrial DNA

According to the endosymbiont theory, plastids and mitochondria in eukaryotes are the descendants of prokaryotic organisms that invaded primitive eukaryotes. Subsequently, they developed a symbiotic relationship and became dependent upon each other. There is much support for this theory. Researchers in 1972 showed homology (genetic similarity) between ribosomal RNA from cyanobacteria and DNA from the chloroplasts of *Euglena gracilis.* This provided support for chloroplasts as the descendants of cyanobacteria. Mitochondria are believed to have come from primitive bacteria and plastids from blue-green algae. Molecular evidence strongly supports the endosymbiotic origin of mitochondria from alpha purple bacteria.

In 1981, Lynn Margulis summarized evidence for this theory. There are many similarities between prokaryotes and organelles: Both have circular DNA and the same size ribosomes, both lack histones and a nuclear membrane, and both show similar response to antibiotics that inhibit protein synthesis. Both also show a primitive mode of translation that begins with formulated methionine. The discovery of promiscuous DNA (DNA segments that have been transferred between organelles or from a mitochondrial genome to the nuclear genome) in eukaryotic cells also lends support to this theory.

Impact and Applications

Genetic investigations have helped tremendously in constructing a genetic map of maize cpDNA. Important features of the map, including two large, inverted, repeat segments containing several rRNA and tRNA genes, are now known. Detection and quantification of mutant mtDNA are essential for the diagnosis of diseases and for providing insights into the molecular basis of pathogenesis, etiology, and ultimately the treatment of diseases. This should help enhance the knowledge of mitochondrial biogenesis. Mitochondrial dysfunction, resulting partly from mutations in mtDNA, may play a central role in organismal aging.

A number of human diseases associated with defects in mitochondrial function have been identified. Large-scale deletions and tRNA point mutations (base changes) in mtDNA are associated with clinical mitochondrial encephalomyopathies. Heteroplasmy (the coexistence of more than two types of mtDNA) has provided experimental systems in which the transmission of mtDNA in animals can be studied. Numerous deleterious point mutations of mtDNA are associated with various types of human disorders involving deficiencies in the mitochondrial oxidative phosphorylation (respiration) apparatus. Leigh disease is caused by a point mutation in mtDNA. Deletions of mtDNA have been associated with diseases such as isolated ocular myopathy, chronic progressive external ophthalmoplegia, Kearn-Sayre syndrome, and Pearson's syndrome.

The influence of the mitochondrial genome and mitochondrial function on nuclear gene expression is poorly understood, but progress is being made toward understanding why a few genes are still sequestered in the mitochondria and toward developing new tools to manipulate mitochondrial genes.

—*Manjit S. Kang*

See also: Ancient DNA; Chloroplast Genes; Genetic Code; Human Genetics; Mitochondrial Diseases; Mitochondrial Genes; Model Organism: *Chlamydomonas reinhardtii*; RNA World.

Further Reading

Attardi, Giuseppe M., and Anne Chomyn, eds. *Methods in Enzymology: Mitochondrial Biogenesis and Genetics.* Vols. 260, 264. San Diego: Academic Press, 1995. One hundred authors contribute to thirty-six chapters, presenting a wealth of new methods and data and covering the significant developments that have expanded the scope of enzyme chemistry.

Cummings, Michael J. *Human Heredity: Principles and Issues.* 5th ed. Pacific Grove, Calif.: Brooks/Cole, 2000. College text that surveys topics such as genetics as a human endeavor; cells, chromosomes, and cell division; transmission of genes from generation to generation; cytogenics; the source of genetic variation; cloning and recombinant DNA; genes and cancer; genetics of behavior; and genes in populations.

Gillham, Nicholas W. *Organelle Genes and Genomes.* London: Oxford University Press, 1997. Contains a comprehensive review of the genetic-molecular aspects of chloroplasts and mitochondria.

Forensic Genetics

Field of study: Human genetics and social issues

Significance: *Forensic genetics uses DNA or the inherited traits derived from DNA to identify individuals involved in criminal cases. Blood tests and DNA testing are used to determine the source of evidence, such as blood stains or semen, left at a crime scene.*

Key terms

ALLELES: alternative versions of genes at a genetic locus that determine an individual's traits

DNA FINGERPRINTING: a DNA test used by forensic scientists to aid in the identification of criminals or to resolve paternity disputes

FORENSIC SCIENCE: the application of scientific knowledge to analyze evidence used in civil and criminal law, especially in court proceedings

Forensic Science and DNA Analysis

Forensic scientists use genetics for two major legal applications: identifying the source of a sample of blood, semen, or other tissue, and establishing the biological relationship between two people in paternity or maternity suits. Forensic scientists are frequently called upon to testify as expert witnesses in criminal trials. One of the most useful sources of inherited traits for forensic science purposes is blood. Such traits include blood type, proteins found in the plasma, and enzymes found in blood cells. The genes in people that determine such inherited traits have many different forms (alleles), and the specific combination of alleles for many of the inherited blood traits can be

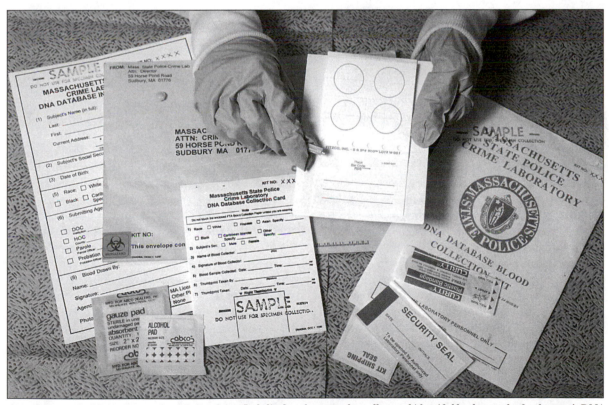

A serologist at the Massachusetts State Police Crime Lab displays forms used to collect and identify bloods samples for the state's DNA database of people convicted of certain crimes. In 1998, a group of prisoners brought a suit against the state to overturn a law requiring blood samples from anyone convicted of any of thirty-three different crimes. (AP/Wide World Photos)

O. J. Simpson and attorneys discuss strategy for cross-examining a forensic scientist during Simpson's 1995 murder trial. Despite DNA evidence that blood found near Simpson's home and in his car matched that of the murder victims, Simpson was acquitted by a jury upon testimony that the evidence might have been contaminated. (AP/Wide World Photos)

used to identify an individual. The number of useful blood group systems is small, however, which means that a number of individuals might have blood groups identical to those of the subject being tested.

The ultimate source of genetic information for identification of individuals is the DNA found in the chromosomes. Using a class of enzymes known as restriction enzymes, technicians can cut strands of DNA into segments, forming bands similar to a supermarket bar code that vary with individuals' family lines. The pattern, termed a DNA "fingerprint" or profile, is inherited much as are the alleles for blood traits. DNA fingerprinting can be used to establish biological relationships (including paternity) with great reliability, because a child cannot have a variation that is not present in one of the parents. Since DNA is stable and can be reliably tested in dried blood or semen even years after a crime has been committed, DNA fingerprinting has revolutionized the solution

of criminal cases in which biological materials are the primary evidence. The likelihood of false matches ranges from one per million to one per billion. These numbers, however, do not include the possibility of mishandling of evidence, laboratory errors, or planting of evidence.

Criminal Cases Involving DNA Evidence

On November 6, 1987, serial rapist Tommy Lee Andrews became the first American ever convicted in a case involving DNA evidence. Samples of semen left at the crime scene by the rapist and blood taken from Andrews were sent to a New York laboratory for testing. Using the techniques of DNA fingerprinting, the laboratory isolated DNA from each sample, compared the patterns, and found a DNA match between the semen and the blood. Andrews was sentenced to twenty-two years in prison for rape, aggravated battery, and burglary.

The 1990-1991 *United States v. Yee* homicide

trial in Cincinnati, Ohio, was the first major case that challenged the soundness of DNA testing methods. DNA analysis by the Federal Bureau of Investigation (FBI) showed a match between blood from the victim's van and from Steven Yee's car. The defense claimed that the matching DNA data were ambiguous or inconsistent, citing what they claimed to be errors, omissions, lack of controls, and faulty analysis. However, after a fifteen-week hearing, the judge accepted the DNA testing as valid.

In 1994, former football star O. J. Simpson was arrested and charged with the murders of his ex-wife, Nicole Brown, and her friend, Ronald Goldman. Blood with DNA that matched Simpson's was found at Brown's home, and blood spots in Simpson's car contained DNA matching Brown's, Goldman's, and Simpson's. Furthermore, blood at Simpson's home contained DNA that matched Brown's and Goldman's. For the most part, the defense admitted the accuracy of the DNA tests and did not scientifically challenge the results of the DNA fingerprinting. Instead, they argued that the biological evidence had been contaminated by shoddy laboratory work and by planting of evidence; the jury found Simpson not guilty of the charges against him.

Impact and Applications

DNA evidence is used in thousands of criminal investigations and tens of thousands of paternity tests annually in the United States. In addition, in numerous cases forensic DNA testing has been used to free previously convicted and incarcerated individuals, some of whom have been in prison for more than a decade. Most states now have data banks containing DNA profiles of people already convicted of sexual or related offenses; when law enforcement officials investigate a crime, they are now able to test DNA collected at the scene to see if it matches that of anyone in the data bank with a history of a similar offense.

—Alvin K. Benson

See also: Biological Determinism; Criminality; DNA Fingerprinting; Eugenics; Eugenics: Nazi Germany; Human Genetics; Insurance; Paternity Tests; Sociobiology; Sterilization Laws.

Further Reading

Burke, Terry, R., Wolf, G. Dolf, and A. Jeffreys, eds. *DNA Fingerprinting: Approaches and Applications.* Boston: Birkhauser, 2001. Describes repetitive DNA and the broad variety of practical applications to law, medicine, politics, policy, and more. Aimed at the layperson.

Coleman, Howard, and Eric Swenson. *DNA in the Courtroom: A Trial Watcher's Guide.* Seattle, Wash.: GeneLex Press, 1994. Gives a good overview of DNA fingerprinting, expert evidence in court, and applications of forensic genetics.

Connors, Edward, et al. *Convicted by Juries, Exonerated by Science: Case Studies in the Use of DNA Evidence to Establish Innocence After Trial.* Washington, D.C.: U.S. Department of Justice, Office of Justice Programs, National Institute of Justice, 1996. Provides case studies in the use of DNA evidence to establish innocence after conviction in a trial.

Fridell, Ron. *DNA Fingerprinting: The Ultimate Identity.* New York: Scholastic, 2001. The history of the technique, from its discovery to early uses. Aimed at younger readers and nonspecialists.

Herrmann, Bernd, and Susanne Hummel, eds. *Ancient DNA: Recovery and Analysis of Genetic Material from Paleographic, Archaeological, Museum, Medical, and Forensic Speciments.* New York: Springer-Verlag, 1994. Written when DNA fingerprinting was just coming to the fore and films such as *Jurrasic Park* were in theaters, this collection of papers by first-generation researchers reflects the broad applications of the technology, including paleontological investigations.

Hummel, Susanne. *Fingerprinting the Past: Research on Highly Degraded DNA and Its Applications.* New York: Springer-Verlag, 2002. Manual about typing ancient DNA.

Jarman, Keith, and Norah Rudin. *An Introduction to Forensic DNA Analysis.* 2d ed. Boca Raton, Fla.: CRC Press, 2001. Emphasizes the advantages and limitations of various DNA techniques used in the analysis of forensic evidence.

Lincoln, Patrick J., and Jim Thompson, eds. *Forensic DNA Profiling Protocols.* Vol. 98. Totowa, N.J.: Humana Press, 1998. Presents techniques of DNA identity-testing, noting exactly what chemicals and procedures are to be used for a variety of situations.

United States National Research Council. *The Evaluation of Forensic DNA Evidence.* Rev. ed. Washington, D.C.: National Academy Press, 1996. Evaluates how DNA is interpreted in the courts, includes developments in population genetics and statistics, and comments on statements made in the original volume that proved controversial or that have been misapplied in the courts.

Web Sites of Interest

Department of Justice, Federal Bureau of Investigation, Handbook of Forensic Services, DNA Examinations. http://www.fbi.gov/ hq/lab/handbook/examsdna.htm. The FBI's step-by-step guide to the collecting, securing, and submitting of DNA evidence from crime scenes.

Earl's Forensic Page. http://members.aol .com/EarlNMeyer/DNA.html. Summarizes how DNA fingerprinting works and its use in crime investigations and in determining paternity.

Fragile X Syndrome

Field of study: Diseases and syndromes
Significance: *There are more than fifty mental retardation disorders associated with the X chromosome, but their frequencies are rare. Fragile X syndrome is the most common inherited form of mental retardation, affecting an estimated one in fifteen hundred males and one in twenty-five hundred females.*

Key terms

SEX CHROMOSOMES: the chromosomes, X and Y, that determine sex; the presence of two X chromosomes codes for females and an X chromosome paired with a Y chromosome codes for males; these chromosomes are received from an individual's parents, each of whom contributes one sex chromosome to their offspring

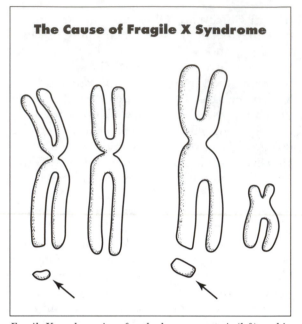

The Cause of Fragile X Syndrome

Fragile X syndrome in a female chromosome pair (left) and in a male pair (right). Note the apparently detached tips of the X chromosomes, the defect that gives the syndrome its name. (Electronic Illustrators Group)

SEX-LINKED TRAIT: Any characteristic controlled by genes on the X or Y chromosome

History of Fragile X Syndrome

In 1969, geneticists studied a family of four mentally retarded brothers who had X chromosomes whose tips appeared to be detached from the rest of the chromosome. It is now recognized that this fragile site occurs in the vicinity of the *FMR1* gene. Males affected with fragile X syndrome have moderate to severe mental retardation and show distinctive facial features, including a long and narrow face, large and protruding ears, and a prominent jaw. Additional features include velvet-like skin, hyperextensible finger joints, and double-jointed thumbs. These features are generally not observed until maturity. Prior to puberty, the only symptoms a child may have are delayed developmental milestones, such as sitting, walking, and talking. Fragile X children may also display an abnormal temperament marked by tantrums, hyperactivity, or autism. A striking feature of most adult fragile X males is an enlarged testicular volume (macroorchidism). This enlargement is not a result of testosterone levels, which are normal. Fragile X men are fertile, and offspring have been documented, but those with significant mental retardation rarely reproduce.

The intelligence quotient (IQ) of the majority of affected males is in the moderate to severely retarded range. Only a few affected males have IQs above seventy-five. Fragile X males frequently show delayed speech development and language difficulties. Repetitive speech patterns may also be present.

Mode of Inheritance

In males, any abnormal gene on the X chromosome is expressed because males have only one X chromosome. In females, two copies of the fragile X chromosome must be present for them to be affected. This is the classic pattern for X-linked, or sex-linked, traits (traits whose genes are located on the X chromosome.)

The pattern of inheritance for fragile X is unusual. Fragile X syndrome increases in severity through successive generations. This is explained by a worsening of the defect in the *FMR1* gene as it is passed from mothers to sons. Since males contribute the Y chromosome to their sons, fathers do not pass the fragile X gene to their sons. They will, however, contribute their X chromosome to their daughters. Because these daughters also receive an X chromosome from their mothers, they generally appear normal or only mildly affected. It is only when these daughters have a son that the condition is expressed.

An explanation for this increasing severity through generations was discovered by analyzing the DNA sequence of the *FMR1* gene. The molecules composing DNA are adenine (A), thymine (T), cytosine (C), and guanine (G) and are referred to collectively as "bases." In fragile X syndrome, a sequence in which the

Seventeen-year-old Jake Porter (wearing Mohawks jersey 45) suffers from fragile X syndrome. Here he is being honored by teammates during halftime at Motor City Bowl in Detroit on December 26, 2002, for scoring a touchdown during an earlier game in McDermott, Ohio. (AP/Wide World Photos)

three bases CGG are repeated over and over was found. The repetitive sequence is found in normal copies of the *FMR1* gene, but in individuals with fragile X syndrome there are many times more copies of the CGG triplet. The longer repetitive sequence in the *FMR1* gene prevents it from being expressed. Individuals not having the fragile X syndrome have a working *FMR1* gene.

—*Linda R. Adkison, updated by Bryan Ness*

See also: Behavior; Chromatin Packaging; Classical Transmission Genetics; Congenital Defects; DNA Replication; Down Syndrome; Intelligence; Repetitive DNA.

Further Reading

Dykens, Elisabeth M., Robert M. Hodapp, and Brenda M. Finucane. *Genetics and Mental Retardation Syndromes: A New Look at Behavior and Interventions.* Baltimore: Paul H. Brookes, 2000. Reviews the genetic and behavioral characteristics of nine mental retardation syndromes, giving in-depth information on genetic causes, prevalence, and physical and medical features of Down, Williams, fragile X, and Prader-Willi syndromes, as well as five other less frequently diagnosed syndromes.

Hagerman, Paul J. *Fragile X Syndrome: Diagnosis, Treatment, and Research.* 3d ed. Baltimore: Johns Hopkins University Press, 2002. Discusses the clinical approach to diagnosing fragile X; the latest research in epidemiology, molecular biology, and genetics; and information on genetic counseling, pharmacotherapy, intervention, and gene therapy.

Hirsch, David. "Fragile X Syndrome." *The Exceptional Parent* 25 (June, 1995). Answers parental concerns about fragile X syndrome genetic testing.

_____. "Fragile X Syndrome: Medications for Aggressive Behavior?" *The Exceptional Parent* 26 (October, 1996). Addresses parental concerns about medications for males displaying aggressive behavior.

Hogenboom, Marga. *Living with Genetic Syndromes Associated with Intellectual Disability.* Philadelphia: Jessica Kingsley, 2001. Gives an accessible introduction to genetics before detailing the ways in which young people are affected by genetic conditions: the extent to which their behavior is determined, the difficulties they face, and the ways in which they can achieve independence and fulfillment.

Parker, James N., and Philip M. Parker, eds. *The 2002 Official Parent's Sourcebook on Fragile X Syndrome.* San Diego: ICON Health, 2002. Draws from public, academic, government, and peer-reviewed research to guide parents where and how to look for information covering virtually all topics related to fragile X syndrome.

Schmidt, Michael A. "Fragile X Syndrome: Diagnosis, Treatment, and Research." *Journal of the American Medical Association* 277 (April 9, 1997). Provides a detailed discussion of fragile X syndrome.

Shannon, Joyce Brennfleck, ed. *Mental Retardation Sourcebook: Basic Consumer Health Information About Mental Retardation and Its Causes, Including Down Syndrome, Fetal Alcohol Syndrome, Fragile X Syndrome, Genetic Conditions, Injury, and Environmental Sources.* Detroit, Mich.: Omnigraphics, 2000. Reviews causes, prevention, family life, education, specific health care issues, and legal and economic concerns for health care consumers.

Warren, Stephen T. "Trinucleotide Repetition and Fragile X Syndrome." *Hospital Practice* 32 (April 15, 1997). Provides detail about CGG repeats in fragile X syndrome.

Web Sites of Interest

Dolan DNA Learning Center, Your Genes Your Health. http://www.ygyh.org. Sponsored by the Cold Spring Harbor Laboratory, this site, a component of the DNA Interactive Web site, offers information on more than a dozen inherited diseases and syndromes, including Fragile X syndrome.

FRAXA Research Foundation. http://www.fraxa.org. The foundation supports research aimed at treatment, and this site offers information on managing the syndrome.

National Fragile X Foundation. Xtraordinary Accomplishments. http://www.nfxf.org. General information about the disorder, with advice for caregivers on testing, medical treatment, education, and life planning.

Gel Electrophoresis

Field of study: Techniques and methodologies

Significance: *Gel electrophoresis is a laboratory technique involving the movement of charged molecules in a buffer solution when an electric field is applied to the solution. The technique allows scientists to separate DNA, RNA, and proteins according to their size. The method is the most widely used way to determine the molecular weight of these molecules and can be used to determine the approximate size of most DNA molecules and proteins.*

Key terms

DENATURING: a method of disrupting the normal three-dimensional structure of a protein or nucleic acid so that it stretches out more or less linearly

GEL: a support matrix formed by interconnecting long polymers into a porous, solid material that retards the movement of molecules entrapped in it

STAINING DYE: a chemical with a high affinity for DNA, RNA, or proteins that causes a visible color to develop that allows the detection of these molecules in the gel

Basic Theory of Electrophoresis

Biologists often need to determine the approximate size of DNA fragments, RNA, or proteins. All of these molecules are much too small to visualize using conventional methods. The size of a piece of DNA capable of carrying all the information needed for a single gene may be only 2 microns long and 20 angstroms wide, while the protein encoded by this gene might form into a globular ball only 2.5 to 10 nanometers in diameter. Therefore, some indirect method of "seeing" the length of these molecules must be used. The easiest and by far most common way to do this is by gel electrophoresis. Electrophoresis is based on the theory that if molecules can be induced to move in the same direction through a tangled web of material, smaller molecules will move farther through the matrix than larger molecules. Thus, the distance a molecule moves will be related to its size, and knowing the basic chemical nature of the molecule will allow an approximation of its relative molecular weight.

As an analogy, imagine a family with two children picnicking by a thick, brushy forest. Their small dog runs into the brush, and the whole family runs in after it. The dog, being the smallest, penetrates into the center of the forest. The six-year-old can duck through many of the branches and manages to get two-thirds of the way in; the twelve-year-old makes it halfway; the mother gets tangled up and must stop after only a short distance; the father, too large to fit in anywhere, cannot enter at all. This is what happens to molecules moving through a gel: Some travel through unimpeded, others are separated into easily visualized size groups, and others cannot even enter the matrix.

The Electrophoresis Setup

The gel is typically composed of a buffer solution containing agarose or acrylamide, two polymers that easily form a gel-like material at room temperature. At first the buffer/polymer solution is liquid and is poured into a casting chamber composed of a special tray or of two plates of glass with a narrow space between them. A piece of plastic with alternating indentations like an oversized comb is pushed into one end of the gel while it is still liquid. When the gel has solidified, the "comb" is removed, leaving small depressions in the matrix (wells) into which the DNA, RNA, or protein sample is applied. The gel is then attached to an apparatus that exposes the ends of the gel to a buffer, each chamber of which is attached to an electric power supply. The buffer allows an even application of the electric field.

Since the molecules of interest are so small, matrices with small pore size must be created. It is important to find a matrix that will properly separate the molecules being studied. The key is to find a material that creates pores large enough to let DNA or proteins enter but small enough to impede larger molecules. By using different concentrations of agarose or acrylamide, anything from very short pieces of DNA that differ only by a single nucleotide to whole chromosomes can be separated.

Agarose is composed of long, linear chains of multiple monosaccharides (sugars). At high temperatures, 95 degrees Celsius (203 degrees Fahrenheit), the agarose will "melt" in a buffer solution. As the gel cools to around 50 degrees Celsius (122 degrees Fahrenheit), the long chains begin to wrap around each other and solidify into a gel. The concentration of agarose determines the pore size, since a larger concentration will create more of a tangle. Agarose is usually used with large DNA or RNA molecules.

Acrylamide is a short molecule made up of a core of two carbons connected through a double bond with a short side-chain with a carboxyl and amino group. When the reactive chemicals ammonium persulfate and TEMED are added, the carbon ends fuse together to create long chains of polyacrylamide. If this were the only reaction, the end result would be much like agarose. However, a small number (usually 5 percent or less) of the acrylamides are the related molecule called bis-acrylamide, a two-headed version of the acrylamide molecule. This allows the formation of interconnecting branch points every twenty to fifty acrylamide residues on the chain, which creates a pattern more like a net than the tangled strands of agarose. This results in a narrower pore size than agarose, which allows the separation of much smaller fragments. Acrylamide is used to separate proteins and small DNA fragments and for sequencing gels in which DNA fragments differing in size by only a single nucleotide must be clearly separated.

Why Nucleic Acids and Proteins Move in a Gel

DNA and RNA will migrate in an electric field since every base has a net negative charge. This means that DNA molecules are negatively charged and will migrate toward the positive pole if placed in an electric field. In fact, since each base contributes the same charge, the amount of negative charge is directly proportional to the length of the DNA. This means that the electromotive force on any piece of DNA or RNA is directly proportional to its length (and therefore its mass) and that the rate of movement of DNA or RNA molecules of the same length should be the same.

The charge on different amino acids varies considerably, and the proportions of the various amino acids vary widely from protein to protein. Therefore, the charge on a protein has nothing to do with its length. To correct for this, proteins are mixed with the detergent sodium dodecyl sulfate, or SDS (the same material that gives most shampoos their suds), before being loaded onto the gel. The detergent coats the protein evenly. This has two important effects. The first is that the protein becomes denatured, and the polypeptide chain will largely exist as a long strand (rather than being compactly bunched, as it normally is). This is important because a tightly balled protein would more easily pass through the polyacrylamide matrix than a linear molecule, and proteins with the same molecular weight might appear to be different sizes. More important, each SDS molecule has a slight negative charge, so the even coating of the protein results in a negative charge that is directly proportional to the size of the protein.

Once the molecules have been subjected to the electric field long enough to separate them in the gel, they must be visualized. This is done by soaking the gel in a solution that contains a dye that stains the molecules. For DNA and RNA, this dye is usually ethidium bromide, a molecule that has an affinity for nucleic acids and slips between the strands or intercalates into the helix. The dye, when exposed to ultraviolet light, glows orange, revealing the location of the nucleic acid in the gel. For proteins, the dye Coomassie blue is usually used, a stain that readily binds to proteins of most types.

—*J. Aaron Cassill, updated by Bryan Ness*
See also: Blotting: Southern, Northern, and Western; DNA Fingerprinting; Genetic Testing; Proteomics; RFLP Analysis; Shotgun Cloning.

Further Reading

Dunn, Michael J., ed. *From Genome to Proteome: Advances in the Practice and Application of Proteomics.* New York: Wiley-VCH, 2000. Reviews advances in proteomics and covers topics such as sample preparation and solubilization, developments in electrophoresis, detection and quantification, mass spectrometry, and proteome data analysis and management.

Hames, B. D., and D. Rickwood, eds. *Gel Electrophoresis of Nucleic Acids: A Practical Approach.* 2d ed. New York: Oxford University Press, 1990. Updated edition of this standard text reviews the advances made in refining established techniques and details many new techniques, including pulse field electrophoresis, gel retardation analysis, and DNA footprinting.

Jolles, P., and H. Jornvall, eds. *Proteomics in Functional Genomics: Protein Structure Analysis.* Boston: Birkhauser, 2000. Discusses a range of topics, including sample preparation, measurement and sequencing techniques, bioinformatics, and equipment issues. Illustrated.

Lai, Eric, and Bruce W. Birren, eds. *Electrophoresis of Large DNA Molecules: Theory and Applications.* Cold Spring Harbor, N.Y.: Cold Spring Harbor Laboratory, 1990. Surveys the technique's biochemical and biophysical foundations and its application to the separation of DNA fragments in a variety of experimental settings.

Link, Andrew J., ed. *2-D Proteome Analysis Protocols.* Totowa, N.J.: Humana Press, 1999. Provides detailed descriptions and helpful illustrations of the techniques that are widely used for the analysis of total cellular proteins.

Pennington, S. R., and M. J. Dunn, eds. *Proteomics: From Protein Sequence to Function.* New York: Springer, 2001. An introductory, illustrated text designed for undergraduates in biochemistry, molecular biology, and genetics that details the study of genomics and proteomics.

Rabilloud, Thierry, ed. *Proteome Research: Two-Dimensional Gel Electrophoresis and Identification Methods.* New York: Springer, 2000. Focuses on the first two phases of proteomics: separation by two-dimensional electrophoresis and microcharacterization of the separated proteins. Illustrated.

Gender Identity

Field of study: Human genetics and social issues

Significance: *Researchers have long sought an understanding of the basis of human gender identity. Discoveries in the field of human genetics have opened the way to examine how genes affect sexual behavior and sexual identity.*

Key terms

HERMAPHRODITE: an individual who has both male and female sex organs

RESTRICTION FRAGMENT LENGTH POLYMORPHISM (RFLP): a technique involving the cutting of DNA with restriction endonucleases (restriction enzymes) that allows researchers to compare genetic sequences from various sources

SEX DETERMINATION: the chromosomal sex of an individual; normal human females have two X chromosomes; normal human males have one X and one Y chromosome

SEXUAL ORIENTATION: the actual sexual behavior exhibited by an individual

Boy or Girl?

The question of what is "male" and what is "female" can have a variety of answers, depending on whether one is thinking of chromosomal (genetic) sex, gonadal sex, phenotypic sex, or self-identified gender. Chromosomal sex is determined at the time of conception. The fertilized human egg has a total of forty-six chromosomes, including one pair of sex chromosomes. If the fertilized egg has a pair of X chromosomes, its chromosomal, or genetic, sex is female. If it has one X chromosome and one Y chromosome, its genetic sex is male. Toward the end of the second month of prenatal development, processes are initiated that lead to the development of the gonadal sex of the individual; the embryo develops testes if male, ovaries if female. Although the chromosomal sex may be XX, the sexual phenotype will not always be female; likewise, if the chromosomal sex is XY, the sexual phenotype does not always turn out to be male. Naturally occurring chromosomal variations or single-gene mutations may interfere with normal development and differentiation, leading to sexual phenotypes that do not correspond to the chromosomal sex.

One such case is that of hermaphrodites, individuals who possess both ovaries and testes. They usually carry both male and female tissue. Some of their cells may be of the female chromosomal sex (XX), and some may be of the male chromosomal sex (XY). Such individuals are called sex chromosome mosaics, and their resulting phenotype may be related to the number and location of cells that are XX and those that are XY. Another example is testicular feminization syndrome, in which a single gene affects sexual differentiation. Individuals with this syndrome have the chromosomal sex of a normal male but have a female phenotype. XY males with this gene, located on the X chromosome, exhibit initial development of the testes and normal production of male hormones. However, the mutant gene prevents the hormones from binding to receptor cells; as a result, female characteristics develop.

Gender Identity Disorder

Gender identity disorder, or transsexualism, is defined by researchers as a persistent feeling of discomfort or inappropriateness concerning one's anatomic sex. The disorder typically begins in childhood and is manifested in adolescence or adulthood as cross-dressing. About one in eleven thousand men and one in thirty thousand women are estimated to display transsexual behavior. Hormonal and surgical sex reassignment are two forms of available treatment for those wanting to take on the physical characteristics of their self-identified gender. Little is known about the causes of gender identity disorder. In some cases, research shows a strong correlation between children who exhibit cross-gender behavior and adult homosexual orientation. Adults with gender identity disorder and adult homosexuals often recall feelings of alienation beginning as early as preschool.

Although some clinical aspects are shared, however, gender identity disorder is different from homosexuality. One definition for homosexuality proposed by Paul Gebhard is "the physical contact between two individuals of the same gender which both recognize as being sexual in nature and which ordinarily results in sexual arousal." Other researchers have underscored the difficulty in defining and measuring sexual orientation. Whatever measure is used, homosexuality is far more common than transsexualism.

Impact and Applications

Biological and genetic links to gender identity have been sought for more than a century. Studies on twins indicate a strong genetic component to sexual orientation. There appears to be a greater chance for an identical twin of a gay person to be gay than for a fraternal twin. Heritability averages about 50 percent in the combined twin studies. The fact that heritability is 50 percent rather than 100 percent, however, may indicate that other biological and environmental factors play a role. One study using restriction fragment length polymorphisms (RFLPs) to locate a gene on the X chromosome associated with male homosexual behavior showed a trend of maternal inheritance. However, not all homosexual brothers had the gene, and some heterosexual brothers shared the gene, indicating that other factors, whether genetic or nongenetic, influence sexual orientation.

Although some genetic factors have been found to influence sexual orientation, most researchers believe that no single gene causes homosexuality. It is also apparent that gender identity and homosexuality are influenced by complexes of factors dictated by biology, environment, and culture. Geneticists and social scientists alike continue to design studies to define how the many factors are interrelated.

—Donald J. Nash

See also: Behavior; Biological Clocks; Hermaphrodites; Homosexuality; Human Genetics; Metafemales; Pseudohermaphrodites; Steroid Hormones; Testicular Feminization Syndrome; X Chromosome Inactivation; XYY Syndrome.

Further Reading

Diamant, L., and R. McAnuity, eds. *The Psychology of Sexual Orientation, Behavior, and Identity:*

A Handbook. Westport, Conn.: Greenwood Press, 1995. Draws from biological and psychological research to provide a comprehensive overview of the major theories about sexual orientation; to summarize recent developments in genetic and neuroanatomic research; to consider the role of social institutions in shaping current beliefs; and to discuss the social construction of gender, sexuality, and sexual identity.

Ettore, Elizabeth. *Reproductive Genetics, Gender, and the Body.* New York: Routledge, 2002. Focuses on prenatal screening to explore how the key concepts of gender and the body are intertwined with the whole process of building genetic knowledge.

Haynes, Felicity, and Tarquam McKenna. *Unseen Genders: Beyond the Binaries.* New York: Peter Lang, 2001. Explores the effects of binary stereotypes of sex and gender on transsexuals, homosexuals, cross-dressers, and transgender and intersex people.

Money, John. *Sex Errors of the Body and Related Syndromes: A Guide to Counseling Children, Adolescents, and Their Families.* 2d ed. Baltimore: P. H. Brookes, 1994. Describes numerous gender variations in order to provide a basis for understanding sexual development anomalies and to enable appropriate counseling.

Zucker, Kenneth J. "Intersexuality and Gender Identity Differentiation." *Annual Review of Sex Research* 10 (1999). An extensive overview of intersexuality, gender identity formation, psychosexual differentiation, concerns about pediatric gender reassignment, hermaphroditism and pseudohermaphroditism, gender socialization, childrearing, and more. Includes a discussion of terminology, a summary, tables, and a bibliography.

Web Sites of Interest

About Gender. http://www.gender.org.uk. A site that looks at the nature versus nurture debate in research on gender roles, identity, and variance, with special emphasis on genetics.

Intersex Society of North America. http://www.isna.org. The society is "a public awareness, education, and advocacy organization which works to create a world free of shame, secrecy, and unwanted surgery for intersex people (individuals born with anatomy or physiology which differs from cultural ideals of male and female)." Includes or links to information on such conditions as clitoromegaly, micropenis, hypospadias, ambiguous genitals, early genital surgery, adrenal hyperplasia, Klinefelter syndrome, androgen insensitivity, and testicular feminization.

Johns Hopkins University, Division of Pediatric Endocrinology, Syndromes of Abnormal Sex Differentiation. http://www.hopkinsmedicine.org/pediatricendocrinology. Site provides a guide to the science and genetics of sex differentiation, including a glossary. Click on "patient resources."

Gene Families

Field of study: Molecular genetics

Significance: *Gene families contain multiple copies of structurally and functionally related genes, derived from duplications of an original gene. Some gene families represent multiple identical copies of an important gene, while others contain different versions of a gene with related functions. Evolution of gene families can lead some genes to take on completely new functions, allowing greater complexity of the genome and perhaps the organism.*

Key terms

CONCERTED EVOLUTION: a process in which the members of a gene family evolve together

PSEUDOGENES: nonfunctional segments of DNA that resemble functional genes

REPETITIVE DNA: a DNA sequence that is repeated two or more times in a DNA molecule or genome

Evolutionary Origin of Gene Families

Gene families are a class of low or moderately repetitive DNA, consisting of structurally and functionally related genes resulting from gene duplication events. Usually, members of gene families are clustered together on a chromosome, but members of a family can be located on more than one chromosome. Several mechanisms can generate tandem copies of genes: chromosome duplication, unequal crossing over, and replication slippage. Duplication of chromosomal segments is often a result of crossing over in inversion heterozygotes and creates tandem repeated segments. Unequal crossing over occurs when homologous segments do not line up correctly during meiosis and one of the crossover products has a duplicated segment. Replication slippage occurs when the DNA polymerase "slips" during DNA replication and copies part of the template strand again. Once there are two copies of a gene in tandem, the latter two mechanisms are more likely to generate additional copies.

A member of a gene family may be functional or functionless. If the gene was not copied completely or further mutations render it nonfunctional, it is called a pseudogene. Further sequence changes in a functional copy may result in a gene with an altered function, such as producing a similar but different form of a protein that can serve some biochemical need or a protein that has a much different function than the original.

Identical Gene Families

Identical gene families contain functional member genes that produce proteins that are identical or very nearly so. These gene families usually contain genes for protein products that need to be found in abundance in the cell because of a crucial function. Multiple copies of the genes allow greater transcription and protein production.

For example, in eukaryotes, ribosomal RNA (rRNA) genes are repeated in tandem several hundred times. In contrast, there are only seven copies of rRNA genes in the prokaryote *Escherichia coli*, and they are dispersed throughout its single chromosome. The rRNA products of these genes make up part of the structure of the ribosome, the organelle responsible for the important process of protein synthesis.

The genes for eukaryotic histone proteins, which are important in maintaining the structure of DNA in chromosomes and in regulating the rate of transcription of many genes, are another example of clustered repeats of the same set of genes. In this case, there are five histone genes, separated by short, unrelated noncoding sequences, repeated several hundred times. The repeats are found in tandem in many invertebrate animal genomes but are dispersed in mammalian genomes.

Nonidentical Gene Families

The human beta-globin gene family is an example of a nonidentical gene family, which has functional member genes that serve different, but usually related, functions. In this case, the different protein products are alternate forms of the same type of protein, perhaps expressed at different times in the organism's development. There are five functional genes and one pseudogene clustered together on chromosome 11. One gene is expressed in the human embryo stage, two in the fetus, and two in the adult. The related alpha-globin gene family, with three genes and four pseudogenes, is a cluster on chromosome 16.

Evolutionary Role of Gene Families

Gene families serve as an example of how genes may be accidentally duplicated by several possible processes, and then by mutation and further duplication the various copies can diverge in function. It is known that long-term genomic evolution (with the exceptions of symbiotic and parasitic genomes) usually involves increases in the number of genes. Although there are a number of mechanisms for this, including polyploidization, it is believed that the formation of gene families can be a first step toward the evolution of "new" genes. Mutations in different members of the gene family cause them to diverge independently, and some may evolve to produce completely different proteins. The presence of gene copies still coding for the original protein allows redundant copies to evolve freely without detrimental changes to cellular physiology.

Although gene family members can evolve to be more different, they may also undergo concerted evolution, in which the various copies evolve together. Unequal crossing over not only changes the number of copies of members of a gene family but also does so by actual duplication, so that some copies are identical. Repeated events of this type can result in all of the genes in the family being identical. In fact, natural selection will sometimes favor this process if it is to the organism's advantage to have multiple identical copies, as with the rRNA and histone identical gene familes.

—Stephen T. Kilpatrick

See also: DNA Replication; Evolutionary Biology; Genomics; Multiple Alleles; Mutations and Mutagenesis; Pseudogenes; Repetitive DNA.

Further Reading

Graur, Dan, and Wen-Hsiung Li. *Fundamentals of Molecular Evolution.* 2d ed. Sunderland, Mass.: Sinauer Associates, 1999. A detailed review of the origin of gene families, with numerous examples.

Holmes, Roger S., and Hwa A. Lim, eds. *Gene Families: Structure, Function, Genetics and Evolution.* River Edge, N.J.: World Scientific, 1996. Contains papers presented at the Eighth International Congress on Isozymes in 1995 in Brisbane, Australia, including molecular evolution, regulation, and developmental roles of gene families in various species.

Rubin, Gerald F., et al. "Comparative Genomics of the Eukaryotes." *Science* 287 (March 24, 2000): 2204-2215. This review article compares the genomes of several model organisms, including yeast, the fruit fly *Drosophila,* and the roundworm *Caenorhabditis elegans.* Based on complete genome sequences, the analysis shows that these species particularly differ in the number and distribution of gene families.

Xue, Gouxiong, and Yongbiao Xue, eds. *Gene Families: Studies of DNA, RNA, Enzymes and Proteins.* Collects key articles by experts in the fields of gene families, DNA, RNA, and proteins, in memory of Clement L. Markert (1917-1999), who developed the concept of isozymes.

Gene Regulation: Bacteria

Fields of study: Bacterial genetics; Cellular biology; Molecular genetics

Significance: *Gene regulation is the process by which the synthesis of gene products is controlled. The study of gene regulation in bacteria has led to an understanding of how cells respond to their external and internal environments.*

Key terms

ALLELE: an alternative form of a gene; for example, $lacI^+$, $lacI^-$, and $lacI^S$ are alleles of the $lacI$ gene

CONTROLLING SITE: a sequence of base pairs to which regulatory proteins bind to affect the expression of neighboring genes

GENE: a sequence of base pairs that specifies a product (either RNA or protein); the average gene in bacteria is one thousand base pairs long

OPERON: one or more genes plus one or more controlling sites that regulate the expression of the genes

TRANSCRIPTION: the use of DNA as the template in the synthesis of RNA

TRANSLATION: the use of an RNA molecule as the guide in the synthesis of a protein

The Discovery of Gene Regulation

In 1961, a landmark paper by French researchers François Jacob and Jacques Monod outlined what was known about genes involved in the breakdown of sugars, the synthesis of amino acids, and the reproduction of a bacterial virus called lambda phage (λ phage). Jacob and Monod described in detail the induction of enzymes that break down the sugar lactose. These enzymes were induced by adding the sugar or, in some cases, structurally related molecules to the media. If these inducer molecules were removed, the enzymes altering lactose were no longer synthesized. Bacteria without the $lacI$ gene ($lacI^-$) produced the enzymes for metabolizing lactose whether or not the inducer was present.

Although bacteria normally have only one copy of each gene locus, they can be given extra copies of selected genes by transforming them with a plasmid containing the genes of interest. Thus, bacteria that are heterozygous at a locus can be produced. When Jacob and Monod produced bacteria heterozygous for the *lacI* gene (*lacI⁻*/*lacI⁺*), they functioned like normal bacteria (*lacI⁺*), indicating that the *lacI⁺* allele was dominant to the *lacI⁻* allele. Certain alleles of the operator site, *lacO*^{C}, result in the synthesis of lactose-altering enzymes whether or not the inducer was present and even when *lacI⁺* was present. These observations suggested that the *lacI⁺* gene specified a repressor that might bind to *lacO⁺* and block transcription of the genes involved in lactose metabolism. Jacob and Monod concluded that inducers interfered with the repressor's ability to bind to *lacO⁺*. This allowed transcription and translation of the lactose operon. In their model, the repressor protein is unable to bind to the altered operator site, *lacO*^{C}. This explained how certain mutations in the operator caused the enzymes for lactose metabolism to be continuously expressed.

Seeing a similarity between the expression of the genes for lactose metabolism, the genes for amino acid synthesis, and the genes for lambda phage proliferation, Jacob and Monod proposed that all genes might be under the control of operator sites that are bound by repressor proteins. An operon consists of the genes that the operator controls. Although the vast majority of operons have operators and are regulated by a repressor, there are some operons without operator sites that are not controlled by a repressor. Generally, these operons are regulated by an inefficient promoter or by transposition of the promoter site, whereas some are inhibited by attenuation, a more complex interaction occurring during transcription and translation. The only controlling site absolutely necessary for gene expression is the promoter site, where RNA polymerase binds.

Lactose Operon: Negatively Controlled Genes

The lactose operon (*lacZYA*) consists of three controlling sites (*lacCRP*, *lacP*$_{ZYA}$, and *lacO*) and three structural genes (*lacZ*, *lacY*, and *lacA*). The lactose operon is controlled by a neighboring operon, the lactose regulatory operon, consisting of a single controlling site

(*lacP*$_I$) and a single structural gene (*lacI*). The order of the controlling sites and structural genes in the bacterial chromosome is *lacP*$_I$, *lacI*, *lacCRP*, *lacP*$_{ZYA}$, *lacO*, *lacZ*, *lacY*, *lacA*. Transcription of the regulatory operon proceeds to the right from the promoter site, *lacP*$_I$. Similarly, transcription of the L-arabinose operon occurs rightward from *lacP*$_{ZYA}$. A cyclic-adenosine monophosphate receptor (CRP) bound by cyclic-adenosine monophosphate (cAMP), referred to as a CRP-cAMP complex, attaches to the *lacCRP* site.

The *lacI* gene specifies the protein subunit of the lactose repressor, a tetrameric protein that binds to the operator site, *lacO*, and blocks transcription of the operon. The *lacZ* gene codes for beta-galactosidase, the enzyme that cleaves lactose into galactose plus glucose. This enzyme also converts lactose into the effector molecule allolactose, which actually binds to the repressor inactivating it. The *lacY* gene specifies the enzyme, known as the "lactose permease," that transports lactose across the plasma membrane and concentrates it within the cell. The *lacA* gene codes for an enzyme called transacetylase, which adds acetyl groups to lactose.

In the absence of lactose, the repressor occasionally diffuses from the operator, allowing RNA polymerase to attach to *lacP*$_{ZYA}$ and make a single RNA transcript. This results in extremely low levels of enzymes called the "basal" level. With the addition of lactose, a small amount of allolactose binding to the repressor induces a conformational change in the repressor so that it no longer binds to *lacO*. The levels of permease and beta-galactosidase quickly increase, and within an hour the enzyme levels may be one thousand times greater than they were before lactose was added.

Normally, cells do not produce levels of lactose messenger RNA (mRNA) or enzymes that are more than one thousand times greater than basal level because the lactose operon is regulated by catabolite repression. As cells synthesize cellular material at a high rate, lactose entrance and cAMP synthesis are inhibited, whereas cAMP secretion into the environment is increased. This causes most of the CRP-cAMP complex to become CRP. CRP is unable to bind to *lacCRP* and promote transcription from *lacP*_{BAD}.

If lactose is removed from the fully induced operon, repressor quickly binds again to *lacO* and blocks transcription. Within a few hours, lactose mRNA and proteins return to their basal levels. Since the lactose operon is induced and negatively regulated by a repressor protein, the operon is classified as an inducible, negatively controlled operon.

Arabinose Operon: Positively Controlled Genes

The L-arabinose operon (*araBAD*) has been extensively characterized since the early 1960's by American researchers Ellis Englesberg, Nancy Lee, and Robert Schleif. This operon is under the control of a linked regulatory operon consisting of (*araC, araO2*) and (*araP*$_C$, *araO1*). The parentheses indicate that the regions overlap: *araO2* is an operator site in the middle of *araC*, whereas *araP*$_C$ and *araO1* represent a promoter site and an operator site respectively, which overlap. The order of the controlling sites and genes for the regulatory operon and the L-arabinose operon is as follows: (*araC, araO2*) (*araP*$_C$, *araO1*), *araCRP, araI1, araI2, araP*$_{BAD}$, *araB, araA, araD*. RNA polymerase binding to *araP*$_C$ transcribes *araC* leftward, whereas RNA polymerase binding to *araP*$_{BAD}$ transcribes *araBAD* rightward.

The *araA* gene specifies an isomerase that converts L-arabinose to L-ribulose, the *araB* gene codes for a kinase that changes L-ribulose to L-ribulose-5-phosphate, and the *araD* gene contains the information for an epimerase that turns L-ribulose-5-phosphate into D-xylulose-5-phosphate. Further metabolism of D-xylulose-5-phosphate is carried out by enzymes specified by genes in other operons.

The *araC* product is in equilibrium between two conformations, one having repressor activity and the other having activator activity. The conformation that functions as an activator is stabilized by the binding of L-arabinose or by certain mutations (*araC*^{C}). In the absence of L-arabinose, almost all the *araC* product is in the repressor conformation; however, in the presence of L-arabinose, nearly all the *araC* product is in the activator conformation.

In the absence of L-arabinose, bacteria will synthesize only basal levels of the lactose regulatory protein and the enzymes involved in the breakdown of L-arabinose. The repressor binding to *araO2* prevents *araC* transcription beginning at *araP*$_C$ from being completed, whereas repressor binding to *araI1* prevents *araBAD* transcription beginning at *araP*$_{BAD}$.

The addition of L-arabinose causes repressor to be converted into activator. Activator binds to *araI1* and *araI2* and stimulates *araBAD* transcription. Activator is absolutely required for the metabolism of L-arabinose since bacterial cells with a defective or missing L-arabinose regulatory protein, *araC*$^-$, only produce basal levels of the L-arabinose enzymes. This is in contrast to what happens to the lactose enzymes when there is a missing lactose regulatory protein, *lacI*$^-$. Because of the absolute requirement for an activator, the L-arabinose operon is considered an example of a positively controlled, inducible operon.

Transcription of the *araBAD* operon is also dependent upon the cyclic-adenosine monophosphate receptor protein (CRP), which exists in two conformations. When excessive adenosine triphosphate (ATP) and cellular constituents are being synthesized from L-arabinose, cAMP levels drop very low in the cell. This results in CRP-cAMP acquiring the CRP conformation and dissociating from *araCRP*. When this occurs, the *araBAD* operon is no longer transcribed. The L-arabinose operon is controlled by catabolite repression very much like the lactose operon.

Tryptophan Operon: Genes Controlled by Attenuation

The tryptophan operon (*trpLEDCBA*) consists of the controlling sites and the genes that are involved in the synthesis of the amino acid tryptophan. The order of the controlling sites and genes in the tryptophan operon is as follows: (*trpP, trpO*), *trpL, trpE, trpD, trpC, trpB, trpA*. RNA polymerase binds to *trpP* and initiates transcription at the beginning of *trpL*.

An inactive protein is specified by an unlinked regulatory gene (*trpR*). The regulatory protein is in equilibrium between its inactive and its repressor conformation, which is stabilized by tryptophan. Thus, if there is a high concentration of tryptophan, the repressor binds to *trpO* and shuts off the tryptophan operon. This operon is an example of an operon that is repressible and negatively regulated.

The tryptophan operon is also controlled by a process called attenuation, which involves the mRNA transcribed from the leader region, *trpL*. The significance of leader region mRNA is that it hydrogen-bonds with itself to form a number of hairpinlike structures. Hairpin-III interacts with the RNA polymerase, causing it to fall off the DNA. Any one of several hairpins can form, depending upon the level of tryptophan in the environment and the cell. When there is no tryptophan in the environment, the operon is fully expressed so that tryptophan is synthesized. This is accomplished by translation of the leader region right behind the RNA polymerase up to a couple of tryptophan codons, where the ribosomes stall. The stalled ribosomes cover the beginning of the leader mRNA in such a way that only hairpin-II forms. This hairpin does not interfere with transcription of the rest of the operon and so the entire operon is transcribed.

When there is too much tryptophan, the operon is turned off to prevent further synthesis of tryptophan. This is accomplished by translation of the leader region up to the end of the leader peptide. Ribosomes synthesizing the leader peptide cover the leader mRNA in such a way that only hairpin-III forms. This hairpin causes attenuation of transcription.

In some cases, the lack of amino acids other than tryptophan can result in attenuation of the tryptophan operon. In fact, cells starved for the first four amino acids (N-formylmethionine, lysine, alanine, and isoleucine) of the leader peptide result in attenuation. When these amino acids are missing, hairpins-I and III both form, resulting in attenuation because of hairpin-III.

Flagellin Operons: Operons Controlled by Transposition

Some pathogenic bacteria change their flagella to avoid being recognized and destroyed by the host's immune system. This change in flagella occurs by switching to the synthesis of another flagellar protein. The phenomenon is known as phase variation. The genes for flagellin are in different operons. The first operon consists of a promoter site, an operator site, and the structural gene for the first flagellin (*flgP*$_{H1}$, *flgO1*, *flgH1*). The first operon is under the negative control of a repressor specified by the second operon. The second operon also specifies the second flagellin and a transposase that causes part of the second operon to reverse itself. This portion of the operon that "flips" is called a "transposon." The promoter sites for the transposase gene (*flgT2*), flagellin gene (*flgH2*), and repressor gene (*flgR2*) are located on either side of the transposase gene in sequences called inverted repeats. Transcription from both promoters in the second operon occurs from left to right: *flgP*$_{T2}$, *flgT2*, *flgP*$_{H2R2}$, *flgH2*, *flgR2*.

When the second operon is active, the repressor binds to *flgO1*, blocking the synthesis of the first flagellin (*flgH1*). Consequently, all bacterial flagella will be made of the second flagellin (*flgH2*). Occasionally, the transposase will catalyze a recombination event between the inverted repeats, which leads to the transposon being reversed. When this occurs, neither *flgH2* nor *flgR2* is transcribed. Consequently, the first operon is no longer repressed by *flgR2*, and *flgH1* is synthesized. All the new flagella will consist of *flgH1* rather than *flgH2*.

Impact and Applications

Many of the genetic procedures developed to study gene regulation in bacteria have contributed to the development of genetic engineering and the production of biosynthetic consumer goods. One of the first products to be manufactured in bacteria was human insulin. The genes for the two insulin subunits were spliced to the lactose operon in different populations of bacteria. When induced, each population produced one of the subunits. The cells were cracked open, and the subunits were purified and mixed together to produce functional human insulin. Many other products have been made in bacteria, yeast, and even plants and animals.

Considerable progress has been made toward introducing genes into plants and animals to change them permanently. In most cases, this is difficult to do because the controlling sites and gene regulation are much more complicated in higher organisms than in bacteria. Nevertheless, many different species of plants have been altered to make them resistant to desiccation, herbicides, insects, and various plant pathogens. Although curing genetic defects by introducing good genes into animals and humans has not been very successful, animals have been transformed so that they produce a number of medically important proteins in their milk. Goats have been genetically engineered to release tissue plasminogen activator, a valuable enzyme used in the treatment of heart attack and stroke victims, into their milk. Similarly, sheep have been engineered to secrete human alpha-1 antitrypsin, useful in treating emphysema. Cattle that produce more than ten times the milk that sheep or goats produce may potentially function as factories for the synthesis of all types of valuable proteins specified by artfully regulated genes.

—*Jaime S. Colomé*

See also: Bacterial Genetics and Cell Structure; Central Dogma of Molecular Biology; Gene Regulation: Eukaryotes; Gene Regulation: *Lac* Operon; Gene Regulation: Viruses; Model Organism: *Escherichia coli*; Molecular Genetics; Transposable Elements.

Further Reading

Inada, Toshifumi, et al. "Mechanism Responsible for Glucose-Lactose Diauxie in *Escherichia coli*: Challenge to the cAMP Model." *Genes to Cells* 1 (March, 1996). Provides an understandable discussion of catabolite repression with numerous diagrams.

Müller-Hill, Benno. *The Lac Operon: A Short History of a Genetic Paradigm.* New York: Walter de Gruyter, 1996. Using a unique combination of personal anecdotes and present-day science, describes the history and present knowledge of a paradigmatic system, the *lac* operon of *Escherichia coli*. Illustrated.

Rasooly, Avraham, and Rebekah Sarah Rasooly. "How Rolling Circle Plasmids Control Their Copy Number." *Trends in Microbiology* 5 (November, 1997). Illustrates how regulatory genes control the rate of synthesis of plasmids in bacteria.

Soisson, Stephen M., et al. "Structural Basis of Ligand-Regulated Oligomerization of AraC." *Science* 276 (April 18, 1997). Explains how two molecules of *AraC* protein interact with the inducer, with each other, and with controlling sites to regulate the expression of the L-arabinose operon.

Gene Regulation: Eukaryotes

Fields of study: Cellular biology; Molecular genetics

Significance: *Gene regulation refers to the processes whereby the information encoded within a DNA sequence is expressed at the required level. For eukaryotes, this primarily pertains to the selective expression of particular proteins during development or in specific tissues.*

Key terms

ANTISENSE TECHNOLOGY: use of antisense oligonucleotides that base pair with mRNA to prevent translation

BASAL TRANSCRIPTION FACTOR: protein that is required for initiation of transcription at all promoters

CHROMATIN REMODELING: any event that changes the nuclease sensitivity (DNA accessibility) of chromatin

CORE PROMOTERS: DNA elements that direct initiation of transcription by the basal RNA polymerase machinery

ENHANCER: a DNA element that serves to enhance transcriptional activity above basal levels

INSULATOR: a DNA element that, when placed between an enhancer and a promoter, prevents activation of that particular gene

RNA INTERFERENCE (RNAI): small, interfering RNAs that cause gene silencing

TRANSCRIPTION FACTOR: a protein that is involved in initiation of transcription but is not part of the RNA polymerase

Nuclear RNA Polymerases

Three nuclear RNA polymerases share the responsibility of transcribing eukaryotic genes.

RNA polymerase I transcribes genes encoding ribosomal RNA (rRNA), RNA polymerase II transcribes protein-coding genes and some small nuclear RNA genes, and RNA polymerase III transcribes genes encoding transfer RNA (tRNA), the 5S rRNA, and some small nuclear RNA. DNA elements known as promoters serve to recruit the RNA polymerases to their transcriptional start sites. The RNA polymerases do not bind their promoters directly. Instead, transcription factors bind to the promoter, and the RNA polymerases are recruited by binding their cognate transcription factors. Promoters for RNA polymerases I and III have limited variability and are recognized by a finite set of transcription factors. In contrast, promoters for RNA polymerase II show significant diversity, and the number of transcription factors involved in recruiting the polymerase is huge. For each polymerase, the core promoter elements are recognized by a set of basal transcription factors which are required for initiation of transcription at all promoters. RNA polymerase, together with its basal transcription factors, constitutes the basal transcription apparatus. Activated or repressed transcription is measured with respect to basal levels.

Transcription can be divided into three phases: Initiation, elongation, and termination. Most regulation occurs at the level of initiation. For RNA polymerases I and III, regulation is generally global and involves a repression of transcription. For RNA polymerase II, regulation is gene-specific, allowing specific regulation of each of thousands of protein-coding genes. RNA polymerase II promoters function only at very low efficiency with the basal transcription factors, and activation is the common mode of regulation. This overview will focus on regulation of protein-coding genes.

Basal Transcription by RNA Polymerase II

RNA polymerase II promoters are modular. The core promoter, which directs transcription by the basal transcription apparatus, typically extends about 35 base pairs upstream or downstream of the transcriptional start site. Core promoters can vary considerably from gene to gene, and there are no universal core promoter elements. Common core promoter elements include the TATA-box, an AT-rich sequence that is located about 25 base pairs upstream of the transcriptional start, and the region immediately surrounding the start site, known as the initiator. The downstream promoter element, DPE, is typically found about 30 base pairs downstream of the transcriptional start, and it is mainly found in genes that do not have a TATA-box. The strength of a given promoter, as defined by the level of basal transcription, depends on which combination of promoter elements is present and on their respective sequences.

The core promoter elements are recognized by basal transcription factors that for RNA polymerase II are named TFIIX, where X is a letter that identifies the individual factor. For example, the TATA-box is bound by the TATA-binding protein, which is a subunit of the transcription factor known as TFIID. A subset of TATA-boxes feature a sequence immediately upstream that serves as a recognition site for TFIIB. TFIIB, in turn, recruits the polymerase. Ultimately, the core promoter is also the target for the factors that regulate transcriptional activity.

Activated Transcription

Transcription is the key step at which gene expression is controlled. Transcriptional initiation is regulated by enhancers, which are DNA elements that function to increase levels of transcription above basal levels. Enhancers may be located on either side of the gene, up to several thousand base pairs from the transcriptional start. Enhancers are recognized by transcriptional activators that mediate an increase in transcriptional activity. Transcriptional activators show great variability in terms of cell type and gene specificity, thus allowing unique regulation of individual genes. Activator proteins are modular, containing both a DNA-binding domain and an activation domain. Some activators function by directly interacting with components of the transcription apparatus to stimulate transcription.

Cellular DNA is not naked but packaged into highly organized and compacted nucleoprotein structures known as chromatin. Packaging of DNA into chromatin can occlude protein-binding sites—for example, interfering with binding of basal transcription factors. Accordingly, activation of transcription may be accomplished by relieving the repression caused by chromatin formation. Indeed, many activators function by recruiting protein complexes whose function is to remodel chromatin to increase DNA accessibility. Decompaction of chromatin at promoters is not always sufficient, as RNA polymerase II may need to transcribe thousands of base pairs. Efficient transcription, therefore, may also depend on specific elongation factors that travel with the RNA polymerase to destabilize chromatin structure.

Since enhancers may activate genes that are located at some distance, mechanisms exist to specify that a certain gene not be the target of a given enhancer. A DNA insulator serves this function: When placed between the enhancer and the promoter, it prevents activation of that particular gene. Unlike enhancers, insulators must therefore be located at a specific positions to work. Recruitment of DNA-binding proteins to these insulator elements prevents activation of a specific gene.

Post-transcriptional Control

The pre-messenger RNA (pre-mRNA) transcript is subject to several types of post-transcriptional processing: Intervening sequences (introns) are removed by splicing, a "cap" structure is added to the 5′ end, and a polyadenosine (poly-A) tail is added to the 3′ end, following cleavage of the transcript. Although historically referred to as post-transcriptional events, this processing occurs during, not after, transcription; the largest RNA polymerase II subunit has a carboxy-terminal domain which serves to recruit proteins involved in mRNA splicing, polyadenylation, and capping, thus securing a tight association between these processes.

Capping and polyadenylation affect both stability of the mRNA and the efficiency of translation. Since most intracellular RNA degradation is in the form of nuclease-mediated degradation from either end, protecting the ends by cap-binding proteins and polyA-binding proteins, respectively, prevents degradation. Short-lived mRNAs often contain elements within the region downstream of the stop codon that explicitly recruit nuclease complexes that degrade the RNA. In general, genes that encode "housekeeping" proteins produce mRNAs with long half-lives, whereas genes whose expression must be rapidly controlled tend to generate mRNAs with short half-lives.

Additional protein diversity may be generated by alternative splicing, a process whereby different combinations of coding sequences, or exons, are incorporated into the final spliced mRNA product. In this fashion, multiple versions of a protein may be made from a single gene.

Finally, the mRNA sequence affects the efficiency with which it is translated. For instance, folding of the mRNA region upstream of the start codon can interfere with binding of the ribosome, and the sequence adjacent to the start codon affects the efficiency of translation initiation. Nucleotide sequences in the untranslated regions of mRNA are also recognized by specific proteins that may anchor the mRNA to specific cellular structures to ensure their translation and accumulation at the appropriate locations.

Changing Gene Expression

Several techniques exist for modification of gene expression. The principle behind antisense technology is the base pairing of a complementary oligonucleotide to a target mRNA that results in the prevention of translation. As natural oligonucleotides (RNA and DNA) are rapidly degraded in the cell, more stable, artificial oligonucleotides with modified backbones are typically used. This strategy has great potential to reduce expression of genes involved in disease states.

RNA interference, or RNAi, also results in sequence-specific gene silencing. The exposure to double-stranded RNA that matches the sequence of coding regions results in loss of the corresponding mRNA. The double-stranded RNA triggers the assembly of a nuclease complex that targets the homologous mRNA for degradation. In plants, for example, RNAi has been suggested to play an important role in resistance to pathogens. RNAi has also evolved into a powerful tool for probing gene activity and for developing gene-silencing therapeutics.

Hundreds of different cell types exist that are specialized to perform unique roles. Since each of these cells contains the same tens of thousands of genes, their specialization requires tightly controlled gene regulation. As summarized above, gene regulation occurs at multiple levels, with DNA sequences from the promoter region to the untranslated mRNA sequences dictating the rates of transcription, pre-mRNA processing, and translation. Although the exact mechanisms are only beginning to be elucidated, knowing the DNA sequence therefore has the potential to reveal which exact modes of regulation are in effect. Understanding normal gene regulation will in turn lead to an understanding of how misregulation may lead to disease.

—*Anne Grove*

See also: Antisense RNA; Bacterial Genetics and Cell Structure; Central Dogma of Molecular Biology; Gene Regulation: Bacteria; Gene Regulation: *Lac* Operon; Gene Regulation: Viruses; Model Organism: *Escherichia coli*; Molecular Genetics; Transposable Elements.

Further Reading

Carey, M., and S. T. Smale. *Transcriptional Regulation in Eukaryotes: Concepts Strategies and Techniques.* Cold Spring Harbor, N.Y.: Cold Spring Harbor Laboratory Press, 2000. Provides strategies for a step-by-step analysis of eukaryotic gene expression. Targeted to students who are just entering the field.

Kadonaga, J. T. "The DPE: A Core Promoter Element for Transcription by RNA Polymerase II." *Experimental and Molecular Medicine* 34 (2002): 259-264. Reviews core promoter elements and how they participate in gene regulation.

Orphanides, G., and D. Reinberg. "A Unified Theory of Gene Expression." *Cell* 108 (2002): 439-451. This review summarizes the different levels of gene regulation and focuses on how the different steps are connected. This entire issue of *Cell* is dedicated to reviews covering various aspects of gene regulation.

Gene Regulation: *Lac* Operon

Fields of study: Bacterial genetics; Cellular biology; Molecular biology

Significance: *Studies of the regulation of the lactose (lac) operon in* Escherichia coli *have led to an understanding of how the expression of a gene is turned on and off through the binding of regulator proteins to the DNA. This has served as the groundwork for understanding not only how bacterial genes work but also how genes of higher organisms are regulated.*

Key terms

ACTIVATOR: a protein that binds to DNA to enhance a gene's conversion into a product that can function within the cell

OPERATOR: a sequence of DNA adjacent to (and usually overlapping) the promoter of an operon; binding of a repressor to this DNA prevents transcription of the genes that are controlled by the operator

OPERON: a group of genes that all work together to carry out a single function for a cell

PROMOTER: a sequence of DNA to which the gene expression enzyme (RNA polymerase) attaches to begin transcription of the genes of an operon

REPRESSOR: a protein that prevents a gene from being made into a functional product when it binds to the operator

Inducible Genes and Repressible Genes

In order for genes or genetic information stored in DNA to be used, the information must first be transcribed into messenger RNA (mRNA); mRNA is synthesized by an enzyme, RNA polymerase, which uses the DNA as a template for making a single strand of RNA that can be translated into proteins. The proteins are the functional gene products that act as enzymes or structural elements for the cell. The process by which DNA is transcribed and then translated is referred to as expression of the genes.

Some genes are always expressed in bacterial cells; that is, they are continually being transcribed into mRNA, which is translated into functional proteins (gene products) of the cell.

The genes involved in using glucose as an energy source are included in this group. Other genes are inducible (expressed only under certain circumstances). The genes for using lactose as an energy source are included in this group. beta-galactosidase, the enzyme that converts lactose into glucose and galactose so that the sugar can be easily metabolized by a cell, is only made in cells when lactose is present. Synthesizing proteins uses a large amount of energy. In order for the cell to conserve energy, it produces proteins only when they are needed. There is no need to make beta-galactosidase if there is no lactose in the cell, so it is synthesized only when lactose is present.

As early as the 1940's François Jacob, Jacques Monod, and their associates were studying the mechanisms by which beta-galactosidase was induced in *Escherichia coli*. They discovered that when there is no lactose in a cell, a repressor protein binds to the DNA of the operon at the operator site. Under these conditions, transcription of genes in the operon cannot occur since the RNA polymerase is physically prevented from binding to the promoter when the repressor is in place. This occurs because the promoter and operator sequences are overlapping. The lactose (*lac*) operon is, therefore, under negative control. When lactose is present, an altered form of the lactose attaches to the repressor in such a way that the repressor can no longer bind to the operator. With the operator sequence vacant, it is possible for the RNA polymerase to begin transcription of the operon genes at the promoter. Lactose (or its metabolite) serves as an inducer for transcription. Only if it is present are the lactose operon genes expressed. The lactose operon is, therefore, an inducible operon under negative control. In 1965, Jacob and Monod were awarded the Nobel Prize in Physiology or Medicine in recognition of their discoveries concerning the genetic control of enzyme synthesis.

Lac Operon Expression in the Presence of Glucose

When a culture of *E. coli* is given equal amounts of glucose and lactose for growth and

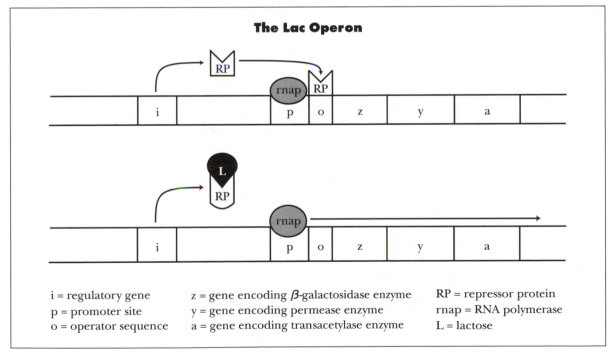

The Lac Operon

i = regulatory gene   z = gene encoding *β*-galactosidase enzyme   RP = repressor protein
p = promoter site   y = gene encoding permease enzyme   rnap = RNA polymerase
o = operator sequence   a = gene encoding transacetylase enzyme   L = lactose

1. In the absence of lactose, the repressor protein binds to the operator, blocking the movement of RNA polymerase. The genes are turned off. 2. When lactose is present, it preferentially binds the repressor protein, freeing up the operator and allowing RNA polymerase to move through the operon. The genes are turned on.

is compared with cultures given either glucose alone or lactose alone, the cells given two sugars do not grow twice as fast, but rather show two distinct growth cycles. Beta-galactosidase is not synthesized initially; therefore, lactose is not used until all the glucose has been metabolized. Laboratory observations show that the presence of lactose is necessary but not a sufficient condition for the lactose (*lac*) operon to be expressed. An activator protein must bind at the promoter in order to unravel the DNA double helix so that the RNA polymerase can bind more efficiently. The activator protein binds only when there is little or no glucose in the cell. If glucose is available, it is preferred over other sugars because it is most easily metabolized to make energy in the form of adenosine triphosphate (ATP). ATP is made through a series of reactions from an intermediate molecule, cyclic adenosine monophosphate (cAMP). The cAMP concentration decreases when ATP is being made but builds up when no ATP synthesis occurs. When the glucose has been used, the concentration of cAMP rises. The cAMP binds to the activator protein to enable it to bind at the operon's promoter. With the activator bound, transcription of the genes, including the beta-galactosidase, occurs.

The activation of a DNA-binding protein by cAMP is a global control mechanism. The lactose operon is only one of many that are controlled in this way. Global control allows bacteria to prevent or turn on transcription of a group of genes in response to a single signal. It ensures that the bacteria always utilize the most efficient energy source if more than one is available. This type of global control only occurs, however, when the operon is also under the control of another DNA-binding protein (the *lac* repressor in the case of the *lac* operon), which makes the operon inducible or repressible or both. Control of transcription through the binding of an activator protein is an example of positive control, since binding of the activator turns on gene expression.

Impact and Applications

Jacob and Monod developed the concept of an operon as a functional unit of gene expression in bacteria. What they learned from studying the *lac* operon has led to a more general understanding of gene transcription and the use of mRNA as an information-bearing intermediate in the process of gene expression. The operon concept has proven to be a universal mechanism by which bacteria organize their genes. Although genes of higher cells (eukaryotes) are not usually organized in operons and although negative control of expression is rare in them, similar positive control mechanisms occur in both bacterial and eukaryotic cells. Studies of the *lac* operon have made possible the understanding of how DNA-binding proteins can attach to a promoter to enhance transcription.

—*Linda E. Fisher*

See also: Bacterial Genetics and Cell Structure; Gene Regulation: Bacteria; Gene Regulation: Eukaryotes; Gene Regulation: Viruses; Model Organism: *Escherichia coli*.

Further Reading

Judson, Horace Freeland. *The Eighth Day of Creation: Makers of the Revolution in Biology.* Rev. ed. New York: Cold Spring Harbor, N.Y.: Cold Spring Harbor Laboratory Press, 1996. A noted text that provides an interesting account of the personalities behind the discoveries that form the basis of modern molecular biology.

Müller-Hill, Benno. *The Lac Operon: A Short History of a Genetic Paradigm.* New York: Walter de Gruyter, 1996. Using a unique combination of personal anecdotes and present-day science, describes the history and present knowledge of a paradigmatic system, the *lac* operon of *Escherichia coli*. Illustrated.

Ptashne, Mark, and Walter Gilbert. "Genetic Repressors." *Scientific American* 222 (June, 1970). Summarizes repression mechanisms that turn genes on and off, using the *lac* operon and the lambda bacterial virus as models.

Tijan, Robert. "Molecular Machines That Control Genes." *Scientific American* 271 (February, 1995). Discusses regulatory proteins that direct transcription of DNA and what happens when they malfunction.

Gene Regulation: Viruses

Fields of study: Cellular biology; Molecular biology; Viral genetics

Significance: *Gene regulation in viruses typically resembles that of the hosts they infect. Because viruses are not alive and are incapable of self-replication, gene regulation at the time of initial infection depends on its host's control systems. Once infection is established, regulation is generally mediated by gene products of the virus's own DNA or RNA.*

Key terms

BACTERIOPHAGE: general term for a virus that infects bacteria

LAMBDA (λ) PHAGE: a virus that infects bacteria and then makes multiple copies of itself by taking over the infected bacterium's cellular machinery

LYSOGENY: a process whereby a virus integrates into a host chromosome as a result of non-lytic, nonproductive, infection

OPERATOR: a sequence of DNA adjacent to (and usually overlapping) the promoter site, where a regulatory protein can bind and either increase or decrease the ability of RNA polymerase to bind to the promoter

PROMOTER: a sequence of DNA to which the gene expression enzyme (RNA polymerase) attaches to begin transcription of the genes of an operon

General Aspects of Regulation

Regardless of the type of organism, DNA is the genetic material that allows species to survive and pass their traits to the next generation. Genes are encoded, along with control sequences that the cell uses to control expression of their associated genes. Although details of these control sequences vary between prokaryotes and eukaryotes, they still function in similar ways. One element common to all genes is a promoter, a sequence that acts as the binding site for RNA polymerase, the enzyme that transcribes the gene into RNA so it can be translated into a protein product. Other control sequences, if present, simply help control whether or not RNA polymerase can bind to the pro-

moter, or they increase or decrease the strength of RNA polymerase binding. These secondary control sequences, therefore, act as switches for turning on or off their associated genes. Some may also act like a dimmer switch, increasing or decreasing the rate at which a gene is expressed.

Viruses are incapable of self-replication and must rely on the host cells they infect. In order to replicate successfully, a virus must be compatible with the host's cell biochemistry and gene-regulation systems. When a virus first enters a host cell, its genes are regulated by the host. Thus, viral promoters and other control elements must be compatible with those of its host. The control elements associated with promoters in prokaryotes are called operators. An operator represents a site where a regulatory protein (a product of yet another gene) can bind and either increase or decrease the ability of RNA polymerase to bind to the promoter of its associated gene or group of genes.

Eukaryotic systems (cells of plants and animals) are more complex and involve a number of proteins called transcription factors, which bind to or near the promoter and assist RNA polymerase binding. There are also enhancer proteins, which bind to other control sequences somewhere upstream from the gene they influence. Because of this greater complexity, viruses that infect eukaryotic cells are also more genetically complex than are viruses infecting prokaryotes.

Viral Genomes

All cells, including bacteria, are subject to infection by parasitic elements such as viruses. Viruses which specifically infect bacteria are known as bacteriophages, from the Greek *phagos*, "to eat." The genetic information in viruses may consist of either RNA or DNA. All forms of viruses contain one or the other, but never both. Regardless of the type of genetic material, gene regulation does have certain features in common.

The size of the viral genome determines the number of potential genes that can be encoded. Among the smallest of the animal viruses are the hepadnaviruses, including hepatitis B virus, the DNA of which consists of some

3 kilobase pairs (3 kbp, or 3,000 base pairs), enough to encode approximately seven proteins. The largest known viruses are the poxviruses, consisting of 200-300 kbp, enough to encode several hundred proteins. Lambda is approximately average in size, with a DNA genome of 48 kbp, enough to encode approximately fifty genes.

Lambda as a Model System: The Lytic Cycle

Following infection of the bacterial host, most bacteriophages replicate, releasing progeny as the cell falls apart, or lyses. Lambda phage is unusual in that, while it can complete a lytic cycle, it is also capable of a nonproductive in-

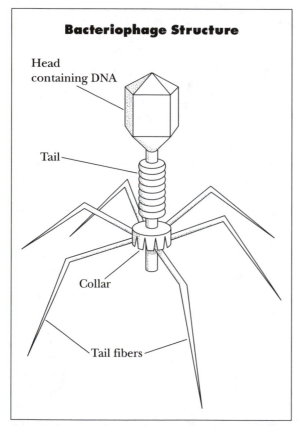

Bacteriophage Structure

Head containing DNA

Tail

Collar

Tail fibers

Bacteriophages, or "phages," are viruses that attach themselves to bacteria and inject their genetic material into the cell. Sometimes, during the assembly of new viral particles, a piece of the host cell's DNA may be enclosed in the viral capsid. When the virus leaves the host cell and infects a second cell, that piece of bacterial DNA enters the second cell, thus changing its genetic makeup. (Electronic Illustrators Group)

fection: following infection, the viral genome integrates into the host chromosome, becoming a prophage in a process known as lysogeny. Such phages are known as temperate viruses.

Most viruses, including lambda, exhibit a temporal control of regulation: Gene expression is sequential. Three classes of proteins are produced, classified based on when after infection they are expressed. "Immediate early" genes are expressed immediately after infection, generally using host machinery and enzymes. "Early" genes are expressed at a later time and generally require proteins expressed from early genes. "Late" genes are expressed following genome replication of the virus. The various temporal classes of gene products may also be referred to as lambda, beta, and gamma proteins.

The lytic cycle of lambda represents a prototype of temporal control. Lambda immediate early gene expression begins following infection of the host cell, *Escherichia coli*. Host cell enzymes catalyze the process. Transcription of lambda DNA begins at a site called a promoter, a region recognized by the host RNA polymerase, which catalyzes transcription. Lambda DNA is circular after entering the cell, and two promoters are recognized: One regulates transcription in a leftward direction (P_L), while the other regulates transcription from the opposite strand in a rightward transcription (P_R).

Among the immediate early genes expressed is one encoding the N protein, expression of which is under the control of P_L. Generally, transcription occurs through a set of genes and is terminated at a specific point. The N protein is an example of an antiterminator, a protein that allows "read-through" of the stop signal for transcription and expression of additional genes. A second protein is encoded by the *cro* gene, the product of which plays a vital role in determining whether the infection is lytic or becomes lysogenic. *Cro* gene expression is controlled through P_R, as are several "early class" genes which regulate viral DNA replication (*O* and *P* genes), repressor synthesis (*cII*), and early gene expression (*Q* gene).

Both the cro and Q proteins are involved in regulating "late" genes, those expressed following DNA replication. Like the N protein, the Q protein is an antiterminator. Late gene products include those that become the structural proteins of the viral capsid. Other late proteins cause cell lysis, releasing progeny phage particles from the cell. The entire process is completed in approximately thirty minutes.

Lambda: The Lysogenic Cycle

Lambda is among those bacterial viruses that can also carry out lysogeny, a nonlytic infection in which the virus integrates within the host chromosome. Lysogeny is dependent on the interaction between two gene products: the repressor, a product of the *cI* gene, and the cro protein.

The cII protein, an early gene product, activates the expression of *cI*, the gene that encodes the repressor. At this point in the cycle, it

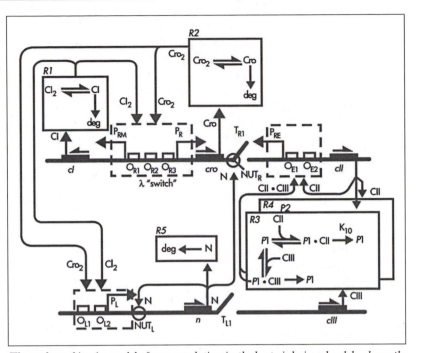

The pathway kinetics model of gene regulation in the bacterial virus lambda shows the "decision circuit" that determines the phage's life cycle: either lytic, in which the virus replicates and destroys its host cell, or lysogenic, in which the viral DNA is incorporated into the host cell's genome and lies dormant. The model, adapted from A. Arkin et al. (Genetics 149, 163348, 1998), was generated using a supercomputer and is consistent with experimental observations. (U.S. Department of Energy Genomes to Life Program, http://doegenomestolife.org)

becomes a "race" (literally) between the activity of the repressor and the cro protein. Each has affinity for the operator regions (O_L and O_R) which control access to the respective promoters, P_L and P_R. If the repressor binds the operator regions before the cro protein, access to these sites by RNA polymerase is blocked, and the virus enters a lysogenic state. If the cro product binds first, repressor action is blocked, and the virus continues in a lytic cycle.

Lambda can remain in lysogeny for an indefinite length of time. Because it is integrated with the host's genome, every time the host reproduces, lambda is also reproduced. Lambda typically remains in the lysogenic phase, unless its host gets into difficulty. For example, if the host is "heat shocked," it produces heat shock proteins that inadvertently destroy the lambda repressor protein. Without the repressor protein to block expression of the early genes, lambda enters the lytic phase. This switch to the

lytic phase allows lambda to reproduce and leave its host before it is potentially destroyed with the host.

Regulation in Other Viral Systems

While lambda is unusual among the complex bacteriophages in carrying out both lytic and lysogenic cycles, regulation among other viruses, including those which infect animals, has certain features in common. Most viruses exhibit a form of temporal control. Regulation in T_{even} bacteriophage infection (T2, T4, or T6) is accomplished by altering the specificity of the RNA polymerase β, resulting in the recognition of alternate promoters at different times after infection. Bacteriophage T7 accomplishes the same task by encoding an entirely new polymerase among its own genes.

The complexity of animal viruses varies significantly; the greater the coding capacity, the more variability in regulation. Some animal viruses, such as the influenza viruses, encode different proteins on unique segments of genetic material, in this case RNA. DNA viruses such as the human herpesviruses (HHV) or poxviruses utilize the same form of temporal control as described above. In place of antiterminators, products of each time frame regulate subsequent gene expression. In some cases, unique polymerase enzymes encoded by the virus carry out transcription of these genes.

Despite their apparent complexity, viruses make useful models in understanding gene expression in general. Control elements resembling operators and promoters are universal among living cells. In addition, an understanding of regulation unique to certain classes of viruses, such as expression of new enzymes, provides a potential target for novel treatments.

—*Richard Adler*

See also: Bacterial Genetics and Cell Structure; Gene Regulation: Bacteria; Gene Regulation: Eukaryotes; Gene Regulation: *Lac* Operon; Genomic Libraries; Viral Genetics; Viroids and Virusoids.

Further Reading

Hendrix, Roger, et al., eds. *Lambda II.* Cold Spring Harbor, N.Y.: Cold Spring Harbor Press, 1983. Description of lambda, growth, and regulation, state of the art for the time. Later work refined the molecular biology, but this volume remains *the* book on the subject.

Ptashne, Mark. *A Genetic Switch: Phage Lambda and Higher Organisms.* 2d ed. Malden, Mass.: Blackwell, 1996. Covers lambda phage as it operates in animals and other eukaryotic organisms.

Ptashne, Mark, and Alexander Gann. *Genes and Signals.* Cold Spring Harbor, N.Y.: Cold Spring Harbor Press, 2002. Summarizes regulation in both prokaryotic and eukaryotic systems, using *Escherichia coli*, lambda phage, and yeast as prototypes.

Ptashne, Mark, et al. "How the Lambda Repressor and Cro Work." *Cell* 19 (1980): 1-11. Reviews factors that determine whether lysis or lysogeny results from infection.

Gene Therapy

Fields of study: Genetic engineering and biotechnology; Human genetics and social issues

Significance: *Gene therapy is the result of a compilation of recombinant DNA technologies, which allows the replacement or supplementation of defective or undesirable genes within a person's genome with functional copies of that gene and/or complementary genes. The primary goal of gene therapy is to reverse the effects of a genetic disease.*

Key terms

EXPRESSION CASSETTE: a synthetic genetic construct that contains the target gene and other DNA elements, which allow the gene to be moved about easily and properly expressed in cells

ONCORETROVIRUS: an RNA-containing viruses that may cause cancerous mutations

VECTOR: a tool for packaging and transferring a gene into a cell

A Brief Background

Gene therapy can be defined quite simply as the use of recombinant DNA technologies to effect a treatment or cure for an inherited (genetic) disease. The term "gene therapy" evokes mixed emotions in scientists and the population at large. In the 1990's, the first positive results using gene therapy to cure genetic diseases in humans began to appear in the medical literature. The topic of gene therapy is alive with scientific, legal, and ethical controversy. By any measure, gene therapy is a very active area of research with tremendous potential to help human beings control previously incurable diseases. However, before the full potential of gene therapy is realized, new scientific technologies will need to be developed, legal and ethical considerations will need to be addressed, and potential risks, many of which are still unforeseen, will need to be minimized to achieve an acceptable risk-benefit ratio.

In many ways, gene therapy is a logical extension of the human desire to improve our surroundings by manipulating evolution, which is a genetically controlled process. People first started altering the process of natural selection many thousands of years ago, when farmers began selectively breeding certain forms of plants and animals found desirable in a process called artificial selection. Artificial selection has been refined over many thousands of years of successful use. In the twentieth century, with the discovery of DNA as the molecule of inheritance and rapid evolution of laboratory methods to isolate and manipulate DNA, it became possible to change the genetic composition of living organisms. The lengthy processes of traditional breeding could theoretically be bypassed, and a major barrier of traditional breeding, generally limited to breeding only within members of the same phylogenetic "family," broke down.

In the broadest sense, gene therapy offers the potential of replacing defective genes within the human genome, be it one or many genes, with new genetic "patches" that can counteract the effect of the defective genes. Additionally, new, beneficial genes that impart desirable characteristics—such as enhanced life span, cancer resistance, or resistance to other diseases—can theoretically be inserted into the human genome even in the absence of defective genes. Finally, what makes gene therapy especially exciting, and simultaneously alarming, is the fact that genes from any living organism, including all animals, bacteria, plants, and even viruses, could potentially be used for gene therapy in humans. No evolutionary boundaries apply in gene therapy.

The Theory of Gene Therapy

The primary goal of gene therapy is to correct a genetic disease by replacing defective genes with functional or supplemental genes that will alleviate the disorder. The driving forces behind gene therapy are recombinant DNA technologies. Recombinant technologies allow the extraction, manipulation, and reinsertion of cellular DNA within and between living organisms. The development of recombinant DNA technologies began in the 1970's, rapidly evolved through the 1980's and the 1990's, and is currently at a level where gene therapy, at least for some conditions, is technically feasible. There have been tremendous advancements in routinely available materials and equipment—including fast, efficient, and affordable laboratory equipment, an explosive proliferation of available biochemicals, and streamlined laboratory procedures. Many key laboratory procedures, which once were very expensive and time-consuming, are now available in reasonably priced and easy-to-use kits, which has greatly increased the speed of biomedical research. The Human Genome Project, completed in April, 2003, offers an abundance of information about the sequence and location of genes within the human genome and will be a tremendous boost to future gene therapy research.

The simplest and most logical targets for gene therapy are hereditary single-gene defects. In these cases, a single faulty gene causes a genetic disease. There are many examples of these single-gene disorders, including certain types of hemophilia, muscular dystrophy, cystic fibrosis, and an immune disorder known as severe combined immunodefiency disorder (SCID). Theoretically, getting a "good copy" of the defective gene into people with these disorders might cure these types of diseases. In reality, however, controlling factors—such as gene insertion, gene expression, gene targeting, and immune response—pose tremendous technical challenges that researchers are currently working to overcome. Gene therapy may also target complex diseases, such as cancer, cardiovascular disease, neurological diseases, and infectious diseases. It is possible that even these complex diseases may one day be treated routinely with gene therapies.

Key Technologies

Although gene therapy and cloning may be employed together in certain scientific and medical research projects, gene therapy is very different from cloning. In the process of cloning, the entire genome of an organism is duplicated to produce a genetically identical organism. In gene therapy, only portions of a genome, usually only one or a few genes, are manipulated at a time, with the goal of correcting a specific genetic disorder. Many of the same legal and ethical questions do apply to both cloning and gene therapy, and both technologies do result in the production of a genetically modified organism (GMO).

Genetic diseases have been studied for many centuries. In many single-gene-defect diseases, the faulty gene has been identified, located, and sequenced. In many cases, the structure and function of the gene product is known in great detail. Through routine molecular biology techniques, functional copies of the gene, suitable for gene therapy, can be isolated from normal human tissues in the laboratory. This functional gene itself may be altered or put together with other genes to create an "expression cassette." An expression cassette is a synthetic genetic construct that contains the target gene and other DNA elements, which allow the gene to be moved about easily and properly expressed in cells.

Once the functional gene is isolated and placed into an expression cassette, the gene is still not ready for use in gene therapy. Because of physical barriers within the human body and the efficiency of the immune system in defending the body from pathogens (disease causing organisms), delivery and expression of foreign gene constructs in the human body are not easy to accomplish. To deliver therapeutic genes into the body, scientists most often harness the power of viruses, since they are very adept at getting around the physical and immune defenses of the body. For safety purposes, most potentially harmful viral genes that might trigger disease or elicit a severe immune response are removed to produce what is called a "disarmed" viral vector.

Viral Vectors

Currently, several classes of virus are used to produce viral vectors for human gene therapy trials. These include oncoretroviruses, such as the Moloney murine leukemia virus (MLV), a virus that causes leukemia in mice; lentiviruses (retroviruses), such as human immunodeficiency virus (HIV), the virus that causes AIDS in humans; adenoviruses, which are extremely infectious viruses that cause cold or flulike symptoms in humans; and herpesviruses, the family of viruses that cause cold sores, genital herpes, and chickenpox in humans. All of these viruses have different applications in human

gene therapy, depending on the specific cells or tissues in the body that are being targeted. For example, herpesvirus vectors have been used for gene therapy research in cells of the nervous system, while oncoretroviruses and lentiviruses have been used for transforming cells of the circulatory system and stem cells.

In addition to being able to transfer "good" genes into the body, vectors must be genetically stable, able to be propagated in cell culture, and able to be purified to a high concentration. After the vector is built, propagated, and purified, the job is still not complete. A growing number of techniques are used to deliver the vector to the correct cells and tissues in the body. In most cases, the cells, tissues, or organs to receive the gene are specifically targeted for delivery. Targeting can be accomplished either by exposing certain cells to the vector outside the body (ex vivo), such as in a culture tube, or by introducing the vector in a targeted way inside the body (in vivo), such as introducing the vector into an organ through a specific blood vessel. Both in vitro and in vivo targeted delivery methods have been used for human gene therapy trials. Targeted delivery appears to be a critical aspect of human gene therapy, in order to increase efficacy and reduce potential risks.

The use of potentially dangerous viruses to transfer genes into the human body is one of the major concerns that surround gene therapy. Even with proper precautions in the design and building of the vector, research and human trials are conducted according to strict biohazard containment procedures in an attempt to prevent the unintentional spread of the gene therapy vector to laboratory and medical personnel.

Clinical Trials

In 1990, the first clinical trial of human gene therapy was conducted in children who were afflicted with severe combined immunodeficiency disorder (SCID). A single defective gene for an enzyme, adenosine deaminase (ADA), had been linked with this fatal disorder, which prevents the immune system from maturing and functioning properly, and so it appeared to be an attractive target for human gene therapy. It was thought that introducing a functional copy of the ADA gene into some cells of the immune system would result in a cure for this genetic disease. Since the ADA gene had already been cloned in 1984 and an early gene therapy vector, derived from a mouse retrovirus, was available, the stage was set for the first test of human gene therapy. Hematopoietic stem cells (very young blood cells) were isolated from young patients with SCID. In laboratory flasks, the stem cells from each patient were exposed to the viral vector containing the good copy of the ADA gene. The goal was to cause genetic transformation of some of the stem cells. The stem cells were then transferred back into the young patients. The good news was that the functional ADA gene

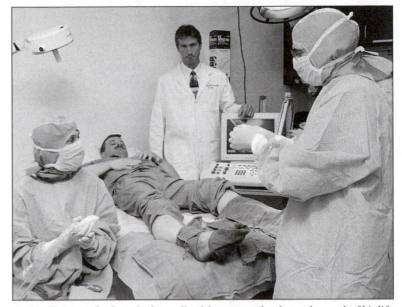

Patient Donovan Decker, who has suffered from muscular dystrophy much of his life, prepares to be injected by Dr. Jerry Mendell with genes to help correct the condition. Donovan was the first person to receive gene therapy for muscular dystrophy, in September, 1999. (AP/Wide World Photos)

did appear in some blood cells of the participants. The bad news was that this genetic modification did not correct the young patients' SCID. As it turned out, ADA production alone was not enough to reverse the SCID disease condition. Overall, this early and heroic attempt at gene therapy—despite the fact that it was not successful in curing the targeted disorder—resulted in useful data and led to tremendous advances in future attempts and eventually to success almost a decade later.

In September, 1999, the first human death attributable to a human gene therapy clinical trial was reported. An eighteen-year-old participant in a human gene therapy trial for hereditary ornithine transcarbamylase (OTC) deficiency died of multiorgan failure caused by a severe immunological reaction to the disarmed adenovirus vector used in the trail. It appears that this patient's immune system may have been sensitized by a previous infection with a wild-type adenovirus and, when exposed to the adenovirus vector, even though it was a disarmed vector, his immune system overreacted, resulting in severe complications and eventually death. This tragic death not only un-

derscored the unforeseen risks associated with human gene therapy trials but also alerted researchers to the need to assess the immune status of gene therapy candidates, especially regarding prior exposure to pathogenic viruses.

In April, 2000, the first successful report of human gene therapy to correct a human genetic disease was published in the journal *Science*. The report contained the details surrounding children afflicted with lethal X-linked severe combined immunodeficiency syndrome (SCID-X). These children were successfully treated at the Necker Hospital for Sick Children in Paris using gene therapy. Eventually, nine of the eleven patients included in this clinical study were cured of their genetic disorder. The techniques used were evolved from the earlier, unsuccessful SCID clinical trials using the ADA gene. In this case, the gene that was introduced into the hematopoietic stem cells was a cytokine receptor gene rather than the ADA gene, and the results were greatly improved.

However, as might be expected in pioneering medical research, unforeseen adverse events soon marred what had been celebrated as an unqualified success. In September, 2002, a three-year-old participant in the SCID-X human gene therapy trials began exhibiting a leukemia-type lymphoproliferative disorder (an inappropriate proliferation of white blood cells). Subsequently another child from the same SCID-X trial showed signs of the same disorder. It was determined that this disorder was most probably due to the nature of the retrovirus vector used. The vector had apparently inserted the therapeutic gene construct into the genome of at least one of the stem cells at a place where it inadvertently activated an oncogene (a cancer-causing gene). These cells with an activated oncogene went on to multiply in the children and cause their leukemia-type disease. Subsequently, another child in the SCID-X trial tested positive for the same oncogene activation, although he did not exhibit the leukemia-type disorder. The combination of successes and unforeseen adverse events underscored the fact that human gene therapy is currently operating in uncharted areas and undoubtedly involves very complex biological factors, some of which are probably still unknown.

Future Prospects: Benefits and Risks

Future prospects are really limited only by imagination and the constraints of currently available technology. The Human Genome Project, which maps all the gene sequences as well as their location in the human genome, will revolutionize the development of human gene therapy. Using these data, scientists will discover targets for gene therapy that can be examined in their native context within the human genome, which should result in the production of much better expression cassettes and allow targeted insertion of therapeutic genes, which in turn should greatly decrease undesirable side effects such as those seen in the SCID-X trial. Gene therapy may within a few decades provide physicians with tools to treat or prevent all sorts of genetic diseases, both simple and complex.

The same technologies developed to correct defective genes may also give scientists the power to insert "desirable genes," possibly from other types of living organisms, to increase life span, impart cancer resistance, provide protec-

tion from environmental toxins, and function as permanent vaccines against infectious disease.

This notion of desirable genes raises the prospect of creating "designer humans"—humans with beneficial or targeted genetic traits, even aesthetic genetic modifications—and all the attendant legal, political, and ethical ramifications.

—Robert A. Sinnott

See also: Bioethics; Bioinformatics; Cloning Vectors; Cystic Fibrosis; DNA Structure and Function; Gene Therapy: Ethical and Economic Issues; Genetic Counseling; Genetic Engineering: Historical Development; Genetic Engineering: Medical Applications; Genetic Engineering: Risks; Genetic Engineering: Social and Ethical Issues; Genetic Screening; Genetic Testing; Genetic Testing: Ethical and Economic Issues; Human Genetics; Human Genome Project; Inborn Errors of Metabolism; Insurance; Knockout Genetics and Knockout Mice; RNA World; Stem Cells; Transgenic Organisms; Tumor-Suppressor Genes.

Further Reading

Fischer, Alain, Salima Hacein-Bey, and Marina Cavazzana-Calvo. "Gene Therapy of Severe Combined Immunodeficiencies." *Nature Reviews (Immunology)* 2 (August, 2002): 615-621. Review of human gene therapy for severe immunological disorders, such as severe combined immunodeficiency disorder (SCID), written by the scientists involved in the first human gene therapy trial in 1990 and the first "successful" human gene therapy interventions, published in 2002.

Habib, Nagy A., ed. *Cancer Gene Therapy: Past Achievements and Future Challenges.* New York: Kluwer Academic/Plenum, 2000. Reviews forty-one preclinical and clinical studies in cancer gene therapy, organized into sections on the vectors available to carry genes into tumors, cell cycle control, apoptosis, tumor-suppressor genes, antisense and ribozymes, immuno-modulation, suicidal genes, angiogenesis control, and matrix metallo proteinase.

Thomas, Clare T., Anja Ehrhardt, and Mark A. Kay. "Progress and Problems with the Use of Viral Vectors for Gene Therapy." *Nature Re-*

views (Genetics) 4 (May, 2003): 346-358. An excellent review of human gene therapy tools and an update of recent successes and failures of human gene therapy trials.

Web Sites of Interest

American Cancer Society, Gene Therapy: Questions and Answers. http://www.cancer.org. Site has searchable information on gene therapy. Topics covered include "What Is Gene Therapy?" and "How Does Gene Therapy Work?"

Genethon. Gene Therapies Research and Applications Center. http://www.genethon .fr/php/index_us.php. Supported by the French Muscular Dystrophy Association, Genethon sponsors research in genetic and cellular therapies for rare diseases. This site offers a section accompanied by computer graphics on the theory of gene therapy.

Gene Therapy: Ethical and Economic Issues

Fields of study: Bioethics; Genetic engineering and biotechnology; Human genetics and social issues

Significance: *Gene therapy has the potential to cure many diseases once viewed as untreatable, such as cystic fibrosis. At the same time, gene therapy presents ethical dilemmas ranging from who decides who will benefit from new therapies to questions of ethics and social policy, such as whether humans should attempt to manipulate natural evolutionary processes. Although there are strong economic incentives for developing new therapies, ethical concerns must be addressed.*

Key terms

GERM CELLS: reproductive cells such as eggs and sperm

GERM-LINE THERAPY: alteration of germ cells resulting in a permanent genetic change in the organism and succeeding generations

INSULIN: a pancreatic hormone that is essential to metabolize carbohydrates, used in the control of diabetes mellitus

RECOMBINANT DNA: genetically engineered DNA prepared by cutting up DNA molecules and splicing together specific DNA fragments, often from more than one species of organism

SOMATIC CELL THERAPY: treatment of specific tissue with therapeutic genes

Gene Therapy

Advances in molecular biology and genetics near the end of the last century have presented tantalizing possibilities for new treatment for medical conditions once viewed as incurable. Gene therapy for the treatment of human genetic diseases can take two forms: somatic cell therapy and germ-line therapy. Somatic cell therapy is less controversial, because it only modifies nonreproductive cells, and therefore the changes cannot be passed on to a person's children. Still, caution is needed, as with any new technology, to be sure the benefits outweigh the risks. Germ-line therapy is more permanent in that the changes include modification of reproductive cells, and thus the changes can be passed on to a person's children. This has led to much greater controversy, because all the same cautions apply to this approach as to somatic cell therapy, with the added problem that any defects introduced by the technology could become permanent features of the human population. Because of this, germ-line gene therapy is currently banned in the United States and in much of the rest of the world.

Somatic Cell Therapy

Somatic cell therapy could provide some clear benefits. For example, it could potentially free insulin-dependent diabetics from reliance on external sources of insulin by restoring the ability of the patient's own body to manufacture it. Scientists have already succeeded in genetically engineering bacteria to grow recombinant insulin, eliminating the need to harvest it from animal pancreatic tissue obtained from slaughterhouses. The next step would seem to be the use of somatic cell therapy to treat individual diabetics.

The primary ethical concerns about treating a disease like diabetes using somatic cell therapy primarily relate to cost and technological

Three-year-old Wilco Conradi at the zoo in Amsterdam on August 18, 2002. After living isolated in a plastic-enclosed space for most of his life, he received gene therapy for the fatal "bubble boy" syndrome, severe combined immunodeficiency disorder (SCID). The results of this treatment appeared promising until it was noted that several children so treated were developing a leukemia-type disorder, likely caused when an oncogene was activated by the vector used to insert the therapeutic gene. Although for children afflicted with SCID, the alternative to no therapy was much worse, such mixed results nevertheless raise ethical concerns. (AP/Wide World Photos)

even adequately assess potential risks? Caution has been advised by many, in light of these concerns.

Assuming that the technological hurdles can be overcome, somatic cell therapy to cure diabetes mellitus appears to offer a fairly clear-cut case for treatment. What about less threatening conditions, such as the insufficient production of growth hormone? A shortage of human growth hormone can result in dwarfism. The use of somatic cell therapy to correct the condition clearly would be beneficial, but growth hormone deficiencies vary, and even otherwise normal children can be shorter than average. In a society in which height is associated with success, wealthy parents have been known to pressure doctors to prescribe human growth hormone to their children who are only slightly smaller than average and not truly suffering from a pituitary gland disorder. If somatic cell gene therapy became widely available for human growth, how many parents would succumb to the temptation to give their children a boost in height? The same potential for abuse is present for any number of perceived defects that might be cured by gene therapy, with only those who are rich being able to afford the technology. When the defect is not life threatening, or even particularly debilitating, do parents have the right to decide that their children receive these treatments?

proficiency. Currently, the potential costs of gene therapy put it out of reach for most people. Is it ethical to develop a technological solution to a problem that will be available to only a few? Of course, this same concern could be directed at many other expensive medical procedures, such as organ transplants, which are often out of reach for most people yet are now well entrenched in medical practice.

A more serious ethical concern, at present, is whether the technology is safe enough to use on humans. Clinical trials of some somatic cell therapies have been halted due to unforeseen complications, including deaths and the development of cancer in some cases. These events have led some ethicists to question whether gene therapy trials should be considered at all. Is it fair to expect individuals who are managing their diabetes with conventional methods to accept the unknown risks inherent in such a complex and poorly understood technology? Do we know so little at this point that we cannot

Germ-Line Therapy

Germ-line gene therapy faces all the same ethical objections as somatic cell therapy, and introduces what some consider more serious ethical concerns. Germ-line therapy changes the characteristics an organism passes on to its offspring. Humans suffer from a variety of inherited diseases, including hemophilia, Huntington's disease (Huntington's chorea), and cystic fibrosis, and physicians have long recognized that certain conditions, such as coronary artery disease and diabetes, seem to run in some families. It is tempting to consider the possibility of eliminating these medical conditions through germ-line therapies: Not only would the person suffering from the disease be cured, but his or her descendants would never have to worry about passing the condition on to their offspring. Eventually, at least in theory, the genes that cause the disease could be eliminated from the general population.

Tempting though it is to see this as a good thing, ethicists believe that such an approach could be extremely susceptible to abuse. They view discussions of human germ-line therapy as an attempt to resurrect the failed agenda of the eugenics movement of the 1920's and 1930's. If scientists are allowed to manipulate human heredity to eliminate certain characteristics, what

---

FDA Limits Gene Therapy Trials

Because of an adverse reaction—a leukemia-like disorder reported in two patients who had undergone successful SCID-X gene therapy—human gene therapy trails are proceeding with caution. The available data suggest that the retrovirus vector used for the SCID-X gene therapy trials, derived from a cancer-causing mouse virus, may be largely responsible. Retrovirus vectors have the ability to insert genes permanently into the human genome, which is desirable to obtain long-term results in gene therapy. A problem occurs, however, if a retrovirus inserts the therapeutic gene near, or in, certain genes called oncogenes or tumor-suppressor genes: Cancerous mutations can develop in the transformed cells. When cells with cancerous mutations replicate over time, cancers can develop. That appears to have happened in at least two of the patients who participated in the SCID-X trials. In response, an advisory committee that monitors data from gene-therapy trials for the U.S. Food and Drug Administration (FDA) recommended that gene therapy for SCID-X be moved to a second-line treatment, meaning that it should be used only in the absence of other medical treatment options, such as a bone marrow transplant from a matched donor.

These results from the SCID-X trials reinvigorated the ethical and legal questions surrounding gene therapy. Moral questions originally arose when scientists became able to alter the human genome and were complicated with the rise of research into embryonic stem cells (cells obtained from human embryos). Embryonic stem cells are an attractive target for researchers in the area of gene therapy because these very young, undifferentiated cells are the progenitors of all the other cells in the human body. Performing gene therapy on embryonic stem cells and then manipulating these cells to develop into specific tissues or organs would allow the quintessential degree of targeted gene therapy. While there is currently no comprehensive ban on the use of embryonic stem cells in gene therapy research, only certain exempted cell lines can be used in federally funded research projects.

The economics of gene therapy may also affect its actual impact on human health care. The technologies involved in gene therapy are currently very expensive and probably will remain so for the foreseeable future. Most gene therapy trails are considered experimental procedures and are therefore not covered by health insurance. These and other real economic conditions, particularly in countries with no national health care policy, may make gene therapy affordable to some and not to others. In this way, gene therapy may increase the disparity in health care services available to people of different socioeconomic groups.

Finally, since the terrorist attacks on the United States of September 11, 2001, any discussion of gene therapy must include the possibility that some of the technologies developed to correct genetic diseases could also be used by people with no moral or legal restraints to cause tremendous human suffering. By using infectious viral vectors developed for gene therapy and incorporating expression cassettes containing harmful or lethal genes, terrorists and others could develop biological weapons with relative ease compared to "traditional" threats such as nuclear weapons. The deliberate spread of these malicious constructs, especially in densely populated areas, could have catastrophic results.

—*Robert A. Sinnott*

is to prevent those same scientists from manipulating the human genome to enhance other characteristics? Would parents be able to request custom-tailored offspring, children who would be tall with predetermined hair color and eyes? Questions concerning class divisions and racial biases have also been raised. Would therapies be equally available to all people who requested them, or would such technology lead to a future in which the wealthy custom-tailor their offspring while the poor must rely on conventional biology? Would those poor people whose parents had been unable to afford germ-line therapy then find themselves denied access to medical care or employment based on their "inferior" or "unhealthy" genetic profiles?

In addition, many ethicists and scientists raise cautionary notes about putting too much faith in new genetic engineering technologies too soon. Most scientists concede that not enough is known about the interdependency of various genes and the roles they play in overall health and human evolution to begin a program to eliminate so-called bad genes. Genes that in one combination may result in a disabling or life-threatening illness may in another have beneficial effects that are not yet known. Germ-line therapy could eliminate one problem while opening the door to a new and possibly worse condition. Thus, while the economic benefits of genetic engineering and gene therapies can be quite tempting, ethicists remind us that many questions remain unanswered. Some areas of genetic research, particularly germ-line therapy, may simply be best left unexplored until a clearer understanding of both the potential social and biological cost emerges.

—*Nancy Farm Männikkö, updated by Bryan Ness*

See also: Bioethics; Bioinformatics; Cloning Vectors; Cystic Fibrosis; DNA Structure and Function; Gene Therapy; Genetic Counseling; Genetic Engineering: Historical Development; Genetic Engineering: Medical Applications; Genetic Engineering: Risks; Genetic Engineering: Social and Ethical Issues; Genetic Screening; Genetic Testing; Genetic Testing: Ethical and Economic Issues; Human Genetics; Human Genome Project; Inborn Errors of Metabolism; Insurance; Knockout Genetics and Knockout Mice; RNA World; Stem Cells; Transgenic Organisms; Tumor-Suppressor Genes.

Further Reading

Anees, Munawar A. *Islam and Biological Futures: Ethics, Gender, and Technology.* New York: Continuum International, 1989. Provides insight into reproductive biotechnologies from the Islamic perspective.

Becker, Gerhold K., and James P. Buchanan, eds. *Changing Nature's Course: The Ethical Challenge of Biotechnology.* Hong Kong: Hong Kong University Press, 1996. Brings together articles based on the November, 1993, symposium, "Biotechnology and Ethics: Scientific Liberty and Moral Responsibility." Topics include environmental and ethical considerations of genetically engineered plants and foods, clinical and ethical challenges of genetic markers for severe human hereditary disorders, and embryo transfer.

Doherty, Peter, and Agneta Sutton, eds. *Man-Made Man: Ethical and Legal Issues in Genetics.* Dublin: Four Courts Press, 1997. Provides an introduction to advances in the field, with topics that include pre-implantation and pre-natal testing; carrier testing with a view to reproductive choice; and somatic gene therapy, germ-line gene therapy, and nontherapeutic genetic interventions.

Harpignies, J. P. *Double Helix Hubris: Against Designer Genes.* Brooklyn, N.Y.: Cool Grove Press, 1996. Examines the moral and ethical aspects of genetic engineering and bioengineering, arguing that these sciences will produce shocking changes to sentient life.

Resnik, David B., Holly B. Steinkraus, and Pamela J. Langer. *Human Germline Gene Therapy: Scientific, Moral, and Political Issues.* Austin, Tex.: R. G. Landes, 1999. Examines the medical, ethical, and social aspects of human reproductive technology.

Rifkin, Jeremy. *The Biotech Century: Harnessing the Gene and Remaking the World.* New York: Jeremy P. Tarcher/Putnam, 1998. Argues that the information and life sciences are fusing into a single powerful technological and economic force that is laying the foundation for the Biotech Century, during which the world is likely to be transformed more fundamentally than in the previous thousand years.

United States. Advisory Committee on Human Radiation Experiments. *Final Report of the Advisory Committee on Human Radiation Experiments.* New York: Oxford University Press, 1996. Describes a variety of experiments sponsored by the U.S. government in which people were exposed to radiation, often without their knowledge or consent.

Walters, LeRoy, and Julie Gage Palmer. *The Ethics of Human Gene Therapy.* Illustrated by Natalie C. Johnson. New York: Oxford University Press, 1997. Surveys the structure and functions of DNA, genes, and cells, and discusses three major types of potential genetic intervention: somatic cell gene therapy, germ-line gene therapy, and genetic enhancements.

Zallen, Doris Teichler. *Does It Run in the Family? A Consumer's Guide to DNA Testing for Genetic Disorders.* New Brunswick, N.J.: Rutgers University Press, 1997. Focuses on the practical aspects of obtaining genetic information, clearly explaining how genetic disorders are passed along in families.

Web Sites of Interest

American Medical Association. http://www.ama-assn.org/ama/pub/printcat/2827.html. The AMA's page on gene therapy, with links to news stories.

Genethon. Gene Therapies Research and Applications Center. http://www.genethon.fr/php/index_us.php. Supported by the French Muscular Dystrophy Association, Genethon sponsors research in genetic and cellular therapies for rare diseases. This site offers a section accompanied by computer graphics on the theory of gene therapy.

National Information Resource on Ethics and Human Genetics. http://www.georgetown.edu/research/nrcbl/nirehg. Site supports links to databases, annotated bibliographies, and articles about the ethics of gene therapy and human genetics in general.

Genetic Code

Field of study: Molecular genetics

Significance: *The molecules of life are made directly or indirectly from instructions contained in DNA. The instructions are interpreted according to the genetic code, which describes the relationship used in the synthesis of proteins from nucleic acid information.*

Key terms

CODON: a three-nucleotide unit of nucleic acids (DNA and RNA) that determines the amino acid sequence of the protein encoded by a gene

NUCLEOTIDES: long nucleic acid molecules that form DNA and RNA, linked end to end; the sequences of these nucleotides in the DNA chain provides the genetic information

READING FRAME: the phasing of reading codons, determined by which base the first codon begins with; certain mutations can also change the reading frame

RNA: ribonucleic acid, a molecule similar to DNA but single-stranded and with a ribose rather than a deoxyribose sugar; RNA molecules are formed using DNA as a template and then use their complementary genetic information to conduct cellular processes or form proteins

TRANSFER RNA (tRNA): molecules that carry amino acids to messenger RNA (mRNA) codons, allowing amino acid polymerization into proteins

TRANSLATION: the process of forming proteins according to instructions contained in an mRNA molecule

Elements of the Genetic Code

Every time a cell divides, each daughter cell receives a full set of instructions that allows it to grow and divide. The instructions are contained within DNA. These long nucleic acid molecules are made of nucleotides linked end to end. Four kinds of nucleotides are commonly found in the DNA of all organisms. These are designated A, G, T, and C for the variable component of the nucleotide (adenine, guanine, thymine, and cytosine, respectively).

The Genetic Code

second position → first position ↓	T	C	A	G	third position ↓
T	Phenylalanine	Serine	Tyrosine	Cysteine	T
	Phenylalanine	Serine	Tyrosine	Cysteine	C
	Leucine	Serine	END CHAIN	END CHAIN	A
	Leucine	Serine	END CHAIN	Tryptophan	G
C	Leucine	Proline	Histidine	Arginine	T
	Leucine	Proline	Histidine	Arginine	C
	Leucine	Proline	Glutamine	Arginine	A
	Leucine	Proline	Glutamine	Arginine	G
A	Isoleucine	Threonine	Asparagine	Serine	T
	Isoleucine	Threonine	Asparagine	Serine	C
	Isoleucine	Threonine	Lysine	Arginine	A
	Methionine	Threonine	Lysine	Arginine	G
G	Valine	Alanine	Aspartic Acid	Glycine	T
	Valine	Alanine	Aspartic Acid	Glycine	C
	Valine	Alanine	Glutamic Acid	Glycine	A
	Valine	Alanine	Glutamic Acid	Glycine	G

The amino acid specified by any codon can be found by looking for the wide row designated by the first base letter of the codon shown on the left, then the column designated by the second base letter along the top, and finally the narrow row marked on the right, in the appropriate wide row, by the third letter of the codon. Many amino acids are represented by more than one codon. The codons TAA, TAG, and TGA do not specify an amino acid but instead signal where a protein chain ends.

The sequence of the nucleotides in the DNA chain provides the information necessary for manufacturing all the proteins required for survival, but information must be decoded.

DNA contains a variety of codes. For example, there are codes for identifying where to start and where to stop transcribing an RNA molecule. RNA molecules are nearly identical in structure to the single strands of DNA molecules. In RNA, the nucleotide uracil (U) is used in place of T and each nucleotide of RNA contains a ribose sugar rather than a deoxyribose sugar. RNA molecules are made using DNA as a template by a process called transcription. The resulting RNA molecule contains the same information as the DNA from which it was made, but in a complementary form. Some RNAs function directly in the structure and activity of cells, but most are used to produce proteins with the help of ribosomes. This latter type is known as messenger RNA (mRNA). The ribosome machinery scans the RNA nucleotide sequence to find signals to start the synthesis of polypeptides, the molecules of which proteins are made. When the start signals are found, the machinery reads the code in the RNA to convert it into a sequence of amino acids in the polypeptide, a process called translation. Translation stops at termination signals. The term "genetic code" is sometimes reserved for the rules for converting a sequence of nucleotides into a sequence of amino acids.

The Protein Genetic Code: General Characteristics

Experiments in the laboratories of Har Gobind Khorana, Heinrich Matthaei, Marshall Nirenberg, and others led to the deciphering of the protein genetic code. They knew that the code was more complicated than a simple one-to-one correspondence between nucleotides and amino acids, since there were about twenty different amino acids in proteins and only four nucleotides in RNA. They found that three adjacent nucleotides code for each amino acid. Since each of the three nucleotide positions can be occupied by any one of four different nucleotides, sixty-four different sets are possible. Each set of three nucleotides is called a codon. Each codon leads to the insertion of one kind of amino acid in the growing polypeptide chain.

Two of the twenty amino acids (tryptophan and methionine) have only a single codon. Nine amino acids are each represented by a pair of codons, differing only at the third position. Because of this difference, the third position in the codons for these amino acids is often called the wobble position. For six amino acids, any one of the four nucleotides occupies the wobble position. The three codons for isoleucine can be considered as belonging to this class, with the exception that AUG is reserved for methionine. Three amino acids (leucine,

arginine, and serine) are unusual in that each can be specified by any one of six codons.

Punctuation

The protein genetic code is often said to be "commaless." The bond connecting two codons cannot be distinguished from bonds connecting nucleotides within codons. There are no spaces or commas to identify which three nucleotides constitute a codon. As a result, the choice of which three nucleotides are to be read as the first codon during translation is very important. For example, if "EMA" is chosen as the first set of meaningful letters in the following string of letters, the result is gibberish:

TH EMA NHI TTH EBA TAN DTH EBA TBI THI M.

On the other hand, if "THE" is chosen as the first set of three letters, the message becomes clear:

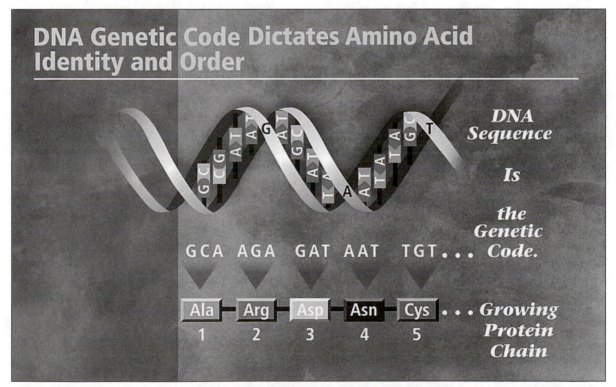

This figure from the Genome Image Gallery of the Department of Energy makes clear the concept of the reading frame and how the genetic code translates into amino acids and hence proteins. (U.S. Department of Energy Human Genome Program, http://www.ornl.gov/hgmis)

THE MAN HIT THE BAT AND THE BAT BIT HIM.

The commaless nature of the code means that one sequence of nucleotides can be read three different ways, starting at the first, second, or third letter. Still, the genetic code does have "punctuation." The beginning of each coding sequence has a start codon, which is always the AUG. Each coding sequence also has a stop codon, which acts like a period at the end of a sentence, denoting the end of the coding sequence.

These ways of reading are called reading frames. A frame is said to be open if there are no stop codons for a reasonable distance. In most mRNAs, only one reading frame is open for any appreciable length. However, in some mRNAs, more than one reading frame is open. Some mRNAs can produce two, rarely three, different polypeptide sequences.

The Near Universality of the Code

The universal genetic code was discovered primarily through experiments with extracts from the bacterium *Escherichia coli* and from rabbit cells. Further work suggested that the code was the same in other organisms. It came to be known as the universal genetic code. The code was deciphered before scientists knew how to determine the sequence of nucleotides in DNA efficiently. After nucleotide sequences began to be determined, scientists could, using the universal genetic code, predict the sequence of amino acids. Comparison with the actual amino acid sequence revealed excellent overall agreement.

Nevertheless, the universal genetic code assignments of codons to amino acids had apparent exceptions. Some turned out to be caused by programmed changes in the mRNA information. In selected codons of some mRNA, a C is changed to a U. In others, an A is changed so that it acts like a G. Editing of mRNA does not change the code used by the ribosomal machinery, but it does mean that the use of DNA sequences to predict protein sequences has pitfalls.

Some exceptions to the universal genetic code are true variations in the code. For exam-

ple, the UGA universal stop codon codes for tryptophan in some bacteria and in fungal, insect, and vertebrate mitochondrial DNA (mtDNA). Ciliated protozoans use UAA and UAG, reserved as stop codons in all other organisms, for the insertion of glutamine residues. Methionine, which has only one codon in the universal genetic code (AUG), is also encoded by AUA in vertebrate and insect mtDNA and in some, but not all, fungal mitochondria. Vertebrate mtDNA also uses the universal arginine codons AGA and AGG as stop codons. AGA and AGG are serine rather than arginine codons in insect mtDNA.

Interpreting the Code

How is the code interpreted? The mRNA codons organize small RNA molecules called transfer RNA (tRNA). There is at least one tRNA for each of the twenty amino acids. They are *L*-shaped molecules. At one end tRNAs have a set of three nucleotides (the anticodon) that can pair with the three nucleotides of themRNA codon. They do not pair with codons for other amino acids. At the other end tRNAs have a site for the attachment of an amino acid.

Special enzymes called aminoacyl tRNA synthetases (RS enzymes) attach the correct amino acids to the correct tRNAs. There is one RS enzyme for each of the twenty amino acids. Interpretation is possible because each RS enzyme can bind only one kind of amino acid and only to tRNA that pairs with the codons for that amino acid. The key to this specificity is a special code in each tRNA located near where the amino acid gets attached. This code is sometimes referred to as the "second genetic code." After binding the correct amino acid and tRNA, the RS enzyme attaches the two molecules with a covalent bond. These charged tRNAs, called aminoacyl-tRNAs, are ready to participate in protein synthesis directed by the codons of the mRNA. Information is stored in RNA in forms other than the triplet code. A special tRNA for methionine exists to initiate all peptide chains. It responds to AUG. However, proteins also have methionines in the main part of the polypeptide chain. Those methionines are carried by a different tRNA that also responds to AUG.

The "Second" Genetic Code

The fidelity of translating codons of messenger RNA (mRNA) into amino acids of the protein product requires that each transfer RNA (tRNA) be attached to the proper amino acid. Twenty distinct aminoacyl tRNA synthetases (RS enzymes) are found in cells; each is specific for a particular amino acid which it attaches to an appropriate tRNA. Because some amino acids (such as isoleucine and valine) are similar in structure, some RS enzymes have an editing feature, which allows them to cleave a mistakenly attached amino acid. The site at which the attachment reaction occurs is distinct from the editing site. The end result is that fewer than one in ten thousand amino acids is attached to the wrong tRNA.

Each RS enzyme must also recognize an appropriate tRNA. One might imagine that the anticodon found in the tRNA would be the recognition site. However, only in a few cases is it the major or sole determinant. Because the anticodon is at one end of the L-shaped tRNA and the amino acid is attached at the other end, this is perhaps not surprising. While tRNA molecules have the same general shape, they typically consist of seventy-six nucleotides, which provide numerous opportunities to distinguish themselves from one another.

The "second" genetic code is sometimes used to refer to the sequence of the tRNA that ensures that the correct one is recognized by its corresponding RS enzyme. Surprisingly, different elements are used by the various RS enzymes. In some cases, elements near the amino acid attachment site are important. This is the case for alanine tRNA, where the primary recognition is a G_3-U_{70} base pair. Incorporating this element into a cysteine tRNA will cause it to accept alanine despite the fact that the anticodon remains that for cysteine. In other cases, structures in the middle of the tRNA molecule are important, such as the variable loop or the D-loop. Usually multiple elements contribute to the recognition and ensure that the correct tRNA is recognized by its respective RS enzyme.

A mutation in the anticodon of a tRNA will usually not restrict its being attached to its designated amino acid. Such a mutation is referred to as a suppressor mutation if it overrides another mutation that leads to a chain termination mutation. For example, a point mutation in the CAG glutamine codon in a gene can convert it to a UAG chain termination codon. This would usually be deleterious because the resultant protein would be shorter than normal. However, if the normal GUA anticodon on tyrosine tRNA is mutated to CUA, it would pair with the UAG in the messenger RNA (mRNA) during protein synthesis; it would suppress the chain termination mutation by inserting tyrosine for the original glutamine in the protein, which may retain its function. This mutated tRNA would, however, insert a tyrosine for the normal UAG chain termination for other genes.

—*James L. Robinson*

The ribosome and associated factors must distinguish an initiating AUG from one for an internal methionine.

Distinction occurs differently in eukaryotes and bacteria. In bacteria, AUG serves as a start codon only if it is near a sequence that can pair with a section of the RNA in the ribosome. Two things are required of eukaryotic start (AUG) codons: First, they must be in a proper context of surrounding nucleotides; second, they must be the first AUG from the mRNA beginning that is in such a context. Context is also important for the incorporation of the unusual amino acid selenocysteine into several proteins. In a limited number of genes, a special UGA stop codon is used as a codon for selenocysteine. Sequences additional to UGA are needed for selenocysteine incorporation. Surrounding nucleotide residues also allow certain termination codons to be bypassed. For example, the mRNA from tobacco mosaic virus encodes two polypeptides, both starting at the same place; however, one is longer than the other. The extension is caused by the reading of a UAG stop codon by tRNA charged with tyrosine.

The production of two proteins with identical beginnings but different ends can also occur by frame shifting. In this mechanism, signals in the mRNA direct the ribosome machinery to advance or backtrack one nucleotide in its reading of the mRNA codons. Frame shifting occurs at a specific sequence in the RNA. Often the code for a frame shift includes a string of seven or more identical nucleotides and a complex RNA structure (a "pseudoknot").

Further codes are embedded in DNA. The linear sequence of amino acids, derived from DNA, has a code for folding in three-dimensional space, a code for its delivery to the proper location, a code for its modification by the addition of other chemical groups, and a code for its degradation. The production of mRNA requires nucleotide codes for beginning RNA synthesis, for stopping its synthesis, and for stitching together codon-containing regions (exons) should these be separated by noncoding regions (introns). RNA also contains signals that can tag them for rapid degradation. DNA has a code recognized by protein complexes for the initiation of DNA replication and signals recognized by enzymes that catalyze DNA rearrangements.

Impact and Applications

A major consequence of the near universality of the genetic code is that biotechnologists can move genes from one species into another and they are still expressed correctly. Since the code is the same in both organisms, the same protein is produced. This has resulted in the large-scale production of specific proteins in bacteria, yeast, plants, and domestic animals. These proteins are of immense pharmaceutical, industrial, and research value.

Scientists developed rapid methods for sequencing nucleotides in DNA in the 1970's. Since the genetic code was known, it suddenly became easier to predict the amino acid sequence of a protein from the nucleotide sequence of its gene than it did to determine the amino acid sequence of the protein by chemical methods. The instant knowledge of the amino acid sequence of a particular protein greatly simplified predictions regarding protein function. This has resulted in the molecular understanding of many inherited human diseases and the potential development of rational therapies based on this new knowledge.

—*Ulrich Melcher, updated by Bryan Ness*
See also: Central Dogma of Molecular Biology; Chromosome Structure; Chromosome

Theory of Heredity; DNA Replication; DNA Structure and Function; Evolutionary Biology; Genetic Code, Cracking of; Genetics, Historical Development of; Mendelian Genetics; Molecular Genetics; One Gene-One Enzyme Hypothesis; Protein Structure; Protein Synthesis; RNA Structure and Function; RNA Transcription and mRNA Processing; RNA World.

Further Reading

Clark, Brian F. C. *The Genetic Code and Protein Biosynthesis.* 2d ed. Baltimore: E. Arnold, 1984. Consists of a brief description of the genetic code.

Clark, David, and Lonnie Russell. *Molecular Biology: Made Simple and Fun.* 2d ed. Vienna, Ill.: Cache River Press, 2000. A detailed and accessible account of molecular biology.

Judson, Horace Freeland. *The Eighth Day of Creation.* Rev. ed. Cold Harbor Spring, N.Y.: Cold Spring Harbor Laboratory Press, 1997. A noted and fascinating history of molecular biology that details the deciphering of the genetic code.

Kay, Lily E. *Who Wrote the Book of Life? A History of the Genetic Code.* Stanford, Calif.: Stanford University Press, 2000. Brings myriad sources together to describe research on the genetic code between 1953 and 1970, the rise of communication technosciences, the intersection of molecular biology with cryptanalysis and linguistics, and the social history of postwar Europe and the United States.

Trainor, Lynn E. H. *The Triplet Genetic Code: The Key to Molecular Biology.* River Edge, N.J.: World Scientific, 2001. Intended for nonspecialists and professionals, surveys the fundamentals of the genetic code and how it has come to revolutionize thinking about living systems as a whole, especially regarding the connection between structure and function.

Web Site of Interest

Oak Ridge National Laboratory, The Genetic Code. http://www.ornl.gov/techresources/human_genome/graphics/slides/images1.html. Site has link to an image of the genetic code, with discussion.

Genetic Code, Cracking of

Field of study: History of genetics;
Molecular genetics

Significance: *The deciphering of the genetic code was a significant accomplishment for molecular biologists. The identification of the "words" used in the code explained how the information carried in DNA can be interpreted, via an RNA intermediate, to direct the specific sequence of amino acids found in proteins.*

Key terms

ANTICODON: a sequence of three nucleotide bases on the transfer RNA (tRNA) that recognizes a codon

CODON: a sequence of three nucleotide bases on the messenger RNA (mRNA) that specifies a particular amino acid

The Nature of the Puzzle

Soon after DNA was discovered to be the genetic material, scientists began to examine the relationship between DNA and the proteins that are specified by the DNA. DNA is composed of four deoxyribonucleotides containing the bases adenine (A), thymine (T), guanine (G), and cytosine (C). Proteins are composed of twenty different building blocks known as amino acids. The dilemma that confronted scientists was to explain the mechanism by which the four bases in DNA could be responsible for the specific arrangement of the twenty amino acids during the synthesis of proteins.

The solution to the problem arose as a result of both theoretical considerations and laboratory evidence. Experiments done in the laboratories of Charles Yanofsky and Sydney Brenner provided evidence that the order, or sequence, of the bases in DNA was important in determining the sequence of amino acids in proteins. Francis Crick proposed that the bases formed triplet "code words." He reasoned that if a single base specified a single amino acid, it would only be possible to have a protein made up of four amino acids. If two bases at a time specified amino acids, it would only be possible to code for sixteen amino acids. If the four bases were used three at a time, Crick proposed, it

would be possible to produce sixty-four combinations, more than enough to specify the twenty amino acids. Crick also proposed that since there would be more than twenty possible triplets, some of the amino acids might have more than one code word. The eventual assignment of multiple code words for individual amino acids was termed "degeneracy." The triplet code words came to be known as codons.

Identifying the Molecules Involved

Since DNA is found in the nuclei of most cells, there was much speculation as to how the codons of DNA could direct the synthesis of proteins, a process that was known to take place in another cellular compartment, the cytosol. A class of molecules related to DNA known as ribonucleic acids (RNAs) were shown to be involved in this process. These molecules consist of ribonucleotides containing the bases A, C, and G (as in DNA) but uracil (U) rather than thymine (T). One type of RNA, ribosomal RNA (rRNA), was found to be contained in structures known as ribosomes, the sites where protein synthesis occurs. Messenger RNA (mRNA) was shown to be another important intermediate. It is synthesized in the nucleus from a DNA template in a process known as transcription, and it carries an imprint of the information contained in DNA. For every A found in DNA, the mRNA carries the base U. For every T in DNA, the mRNA carries an A. The Gs in DNA become Cs in mRNA, and the Cs in DNA become Gs in mRNA. The information in mRNA is found in a form that is complementary to the nucleotide sequence in DNA. The mRNA is transported to the ribosomes and takes the place of DNA in directing the synthesis of a protein.

Deciphering the Code

The actual assignment of codons to specific amino acids resulted from a series of elegant experiments that began with the work of Marshall Nirenberg and Heinrich Matthaei in 1961. They obtained a synthetic mRNA consisting of polyuridylic acid, or poly (U), made up of a string of Us. They added poly (U) to a cell-free system that contained ribosomes and all other ingredients necessary to make proteins in vitro. When the twenty amino acids were added to the system, the protein that was produced contained a string of a single amino acid, phenylalanine. Since the only base in the synthetic mRNA was U, Nirenberg and Matthaei had discovered the code for phenylalanine: UUU. Because UUU in mRNA is complementary to AAA in DNA, the actual DNA bases that direct the synthesis of phenylalanine are AAA. By convention, the term "codon" is used to designate the mRNA bases that code for specific amino acids. Therefore UUU, the first code word to be discovered, was the codon for phenylalanine.

Using cell-free systems, other codons were soon discovered by employing other synthetic mRNAs. AAA was shown to code for lysine, and CCC was shown to code for proline. Scientists working in the laboratory of Severo Ochoa began to synthesize artificial mRNAs using more than one base. These artificial messengers produced proteins with various proportions of amino acids. Using this technique, it was shown that a synthetic codon with twice as many Us as Gs specified valine. It was not clear, however, if the codon was UUG, UGU, or GUU. Har Gobind Khorana and his colleagues began to synthesize artificial mRNA with predictable nucleotide sequences, and the use of this type of mRNA contributed to the assignment of additional codons to specific amino acids.

In 1964, Philip Leder and Nirenberg developed a cell-free protein-synthesizing system in which they could add triplet codons of known sequence. Using this new system, as well as Khorana's synthetic messengers, scientists could assign GUU to valine and eventually were able to assign all but three of the possible codons to specific amino acids. These three codons, UAA, UAG, and UGA, were referred to as "nonsense" codons because they did not code for any of the twenty amino acids. The nonsense codons were later found to be a type of genetic punctuation mark; they act as stop signals to specify the end of a protein.

There is no direct interaction between the mRNA codon and the amino acid for which it codes. Yet another type of RNA molecule was found to act as a bridge or, in Crick's terminology, an "adaptor" between the mRNA codon and the amino acid. This type of RNA is a small

The assignment of codons to specific amino acids resulted from a series of elegant experiments that began with the work of Marshall Nirenberg (above) and Heinrich Matthaei in 1961. (Jim Willier-Stokes Imaging)

molecule known as transfer RNA (tRNA). Specific enzymes connect the amino acids to their corresponding tRNA; the tRNA then carries the amino acid to the appropriate protein assembly location specified by the codon. The tRNA molecules contain recognition triplets known as anticodons, which are complementary to the codons on the mRNA. Thus, the tRNA that carries phenylalanine and recognizes UUU contains an AAA anticodon.

By 1966, all the codons had been discovered. Since some codons had been identified as "stop" codons, scientists had begun searching for one or more possible "start" codons. Since all proteins were shown to begin with the amino acid methionine or a modified form of methionine (which is later removed), the methionine codon, AUG, was identified as the start codon for most proteins. It is interesting that AUG also codes for methionine when this amino acid occurs at other sites within the protein.

The cracking of the genetic code gave scientists a valuable genetic tool. Once the amino acid sequence was known for a protein, or for even a small portion of a protein, knowledge of the genetic code allowed scientists to search for the gene that codes for the protein or, in some cases, to design and construct the gene itself. It also became possible to predict the sequence of amino acids in a protein if the sequence of nucleotide bases in a gene were known. Knowledge of the genetic code became invaluable in understanding the genetic basis of mutation and in attempts to correct these mutations by gene therapy. The discovery of the genetic code was therefore key to the development of genetics in the late twentieth century, perhaps outshined only by the discovery of DNA's double-helical structure in 1953 and the completion of the Human Genome Project in 2003.

—*Barbara Brennessel*

See also: Central Dogma of Molecular Biology; Chromosome Structure; Chromosome Theory of Heredity; DNA Replication; DNA Structure and Function; Evolutionary Biology; Genetic Code; Genetics, Historical Development of; Human Genome Project; Mendelian Genetics; Molecular Genetics; One Gene-One Enzyme Hypothesis; Protein Structure; Protein Synthesis; RNA Structure and Function; RNA Transcription and mRNA Processing; RNA World.

Further Reading

Crick, Francis H. C. "The Genetic Code III." *Scientific American* 215 (October, 1966). Reprinted in *The Chemical Basis of Life: An Introduction to Molecular and Cell Biology*. San Francisco: W. H. Freeman, 1973. The co-discoverer of DNA's double helical structure summarizes the story of the genetic code.

_____. "The Genetic Code: Yesterday, Today, and Tomorrow." *Cold Spring Harbor Symposia on Quantitative Biology* 31 (1966). Summarizes how the genetic code was solved and serves as an introduction to papers presented during a symposium on the genetic code.

Edey, Maitland A., and Donald C. Johnson. *Blueprints: Solving the Mystery of Evolution.* Reprint. New York: Viking, 1990. Focuses on evolution from the molecular genetic perspective and emphasizes the process of scientific discovery; three chapters are devoted to the genetic code.

Judson, Horace Freeland. *The Eighth Day of Creation.* Rev. ed. Cold Harbor Spring, N.Y.: Cold Spring Harbor Laboratory Press, 1997. A noted and fascinating history of molecular biology that details the deciphering of the genetic code.

Kay, Lily E. *Who Wrote the Book of Life? A History of the Genetic Code.* Stanford, Calif.: Stanford University Press, 2000. Kay brings myriad sources together to describe research on the genetic code between 1953 and 1970, the rise of communication technosciences, the intersection of molecular biology with cryptanalysis and linguistics, and the social history of postwar Europe and the United States.

Portugal, Franklin H., and Jack S. Cohn. *A Century of DNA: A History of the Discovery of the Structure and Function of the Genetic Substance.* Cambridge, Mass.: MIT Press, 1977. Provides a comprehensive historical background and identifies many of the scientists who worked to solve the genetic code.

Trainor, Lynn E. H. *The Triplet Genetic Code: The Key to Molecular Biology.* River Edge, N.J.: World Scientific, 2001. Intended for nonspecialists as well as professionals, surveys the fundamentals of the genetic code and how it has come to revolutionize thinking about living systems as a whole.

Web Site of Interest

Cracking the Code of Life. http://www.pbs.org/wgbh/nova/genome. The companion Web site to the 2001 PBS broadcast of the same name. Discusses heredity, gene manipulation, DNA sequencing, a "journey into DNA," and more.

Genetic Counseling

Field of study: Human genetics and social issues

Significance: *Genetic counseling involves helping individuals or families cope with genetic syndromes or diseases that exist, or could potentially occur, in a family setting. Genetic counselors pro-*

vide information regarding the occurrence or risk of occurrence of genetic disorders, discuss available options for dealing with those risks, and help families determine their best course of action.

Key terms

GENETIC SCREENING: the process of investigating a specific population of people to detect the presence of genetic defects

NONDIRECTIVE COUNSELING: a practice that values patient autonomy and encourages patients to reach a decision that is right for them based upon their personal beliefs and values

PEDIGREE ANALYSIS: analysis of a family's history by listing characteristics such as age, sex, and state of health of family members, used to determine the characteristics of a genetic disease and the risk of passing it on to offspring

PRENATAL DIAGNOSIS: the process of detecting a variety of birth defects and inherited disorders before a baby is born by various imaging technologies, genetic tests and biochemical assays

The Establishment of Genetic Counseling

Historically, people have long understood that some physical characteristics are hereditary and that particular defects are often common among relatives. This concept was widely accepted by expectant parents and influenced the thinking of many scientists who experimented with heredity in plants and animals. Many efforts were made to understand, predict, and control the outcome of reproduction in humans and other organisms. Gregor Mendel's experiments with garden peas in the mid-1800's led to the understanding of the relationship between traits in parents and their offspring. During the early twentieth century, Walter Sutton proposed that newly discovered hereditary factors were physically located on complex structures within the cells of living organisms. This led to the chromosome theory of inheritance, which explains mechanically how genetic information is transmitted from parent to offspring in a regular, orderly manner. In 1953, James Watson and Francis Crick (along with Maurice Wilkins and Rosalind Franklin)

discovered the double-helix structure of DNA, the molecule that carries the genetic information in the cells of most living organisms. Three years later, human cells were found to contain forty-six chromosomes each.

These discoveries, along with other developments in genetics, periodically generated efforts (often misguided) to control the existence of "inferior" genes, a concept known as eugenics. Charles F. Dight, a physician influenced by the eugenics movement, left his estate in 1927 "To Promote Biological Race Betterment—betterment in Human Brain Structure and Mental Endowment and therefor[e] in Behavior." In 1941 the Dight Institute for Human Genetics began to shift their emphasis from eugenics to genetic studies of individual families. In 1947, Sheldon Reed began working at the Dight Institute as a genetic consultant to individual families. Reed believed that his profession should put the clients' needs before all other considerations and that it should be separated from the concept of eugenics. He rejected the older names for his work, such as "genetic hygiene," and substituted "genetic counseling" to describe the type of social work contributing to the benefit of the family. As a result, the field of genetic counseling was born and separated itself from the direct concern of its effect upon the state or politics. In fact, Reed predicted that genetic counseling would have been rejected had it been presented as a form of eugenics.

Genetic counseling developed as a preventive tool and became more diagnostic in nature as it moved from academic centers to the major medical centers. In 1951, there were ten genetic counseling centers in the United States employing academically affiliated geneticists. Melissa Richter and Joan Marks were instrumental in the development of the first graduate program in genetic counseling at Sarah Lawrence College in New York in 1969. By the early 1970's, there were nearly nine hundred genetic counseling centers worldwide. By 2002 there were approximately two thousand genetic counselors in the United States not only working with individual families concerning genetic conditions but also involved in teaching, research, screening programs, public health, and the coordination of support groups. In 1990, the Human Genome Project began as a fifteen-year effort coordinated by the U.S. Department of Energy and the National Institutes of Health to map and sequence the entire human genome, prepare a model of the mouse genome, expand medical technologies, and study the ethical, legal, and social implications of genetic research.

The Training of the Genetic Counselor

Most genetic counseling students have undergraduate degrees in genetics, nursing, psychology, biology, social work, or public health. Training programs for genetic counselors are typically two-year masters-level programs and include field training in medical genetics and counseling in addition to a variety of courses focusing on genetics, psychosocial theory, and counseling techniques. During the two-year program, students obtain an in-depth background in human genetics and counseling through coursework and field training at genetic centers. Coursework incorporates information on specific aspects of diseases, including the prognoses, consequences, treatments, risks of occurrence, and prevention as they relate to individuals or families. Field training at genetic centers enables students to develop research, analytical, and communication skills necessary to meet the needs of individuals at risk for a genetic disease.

Many genetic counselors work with M.D. or Ph.D. geneticists and may also be a part of a health-care team that may include pediatricians, cardiologists, psychologists, endocrinologists, cytologists, nurses, and social workers. Other genetic counselors are in private practice or are engaged in research activities related to the field of medical genetics and genetic counseling. Genetic counseling most commonly takes place in medical centers, where specialists work together in clinical genetics units and have access to diagnostic facilities, including genetic laboratories and equipment for prenatal screening.

The Role of the Genetic Counselor

Prior to the 1960's, most genetic counselors were individuals with genetic training who consulted with patients or physicians about specific risks of occurrence of genetic diseases. It was not until 1959, when French geneticist Jérôme Lejeune discovered that children with Down syndrome have an extra chromosome 21, that human genetics was finally brought to the attention of ordinary physicians. Rapid growth in knowledge of inheritance patterns, improvements in the ability to detect chromosomal abnormalities, and the advent of screening programs for certain diseases in high-risk populations all contributed to the increased interest in genetic counseling. Development of the technique of amniocentesis, which detects both chromosomal and biochemical defects in fetal cells, led to the increased specialization of genetic counseling. By the 1970's, training of genetic counselors focused on addressing patients' psychosocial as well as medical needs. Genetic counseling thus became a voluntary social service intended exclusively for the benefit of the particular family involved.

Genetic counselors provide information and support to families who have members with genetic disorders, individuals who themselves are affected with a genetic condition, and families who may be at risk for a variety of inherited genetic conditions, including Huntington's disease (Huntington's chorea), cystic fibrosis, and Tay-Sachs disease. The counselor obtains the family medical history and medical records in order to interpret information about the inherited genetic abnormality. Genetic counselors analyze inheritance patterns, review risks of recurrence, and offer available options for the genetic condition. Other functions of genetic counselors include discussing genetic risks with blood-related couples considering marriage, contacting parents during the crisis following fetal or neonatal death, preparing a community for a genetic population screening program, and informing couples about genetically related causes of their infertility. A pregnant patient is most commonly referred to a genetic counselor by an obstetrician because of her advanced age (thirty-five years or older).

In addition to obtaining accurate diagnosis of the genetic abnormality, genetic counselors strive to explain the genetic information as clearly as possible, making sure that the individual or family understands the information fully and accurately. The genetic counselor must evaluate the reliability of the diagnosis and the risk of occurrence of the genetic disease. Because the reliability of various tests will affect a patient's decision about genetic testing and abortion, the counselor must give the patient a realistic understanding of the meaning and inherent ambiguity of test results. Most genetic counselors practice the principle of nondirectiveness and value patient autonomy. They present information on the benefits, limitations, and risks of diagnostic procedures without recommending a course of action, encouraging patients to reach their own decisions based on their personal beliefs and values. This attitude reflects the historical shift of genetic counseling away from eugenics toward a focus on the individual family. The code of ethics of the National Society of Genetic Counselors states that its members strive to "respect their clients' beliefs, cultural traditions, inclinations, circumstances, and feelings as well as provide the means for their clients to make informed independent decisions, free of coercion, by providing or illuminating the necessary facts and clarifying the alternatives and anticipated consequences."

Diagnosis of Genetic Abnormalities

In the latter half of the twentieth century, discoveries in genetics and developments in reproductive technology contributed to the advancements in prenatal diagnosis and genetic counseling. Prenatal diagnostic procedures eventually became an established part of obstetrical practice with the development of amniocentesis in the 1960's, followed by ultrasound, chorionic villus sampling (CVS), and fetal blood sampling. Amniocentesis, CVS, and fetal blood sampling are ways to obtain fetal cells for analysis and detection of various types of diseases. Amniocentesis, a cytogenetic analysis of the cells within the fluid surrounding the fetus, is performed between the fifteenth and twentieth weeks of gestation and detects possible chromosomal abnormalities such as Down syndrome and trisomy 18. The information obtained from CVS is similar to that obtained from amniocentesis, except the testing can be performed earlier in the pregnancy (during the tenth to twelfth weeks of gestation). Fetal blood sampling can be performed safely only after eighteen weeks of pregnancy. An ultrasound, offered to all pregnant women, uses high-frequency sound waves to create a visual image of the fetus and detects anatomical defects such as spina bifida, cleft lip, or certain heart malformations. Pedigree analysis may also be used for diagnostic purposes and to determine the risk of passing a genetic abnormality on to future generations. A pedigree of the family history is constructed, listing the sex, age, and state of health of the patient's close relatives; from that, recurrent miscarriages, stillbirths, and infant deaths are explored.

Prenatal diagnostic techniques are used to identify many structural birth defects, chromosomal abnormalities, and more than five hundred specific disorders. Genetic counselors who believe that their client is at risk for passing on a particular disease may suggest several genetic tests, depending on the risk the patient may face. Screening of populations with high frequencies of certain hereditary conditions, such as Tay-Sachs disease among Ashkenazi Jews, is encouraged so that high-risk couples can be identified and their pregnancies monitored for affected fetuses. Pregnant women may also be advised to undergo testing if an abnormality has been found by the doctor, the mother will be thirty-five years of age or older at the time of delivery, the couple has a family history of a particular genetic abnormality, the mother has a history of stillbirths or miscarriages, or the mother is a carrier of metabolic disorders (for example, hemophilia) that can be passed from mothers to their sons.

The Human Genome Project is expected to have a dramatic impact on presymptomatic diagnosis of individuals carrying specific diseases, multigene defects involved in common diseases such as heart disease and diabetes, and individual susceptibility to environmental factors that interact with genes to produce diseases. The isolation and sequencing of genes associated with genetic abnormalities such as cystic fibrosis, kidney disease, Alzheimer's disease, and Huntington's disease (Huntington's chorea) allow for individuals to be tested for those specific conditions. Many genetic tests have been developed so that the detection of genetic conditions can be made earlier and with more precision.

Ethical Aspects of Genetic Counseling

With advancements in human genetics and reproductive technology, fundamental moral and ethical questions may arise during difficult decision-making processes involving genetic abnormalities for which families may be unprepared. Diagnosis of a particular genetic disease may allow individuals or families to make future plans and financial arrangements. However, improvements in the capability to diagnose numerous hereditary diseases often exceed the ability to treat such diseases. The awareness that an unborn child is genetically predisposed toward a disease with no known cure may lead to traumatic anxiety and depression. The psychological aspects of genetic counseling and genetic centers must therefore continue to be explored in genetic centers throughout the world.

Questions about who should have access to the data containing patients' genetic makeup must also be considered as the ability to screen for genetic diseases increases. Violating patients' privacy could have devastating consequences, such as genetic discrimination in job hiring and availability of health coverage. Employers and insurance companies have already denied individuals such opportunities based on information found through genetic testing. Disclosure of genetic information not only contributes to acts of discrimination but also may result in physical and psychological harm to individuals.

With data derived from the Human Genome Project increasing rapidly, problems arising from the application of new genetic knowledge in clinical practice must be addressed. The norm of nondirective counseling will be challenged, raising questions of who provides and who receives information and how it is given. Many believe that genetic counseling is beneficial to those faced with genetic abnormalities, while others fear that genetic counseling is a form of negative eugenics, an attempt to "improve" humanity as a whole by discouraging the birth of children with genetic defects. Since most genetic conditions can be neither treated nor modified in pregnancy, abortion is often the preventive measure used. Thus, ethical issues concerning the respect for autonomy of the unborn child must also be considered.

—Jamalynne Stuck and Doug McElroy

See also: Amniocentesis and Chorionic Villus Sampling; Bioethics; Gene Therapy; Gene Therapy: Ethical and Economic Issues; Genetic Screening; Genetic Testing; Genetic Testing: Ethical and Economic Issues; Hereditary Diseases; Human Genetics; In Vitro Fertilization and Embryo Transfer; Insurance; Linkage Maps; Pedigree Analysis; Prenatal Diagnosis.

Further Reading

Leroy, Bonnie, Dianne M. Bartels, and Arthur L. Caplan, eds. *Prescribing Our Future: Ethical Challenges in Genetic Counseling*. New York: Aldine de Gruyter, 1993. Offers ethical insights into the implications of genetic counseling, including the issue of neutrality, the potential impact of the Human Genome Project, workplace ideology of counselors, and the role of public policy.

Resta, Robert G., ed. *Psyche and Helix: Psychological Aspects of Genetic Counseling*. New York: Wiley-Liss, 2000. Articles reprinted from myriad sources cover topics pertaining to the medical, social, psychological, and emotional effects of genetic diseases, including the management of guilt and shame, patient care, and a detailed analysis of a genetic counseling session.

Rothman, Barbara Katz. *The Tentative Pregnancy: How Amniocentesis Changes the Experience of Motherhood*. Rev. ed. New York: Norton, 1993. Provides a discussion of decisions faced by patients who seek genetic counseling.

Schneider, Katherine A. *Counseling About Cancer: Strategies for Genetic Counseling*. 2d ed. New York: Wiley-Liss, 2002. A thorough resource to help genetic counselors and other health care providers effectively assist patients and families in managing hereditary cancer. Gives clinical features of thirty cancer syndromes, tables listing major cancer syndromes by cancer type, and many case studies.

Weil, Jon. *Psychosocial Genetic Counseling.* New York: Oxford University Press, 2000. Examines the psychosocial components of counseling interactions, including the role of emotions such as anxiety and guilt, and the complex process of decision making. Illustrated.

Wexler, Alice. *Mapping Fate: A Memoir of Family, Risk, and Genetic Research.* Berkeley: University of California Press, 1996. Describes Wexler's personal quest to discover the genetic basis for Huntington's disease.

Young, Ian D. *Introduction to Risk Calculation in Genetic Counseling.* 2d ed. New York: Oxford University Press, 1999. Designed for professionals, but useful to consumers in understanding the different types of quantitative risk assessment. Illustrated.

Web Sites of Interest

American Board of Genetic Counseling. http://www.faseb.org/genetics/abgc/abgcmenu.htm. A professional organization that educates, administers examinations, and certifies genetic counselors.

Human Genome Project Information, Genetic Counseling. http://www.doegenomes.org. Site links to genetic counseling information and related resources.

National Society of Genetic Counselors. http://www.nsgc.org. Offers a search engine for locating genetic counselors in the United States and a newsroom with press releases and fact sheets about the counseling services.

Genetic Engineering

Field of study: Genetic engineering and biotechnology

Significance: *The development of the tools of recombinant DNA technology used in genetic engineering has generated unprecedented inquiry into the nature of the living system and has revolutionized the study of genetics. The implications of this research are far-reaching, ranging from a better understanding of basic biological principles and molecular mechanisms to pharmacological, diagnostic, and therapeutic applications that promise to help prevent and treat a wide range of genetic diseases.*

Key terms

BIOTECHNOLOGY: the application of recombinant DNA technology to the development of specific products and procedures

CLONING: the process by which large amounts of a single gene or genome (the entire genetic content of a cell) are reproduced

COMPLEMENTARY BASE PAIRING: hydrogen bond formation that only occurs between adenine and thymine or cytosine and guanine

DNA SEQUENCE ANALYSIS: chemical methods that permit the determination of the order of nucleotide bases in a DNA molecule

GENOMIC LIBRARY: a collection of clones that includes the entire genome of a single species as fragments ligated to vector DNA

PROBE HYBRIDIZATION: a method that permits the identification of a unique sequence of DNA bases using a single-stranded DNA segment complementary to the unique sequence and carrying a molecular tag allowing identification

TRANSGENIC ORGANISM: a species in which the genome has been modified by the insertion of genes obtained from another species

VECTOR: a segment of DNA, usually derived from viruses, bacteria, or yeast, that contains regulatory sequences that permit the amplification of single genes or genetic segments

Restriction Enzymes

Many of the methods used in genetic engineering represent adaptations of naturally occurring genetic processes. One of the earliest and most significant discoveries was the identification of a family of DNA enzymes called restriction endonucleases, more commonly called restriction enzymes. Restriction enzymes are DNA-modifying enzymes produced by microorganisms as a protection against viral infection; their uniqueness and utility in recombinant DNA technology reside in their ability to cleave DNA at precise recognition sites based on DNA sequence specificity. Several hundred restriction enzymes have been isolated, and many recognize unique DNA segments and initiate DNA cleavage only at these sites. The site-specific cleavages generated by restriction enzymes can be used to produce a unique set of DNA segments that can be used to "map" individual genes and distinguish them from all other genes. This type of genetic analysis, based on differences in the sizes of DNA segments from different genes or different individuals when cleaved with restriction enzymes, is referred to as restriction fragment length polymorphism (RFLP) analysis.

If genes or DNA segments from different sources or species are cleaved with the same restriction enzyme, the DNA segments produced, though genetically unrelated, can be mixed together to produce recombinant DNA. This occurs because most restriction enzymes produce complementary, linear, single-stranded DNA ends that can join together. An additional enzyme called DNA ligase is used to seal the link between the DNA molecules with covalent bonds. This procedure, developed in the 1970's, is at the core of recombinant DNA technology and can be used to analyze the structure and function of the genome at the molecular level.

Another key development has been the use of vectors to amplify DNA fragments. Vectors are specially designed DNA molecules derived from viruses, bacteria, or other microorganisms, such as yeast, that contain regulatory sequences permitting the amplification or expression of a DNA fragment or gene. Vectors are available for numerous applications.

Vectors

Plasmids are small, circular DNAs that have been isolated from many species of bacteria. These naturally occurring molecules often encode antibiotic resistance genes that can be transferred from one bacterial cell to another in a process called transformation. In the laboratory, plasmids can be used as vectors in the amplification of genes inserted by restriction enzyme treatment of both vector and insert DNA, followed by DNA ligation to produce recombinant plasmids. The recombinant DNA is then inserted into host bacterial cells by transformation, a routine process in which bacterial cells are made "competent," that is, able to take up DNA from their surroundings. Once inside

the host cell, the recombinant plasmid will be replicated by the host cell, along with the host's own genome. Bacterial cells reproduce rapidly and generate large colonies of cells, each cell containing a copy of the recombinant plasmid. By this process the fragment of DNA in the recombinant vector is "cloned."

The cloned DNA can then be isolated from the bacterial cells and used for other applications or studies. Plasmids are useful for cloning small genes or DNA fragments; larger fragments can be cloned using viral vectors such as the bacterial virus (bacteriophage) lambda (phage λ). This virus can infect bacterial cells and reproduce to high copy number. If nonessential viral genes are removed, recombinant viruses containing genes of interest can be produced. Synthetic recombinant vectors incorporating bacterial and viral components, called cosmids, have also been developed. In addition, synthetic minichromosomes called yeast artificial chromosomes (YACs), which incorporate large segments of chromosomal DNA and which are capable of replication in bacterial or eukaryotic systems, have been developed.

DNA Sequence Analysis

A further key discovery in genetic engineering has been the development of chemical methods of DNA sequence analysis. These methods permit a determination of the linear sequence of nucleotide bases in DNA. DNA sequence analysis permits a direct determination of gene structure with respect to regulatory and protein-coding regions and can be used to predict the structure and function of proteins encoded by specific genes.

There are many important applications of the basic principles of genetic engineering. Notable examples include the Human Genome Project, the identification and characterization of human disease genes, the production of large amounts of proteins for therapeutic or industrial purposes, the creation of genetically engineered plants that are disease-resistant and show higher productivity, the creation of genetically engineered microorganisms that can help clean up pollution, and the treatment of genetic disorders using gene therapy.

Gene Cloning

The ability to clone DNA fragments has directly facilitated DNA sequence analysis. In addition to allowing the better understanding of specific genes, cloning was an integral tool in the Human Genome Project, an international effort to elucidate the structure of the entire human genome. The Human Genome Project offers the promise of greatly increasing the understanding of the genes responsible for inherited single-gene disorders as well as the involvement of specific genes in multifactorial disorders such as coronary heart disease.

The underlying genetic defects for a number of disease-causing genes have been identified, including sickle-cell disease (which results from a single nucleotide base substitution in one of the globin genes), Duchenne muscular dystrophy (caused by deletions in the muscle protein gene for dystrophin), and cystic fibrosis (caused by a variety of mutations in the gene for the chloride channel conductance protein). The identification of these disease genes has permitted the design of diagnostic tests and in some cases therapeutic strategies, including attempts to replace defective genes.

The analysis of gene function has been made possible by a process called site-directed mutagenesis, in which specific mutations can be introduced into cloned genes. These mutant genes can then be inserted into expression vectors, where the faulty protein can be produced and studied. Alternatively, the mutant genes can be introduced into animals, such as mice, to explore the effects of specific mutations on development and cell function.

Transgenic Organisms

One of the earliest successes in producing transgenic organisms was when *Escherichia coli* bacteria were engineered to produce human insulin for the treatment of diabetes. The tech-

Among less well known genetic engineering projects is the work of Oregon State University professor Steve Strauss and his colleagues, who are genetically modifying poplar trees to grow larger leaves in order to find genes that affect growth. (AP/Wide World Photos)

nology involved the cloning of the human insulin gene and its insertion into bacterial expression vectors. Subsequently, many gene products have been produced by genetically engineered microorganisms, including clotting factors (used in the treatment of hemophilia), growth factors such as epidermal growth factor (used to accelerate wound healing) and colony-stimulating factors (used to stimulate blood cell formation in the bone marrow), and interferons (used in the treatment of immune system disorders and certain types of cancer). The advantages of using genetically engineered products are enormous: Therapeutic proteins or hormones can be produced in much larger amounts than could be obtained from tissue isolation, and the genetically engineered products are free of viruses and other contaminants.

Introduction of foreign genes into the fertilized eggs of host animals is called germ-line transformation and involves the insertion of individual genes into fertilized eggs. After the eggs are implanted in foster mothers, the resulting transgenic offspring will have the mutated gene in all their cells and will be able to pass the gene on to their future offspring.

Many of the methods for introducing foreign genes into host cells take advantage of the naturally occurring processes facilitated by viruses. Genetically engineered retroviruses, for example, can be used to insert a foreign gene into a recipient cell following viral infection. Foreign genes may also be incorporated into lipid membranes to form liposomes, which then can bind to the target cell and insert the gene. Chemical methods of gene transfer include the use of calcium phosphate or dextran sulfate to generate pores in the recipient cell membrane through which the foreign DNA enters the cell. Microinjection involves the use of microscopic needles to insert foreign DNA directly into the nucleus of the target cell and is often used to insert genetic material into fertilized eggs. Electroporation involves the use of an electric current to open pores in the cell membrane, permitting DNA uptake by the recipient cell. Finally, particle bombardment represents a method of gene transfer in which metal pellets coated with DNA are transferred into target cells under high pressure using "gene guns." This method is particularly useful for inserting genes into plant cells that are resistant to DNA uptake because of thick cell walls.

Genetically engineered transgenic species have many biological uses. Transgenic animals have been used to analyze the functions of specific genes in development and to generate animal models of human diseases. For example, a transgenic mouse strain incorporating a human breast cancer gene has been developed to explore the mechanisms by which this disease occurs. In addition, transgenic mice have been used to analyze the normal functions of specific genes by producing "knockout" mice, whose genomes contain mutated, nonfunctional copies of the genes of interest. This technology, developed by Mario Capecchi, uses homologous recombination, in which only complementary nucleotide base pairs carry out the genetic exchange within the host chromosome. Thus, the effects of the inserted gene, or transgene, on development and physiology can be examined. Knockout mice lacking a functional adenosine deaminase (*ADA*) gene, for example, show disease characteristics comparable to those of humans with severe combined immunodeficiency disorder (SCID). These mice have been very useful for determining the efficacy of novel treatments, including the replacement of the faulty gene by gene therapy.

Transgenic animals have also been developed to produce therapeutic gene products in large quantities. For example, transgenic sheep have been developed that secrete the human protein alpha-1 antitrypsin (AAT) in their milk. AAT is used to treat an inherited form of emphysema. The process involves the microinjection of fertilized sheep eggs with the human *AAT* gene linked to regulatory sequences that allow the gene to be actively expressed in the mammary tissue. Although the process of generating transgenic animals is inefficient, individual transgenic animals can produce tremendous amounts of gene products that can be readily purified from the milk. Additional transgenic livestock have been engineered to produce tissue plasminogen activator (used in the treatment of blood clots), hemoglobin (used as a blood substitute), erythropoietin (used to stimulate red blood cell formation in kidney dialysis patients), human growth hormone (used to treat pituitary dwarfism), and factor VIII (used to treat hemophilia).

Transgenic plants have also been produced, using the Ti (tumor-inducing) plasmid. This plasmid is found naturally in the bacterium *Agrobacterium tumefaciens*. The Ti plasmid has been used to transfer a toxin gene from the bacterium *Bacillus thuringiensis* that kills insect pests, thereby avoiding the use of pesticides.

Genetically Engineered Viruses

An additional medical application involves the use of genetically engineered viruses in the treatment of genetic diseases. Retroviruses are the most important group of viruses used for these purposes, since the life cycle of the virus involves the incorporation of the viral genome into host chromosomes. Removal of most of the virus's own structural genes removes its ability to cause disease, while the regulatory genes are retained and ligated to the therapeutic gene. The recombinant retrovirus then becomes harmless; however, it can still enter a cell and become integrated into the host cell genome, where it can direct the expression of the therapeutic gene. The first successful clinical application was the use of genetically engineered retroviruses in the treatment of severe combined immunodeficiency disorder (SCID). Viruses with a functional copy of the *ADA* gene were able to reverse SCID. However, in 2002 researchers in France and the United States discovered that this treatment appears to lead to a greatly increased risk of developing leukemia, and clinical trials were suspended.

Similar methods have been used to develop recombinant vaccines. For example, a recombinant vaccinia virus has been produced by the insertion of genes from other viruses. During the process of infection, the recombinant vaccinia virus produces proteins from the foreign genes, which act as antigens which lead to immunity following vaccination. This strategy is particularly useful in the development of vaccines against viruses that are highly pathogenic such as the human immunodeficiency virus (HIV), in which it is not possible to use a whole killed or attenuated (weakened) live viral vaccine because of the risk of developing the disease from the vaccination. Genetically engineered viruses may also be useful in the treatment of diseases such as cancer since they could be designed to target specific cells with abnormal cell surface receptors. Recombinant adenoviruses containing a single gene mutation have been engineered that are capable of lethal infection in cancer cells but not in normal tissues of the body.

Impact and Applications

The methods of recombinant DNA technology have revolutionized our understanding of the molecular basis of life and have led to a variety of useful applications. Some of the most important discoveries have involved an increased understanding of the molecular basis of disease processes, which has led to new methods of diagnosis and treatment. Genetically engineered animals can be used to produce unlimited amounts of therapeutic gene products and can also serve as genetic models to enhance understanding of the physiological basis of disease. Plants can be genetically engineered for increased productivity and disease resistance. Genetically engineered viruses have been developed as vaccines against infectious disease. The methods of recombinant DNA technology were originally developed from natural products and processes that occur within the living system. The ultimate goals of this research must involve applications that preserve the integrity and continuity of the living system.

—*Sarah Crawford Martinelli,*
updated by Bryan Ness

See also: Animal Cloning; Biopharmaceuticals; Cloning; Cloning: Ethical Issues; Cloning Vectors; DNA Replication; DNA Sequencing Technology; Gene Therapy; Gene Therapy: Ethical and Economic Issues; Genetic Engineering: Agricultural Applications; Genetic Engineering: Historical Development; Genetic Engineering: Industrial Applications; Genetic Engineering: Medical Applications; Genetic Engineering: Risks; Genetic Engineering: Social and Ethical Issues; Genetically Modified (GM) Foods; High-Yield Crops; Knockout Genetics and Knockout Mice; Polymerase Chain

Reaction; Restriction Enzymes; Reverse Transcriptase; Shotgun Cloning; Synthetic Genes; Transgenic Organisms; Xenotransplants.

Further Reading

Altieri, Miguel A. *Genetic Engineering in Agriculture.* Chicago: LPC Group, 2001. Raises serious questions about the drive toward genetically engineered crops.

Anderson, Luke. *Genetic Engineering, Food, and Our Environment.* White River Junction, Vt.: Chelsea Green, 1999. Introduces issues surrounding genetic engineering, including the social, environmental, and health implications arising from the commercial use of this technology in food and farming.

Boylan, Michael, and Kevin E. Brown. *Genetic Engineering: Science and Ethics on the New Frontier.* Upper Saddle River, N.J.: Prentice Hall, 2001. Written by a biologist and philosopher, this text includes discussions on the professional and practical principles of conduct, the biology of genetic therapy, the limits of science, somatic gene therapy, enhancement, cloning, and germ line therapy. Illustrated.

Drlica, Karl. *Understanding DNA and Gene Cloning: A Guide for the Curious.* Rev. ed. New York: Wiley, 2003. An excellent introduction to the basic properties of DNA and its modern applications. Consists of four sections: basic molecular biology, manipulation of DNA, insights gained through the use of gene cloning (including a chapter on retroviruses), and human genetics.

Hill, Walter E. *Genetic Engineering: A Primer.* Newark, N.J.: Harwood Academic, 2000. Written to "help those with little scientific background become conversant with the area generally called genetic engineering." Illustrations, glossary, index.

Le Vine, Harry, III. *Genetic Engineering: A Reference Handbook.* Santa Barbara, Calif.: ABC-CLIO, 1999. Part of the series Contemporary World Issues, covers the basics of genetic engineering. Illustrated.

Nicholl, Desmond S. T. *An Introduction to Genetic Engineering.* 2d ed. New York: Cambridge University Press, 2002. Three sections detail basic molecular biology, methods used to manipulate genes, and modern applications of genetic engineering. Illustrated.

Steinberg, Mark, and Sharon D. Cosloy, eds. *The Facts On File Dictionary of Biotechnology and Genetic Engineering.* New ed. New York: Checkmark Books, 2001. Collects sixteen hundred medical, chemical, and engineering terms relating to plant and animal biology and molecular genetics and genetic engineering. Four appendices cover acronyms and abbreviations, the genetic code, purine and pyrimidine bases, and amino acid side chains.

Walker, Mark, and David McKay. *Unravelling Genes: A Layperson's Guide to Genetic Engineering.* St. Leonards, N.S.W.: Allen & Unwin, 2000. Explains the core concepts of genetic engineering, including the scientific principles and technological advances that have made gene therapy, cloning, and genetically modified food products available. Special focus is given to gene therapy treatments for Alzheimer's disease, cystic fibrosis, and hemophilia.

Williams, J. G., A. Ceccarelli, and A. Wallace. *Genetic Engineering.* 2d ed. New York: Springer, 2001. Surveys some of the techniques which have made recent advances in genetic engineering possible and shows how they are being applied to clinical problems.

Yount, Lisa. *Biotechnology and Genetic Engineering.* New York: Facts On File, 2000. Gives background on controversial genetic engineering technologies and the social, political, ethical, and legal issues they raise. Includes a chronology from the birth of agriculture to recent findings.

Web Sites of Interest

Centers for Disease Control, Office of Genomics and Disease Prevention. http://www.cdc.gov/genomics/default.htm. Offers information on the genetic discoveries and prevention of diseases in humans. Includes links to related resources.

U.S. Department of Agriculture, Biotechnology: An Information Resource. http://www.nal.usda.gov/bic. A government site that offers dozens of links to information on genetic engineering (biotechnology). glossary, and an annotated bibliography.

Genetic Engineering: Agricultural Applications

Field of study: Genetic engineering and biotechnology

Significance: *Genetic engineering is the deliberate manipulation of an organism's DNA by introducing beneficial or eliminating specific genes in the cell. For agricultural applications, the technology enables scientists to isolate, modify, and insert genes into the same or different crop, clone an adult plant from a single cell of a parent plant, and create genetically modified (GM) foods.*

Key terms

CLONING: regeneration of full-grown adult group of organisms from some form of asexual reproduction—for example, from protoplasts

EXOGENOUS GENE: a gene produced or originating from outside an organism

GENOME: the collection of all the DNA in an organism

PLASMID: a small, circular DNA molecule that occurs naturally in some bacteria and yeasts

PROTOPLASTS: plant cells whose cell walls have been removed by enzymatic digestion

RECOMBINANT DNA: a molecule of DNA formed by the joining of DNA segments from different sources

TRANSGENIC CROP PLANT: a crop plant that contains a gene or genes that have been artificially inserted into its genome

VECTOR: a carrier organism, or a DNA molecule used to transmit genes in a transformation procedure

Producing Transgenic Crop Plants

To produce a transgenic crop, a desirable gene from another organism, of the same or a different species, must first be spliced into a vector such as a virus or a plasmid. In some cases additional modification of the gene may be attempted in the laboratory. The most common vector used for producing transgenic plants is the "Ti" plasmid, or tumor-inducing plasmid, found in the cells of the bacterium called *Agrobacterium tumefaciens*. *A. tumefaciens* infection causes galls or tumorlike growths to

develop on the tips of the plants. Botanists use the infection process to introduce exogenous genes of interest into host plant cells in order to generate entire crop plants that express the novel gene.

Unfortunately, it was discovered that *A. tumefaciens* could infect only dicotyledons such as potatoes, apples, pears, roses, tobacco, and soybeans. Monocotyledons like rice, wheat, corn, barley, and oats could not be infected with the bacterium. Three primary methods are used to overcome this problem: particle bombardment, microinjection, and electroporation. Particle bombardment is a process in which microscopic DNA-coated pellets are shot through the cell wall using a gene gun. Microinjection involves the direct injection of DNA material into a host cell using a finely drawn micropipette needle. In electroporation, the recipient plant cell walls are removed with hydrolyzing enzymes to make protoplasts, and a few pulses of electricity are used to produce membrane holes through which some DNA can randomly enter.

Reducing Damage from Pests, Predators, and Disease

Geneticists have identified many of the genes for resistance to insect predation and damage caused by viral, bacterial, and fungal diseases in agricultural plants. For instance, seeds of common beans produce a protein that blocks the digestion of starch by two insect pests, cowpea weevil and Azuki bean weevil. The gene for this protein has now been transferred to the garden pea to protect stored pea seeds from pest infestation.

Bacillus thuringiensis (*Bt*), a common soil bacterium, produces an endotoxin called the *Bt* toxin. The *Bt* toxin, considered an environmentally safe insecticide, is toxic to a number of caterpillars, including the tobacco hornworm and gypsy moth. An indirect approach to pest management completely bypasses the problem of plant transformation. This involves inserting the *Bt* gene into the genome of a bacterium that colonizes the leaf, synthesizes, and secrets the pesticide on the leaf surface. Transgenic corn and cotton have also been modified with the *Bt* gene, enabling the plants to manufacture their own pesticide, which is nontoxic to humans.

Glyphosate, the most widely used nonselective herbicide, and other broad-spectrum herbicides are often toxic to crop plants, as well as the weeds they are intended to kill. A major thrust is to identify and transfer herbicide resistance genes into crop plants. Cotton plants, for example, have been genetically engineered to be resistant to certain herbicides.

Improving Crop Yield and Food Quality

Genetic engineering is now being used to modify crops, to improve the quality of food taste, fatty acid profile, protein content, sugar composition, and resistance to spoilage. New, useful or attractive horticultural varieties are also produced, by transforming plants with new or altered genes. For example, plants have been engineered that have additional genes for enzymes that produce anthocyanins, which has resulted in flowers with unusual colors and patterns.

Cereals are the staple food and major source of protein for a large percentage of the earth's population, and contain 10 percent protein in the dry weight. Grains unfortunately lack one or more essential amino acids and therefore provide incomplete nutrition. There are efforts to engineer missing amino acids into cereal protein and to insert genes for higher yields. The development of a high-yielding dwarf rice plant so dramatically helped the nutritional status of millions of people in Southeast Asia that it has been called the "miracle rice."

Researchers based at the Swiss Federal Institute of technology in Zurich have genetically engineered a more nutritious type of rice by inserting three genes into rice that make the plant produce beta-carotene or pro-vitamin A. The color that is imparted to the rice by the vitamin gives it the name "golden rice." Mammals, including humans use beta-carotene, from their food to produce vitamin A, which is necessary for good eye-

sight. It is estimated in 2003 that about 124 million children in the world lack vitamin A, which puts them at risk of permanent blindness and other serious diseases. Golden rice could help alleviate the serious problem of vitamin A deficiency. Iron deficiency is the world's worst nutrition disorder, causing anemia that affects 2 billion people worldwide. The scientists have also managed to insert genes into rice to make it iron-rich.

For improved quality of fruit after harvest, genetic engineers are inserting genes to slow the rate of senescence (aging) and thus slow spoilage of harvested crops, especially fruits. For example, scientists at Calgene (Davis, California) have inserted a gene into tomato plants that blocks the synthesis of the enzyme polygalacturonase, which causes tomato softening, thereby delaying aging (hence rotting) of the fruit.

Improved tolerance to environmental stress for agricultural plants is also being explored by biotechnology, especially for drought, saline conditions, chilling temperatures, high light intensities, and extreme heat. Some plants have genes that enable them to adapt naturally to harsh environments, and genetic engineers are

Tobacco engineered to have no nicotine became economically important to this Amish farmer during the drought of 2002. (AP/Wide World Photos)

A demonstration in Seattle against Starbucks' usage of genetically engineered ingredients, dubbed "frankenfoods" by protesters. Although many in the developed nations of Europe and North American are concerned over unintended consequences, genetic engineering in agriculture has made it possible to breed varieties of desirable crop plants with a wider range of tolerance for climatic and soil conditions, as well as pests. Such crops offer hope that poorer nations will be able to feed their growing populations. (AP/Wide World Photos)

using these genes to produce similar effects in crop plants.

Biotechnology has produced a marked increase in crop productivity worldwide. In 1999, about 50 percent of the soybean, 33 percent of the corn, and 35 percent of the cotton crops in the United States and 62 percent of the canola crop in Canada were planted with genetically modified seed. In 1996, genetically engineered corn and soybeans were first grown commercially on 1.7 million hectares (4.2 million acres). The land planted in these crops had swelled to 39.9 million hectares (98.8 million acres) by 2003.

Impact and Implications

The various applications of genetic engineering to agriculture have made it possible to alter genes and modify crops for the benefit of humankind, in addition to industrial and medical applications. This should be a subject of interest to everyone because it impacts every aspect of our daily living, and calls for ideas to be tapped from all sectors of our communities. It is a modern innovative trend that has become a major thrust in agriculture by production of genetically modified (GM) foods that are more nutritious and better preserved, but raises concerns as well because of potential dangers of microbial infections and chemical hazards.

Many nonscientists, and some scientists, are leery of GM foods, feeling that too little is understood about the environmental effects of growing GM plants and the potential health dangers of eating GM foods. Resistance to GM foods is widespread in Europe and parts of Asia, and a number of environmental groups strongly oppose all GM crops. Some have gone so far as to call them "frankenfoods." So far they appear to be safe and successful, holding out great promise to solve the problem of world hunger. They make it possible to breed varieties of desirable crop plants with a wider range of tolerance for different climatic and soil conditions, offering hope for the promotion of global agriculture to feed poorer nations. Genetic engineering must be seen as an indispensable component of modern scientific advancement and social development for every nation, if handled wisely without exposing living organisms to harmful microorganisms and toxic chemicals in the process.

—*Samuel V. A. Kisseadoo*

See also: Animal Cloning; Biofertilizers; Biological Weapons; Biopesticides; Biopharmaceuticals; Cloning; Cloning: Ethical Issues; Cloning Vectors; DNA Replication; DNA Sequencing Technology; Gene Therapy; Gene Therapy: Ethical and Economic Issues; Genetic Engineering; Genetic Engineering: Historical Development; Genetic Engineering: Industrial Applications; Genetic Engineering: Medical Applications; Genetic Engineering: Risks; Genetic Engineering: Social and Ethical Issues; Genetically Modified (GM) Foods; High-Yield Crops; Knockout Genetics and Knockout Mice; Polymerase Chain Reaction; Restriction Enzymes; Reverse Transcriptase; Shotgun Cloning; Synthetic Genes; Transgenic Organisms; Xenotransplants.

Further Reading

Borlaug, Norman E. "Ending World Hunger: The Promise of Biotechnology and the Threat of Antiscience Zealotry." *Plant Physiology* 124 (2000): 487-490. The father of Green Revolution and Nobel Peace Prize winner speaks of his unwavering support for GMOs.
Potrykus, Ingo. "Golden Rice and Beyond." *Plant Physiology* 125 (2001): 1157-1161. The originator of the wonder rice presents scientific, ethical, intellectual, and social challenges of developing and using the GMOs. Illuminating and insightful.
Raven, Peter H., et al. *Biology of Plants.* 6th ed. New York: W. H. Freeman/Worth, 1999. Includes discussion of all the essential features and processes in plants, genetic engineering, adaptations, and plant uses. Excellent illustrations, colored photos, appendix, glossary, and index.
Rost, Thomas L., et al. *Plant Biology.* New York: Wadsworth, 1998. Vital botanical information on all aspects of plant biology plus genetics. Excellent photographs and illustrations, summaries, questions, further readings, glossary, and index.
Simpson, Beryl Brintnall, and Molly Conner Ogorzaly. *Economic Botany.* 3d ed. Boston, New York: McGraw Hill, 2001. Good account of the most important crop plants of the world especially their genetic and agricultural diversities. Useful illustrations, photographs, glossary, additional readings, and index.
Starr, Cecie. *Biology: Concepts and Applications.* 5th ed. United States, Canada: Brooks/Cole, 2003. Current detailed biological information including genetics and evolution. Contains excellent photos, illustrations, quizzes, appendix, glossary, and index.

Web Sites of Interest

Food and Agriculture Organization of the United Nations. Biotechnology in Food and Agriculture. http://www.fao.org. Addresses the role of biotechnology in worldwide food production.
National Academic of Sciences. Transgenic Plants and World Agriculture. http://www.nap.edu/html/transgenic. An online, downloadable pamphlet, published in July of 2000 by a consortium of leading research societies around the world, assesses the need to modify crops genetically in order to feed the increasing world population and then discusses examples of the technology, safety, effect on the environment, funding sources, and intellectual property issues.

Genetic Engineering: Historical Development

Field of study: Genetic engineering and biotechnology; History of genetics
Significance: *Genetic engineering, or biotechnology, is the use of biology, genetics, and biochemistry to manipulate genes and genetic materials in a highly controlled fashion. It has led to major advancements in the understanding of the molecular*

organization, function, and manipulation of genes. The methods have been used to identify causes and solutions to many different human genetic diseases and have led to the development of many new medicines, vaccines, plants, foods, animals, and environmental cleanup techniques.

Key terms
CLONE: a group of genetically identical cells
PLASMIDS: small rings of DNA found naturally in bacteria and some other organisms
RECOMBINANT DNA: a DNA molecule made up of two or more sequences derived from different sources

Foundations of Genetic Engineering
Microbial genetics, which emerged in the mid-1940's, was based upon the principles of heredity that were originally discovered by Gregor Mendel in the middle of the nineteenth century and the resulting elucidation of the principles of inheritance and genetic mapping during the first forty years of the twentieth century. Between the mid-1940's and the early 1950's, the role of DNA as genetic material became firmly established, and great advances occurred in understanding the mechanisms of gene transfer between bacteria. A broad base of knowledge accumulated from which later developments in genetic engineering would emerge. The discovery of the structure of DNA by James Watson and Francis Crick in 1953 provided the stimulus for the development of genetics at the molecular level, and, for the next few years, a period of intense activity and excitement evolved as the main features of the gene and its expression were determined. This work culminated with the establishment of the complete genetic code in 1966, which set the stage for advancements in genetic engineering.

Initially, the term "genetic engineering" included any of a wide range of techniques for the manipulation or artificial modification of organisms through the processes of heredity and reproduction, including artificial selection, control of sex type through sperm selection, extrauterine development of an embryo, and development of whole organisms from cultured cells. However, during the early 1970's, the term came to be used to denote the nar-

rower field of molecular genetics, involving the manipulation, modification, synthesis, and artificial replication of DNA in order to modify the characteristics of an individual organism or a population of organisms.

The Development of Genetic Engineering
Molecular genetics originated during the late 1960's and early 1970's in experiments with bacteria, viruses, and free-floating rings of DNA found in bacteria known as plasmids. In 1967, the enzyme DNA ligase was isolated. This enzyme can join two strands of DNA together, acting like a molecular glue. It is the prerequisite for the construction of recombinant DNA molecules, which are DNA molecules that are made up of sequences not normally joined together in nature.

The next major step in the development of genetic engineering came in 1970, when researchers discovered that bacteria make special enzymes called restriction endonucleases, more commonly known as restriction enzymes. Restriction enzymes recognize particular sequences of nucleotides arranged in a specific order and cut the DNA only at those specific sites, like a pair of molecular scissors. Whenever a particular restriction enzyme or set of restriction enzymes is used on DNA from the same source, the DNA is cut into the same number of pieces of the same length and composition. With a molecular tool kit that included isolated enzymes of molecular glue (ligase) and molecular scissors (restriction enzymes), it became possible to remove a piece of DNA from one organism's chromosome and insert it into another organism's chromosome in order to produce new combinations of genes (recombinant DNA) that may not exist in nature. For example, a bacterial gene could be inserted into a plant, or a human gene could be inserted into a bacterium.

The first recombinant DNA molecules were generated by Paul Berg at Stanford University in 1971, and the methodology was extended in 1973 when DNA fragments were joined to *Escherichia coli* (*E. coli*) plasmids. These recombinant molecules could replicate when introduced into *E. coli* cells, and a colony of identical cells, or clones, could be grown on agar plates.

This development marked the beginning of the technology that has come to be known as gene cloning, and the discoveries of 1972 and 1973 triggered what became known as "the new genetics." The use of the new technology spread very quickly, and a sense of urgency and excitement prevailed. However, because of rising concerns about the morality of manipulating the genetic material of living organisms, as well as the fear that potentially harmful organisms might accidentally be produced, U.S. biol-

ogists called for a moratorium on recombinant DNA experiments in 1974, and the National Institutes of Health (NIH) issued safety guidelines in 1976 to control laboratory procedures for gene manipulation.

In 1977, the pioneer genetic engineering company Genentech produced the human brain hormone somatostatin, and, in 1978, Genentech produced human insulin in *E. coli* by the plasmid method of recombinant DNA. Human insulin was the first genetically engi-

The Asilomar Conference

Rising concerns related to safety and ethical issues surrounding experiments involving recombinant DNA technology led the National Institutes of Health (NIH) and the National Institute of Medicine (NIM) to appoint the Recombinant DNA Advisory Committee (RAC) to study the matter in 1973. RAC consisted of twelve experts from the areas of molecular biology, genetics, virology, and microbiology. Not only was there adverse public opinion in reaction to recombinant DNA (recombinant DNA) experiments, but many specialists in the field of genetic engineering were beginning to doubt their own ability to make important decisions that could impact society.

In February, 1975, the Asilomar Conference was convened under the direction of the NIH in Pacific Grove, California, to address the relevant issues. A total of 140 prominent international researchers and academicians, including Dr. Phillip Sharp, Nobel Laureate Professor at the Massachusetts Institute of Technology's Center for Cancer Research, met to discuss their opinions about recombinant DNA experiments.

Some of the issues debated at the Asilomar Conference included whether or not genetically altered microorganisms that posed a health hazard to humans and other living things might escape from lab facilities, how different genetically tailored recombinant DNA organisms should be classified, and what guidelines should be established to regulate recombinant DNA technology. The scientists concluded that only "safe" bacteria and plasmids that could not escape from the laboratory should be developed. They called for a moratorium on recombinant DNA experiments and demanded that the federal government establish guidelines regulating these experi-

ments. Appropriate safeguards on both physical and biological contaminant procedures would have to be in effect before recombinant DNA experiments continued. Within a year, the NIH had developed guidelines based upon the recommendations made at the Asilomar Conference.

Many positive outcomes resulted from the Asilomar Conference. Scientists demonstrated to the public their genuine concern for the development of safe scientific technology. It marked the first time in history that scientists themselves halted scientific research until the potential hazards could be properly assessed. It also became clear that for future meetings on recombinant DNA technology it would be wise to include scientists with training in infectious diseases, epidemiology, and public health, as well as people from other disciplines, in order to establish a more complete picture of the potential problems and solutions. As a result, a variety of scientists and nonscientists became part of national and local review boards on biotechnology.

Conferences that followed focused on "worst case scenarios" of recombinant DNA experiments. For the first time, debate of scientific issues spread beyond the scientific community to include the general public. Broad social, ethical, environmental, and ecological issues became part of conference agendas and discussions. The RAC membership was changed to include experts in epidemiology, infectious diseases, botany, tissue culture, and plant pathology, as well as nonscientists. NIH guidelines for federally funded research involving recombinant DNA molecules were published on June 23, 1976. As recombinant DNA research continued to progress, appropriate modifications to the NIH guidelines were made.

—*Alvin K. Benson*

neered product to be approved for human use. By 1979, small quantities of human somatostatin, insulin, and interferon were being produced from bacteria by using recombinant DNA methods. Because such research was proven to be safe, the NIH gradually relaxed the guidelines on gene splicing between 1978 and 1982. The 1978 Nobel Prize in Physiology or Medicine was shared by Hamilton O. Smith, the discoverer of restriction enzymes, and Daniel Nathans and Werner Arber, the first people to use these enzymes to analyze the genetic material of a virus.

By the early 1980's, genetic engineering techniques could be used to produce some biomolecules on a large scale. In December, 1980, the first genetically engineered product was used in medical practice when a diabetic patient was injected with human insulin generated in bacteria; in 1982, the Food and Drug Administration (FDA) approved the general use of insulin produced from bacteria by recombinant DNA procedures for the treatment of people with diabetes. During the same time period, genetically engineered interferon was tested against more than ten different cancers. Methods for adding genes to higher organisms were also developed in the early 1980's, and genetic researchers succeeded in inserting a human growth hormone gene into mice, which resulted in the mice growing to twice their normal size. By 1982, geneticists had proven that genes can be transferred between plant species to improve nutritional quality, growth, and resistance to disease.

In 1985, experimental guidelines were approved by the NIH for treating hereditary defects in humans by using transplanted genes; the more efficient polymerase chain reaction (PCR) cloning procedure for genes, which produces two double helixes in vitro that are identical in composition to the original DNA sample, was also developed. The following year, the first patent for a plant produced by genetic engineering, a variety of corn with increased nutritional value, was granted by the U.S. Patent and Trademark Office. In 1987, a committee of the National Academy of Sciences concluded that no serious environmental hazards were posed by transferring genes between species of organisms, and this action was followed in 1988 by the U.S. Patent and Trademark Office issuing its first patent for a genetically engineered higher animal, a mouse that was developed for use in cancer research.

Impact and Applications

The application of genetic engineering to gene therapy (the science of replacing defective genes with sound genes to prevent disease) took off in 1990. On September 14 of that year, genetically engineered cells were infused into a four-year-old girl to treat her adenosine deaminase (ADA) deficiency, an inherited, life-threatening immune deficiency called severe combined immunodeficiency disorder (SCID). In January, 1991, gene therapy was used to treat skin cancer in two patients. In 1992, small plants were genetically engineered to produce small amounts of a biodegradable plastic, and other plants were manufactured to produce antibodies for use in medicines.

By the end of 1995, mutant genes responsible for common diseases, including forms of schizophrenia, Alzheimer's disease, breast cancer, and prostate cancer, were mapped, and experimental treatments were developed for either replacing the defective genes with working copies or adding genes that allow the cells to fight the disease. In February, 1997, a lamb named Dolly was cloned from the DNA of an adult sheep's mammary gland cell; it was the first time scientists successfully cloned a fully developed mammal. By the end of 1997, approximately fifty genetically engineered products were being sold commercially, including human insulin, human growth hormone, alpha interferon, hepatitis B vaccine, and tissue plasminogen activators for treating heart attacks. In 1998, strong emphasis was placed on research involving gene therapy solutions for specific defects that cause cancer, as well as on a genetically engineered hormone that can help people with damaged hearts grow their own bypass vessels to carry blood around blockages.

In spite of the many successes and optimism that prevailed for many years, there have also been some serious setbacks. In 1999 a healthy eighteen-year-old participating in a gene therapy clinical trial at the University of Pennsylvania died unexpectedly, casting doubt on the safety of some types of gene therapy. In another set of clinical trials in France in 2002, involving the treatment of children with SCID, two of the children developed leukemia, raising doubts about the safety of yet another gene therapy protocol. As a result of these events, gene therapy trials of many types were put on hold and extensive discussions and investigations ensued. Still, scientists are hopeful that these kinds of obstacles can be overcome, leading to much greater availability of cures for genetic diseases.

—*Alvin K. Benson*

See also: Animal Cloning; Biofertilizers; Biological Weapons; Biopesticides; Biopharmaceuticals; Cloning; Cloning: Ethical Issues; Cloning Vectors; DNA Replication; DNA Sequencing Technology; Gene Therapy; Gene Therapy: Ethical and Economic Issues; Genetic Engineering; Genetic Engineering: Agricultural Applications; Genetic Engineering: Industrial Applications; Genetic Engineering: Medical Applications; Genetic Engineering: Risks; Genetic Engineering: Social and Ethical Issues; Genetically Modified (GM) Foods; High-Yield Crops; Knockout Genetics and Knockout Mice; Polymerase Chain Reaction; Restriction Enzymes; Reverse Transcriptase; Shotgun Cloning; Synthetic Genes; Transgenic Organisms; Xenotransplants.

Further Reading

Fredrickson, Donald S. *The Recombinant DNA Controversy, a Memoir: Science, Politics, and the Public Interest, 1974-1981.* Washington, D.C.: ASM Press, 2001. An overview of the initial concerns about potential hazards of recombinant DNA cloning.

Grace, Eric S. *Biotechnology Unzipped: Promises and Reality.* Washington, D.C.: National Academy Press, 1997. Provides a nontechnical history and explanation of biotechnology for general readers.

Judson, Horace Freeland. *The Eighth Day of Creation.* Rev. ed. Cold Harbor Spring, N.Y.: Cold Spring Harbor Laboratory Press, 1997. A noted and fascinating history of molecular biology that details the deciphering of the genetic code.

Portugal, Franklin H., and Jack S. Cohn. *A Century of DNA: A History of the Discovery of the Structure and Function of the Genetic Substance.* Cambridge, Mass.: MIT Press, 1977. Provides a comprehensive historical background and identifies many of the scientists who worked to solve the genetic code.

Shannon, Thomas A., ed. *Genetic Engineering: A Documentary History.* Westport, Conn.: Greenwood Press, 1999. A variety of scientific, social, and ethical perspectives on genetic engineering.

Web Site of Interest

National Health Museum, Biotech Chronicles. http://www.accessexcellence.org/ab/bc. Site discusses the history of biotechnology and includes a time line, from 6000 B.C.E. to the present, with key figures and links.

Genetic Engineering: Industrial Applications

Field of study: Genetic engineering and biotechnology

Significance: *Industrial applications of genetic engineering include the production of new and better fuels, medicines, products to clean up existing pollution, and tools for recovering natural resources. Associated processes may maximize the use and production of renewable resources and biodegradable materials, while minimizing the generation of pollutants during product manufacture and use.*

Key terms

BIOMASS: any material formed either directly or indirectly by photosynthesis, including plants, trees, crops, garbage, crop residue, and animal waste

BIOREMEDIATION: biologic treatment methods to clean up contaminated water and soils

CLONING VECTOR: a DNA molecule that maintains and replicates a foreign piece of DNA in a cell type of choice, typically the bacterium *Escherichia coli*

GENETIC TRANSFORMATION: the transfer of extracellular DNA among and between species

NANOTECHNOLOGY: development and use of devices that have a size of only a few billionths of a meter

PLASMIDS: small rings of DNA found naturally in bacteria and some other organisms, used as cloning vectors

RECOMBINANT DNA: a DNA molecule made up of sequences that are not normally joined together

Foundations in Medical Advancements

Since the 1970's, numerous industrial processes have involved applications of genetic engineering and biotechnology, ranging from the production of new medicines and foods to the manufacture of new materials for cleaning up the environment and enhancing natural resource recovery. With these applications, a primary focus has been the development of industrial processes that reduce or eliminate the production of waste products and consume low amounts of energy and nonrenewable resources. The chemical, plastic, paper, textile, food, farming, and pharmaceutical industries are positively impacted by biotechnology.

The dawn of the age of genetic engineering was 1971, when Herbert Boyer and Stanley Cohen successfully spliced a toad gene between two recombined ends of bacterial DNA. After further experimentation and resulting successes, Boyer and Robert Swanson in 1976 formed Genentech, a company devoted to the development and promotion of biotechnology and genetic engineering applications. In 1978, Boyer discovered a synthetic version of the human insulin gene and inserted it into *Escherichia coli* (*E. coli*) bacteria. The *E. coli* served as cloning vectors to maintain and replicate large amounts of human insulin. This application of recombinant DNA technology to produce human insulin for diabetics was a foundation for the future of industrial applications of genetic engineering and biotechnology. The Eli Lilly company began manufacturing large quantities of human insulin by vector cloning in 1982. Growth hormones for children and antibodies for cancer patients were soon being similarly cloned in bacteria. The pharmaceutical industry was revolutionized.

Later applications of genetic engineering to the medical industry include the production of new vaccines for use in fighting a variety of diseases. One approach is to use genetically altered viruses to insert manufactured vaccines directly into the cells of diseased animals and humans. In other cases, antigens that invoke immunity to certain diseases are being produced from genetically modified viruses, bacteria, fungi, and other disease-causing microorganisms.

Cleaning up Waste

Genetic engineering methods are being employed in myriad applications to help clean up waste and pollution worldwide. The idea had its beginning in 1972, when Ananda Chakrabarty, a researcher at General Electric (who would later join the college of medicine at the University of Illinois at Chicago), applied for a patent on a genetically modified bacterium that could partially degrade crude oil. Other scientists quickly recognized that toxic wastes might be cleaned up by pollution-eating microorganisms. After a financial downturn for a number of years, a resurgence in bioremediation technology occurred in the late 1980's and early 1990's, when genetically engineered bacteria were produced that could accelerate the breakdown of oil, as well as a diversity of unnatural and synthetic compounds, such as plastics, chlorinated insecticides, herbicides, and fungicides. In 1987 and 1988, bacterial plasmid transfer was used to degrade a variety of hydrocarbons found in crude oil. In the 1990's, naturally occurring and genetically altered bacteria were employed to degrade crude oil spills, such as the major spill that occurred in Alaska's Prince William Sound after the *Exxon Valdez* accident.

Some genetically altered bacteria have been designed to concentrate or transform toxic metals into less toxic or nontoxic forms. In 1998, a gene from *E. coli* was successfully transferred into the bacterium *D. radiodurans*, allowing this microbe to resist high levels of radioactivity and convert toxic mercury II into less toxic elemental mercury. Other altered microbial genes have been added to this bacterium, allowing it to metabolize the toxic organic chemical toluene, a carcinogenic constituent of gasoline. Genetically altered plants have

A genetically engineering enzyme developed from a hybrid poplar tree, shown here by researchers Arun Goyal (left) and Neil Nelson, could reduce the cost of manufacturing paper by replacing chlorine used for pulp bleaching, and might also become a component of animal feed and a means of decomposing harmful toxic pollution. (AP/Wide World Photos)

been produced that absorb toxic metals, including lead, arsenic, and mercury, from polluted soils and water. At Michigan State University, naturally occurring bacteria have been combined with genetically modified bacteria to degrade polychlorinated biphenyls (PCBs). A genetically altered fungus that helps clean up toxic substances discharged when paper is manufactured also produces methane as a by-product, which can be used as a fuel.

Biomass and Materials Science

Genetically altered microorganisms can transform animal and plant wastes into materials usable by humans. Bioengineered bacteria and fungi are being developed to convert biomass wastes, such as sewage solid wastes (paper, garbage), agricultural wastes (seeds, hulls, corn cobs), food industry by-products (cartilage, bones, whey), and products of biomass, such as sugars, starch, and cellulose, into useful products like ethanol, hydrogen gas, and methane.

Commercial amounts of methane are being generated from animal manure at cattle, poultry, and swine feed lots, sewage treatment plants, and landfills. Biofuels will be cleaner and generate less waste than fossil fuels. In a different application involving fuel technology, genetically modified microbes are being used to reduce the pollution associated with fossil fuels by eating the sulfur content from these fuels.

In applications involving the generation of new materials, a gene generated in genetically modified cotton can produce a polyester-like substance that has the texture of cotton, is even warmer, and is biodegradable. Other genetically engineered biopolymers are being produced to replace synthetic fibers and fabrics. Polyhydroxybutyrate, a feedstock used in producing biodegradable plastics, is being manufactured from genetically modified plants and microbes. Natural protein polymers, very similar to spider silk and the adhesives generated by barnacles, are being produced from the fermentation of genetically engineered microbes. Sugars produced by genetically altered field corn are being converted into a biodegradable polyester-type material for use in manufacturing packaging materials, clothing, and bedding products. Genetically tailored yeasts can produce a variety of plastics. Such biotechnological advancements will help eliminate the prevalent use of petroleum-based chemicals that has been necessary in the creation of plastics and polyesters.

The fields of biotechnology and nanotechnology are being merged in some materials science applications. Genetic codes discovered in microorganisms can be used as codes for nanostructures, such as task-specific silicon chips and microtransistors. Nanotech production of bioactive ceramics may provide new ways to purify water, since bacteria and viruses stick to these ceramic fibers. Recombinant DNA technology combined with nanotechnology provides the promise for the production of a variety of commercially useful polymers. Carbon nanotubes possessing great tensile strength may be used as computer switches and hydrogen energy storage devices for vehicles. When these nanotubes are coated with reaction specific biocatalysts, many other specialized applications are apparent. In the future, DNA fragments themselves may be used as electronic switching devices.

Natural Resource Recovery

Bioengineered microbes are being developed to extract and purify metals from mined ores and from seawater. The microbes obtain energy by oxidizing metals, which then come out of solution. Chemolithotrophic bacteria, such as *Bacillus cereuss*, are energized when they oxidize nickel, cobalt, and gold. They may be used to filter out and concentrate precious metals from seawater. Iron and sulfur-oxidizing bacteria can also concentrate and release precious metals from seawater. Genetically modified thermophilic bacteria are being produced to extract precious metals from sands. Some genetically altered microorganisms can withstand extreme environments of high salinity, acidity, heavy metals, temperature, and/or pressure, like those that exist around hydrothermal vents where precious minerals are present near the bottom of the ocean.

Genetically engineered strains of the bacteria *Pseudomonas* and *Bacillus* are being produced that can extract oil from untapped reservoirs and store it rather than digest it. These bacteria can be extracted and processed to recover the oil. Other strains are being developed to absorb oil from the vast supplies of oil shale in North America. The process would involve drilling into the oil shale and breaking it into pieces with chemical explosives. A solution of the bioengineered microbes would then be injected through a well into the rock fragments, where they would grow and absorb the oil. The solution would be pumped back to the surface through another well and the bacteria processed to remove the oil. Since this process would eliminate the need for large, open-pit oil shale mines, as well as the need to store oil shale at the surface, the negative environmental impact of oil recovery from shale would be greatly reduced.

—*Alvin K. Benson*

See also: Animal Cloning; Biofertilizers; Biological Weapons; Biopesticides; Biopharmaceuticals; Cloning; Cloning: Ethical Issues; Cloning Vectors; DNA Replication; DNA Sequencing Technology; Gene Therapy; Gene Therapy: Ethical and Economic Issues; Genetic Engineering; Genetic Engineering: Agricultural Applications; Genetic Engineering: Historical Development; Genetic Engineering: Medical Applications; Genetic Engineering: Risks; Genetic Engineering: Social and Ethical Issues; Genetically Modified (GM) Foods; High-Yield Crops; Knockout Genetics and Knockout Mice; Polymerase Chain Reaction; Restriction Enzymes; Reverse Transcriptase; Shotgun Cloning; Synthetic Genes; Transgenic Organisms; Xenotransplants.

Further Reading

Evans, Gareth M. *Environmental Biotechnology: Theory and Application.* Hoboken, N.J.: Wiley, 2003. Describes basic principles and methods involved in the remediation of contaminated soils and groundwater through applications of biotechnology and natural processes.

Krimsky, Sheldon. *Biotechnics and Society: The Rise of Industrial Genetics.* New York: Praeger, 1991. Includes an overview of applied biotechnology to industrial processes and a description of some industrial products produced by applications of genetic engineering; also addresses concerns over the environmental release of genetically engineered organisms.

Nicholl, Desmond S. T. *An Introduction to Genetic Engineering.* 2d ed. New York: Cambridge University Press, 2002. A basic introduction to the ideas of genetic engineering, including a description of many of the technological applications.

Sofer, William. *Introduction to Genetic Engineering.* Boston: Butterworth-Heinemann, 1991. Contains the general principles of molecular biology and molecular cloning and how genetic engineering pieces together genes from different organisms to produce new products.

Genetic Engineering: Medical Applications

Fields of study: Genetic engineering and biotechnology; Human genetics and social issues

Significance: *Genetic engineering has produced a wide range of medical applications, including recombinant DNA drugs, transgenic animals that produce pharmaceutically useful proteins, methods for the diagnosis of disease, and gene therapy to introduce a functional gene to replace a defective one.*

Key terms

CLONE: in recombinant DNA technology, a piece of DNA into which a gene of interest has been inserted to obtain large amounts of that gene

GENE TARGETING: the process of introducing a gene that replaces a resident gene in the genome

GENE THERAPY: any procedure to alleviate or treat the symptoms of a disease or condition by genetically altering the cells of the patient

GERM-LINE GENE THERAPY: a genetic change in gametes or fertilized ova so all cells in the organism will have the change and the change will be passed on to offspring

KNOCKOUT: the inactivation of a specific gene within a cell (or whole organism, as in the case of knockout mice), to determine the effects of loss of function of that gene

SOMATIC GENE THERAPY: a genetic change in a specific somatic tissue of an organism, which will not be passed on to offspring

STEM CELL: a an undifferentiated cell that retains the ability to give rise to other, more specialized cells

TRANSGENIC ANIMAL: an animal in which introduced foreign DNA is stably incorporated into the germ line

Multiple Applications: Drug Production

Genetic engineering, the manipulation of DNA to obtain a large amount of a specific gene, has produced numerous medical applications. As a result of the completion in 2003 of the Human Genome Project—the determination of the DNA sequences of all the chromosomes in humans—genetic engineering will continue at an accelerated pace and result in even more important medical applications.

Recombinant DNA technology can be used to mass-produce protein-based drugs. The gene for the protein of interest is cloned and expressed in bacteria. For example, insulin needed for people with Type I diabetes mellitus was isolated from the pancreases of cattle or pigs in slaughterhouses, an expensive and far from ideal process. There are some small chemical differences between human and cow and pig insulin. About 5 percent of those receiving cow insulin have an allergic reaction to it and therefore need insulin from other animals or human cadavers. In 1982, the human gene for insulin was isolated, and a transgenic form called Humulin was successfully produced using *Escherichia coli* bacteria grown in a controlled environment by pharmaceutical companies.

Many other protein-based drugs are produced in bacteria using recombinant DNA technology. Among these are human growth hormone, to treat those deficient in the hormone; factor VIII, to promote blood clotting in hemophiliacs; tissue plasminogen activator, to dissolve blood clots in heart attack and stroke victims; renin inhibitor, to lower blood pressure; fertility hormones, to treat infertility; epidermal growth factor, to increase the rate of healing in burn victims; interleukin-2, to treat kidney cancer; and interferons, to treat certain leukemias and hepatitis.

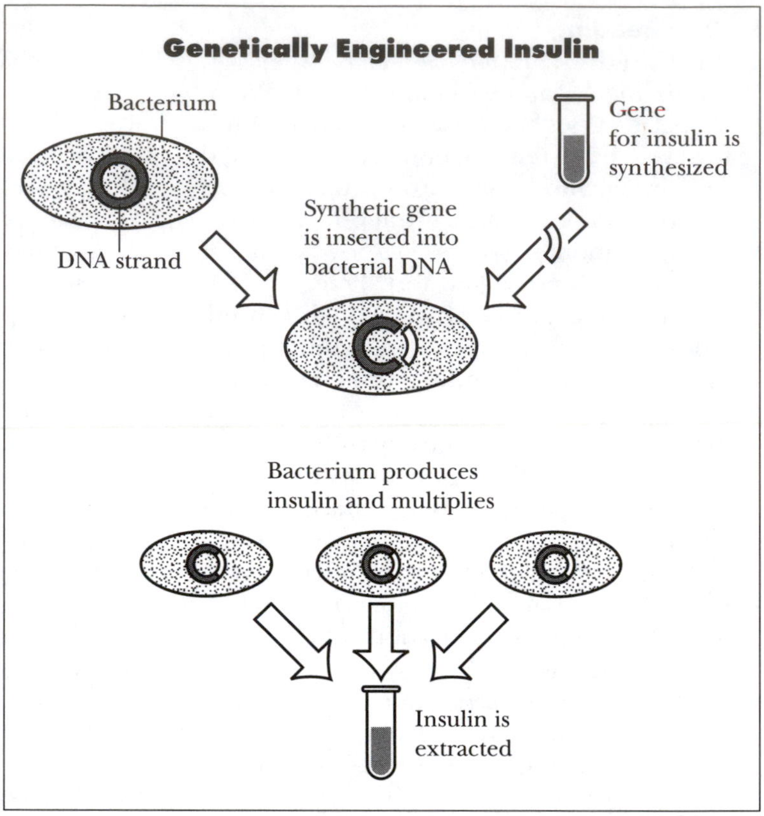

Transgenic Pharming

Sometimes a protein from a higher organism that is expressed in bacteria does not function properly because bacteria cannot perform certain protein modifications. In such cases, the protein can be produced in a higher organism. In transgenic pharming, a gene that codes for a pharmaceutically useful protein is introduced into an animal such as a cow, pig, or sheep. For example, a transcriptional promoter from a sheep gene that is expressed in sheep's milk is spliced to the gene of interest, such as for alpha-1-antitrypsin, ATT, a glycoprotein (a protein modified with sugar groups) in blood serum that helps the microscopic air sacs of the lungs function properly. People who lack ATT are at risk for developing emphysema. This sheep promoter and *ATT* gene are injected into the nuclei of fertilized sheep ova that are implanted in surrogate mother sheep. The offspring are examined, and if the procedure is successful, a few of the female lambs will produce the ATT protein in their milk. Once a transgenic animal is created that expresses the *ATT* gene, transgenic animals expressing the

Genetically Engineered Insulin

Bacterium

DNA strand

Gene for insulin is synthesized

Synthetic gene is inserted into bacterial DNA

Bacterium produces insulin and multiplies

Insulin is extracted

Genetic engineering is being used to synthesize large quantities of drugs and hormones such as insulin for therapeutic use. (Hans & Cassidy, Inc.)

gene can be bred to each other to produce a whole flock of sheep making ATT—an easier way to obtain ATT than isolating it from donated human blood.

Vaccines

Recombinant DNA methods can be used to produce DNA vaccines that are safer than vaccines made from live viruses. Edible vaccines have also been created by introducing into plants genes that will cause a specific immune response. For example, a vaccine for hepatitis has been made in bananas. The idea is that by eating the fruit, individuals will be vaccinated.

Diagnosis

Recombinant DNA methods are used in the diagnosis, as well as treatment, of diseases. Oligonucleotide DNA sequences specific for, and which will only bind to, a particular mutation are used to show if that particular mutation is present. Also, DNA microarrays are important for gene expression profiling, to aid in cancer diagnosis. For example, oligonucleotides representing portions of many different human genes can be fixed to special "chips" in an array. Messenger RNAs from a cancer patient are bound to the array to show which genes are expressed in that cancer. A certain subtype of cancers express a certain group of genes. This knowledge can be used to design specific treatment regimens for each subtype of cancer.

Mice and other animals are used as models for human diseases. Through recombinant DNA technology, a specific gene is "knocked out" (inactivated) to study the effect of the loss of that gene. Mice models are particularly useful in the study of diseases such as diabetes, Parkinson's disease, and severe combined immunodeficiency disorder (SCID).

Gene Therapy

In gene therapy, a cloned functional copy of a gene is introduced into a person to compensate for the person's defective copy. Due to ethical concerns, germ-line gene therapy is not being conducted. Many geneticists and bioethicists oppose germ-line therapy because any negative consequences of the therapy would be passed on to future generations. Therefore, germ-line therapy must wait until scientists, policymakers, and legislators are more confident of consistently positive outcomes. In general, there is support for somatic gene therapy, where the somatic tissue of an individual is modified to produce the correct gene product.

Gene therapy has been attempted for a number of diseases, including SCID and hemophilia. Gene therapy trials have been under close scrutiny, however, during clinical trials

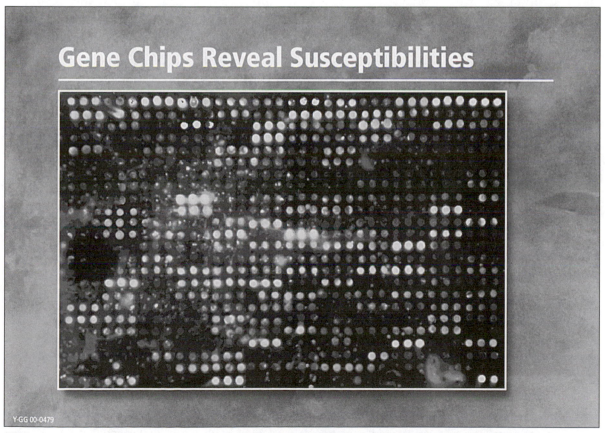

DNA microarrays such as the one above can show which genes are expressed in a cancer, knowledge that can be used to design specific treatment regimens for each subtype of cancer. (Mitch Doktycz, Life Sciences Division, Oak Ridge National Laboratory; U.S. Department of Energy Human Genome Program, http://www.ornl.gov/hgmis)

for gene therapy, one young man died in 1999 and two cases of leukemia in children were detected. These trials used inactivated viruses as vectors, which may have played a role in the death and leukemia cases. Efforts are therefore focusing on the development of DNA delivery systems that do not use viruses.

Future Prospects

In the future, stem cells may be used to generate tissues to replace defective tissues. Catalytic RNAs (ribozymes) may be used to repair genetically defective messenger RNAs. RNA-mediated interference may be used to partially inactivate, rather than knock out, genes to determine the genes' functions in the cell. With the completion of the DNA sequence of the human genome, more genes will inevitably be identified and their functions determined, leading to many more applications to medical diagnosis and therapy.

—*Susan J. Karcher*

See also: Animal Cloning; Biofertilizers; Biological Weapons; Biopesticides; Biopharmaceuticals; Cloning; Cloning: Ethical Issues; Cloning Vectors; DNA Replication; DNA Sequencing Technology; Gene Therapy; Gene Therapy: Ethical and Economic Issues; Genetic Engineering; Genetic Engineering: Agricultural Applications; Genetic Engineering: Historical Development; Genetic Engineering: Industrial Applications; Genetic Engineering: Risks; Genetic Engineering: Social and Ethical Issues; Genetically Modified (GM) Foods; High-Yield Crops; Knockout Genetics and Knockout Mice; Polymerase Chain Reaction; Restriction Enzymes; Reverse Transcriptase; Shotgun Cloning; Synthetic Genes; Transgenic Organisms; Xenotransplants.

Further Reading

Botstein, David, and Neil Risch. "Discovering Genotypes Underlying Human Phenotypes: Past Successes for Mendelian Disease, Future Approaches for Complex Disease." *Nature Genetics*, supp. 33 (March 2003): 228-237. Discusses how human genome sequence analysis is helping to identify complex diseases.

Epstein, Richard J. *Human Molecular Biology; An Introduction to the Molecular Basis of Health and Disease*. Cambridge, England: Cambridge University Press, 2003. Focuses on molecular biology and clinical information about human diseases. Includes chapters on genetic engineering, gene knockouts, and gene therapy. Illustrations, color photographs.

Langer, Robert. "Delivering Genes." *Scientific American* 288 (April, 2003): 56. Discusses alternatives to viruses for introducing genes into cells.

Langridge, William H. R. "Edible Vaccines." *Scientific American* 283 (September, 2000): 66-71. Describes the making of vaccines in plants.

Lewis, Ricki. *Human Genetics: Concepts and Applications*. 5th ed. Boston: McGraw-Hill, 2003. A well-written introductory text. Includes chapters on genetically modified organisms, gene therapy, and the Human Genome Project. Illustrations, color photos, problems, glossary, index. Lists links to Web sites.

Service, Robert F. "Recruiting Genes, Proteins for a Revolution in Diagnostics." *Science* 300 (April 11, 2003): 236-239. Overview of the use of DNA microarrays to diagnose diseases.

Strachan, Tom, and Andrew P. Read. *Human Molecular Genetics*. New York: Wiley-Liss, 1999. An advanced text with a chapter on gene therapy and genetic-based therapeutic approaches to treating diseases. Illustrations, photos, glossary, index.

Web Sites of Interest

American Medical Association. http://ama-assn.org. The AMA includes information on genetic diseases and disorders as well as links to affiliated professional organizations and other resources.

Centers for Disease Control, Office of Genomics and Disease Prevention. http://www.cdc.gov/genomics/default.htm. Offers information on the genetic discoveries and prevention of diseases in humans. Includes links to related resources.

Dolan DNA Learning Center, Your Genes Your Health. http://www.ygyh.org. Sponsored by the Cold Spring Harbor Laboratory, this site, a component of the DNA Interactive Web site, offers information on more than a dozen inherited diseases and syndromes.

National Center for Biotechnology Information. Online Mendelian Inheritance in Man. http://www.ncbi.nlm.nih.gov/Omim. A catalog of human genes and genetic disorders for scientists, offering maps of genes and diseases, statistical summaries, and links to similar sites devoted to medical literature and biotechnology.

Genetic Engineering: Risks

Fields of study: Bioethics; Genetic engineering and biotechnology

Significance: *The application of biotechnology, specifically genetic engineering, creates real and foreseeable risks to humans and to the environment. Furthermore, like any new technology, it may cause unforeseen problems. Predicting the occurrence and severity of both anticipated and unexpected problems resulting from biotechnology is a subject of much debate in the scientific community.*

Key terms

FITNESS: the probability of a particular genotype surviving to maturity and reproducing

GENOME: the genetic content of a single set of chromosomes

GENOTYPE: the genetic makeup of an individual, referring to some or all of its specific genetic traits

SELECTION: a natural or artificial process that removes genotypes of lower fitness from the population and results in the inheritance of traits from surviving individuals

TRANSGENIC ORGANISM: an organism that has had its genome deliberately modified using genetic engineering techniques and that is usually capable of transmitting those changes to offspring

The Nature of Biotechnological Risks

Most of the potential risks of biotechnology center on the use of transgenic organisms. Potential hazards can result from the specific protein products of newly inserted or modified genes; interactions between existing, altered, and new protein products; the movement of transgenes into unintended organisms; or changes in the behavior, ecology, or fitness of transgenic organisms. It is not the process of removing, recombining, or inserting DNA that usually causes problems. Genetically modifying an organism using laboratory techniques creates a plant, animal, or microbe that has DNA and RNA that is fundamentally the same as that found in nature.

Risks to Human Health and Safety

The problem most likely to result from ingesting genetically modified (GM) foods is unexpected allergenicity. Certain foods such as milk or Brazil nuts contain allergenic proteins that, if placed into other foods using recombinant DNA technology, could cause the same allergic reactions as the food from which the allergenic protein originally came. Scientists and policymakers will, no doubt, guard against or severely restrict the movement of known allergens into the food supply. New or unknown allergens, however, could necessitate extensive testing of each GM food product prior to general public consumption. Safety testing will be especially important for proteins that have no known history of human consumption.

Unknown, nonfunctional genes that produce compounds harmful or toxic to humans and animals could become functional as a result of the random insertion of transgenes into an organism. Unlike traditional breeding methods, recombinant DNA technology provides scientists with the ability to introduce specific genes without extra genetic material. These methods, however, usually cannot control where the gene is inserted within the target genome. As a result, transgenes are randomly placed among all the genes that an organism possesses, and sometimes "insertional mutagenesis" occurs. This is the disruption of a previously functional gene by the newly inserted gene. This same process may also activate previously inactive genes residing in the target genome. Early testing of transgenic organisms would easily reveal those with acute toxicity problems; however, testing for problems caused by the long-term intake of new proteins is difficult.

Many human and animal disease organisms are becoming resistant to antibiotics. Some

Greenpeace has been active in protesting genetically engineered organisms, especially for use as food. Near Live Oak, California, one protester warns passersby of a rice "pharm" crop that has been engineered to produce human proteins for drug production. Environmentalists fear the effects such experimentation might have on the food supply and wild-type species. (AP/Wide World Photos)

scientists worry that biotechnology may accelerate that process. Recombinant DNA technologies usually require the use of antibiotic resistance genes as "reporter" genes in order to identify cells that have been genetically modified. Consequently, most transgenic plants contain antibiotic resistance genes that are actively expressed. Although unlikely, it is possible that resistance genes could be transferred from plants to bacteria or that the existence of plants carrying active antibiotic resistance genes could encourage the selection of antibiotic-resistant bacteria. As long as scientists continue using nat-

urally occurring antibiotic resistance genes that are already commonly found in native bacterial populations, there is little reason to believe that plants with these genes will affect the rate of bacteria becoming resistant to antibiotics.

Another possible problem associated with antibiotic resistance genes is the reduction or loss of antibiotic activity in individuals who are taking antibiotic medication while eating foods containing antibiotic resistance proteins. Would the antibiotic be rendered useless if transgenic foods were consumed? Scientists have found that this is not the case for the most commonly used resistance gene, NPTII (neomycin phosphotransferase II), which inactivates and provides resistance to kanamycin and neomycin. Studies have shown this protein to be completely safe to humans, to be broken down in the human gut, and to be present in the current food supply. Each person consumes, on average, more than one million kanamycin-resistant bacteria daily through the ingestion of fresh fruit and vegetables. These results are probably similar for other naturally occurring resistance genes of bacterial origin.

Risks to the Environment

If environmentally advantageous genes are added to transgenic crops, then those crops, or crop-weed hybrids, may become weeds, or their weediness may increase. For example, tolerance to high-salt environments is a useful and highly desirable trait for many food crops. The addition of transgenes for salt tolerance may allow crop-weed hybrids to displace naturally occurring salt-tolerant species in high-salt environments. Most crop plants are poor competitors in natural ecosystems and probably would not become weeds even with the addition of one or a few genes conferring some competitive advantage. Hybrids between crops and related weed species, however, can show increased weediness, and certain transgenes may also contribute to increased weediness.

Biotechnology may accelerate the development of difficult-to-control pests. Crops and domesticated animals are usually protected from important diseases and insect pests by specific host resistance genes. Genetic resistance is the most efficient, effective, and environmentally friendly means for controlling and preventing agricultural losses caused by pests. Such genes are bred into plants and animals by mating desirable genotypes to those that carry genes for resistance. This method is limited to those species that can interbreed. Biotechnology provides breeders with methods for moving resistance genes across species barriers, which was not possible prior to the 1980's. Bacteria and viruses, however, have been moving bits of DNA in a horizontal fashion (that is, across species and kingdom barriers) since the beginning of life. The widespread use of an effective, specific host resistance gene in domesticated species historically has led to adaptation in the pest population eventually making the resistance gene ineffective. Recombinant methods will likely accelerate the loss of resistance genes as compared with traditional methods because one resistance gene can be expressed simultaneously in many species, is often continuously expressed at high levels within the host, and will more likely be used over large areas because of the immediate economic benefits such a gene will bring to a grower or producer.

Hybrid plants carrying genes that increase fitness (through, for example, disease resistance or drought tolerance) may decrease the

Gene Flow from Crop Plants to Wild Relatives

Crop plants commonly exchange genes with related wild plants that are growing nearby, in a process known as gene flow. Pollen seems to be the most effective agent for gene flow, introducing genes of the parent plant to the recipient plant through fertilization of egg cells. Concern has arisen that genes engineered into crop plants, called transgenes, might spread to their nondomesticated relatives. As bioengineered varieties continue to be developed and as farmers grow the resulting transgenic plants on a commercial scale, the chances of transgenes escaping both to other crop plants and to nondomesticated, wild relatives will increase.

Agriculturally useful traits engineered into crop plants include resistance to herbicides, insects, and pathogens, and tolerance of harsh environmental conditions such as cold, drought, and high salinity. These traits not only give the crops a survival edge under appropriate conditions but also might do the same for nearby wild relatives that acquire the transgenes. As a result, farmers face the possibility that wild plants invigorated by transgenes coding for herbicide resistance could turn into "superweeds," requiring more expensive or more environmentally harmful herbicides.

Further, if transgenes permit a crop to be grown closer to locally rare, wild relatives because it can tolerate an environmental stress that it could not tolerate before, the previously isolated species might hybridize. If hybridization occurs repeatedly, the risk of extinction for the wild population increases.

Another fear is that the spread of transgenes could diminish the genetic diversity of agronomically important native plants. For example, in Mexico, which is located in the evolutionary cradle of corn, concerns about the spread of transgenes to ancient, native corn varieties, which conventional corn breeders value as genetic reservoirs, led the Mexican government to outlaw the planting of bioengineered corn in 1998.

In addition, wild plants that acquire transgenes for insecticidal properties could harm insects that the crop bioengineers had not targeted. For example, moth and butterfly species, whose larvae depend for food primarily on these wild plants, might be vulnerable if acquired transgenes endow their food plants with insect-killing abilities.

The potential for transgene flow from crops to wild relatives varies with the crop and the geographic location. Most cultivated plants spontaneously mate with one or more wild cousins somewhere in their agricultural distributions. In the United States, some of the major genetically engineered crops, including corn and soybeans, generally have no nearby, wild relatives. About twenty other U.S. crops (some already having transgenic varieties), however, are grown near nondomesticated kin. These crops include rice, sorghum, canola, strawberries, and turf grasses. The hazards from transgene flow to wild relatives, though, could prove lower than the risks of crop-to-crop gene flow, because of the prospect that transgenes for production of pharmaceuticals or other industrial chemicals could make their way into food crops.

—*Jane F. Hill*

native genetic diversity of a wild population through competitive or selection advantage. As new genes or genes from unrelated species are developed and put into domesticated species, engineered genes may move, by sexual outcrossing, into related wild populations. Gene flow from nontransgenic species into wild species has been taking place ever since crops were first domesticated, and there is little evidence that such gene flow has decreased genetic diversity. In most situations, transgene flow will likewise have little or no detrimental effects on the genetic diversity of wild populations; however, frequent migration of transgenes for greatly increased fitness could have a significant impact on rare native genes in the world's centers of diversity. A center of diversity harbors most of the natural genetic resources for a given crop and is a region in which wild relatives of a crop exist in nature. These centers are vital resources for plant breeders seeking to improve crop plants. The impact of new transgenes on such centers should be fully investigated before transgenic crops are grown near their own center of diversity.

Impact and Applications

The risks associated with genetically modified organisms have been both overstated and understated. Proponents of biotechnology have downplayed likely problems while opponents have exaggerated the risks of the unknown. As with any new technology, there will be unforeseen problems; however, as long as transgenic organisms are scientifically and objectively evaluated on a case-by-case basis prior to release or use, society should be able to avoid the obvious or most likely problems associated with biotechnology and benefit from its application.

—*Paul C. St. Amand*

See also: Animal Cloning; Biofertilizers; Biological Weapons; Biopesticides; Biopharmaceuticals; Cloning; Cloning: Ethical Issues; Cloning Vectors; DNA Replication; DNA Sequencing Technology; Gene Therapy; Gene Therapy: Ethical and Economic Issues; Genetic Engineering; Genetic Engineering: Agricultural Applications; Genetic Engineering: Historical Development; Genetic Engineering: Industrial Applications; Genetic Engineering: Medical Applications; Genetic Engineering: Social and Ethical Issues; Genetically Modified (GM) Foods; High-Yield Crops; Knockout Genetics and Knockout Mice; Polymerase Chain Reaction; Restriction Enzymes; Reverse Transcriptase; Shotgun Cloning; Synthetic Genes; Transgenic Organisms; Xenotransplants.

Further Reading

Engel, Karl-Heinz, et al. *Genetically Modified Foods: Safety Aspects.* Washington, D.C.: American Chemical Society, 1995. Details the policy and safety issues regarding food biotechnology.

Krimsky, Sheldon, et al., eds. *Agricultural Biotechnology and the Environment: Science, Policy, and Social Issues.* Urbana: University of Illinois Press, 1996. Covers biotechnology risks related to agriculture.

Nottingham, Stephen. *Genescapes: The Ecology of Genetic Engineering.* New York: Zed Books, 2002. Provides a framework for assessing the environmental impacts of genetically modified organisms and warns about the risks. Topics include microorganisms, transgenic crops, invasion, genetic pollution, impact on nontarget species, and the possibilities for engineered solutions.

Thomas, John, et al., eds. *Biotechnology and Safety Assessment.* 2d ed. Philadelphia: Taylor & Francis, 1999. Covers a wide range of topics related to safety in biotechnology.

Web Sites of Interest

Physicians and Scientists for Responsible Application of Science and Technology. http://www.psrast.org. Developed for the general reader, this site discusses the risks of genetically modified foods. Topics include a general introduction to the topic, articles about environmental and health risks, global implications, and suggested alternatives.

The Edmonds Institute. http://www.edmonds-institute.org. A site that offers "A Brief History of Biotechnology Risk Debates and Policies in the United States" and "Manual for Assessing Ecological and Human Health Effects of Genetically Engineered Organisms."

Genetic Engineering: Social and Ethical Issues

Fields of study: Bioethics; Genetic engineering and biotechnology; Human genetics and social issues

Significance: *New technologies for manipulating the genetic makeup of living organisms raise serious questions about the social desirability of controlling genes and the moral right of humans to redesign living beings.*

Key terms

BIODIVERSITY: the presence of a wide variety of forms of life in an environment

BIOTECHNOLOGY: the technological manipulation of living organisms; genetic engineering is the most common form of biotechnology

RECOMBINANT DNA: a new combination of genes spliced together on a single piece of DNA; recombinant DNA is the basis of genetic engineering technology

TRANSGENIC ORGANISM: a organism into which the DNA of another species has been inserted

Genetic Engineering as a Social and Ethical Problem

English author Mary Shelley's 1818 horror novel *Frankenstein,* about a scientist who succeeds in bringing a creature to life, expressed anxiety about the possibility of human control over the basic mysteries of existence. The novel's continuing popularity and the many films and other works based on it attest to deep-seated feelings that unrestrained science may violate essential principles of nature and religion and that human powers may grow to exceed human wisdom. With the rise of genetic engineering in the 1970's, many serious philosophers and social critics feared that the Frankenstein story was moving from the realm of science fiction into reality.

The basic blueprint of all living beings was found in 1953, when Francis Crick and James Watson discovered the structure of DNA. A little less than two decades later, in 1970, it became possible to conceive of redesigning this blueprint when Hamilton Smith and Daniel Nathans of The Johns Hopkins University discovered a class of "restriction" enzymes that could be used as scissors to cut DNA strands at specific locations. In 1973, two researchers in California, Stanley Cohen and Herbert Boyer, spliced recombinant DNA strands into bacteria that reproduced copies of the foreign DNA. This meant that it would be possible to combine genetic characteristics of different organisms. In 1976, Genentech in San Francisco, California, became the first corporation formed to develop genetic engineering techniques for commercial purposes.

By the 1990's, genetic engineering was being used on plants, animals, and humans. The Flavr Savr tomato, the first genetically modified (GM) food to be approved by the U.S. government, was developed when biotechnologists inserted a gene that delayed rotting in tomatoes. Transgenic animals (containing genes from humans and other animals) became commonplace in laboratories by the middle of the 1990's. The year 1990 saw the first successful use of genetic engineering on humans, when doctors used gene therapy to treat two girls suffering from an immunodeficiency disease. The long-felt discomfort over scientific manipulation of life, the suddenness of the development of the new technology, and the application of the technology to humans all combined to make many people worry about the social and ethical implications of genetic engineering. The most serious concerns were over genetic manipulation of humans, but some critics also pointed out possible problems with the genetic engineering of plants and animals.

Engineering of Plants and Animals

According to a Harris Poll survey conducted for the U.S. Office of Technology in the fall of 1968, a majority of Americans were not opposed to using recombinant DNA techniques to produce hybrid agricultural plants. Some social critics, such as Jeremy Rifkin, have argued that such ready acceptance of the genetic engineering of plants is shortsighted. These critics question the wisdom of intervening in the ecological balance of nature. More specifically, they maintain that manipulating the genetic structure of plants tends to lead to a reduction in the diversity of plant life, making plants less resistant to disease. It could also lead to the spread of diseases from one plant species to another, as genes of one species are implanted in another. Furthermore, new and unnatural varieties of food plants could have unforeseen health risks for human beings.

Since genetic engineering is a highly technical procedure, those who control technology have great power over the food supply. Thus, both corporate power over consumers and the power of more technologically advanced nations over less technologically advanced nations could be increased as GM foods fill the marketplace. In addition, plants that are genetically engineered to produce more often re-

quire more fertilizer and greater amounts of irrigated water than ordinary plants. The technology would therefore serve the interests of corporate agribusiness at the expense of small-scale, low-income farmers.

Many of the concerns about the genetic engineering of animals are similar to those about the engineering of plants. Loss of biodiversity, vulnerability to disease, and business control over livestock are all frequently mentioned objections to the genetic manipulation of animals. Moral issues tend to become more important, though, when opponents of genetic engineering discuss its use with animals. Many religious beliefs hold that the order of the world, including its division into different types of creatures, is divinely ordained. From the perspective of such beliefs, the relatively common experimental practice of injecting human growth genes into mice could be seen as the sacrilegious creation of monsters. Opponents of the genetic alteration of animals argue, further, that animals will suffer. They point out that selective breeding, a slow process, has led to about two hundred diseases of genetic origin in purebred dogs. Genetic engineering brings about change much faster than breeding, increasing the probability of genetic diseases.

Engineering of Humans

Some of the greatest ethical and social problems with genetic engineering involve its use on humans. Gene therapy seeks to cure inherited diseases by altering the defective genes that cause them. Those who favor gene therapy maintain that it can be a powerful tool to overcome human misery. Those who oppose this type of medical procedure usually focus on three major ethical issues. First, critics maintain that this technology raises the problem of ownership of human life. In the early 1990's, the National Institutes of Health (NIH) began filing for patents on human genes, meaning that the blueprints for human life could actually be owned. Because all human DNA comes from human tissue, the question arises of whether participants in genetic experiments own their own DNA or it belongs to the researchers who have extracted it.

The second problem involves eugenic impli-

In Battle Creek, Michigan, a demonstration outside the headquarters of Kellogg highlights the company's use of genetically altered crops without labeling. Those in favor of labeling the use of genetically altered ingredients maintain that the public has a right to know about the use of such ingredients so they can make informed purchase decisions. (AP/Wide World Photos)

cations. Eugenics is the practice of trying to produce "better" humans. If scientists can alter genes to produce humans with more desirable health characteristics, then scientists can also alter genes to produce humans with more desirable characteristics of personality or physical appearance. In this way, genetic engineering poses the risk of becoming an extreme and highly technological form of discrimination. The third problem is related to both of the first two: the reduction of humans to objects. When human life becomes something that can be owned and redesigned at will, some ethicists claim, it will cease to be seen as a sacred mystery and will become simply another piece of biological machinery. As objects, people will gradually lose the philosophical justification for their political and moral rights.

Impact and Applications

Concerns about the social and ethical implications of genetic engineering have led to a number of attempts to limit or control the technology. The environmental group Greenpeace has campaigned against GM agricultural products and called for the clear labeling of all foods produced by genetic manipulation. In September, 1997, Greenpeace filed a legal petition against the U.S. Environmental Protection Agency (EPA), objecting to the EPA's approval of GM plants.

Activist Jeremy Rifkin became one of the most outspoken opponents of all forms of genetic engineering. Rifkin and his associates called on the U.S. NIH to stop government-funded transgenic animal research. A number of organizations, such as the Boston-based Council for Responsible Genetics (CRG), lobbied to increase the legal regulation of genetic engineering. In 1990, in response to pressure from critics of genetic engineering, the Federal Republic of Germany enacted a gene law to govern the use of biotechnology. In the United States, the federal government and many state governments considered laws regarding genetic manipulation. A 1995 Oregon law, for example, granted ownership of human tissue and genetic information taken from human tissue to the person from whom the tissue was taken.

—*Carl L. Bankston III*

See also: Animal Cloning; Biological Weapons; Biopharmaceuticals; Cloning: Ethical Issues; Eugenics; Cloning Vectors; Gene Therapy; Gene Therapy: Ethical and Economic Issues; Genetic Engineering; Genetic Engineering: Agricultural Applications; Genetic Engineering: Historical Development; Genetic Engineering: Industrial Applications; Genetic Engineering: Medical Applications; Genetic Engineering: Risks; Genetically Modified (GM) Foods; High-Yield Crops; Knockout Genetics and Knockout Mice; Organ Transplants and HLA Genes; Paternity Tests; Patents on Life-Forms; Shotgun Cloning; Synthetic Genes; Transgenic Organisms; Xenotransplants.

Further Reading

Boylan, Michael, and Kevin E. Brown. *Genetic Engineering: Science and Ethics on the New Frontier.* Upper Saddle River, N.J.: Prentice Hall, 2001. Written by a biologist and philosopher, text includes discussions on the professional and practical principles of conduct, the biology of genetic therapy, the limits of science, somatic gene therapy, enhancement, cloning, and germ line therapy. Illustrated.

Evans, John Hyde. *Playing God? Human Genetic Engineering and the Rationalization of Public Bioethical Debate.* Chicago: University of Chicago Press, 2002. Chapters include "Framework for Understanding the Thinning of a Public Debate," "The Eugenicists and the Challenge from the Theologians," "Gene Therapy, Advisory Commissions, and the Birth of the Bioethics Profession," and "The President's Commission: The 'Neutral' Triumph of Formal Rationality."

Gonder, Janet C., Ernest D. Prentice, and Lilly-Marlene Russow, eds. *Genetic Engineering and Animal Welfare: Preparing for the Twenty-first Century.* Greenbelt, Md.: Scientists Center for Animal Welfare, 1999. Covers ethics and the well-being of animals used in genetic engineering and xenotransplantation. Illustrated.

Hubbell, Sue. *Shrinking the Cat: Genetic Engineering Before We Knew About Genes.* Illustrations by Liddy Hubbell. Boston: Houghton

Mifflin, 2001. Discusses the way genes have been altered by humans for centuries by focusing on corn, silkworms, domestic cats, and apples and notes some of the mistakes that were made in the quest for improvements.

Kass, Leon R. *Life, Liberty, and the Defense of Dignity: The Challenge for Bioethics.* San Francisco: Encounter Books, 2002. Examines genetic research, cloning, and active euthanasia, and argues that biotechnology has left humanity out of its equation, often debasing human dignity rather than celebrating it.

Lambrecht, Bill. *Dinner at the New Gene Cafe: How Genetic Engineering Is Changing What We Eat, How We Live, and the Global Politics of Food.* New York: Thomas Dunne Books, 2001. Chronicles the growing debate over genetically altered food in the United States between corporate profiteers and consumers, farmers, and environmentalists. Illustrated.

Long, Clarisa, ed. *Genetic Testing and the Use of Information.* Washington, D.C.: AEI Press, 1999. Chapters include "Genetic Privacy, Medical Information Privacy, and the Use of Human Tissue Specimens in Research," "The Social Implications of the Use of Stored Tissue Samples: Context, Control, and Community," and "Genetic Discrimination."

Reiss, Michael J., and Roger Straughan, eds. *Improving Nature? The Science and Ethics of Genetic Engineering.* New York: Cambridge University Press, 2001. Elucidates the ethical issues surrounding genetic engineering for the nonbiologist. Chapters examine genetic engineering in microorganisms, plants, animals, and humans.

Rifkin, Jeremy. *The Biotech Century: Harnessing the Gene and Remaking the World.* New York: Putnam, 1998. One of the best-known critics of biotechnology warns that procedures such as cloning and genetic engineering could be disastrous for the gene pool and for the natural environment.

Yount, Lisa, ed. *The Ethics of Genetic Engineering.* San Diego: Greenhaven Press, 2002. Essays written by scientists, science writers, ethicists, and consumer advocates present the growing controversy over genetically modi-

fying plants and animals, altering human genes, and cloning humans.

Veatch, Robert M. *The Basics of Bioethics.* 2d ed. Upper Saddle River, N.J.: Prentice Hall, 2003. In a textbook designed for students, Veatch presents an overview of the main theories and policy questions in biomedical ethics. Includes diagrams, case studies, and definitions of key concepts.

Web Sites of Interest

American Medical Association. http://ama-assn.org. The AMA has posted it guidelines on the ethics of genetic engineering.

Council for Responsible Genetics. http://www.gene-watch.org. An organization that encourages debate on issues concerning genetic technologies.

National Information Resource on Ethics and Human Genetics. http://www.georgetown.edu/research/nrcbl/nirehg. Site supports links to databases, annotated bibliographies, and articles about the ethics of genetic engineering and human genetics.

Genetic Load

Field of study: Population genetics

Significance: *Genetic load is a measure of the number of recessive deleterious (lethal or sublethal) alleles in a population. These alleles are maintained in populations at equilibrium frequencies by mutation (which introduces new alleles into the gene pool) and selection (which eliminates unfavorable alleles from the gene pool). Genetic load is one of the causes of inbreeding depression, the reduced viability of offspring from closely related individuals. For this reason, genetic load is a primary concern in the fields of agriculture, animal husbandry, conservation biology, and human health.*

Key terms

DELETERIOUS ALLELES: alternative forms of a gene that, when expressed in the homozygous condition in diploid organisms, may be lethal or sublethal—in the latter case typically resulting in an aberrant phenotype with low fitness

INBREEDING DEPRESSION: reduced fitness of an individual or population arising as the result of decreased heterozygosity across loci

Genetic Load in Diploid Populations

Genetic diversity is a measure of the total number of alleles within a population and it is mutation, the ultimate source of all genetic variation, that gives rise to new alleles. Favorable mutations are rare and are greatly outnumbered by mutations that are selectively neutral or deleterious (that is, lethal or sublethal). In diploid organisms most mutant (deleterious) alleles are hidden from view because they are masked by a second, normal, or wild-type, allele; that is they are typically (but not always) recessive. On the other hand, in haploid organisms lethal and deleterious genes are immediately exposed to differential selection.

Genetic load is defined as an estimate of the number of deleterious alleles in a population. Total genetic load is therefore the sum of two major components, the lethal load (L) and the detrimental, but nonlethal load (D). Empirical and theoretical studies suggest detrimental alleles rather than lethals constitute the greater majority of the genetic load in natural populations. When expressed in the homozygous condition, the primary effect of deleterious alleles within the gene pool on individuals is straightforward: death or disability accompanied by lower fitness. However, the impact of lethal and sublethal alleles on the mean fitness of populations, as opposed to individuals, is dependent upon many factors, such as their frequency within the gene pool, the number of individuals in the population, and whether or not those individuals are randomly mating.

How and why are recessive alleles maintained within a population at all? Why are they not eliminated by natural selection? First, recessive deleterious alleles must obtain a sufficient frequency before homozygous individuals occur

The April 25, 1986, accident at the Chernobyl nuclear power plant in the Ukraine released 5 percent of the radioactive reactor core into the atmosphere, contaminating large areas of Belarus, Ukraine, and Russia and quite possibly increasing genetic loads in affected populations. (AP/Wide World Photos)

in sufficient number to be detected. Second, in some situations recessive alleles that are deleterious or lethal in the homozygous state are advantageous in heterozygotes. Third, new deleterious alleles are constantly introduced into the population by mutation or are reintroduced by back mutation. Finally, the rate at which deleterious genes are purged from the population critically depends upon the "cost of selection" against them, and selection coefficients may vary considerably depending upon the allele and intra- or extracellular environments. In large randomly mating diploid populations, genetic load theoretically reaches an "equilibrium value" maintained by a balance between the mutation rate and the strength of selection. Finally, it should be borne in mind that nonlethal alleles that are not advantageous under present circumstances nevertheless constitute a pool of alleles that may be advantageous in a different (or changing) environment or in a different genetic background. In other words, some neutral and nonlethal mutations may have unpredictable "remote consequences."

Population Size, Inbreeding, and Genetic Load

As it is used among population geneticists, genetic load is most appropriately defined as the proportionate decrease in the average fitness of a population relative to that of the optimal genotype. The "proportionate decrease in the average fitness" is, of course, due to the presence of lethal and deleterious nonlethal alleles that are maintained in equilibrium by mutation and selection. Genetic load within populations may be substantially increased under certain circumstances. Small populations, species whose mating system involves complete or partial inbreeding, and populations with increased mutation rates all are expected to accumulate load at values exceeding that of large outbreeding populations. Small populations face multiple genetic hazards, including inbreeding depression.

Inbreeding decreases heterozygosity across loci and, relative to randomly mating populations, the fitness of inbred individuals is typically depressed. Inbreeding causes rare recessive alleles to occur more frequently in the homozygous condition, increasing the frequency of aberrant phenotypes that are observed. Complete or partial inbreeding (or, in plants, self-fertilization) leads to the accumulation of deleterious mutations that increase genetic load. Paradoxically, continued inbreeding results in lower equilibrium frequencies of deleterious alleles because they are expressed with greater frequency in the homozygous state. Thus, inbreeding populations may eliminate, or "purge," some proportion of their genetic load via selection against deleterious recessive alleles. Nevertheless it is true that, compared to large genetically diverse populations, small inbred populations with reduced genetic diversity are more likely to go extinct. For these reasons, population sizes, inbreeding, and genetic load are among the primary concerns of conservation biologists working to ensure the survival of rare or endangered species.

As previously mentioned, increased muta-tion rates may also increase genetic load. For example, the rate of nucleotide substitution in mammalian mitochondrial DNA (mtDNA) is nearly ten times that of nuclear DNA. The tenfold mutation rate difference is postulated to be due to highly toxic, mutagenic reactive oxygen species produced by the mitochondrial electron transport chain and/or relatively inefficient DNA repair mechanisms. Thus, mitochondrial genomes accumulate fixed nucleotide changes rapidly via "Müller's ratchet." Mutation rates and genetic load may also be increased by exposure to harmful environments. For example, an accident on April 25, 1986, at the Chernobyl nuclear power plant in the Ukraine released 5 percent of the radioactive reactor core into the atmosphere, contaminating large areas of Belarus, Ukraine, and Russia. Radiation exposure of this kind and toxic chemicals (such as heavy metals) in watersheds pose significant human health risks that, over time, may be associated with increased genetic loads in affected populations.

—*J. Craig Bailey*

See also: Consanguinity and Genetic Disease; Hardy-Weinberg Law; Heredity and Environment; Inbreeding and Assortative Mating; Lateral Gene Transfer; Natural Selection; Pedigree Analysis; Polyploidy; Population Genetics; Punctuated Equilibrium; Quantitative Inheritance; Sociobiology; Speciation.

Further Reading

Charlesworth, D., and B. Charlesworth. "Inbreeding Depression and Its Evolutionary Consequences." *Annual Review of Ecology and Systematics* 18 (1987): 237-268. A review of empirical studies of genetic load and its short- and long-term affects on the evolutionary potentialities of inbred populations.

Thornhill, Nancy Wilmsen, ed. *The Natural History of Inbreeding and Outbreeding: Theoretical and Empirical Perspectives.* Chicago: University of Chicago Press, 1993. Several articles in this book expertly consider the complex relationship between the costs and benefits associated with many different mating systems (totally outbreeding, inbreeding, partial selfing, and haplodiploidy) in relation to total genetic load.

Wallace, Bruce. *Genetic Load: Its Biological and Conceptual Aspects.* Englewood Cliffs, N.J.: Prentice-Hall, 1970. This 116-page treatise provides an introduction to the concept of genetic load in individuals and populations and discusses how genetic load is calculated. It also provides a discussion of how the interplay among mutation rates, selection, and inbreeding influences the dynamics of genetic load within populations.

Genetic Screening

Field of study: Human genetics and social issues

Significance: *Genetic screening is a preventive health measure that involves the mandatory or voluntary testing of certain individuals for the purpose of detecting genetic disorders or identifying defective genes that can be transmitted to offspring. The primary goals of genetic screening include the prevention and/or treatment of genetic disorders and the option to make informed and rational decisions about conception and birth. It has raised concerns about confidentiality, discrimination, and the right to privacy.*

Key terms

ALLELE: a form of a gene at a locus; each locus in an individual's chromosomes has two alleles, which may be the same or different

CARRIER: a healthy individual who has one normal allele and one defective allele at the same gene locus

GENETIC DISORDER: a disorder caused by a change in a gene or chromosome

INBORN ERROR OF METABOLISM: an inherited disease caused by a mutation in a gene that codes for an enzyme important in a metabolic pathway

LOCUS (*pl.* LOCI): the actual location of a gene on a chromosome

Neonatal Screening

The most widespread use of genetic screening is the testing of newborn babies, called neonatal screening. Every year, millions of newborn babies are tested for inborn errors of metabolism. The purpose of this kind of screening is to provide immediate treatment after birth if a defect is detected, so that the newborn has a chance of having a normal life. A classic example of neonatal screening in the United States and many countries is the mandatory mass screening of newborn babies for phenylketonuria (PKU), a disorder that causes irreversible brain damage when not treated. Individuals with PKU lack the enzyme phenylalanine hydroxylase, which converts the essential amino acid phenylalanine into the amino acid tyrosine. PKU results in accumulation of phenylalanine in the body, which causes the brain damage.

Newborn babies are screened for PKU using the Guthrie test, named after its inventor, Robert Guthrie. The Guthrie test detects high levels of phenylalanine in the blood of newborns. Blood samples are taken from the heels of newborn babies in the hospital nursery, placed on

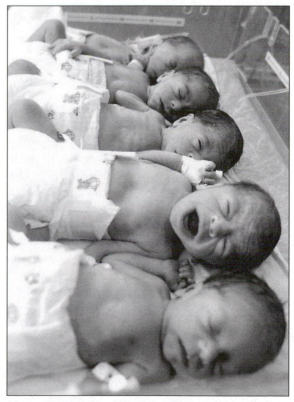

Many states mandate that newborns be screened for phenylketonuria (PKU), hypothyroidism, and, in some cases, other inborn errors of metabolism such as galactosemia. (AP/Wide World Photos)

filter papers as dried spots, and sent off to appropriate laboratories for analysis. Newborns with positive results can be effectively treated with a diet low in phenylalanine (low-protein foods). The level of phenylalanine in the blood is regularly monitored. If treatment is not initiated within the first two months of life, irreversible brain damage will occur.

In the United States, only one other neonatal tests is performed routinely in all states: for hypothyroidism. There is no federal mandate for which tests must be provided to everyone, so each state decides independently. Many states also screen for galactosemia, an inherited disorder characterized by seizures, mental retardation, vomiting, and liver disease, caused by the accumulation of galactose in the blood. Many states also screen for maple syrup urine disease, homocystinuria, and biotinidase deficiency. A majority of states also screen for sickle-cell disease, an inherited blood disorder characterized by anemia, pain in the abdomen and joints, and damage to organs.

Carrier Screening

Carrier screening is the voluntary testing of healthy individuals of reproductive age who may be carriers of heritable, defective genes, the purpose being to inform couples of their risk of having a child with a genetic disorder. In the United States, screening has been limited to some ethnic groups known to have a high incidence of a specific genetic disorder. In the 1970's, for example, Tay-Sachs screening of Ashkenazi Jews of reproductive age was successfully implemented. Tay-Sachs disease is an inherited, progressive disease in infants characterized by a startle response to noise, blindness, seizures, paralysis, and death in infancy caused by the absence of an enzyme called hexosaminidase A. Tay-Sachs screening measures the level of hexosaminidase A in the blood. People with Tay-Sachs disease have no detectable level of the enzyme, while carriers have a below-normal level of the enzyme.

In the early 1970's, mandatory, large-scale screening of African American couples and some schoolchildren was implemented in an effort to identify carriers of the gene for sickle-cell disease. Blood samples taken from individuals were tested for the presence of distorted or sicked-shaped red blood cells caused by the production of abnormal hemoglobin, the molecule that transports oxygen in the body. The laws mandating screening were later repealed amid charges of racial discrimination.

After successful identification of the cystic fibrosis gene in 1989, the scientific and medical community began debating the costs and benefits of screening millions of carriers of the gene in the United States. Cystic fibrosis is a common inherited disorder characterized by accumulation of mucus in the lungs and pancreas; it affects Caucasian children and young adults. The proportion of carriers in the population varies by ethnicity. Northern Europeans have the highest proportion of carriers (1 in 22), while African Americans have a much lower proportion (1 in 65). Some companies have begun voluntary screening for the cystic fibrosis gene in couples with a family history of the disorder.

Impact and Applications

The benefits of genetic screening include early intervention and treatment, detection of new mutations by researchers, and the education of people about genetic disorders so that they are able to make informed and responsible decisions about reproductive issues. However, screening for genetic defects has raised ethical and social issues over confidentiality, discrimination, and the right to privacy.

One example is the sickle-cell screening program of the early 1970's. Screening results were not kept in strictest confidence; consequently, many healthy African Americans who were carriers of the sickle-cell gene were stigmatized and discriminated against in terms of employment and insurance coverage. There were also charges of racial discrimination because carriers were advised against bearing children.

Although the sickle-cell disease screening programs were unsuccessful, there were some successes in other screening programs. For example, the number of newborns with Tay-Sachs disease has dropped dramatically as a result of carrier screening programs. The screening programs succeeded because of the tremendous effort put forth to educate Ashkenazi Jewish communities about the disorder and the consequent acceptance of the programs by these communities.

Genetic screening may be even more useful in the future, when genetic disorders that are currently incurable might be treated by gene therapy (the replacement of a defective or missing gene with a normal, functional copy of the gene). The first clinical trials of gene therapy began in 1990 for severe combined immunodeficiency disorder (SCID), a lethal genetic disorder in which a person has no functional immunity. Much attention has been focused on gene therapy as a significant treatment option for patients with disorders such as cystic fibrosis. So far, gene therapy has garnered mixed results, and much progress will be needed before it becomes possible, much less routinely feasible. However, although it may well be decades or more before that occurs, future generations may well enjoy the results of today's research.

—*Oluwatoyin O. Akinwunmi*

See also: Amniocentesis and Chorionic Villus Sampling; Bioethics; Cystic Fibrosis; Down Syndrome; Gene Therapy; Gene Therapy: Ethical and Economic Issues; Genetic Counseling; Genetic Testing; Genetic Testing: Ethical and Economic Issues; Hereditary Diseases; Human Genetics; In Vitro Fertilization and Embryo Transfer; Inborn Errors of Metabolism; Insurance; Linkage Maps; Phenylketonuria (PKU); Prenatal Diagnosis; Sickle-Cell Disease; Tay-Sachs Disease.

Further Reading

Anderson, W. French. "Gene Therapy." *Scientific American* 273 (September, 1995). Provides an overview for the general reader.

Chadwick, Ruth, et al., eds. *The Ethics of Genetic Screening.* Boston: Kluwer Academic, 1999. Covers three aspects of the topic: ethical principles that guide genetic screening and testing programs, descriptions of genetic screening programs in European nations, and analysis of social and historical conditions that influence national programs.

Evans, Mark I., ed. *Metabolic and Genetic Screening.* Philadelphia: W. B. Saunders, 2001. Covers principles of screening, screening for neural tube defects, second-trimester biochemical screening, prenatal genetic screening in the Ashkenazi Jewish population, cystic fibrosis, identifying and managing hereditary risk of breast and ovarian cancer, and genetic implications for newborn screening for phenylketonuria.

Heyman, Bob, and Mette Henriksen. *Risk, Age, and Pregnancy: A Case Study of Prenatal Genetic Screening and Testing.* New York: Palgrave, 2001. Provides a detailed case study of a prenatal genetic screening and testing system in a British hospital, giving perspectives of pregnant women, hospital doctors, and midwives, and elucidating the communication between women and the hospital doctors who advise them.

Pierce, Benjamin A. *The Family Genetic Sourcebook.* New York: John Wiley & Sons, 1990. An introduction to the principles of heredity and a catalog of more than one hundred human traits. Topics include heredity, inheritance patterns, chromosomes and chromosomal abnormalities, genetic risks, genetic counseling, and family history. Written for the general reader, with short descriptions, and includes suggested readings, appendices, glossary, and index.

Roberts, Leslie. "To Test or Not to Test?" *Science* 247 (January, 1990). Although a bit dated, outlines well the basic issues surrounding cystic fibrosis screening.

Shannon, Joyce Brennfleck, ed. *Medical Tests Sourcebook.* Detroit, Mich.: Omnigraphics, 1999. As the lengthy subtitle puts it, a guide to "basic consumer health information about medical tests, including periodic health exams, general screening tests, tests you can do at home, findings of the U.S. Preventive Services Task Force, X-ray and radiology tests, electrical tests, tests of blood and other body fluids and tissues, scope tests, lung tests, genetic tests, pregnancy tests, newborn screening tests, sexually transmitted disease tests, and computer aided diagnoses, along with a section on paying for medical tests, a glossary, and resource listings." Illustrated.

Web Sites of Interest

American Medical Association. http://ama-assn.org. The AMA's guidelines on the ethics of genetic screening.

Centers for Disease Control, Genomics and Disease Prevention. http://www.cdc.gov/genomics/info/reports/program/population.htm. A journal article on genetic screening, entitled "Population Screening in the Age of Genomic Medicine."

Genetic Testing

Field of study: Human genetics and social issues

Significance: *Genetic testing comprises any procedure used to detect the presence of a genetic disorder or a defective gene in a fetus, newborn, or adult. The results of genetic tests can be useful in family planning, treatment decisions and medical research. Genetic testing has significant implications with respect to reproductive choices, privacy, insurance coverage, and employment.*

Key terms

GENETIC DISORDER: a disorder caused by a mutation in a gene or chromosome

GENETIC MARKER: a distinctive DNA sequence that shows variation in the population and can therefore potentially be used for identification of individuals and for discovery of disease genes

Prenatal Diagnosis

Prenatal diagnosis is the testing of a developing fetus in the womb, or uterus, for the presence of a genetic disorder. The purpose of this type of genetic testing is to inform a pregnant woman of the chances of having a baby with a genetic disorder. Prenatal diagnosis is limited to high-risk individuals and is usually recommended only if a woman is thirty-five years of age or older, if she has had two or more spontaneous abortions, or if there is a family history of a genetic disorder. Hundreds of genetic disorders can be tested in a fetus. One of the most common genetic disorders screened for is Down syndrome, or trisomy 21, a form of mental retardation caused by having an extra copy of chromosome 21. The incidence of Down syndrome increases sharply in children born to women over the age of forty.

The technique most commonly used for prenatal diagnosis is amniocentesis. It is performed between the sixteenth and eighteenth week of pregnancy. Amniocentesis involves the insertion of a hypodermic needle through the abdomen into the uterus of a pregnant woman. The insertion of the needle is guided by ultrasound, a technique that uses high-frequency sound waves to locate a developing fetus or internal organs and presents a visual image on a video monitor. A small amount of amniotic fluid, which surrounds and protects the fetus, is withdrawn. The amniotic fluid contains fetal secretions and cells sloughed off the fetus that are analyzed for genetic abnormalities. Chromosomal disorders such as Down syndrome, Edwards' syndrome (trisomy 18), and Patau syndrome (trisomy 13) can be detected by examining the chromosome number of the fetal cells. Certain biochemical disorders such as Tay-Sachs disease, a progressive disorder characterized by a startle response to sound, blindness, paralysis, and death in infancy, can be determined by testing for the presence or absence of a specific enzyme activity in the amniotic fluid. Amniocentesis can also determine the sex of a fetus and detect common birth defects such as spina bifida (an open or exposed spinal cord) and anencephaly (partial or complete absence of the brain) by measuring levels of alpha fetoprotein in the amniotic fluid. The limitations of amniocentesis include inability to detect most genetic disorders, possible fetal injury or death, infection, and bleeding.

Chorionic villus sampling (CVS) is another technique used for prenatal diagnosis. It is performed earlier than amniocentesis (between the eighth and twelfth week of pregnancy). Under the guidance of ultrasound, a catheter is inserted into the uterus via the cervix to obtain a sample of the chorionic villi. The chorionic villi are part of the fetal portion of the placenta, the organ that nourishes the fetus. The chorionic villi can be analyzed for chromosomal and biochemical disorders but not for congenital birth defects such as spina bifida and anencephaly. The limitations of this technique are inaccurate diagnosis and a slightly higher chance of fetal loss than in amniocentesis.

Neonatal Testing

The most widespread genetic testing is the mandatory testing of every newborn infant for the inborn error of metabolism (a biochemical disorder caused by mutations in the genes that code for the synthesis of enzymes) phenyl-ketonuria (PKU), a disorder in which the enzyme for converting phenylalanine to tyrosine is nonfunctional. The purpose of this type of testing is to initiate early treatment of infants. Without treatment, PKU leads to brain damage and mental retardation. A blood sample is taken by heel prick from a newborn in the hospital nursery, placed on filter papers as dried spots, and subsequently tested, using the Guthrie test, for abnormally high levels of phenylalanine. In infants who test positive for PKU, a diet low in phenylalanine is initiated within the first two months of life. Newborns can be tested for many other disorders such as sickle-cell disease and galactosemia (accumulation of galactose in the blood), but the cost benefit ratio is only acceptable in the more common genetic diseases, and most tests are only per-formed if there is a family history of the genetic disease or some other reason to suspect its presence.

Carrier Testing

A healthy couple contemplating having children can be tested voluntarily to determine if they carry a defective gene for a disorder that runs in the family. This type of testing is known as carrier testing because it is designed for carriers (individuals who have a normal gene paired with a defective form of the same gene but have no symptoms of a genetic disorder). Carriers of the genes responsible for Tay-Sachs disease, cystic fibrosis (accumulation of mucus in the lungs and pancreas), Duchenne muscular dystrophy (wasting away of muscles), and hemophilia (uncontrolled bleeding caused by lack of blood clotting factor) can be detected by DNA analysis.

When the gene responsible for a specific genetic disorder is unknown, the location of the gene on a chromosome can be detected indirectly by linkage analysis. Linkage analysis is a

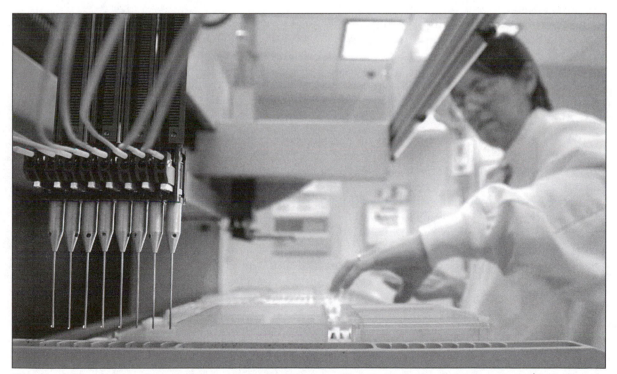

DNA samples from patients are removed by an eight-needle apparatus and deposited into a tray for genetic testing at Myriad Genetics in Salt Lake City. (AP/Wide World Photos)

technique in which geneticists look for consistent patterns in large families where the mutated gene and a genetic marker always appear together in affected individuals and those known to be carriers. If a genetic marker lies close to the defective gene, it is possible to locate the defective gene by looking for the genetic marker. The genetic markers used commonly for linkage analysis are restriction fragment length polymorphisms (RFLPs). When human DNA is isolated from a blood sample and digested at specific sites with special enzymes called restriction endonucleases, RFLPs are produced. RFLPs are found scattered randomly in human DNA and are of different lengths in different people, except in identical twins. They are caused by mutations or the presence of varying numbers of repeated copies of a DNA sequence and are inherited. RFLPs are separated by gel electrophoresis, a technique in which DNA fragments of varying lengths are separated in an electric field according to their sizes. The separated DNA fragments are blotted onto a nylon membrane, a process known as Southern blotting. The membrane is probed and then visualized on X-ray film. The characteristic pattern of DNA bands visible on the film is similar in appearance to the bar codes on grocery items.

An early successful example of linkage analysis involved the search for the gene that causes Huntington's disease, an always fatal neurological disease that typically shows onset after 35 or 40 years of age. In 1983, James Gusella, Nancy Wexler, and Michael Conneally reported a correlation between one specific RFLP they named *G8* and Huntington's disease (Huntington's chorea). After studying numerous RFLPs of generations of an extended Venezuelan family with a history of Huntington's disease, they discovered that *G8* was present in members afflicted with the genetic disorder and was absent in unaffected members.

High-risk individuals or families can be tested voluntarily for the presence of a mutated gene that may indicate a predisposition to a late-onset genetic disorder such as Alzheimer's disease or to other conditions such as hereditary breast, ovarian, and colon cancers. This type of testing is called predictive testing. Un-like tests for many of the inborn errors of metabolism, predictive testing can only give a rough idea of how likely an individual may be to develop a particular genetic disease. It is not always clear how such information should be used, but at least in some cases lifestyle or therapeutic changes can be instituted to lessen the likelihood of developing the disease.

Impact and Applications

Genetic testing has had a significant impact on families and society at large. It provides objective information to families about genetic disorders or birth defects and provides an analysis of the risks for genetic disorders through genetic counseling. Consequently, many prospective parents are able to make informed and responsible decisions about conception and birth. Some choose not to bear children, some terminate pregnancy after prenatal diagnosis, and some take a genetic gamble and hope for a normal child. Genetic testing can have a profound psychological impact on an individual or family. A positive genetic test could cause a person to experience depression, while a negative test result may eliminate anxiety and distress. Questions have been raised in the scientific and medical community about the reliability and high costs of tests. There is concern about whether genetic tests are stringent enough to ensure that errors are not made. DNA-based diagnosis can lead to errors if DNA samples are contaminated. Such errors can be devastating to families. People at risk for late-onset disorders such as Huntington's disease can be tested to determine if they are predisposed to developing the disease. There is, however, controversy over whether it is ethical to test for diseases for which there are no known cures or preventive therapies. The question of testing also creates a dilemma in many families. Unlike other medical tests, predictive testing involves the participation of many members of a family. Some members of a family may wish to know their genetic status, while others may not.

While there has been great enthusiasm over genetic testing, there are also social, legal, and ethical issues such as discrimination, confidentiality, reproductive choice, and abuse of genetic information. Insurance companies and employers may require prospective customers and employees to submit to genetic testing or may inquire about a person's genetic status. Individuals may be denied life and health insurance coverage because of their genetic status, or a prospective customer may be forced to pay exorbitant insurance premiums. The potential for discrimination with respect to employment and promotions also exists. For example, as a result of the sickle-cell screening programs of the early 1970's, many African Americans with sickle-cell disease were denied employment and insurance coverage and some were denied entry into the U.S. Air Force. The Americans with Disabilities Act, signed into federal law in 1990, contained provisions safeguarding employees from genetic discrimination by employers. By 1994, companies with fifteen or more employees had to comply with the law, which prohibits employment discrimination because of genetic status and also prohibits genetic testing by employers.

As genetic testing becomes standard practice, the potential for misuse of genetic tests and genetic information will become greater. Prospective parents may potentially use prenatal diagnosis as a means to ensure the birth of a "perfect" child. Restriction fragment length polymorphism analysis, used in genetic testing, has applications in DNA fingerprinting or DNA typing. DNA fingerprinting is a powerful tool for identification of individuals used to generate patterns of DNA fragments unique to each individual based on differences in the sizes of repeated DNA regions in humans. It is used to establish identity or nonidentity in immigration cases and paternity and maternity disputes; it is also used to exonerate the innocent accused of violent crimes and to link a suspect's DNA to body fluids or hair left at a crime scene. Several states in the United States have been collecting blood samples from a variety of sources, including newborn infants during neonatal testing and individuals convicted of violent crimes, and have been storing genetic information derived from them in DNA databases for future reference. Such information could be misused by unauthorized people.

—*Oluwatoyin O. Akinwunmi,*
updated by Bryan Ness

See also: Amniocentesis and Chorionic Villus Sampling; Bioethics; Breast Cancer; Cystic Fibrosis; DNA Fingerprinting; Down Syndrome; Gene Therapy; Gene Therapy: Ethical and Economic Issues; Genetic Counseling; Genetic Screening; Genetic Testing: Ethical and Economic Issues; Hemophilia; Hereditary Diseases; Human Genetics; Huntington's Disease; In Vitro Fertilization and Embryo Transfer; Inborn Errors of Metabolism; Insurance; Linkage Maps; Paternity Tests; Phenylketonuria (PKU); Prenatal Diagnosis; RFLP Analysis; Sickle-Cell Disease; Tay-Sachs Disease.

Further Reading

Anderson, W. French. "Gene Therapy." *Scientific American* 273 (September, 1995). Provides an overview for the general reader.

Genetic Testing for Breast Cancer Risk: It's Your Choice. NIH Publication 99-4252 4008061621. DHHS Publication 99-4252 4008061622. Washington, D.C.: National Action Plan on Breast Cancer, Public Health Service's Office on Women's Health, Department of Health and Human Services, 1999. Examines the diagnosis of human chromosome abnormalities and the genetic aspects of breast and ovarian cancer. Illustrated.

Heller, Linda. "Genetic Testing." *Parents* 70 (November, 1995). Explores the risks of some tests.

Heyman, Bob, and Mette Henriksen. *Risk, Age, and Pregnancy: A Case Study of Prenatal Genetic Screening and Testing.* New York: Palgrave, 2001. Provides a detailed case study of a prenatal genetic screening and testing system in a British hospital, giving perspectives of pregnant women, hospital doctors, and midwives, and elucidating the communication between women and the hospital doctors who advise them.

Jackson, J. F., H. F. Linskens, and R. B. Inman, eds. *Testing for Genetic Manipulation in Plants.* New York: Springer, 2002. Surveys the developing methods for detecting and characterizing genetic manipulation in plants and plant products such as seeds and foods. Figures, tables.

Shannon, Joyce Brennfleck, ed. *Medical Tests Sourcebook.* Detroit, Mich.: Omnigraphics,

1999. Gives lay readers insight into basic consumer health information about a range of medical tests. Topics covered include general screening tests, medical imaging tests, genetic testing, newborn screenings, and sexually transmitted disease tests, as well as Medicare, Medicaid, and other information on paying for medical tests.

Web Site of Interest

National Institutes of Health, National Library of Medicine. http://www.nlm.nih.gov. This site links to comprehensive information on genetic testing, research, and more. Also available in Spanish.

Genetic Testing: Ethical and Economic Issues

Field of study: Bioethics; Human genetics and social issues

Significance: *Using a suite of molecular, biochemical, and medical techniques, it is now possible to identify carriers of a number of genetic diseases and to diagnose some genetic diseases even before they display physical symptoms. In addition, numerous genes that predispose people to particular diseases such as cancer, alcoholism, and heart disease have been identified. These technologies raise important ethical questions about who should be tested, how the results of tests should be used, who should have access to the test results, and what constitutes normality.*

Key terms

DOMINANT TRAIT: a genetically determined trait that is expressed when a person receives the gene for that trait from either or both parents

RECESSIVE TRAIT: a genetically determined trait that is expressed only if a person receives the gene for the trait from both parents

The Dilemmas of Genetic Testing

Historically, it was impossible to determine whether a person was a carrier of a genetic disease or whether a fetus was affected by a genetic disease. Now both of these things and much

more can be determined through genetic testing. Although there are obvious advantages to acquiring this kind of information, there are also potential ethical problems. For example, if two married people are both found to be carriers of cystic fibrosis, each child born to them will have a 25 percent chance of having cystic fibrosis. Using this information, they could choose not to have any children, or, under an oppressive government desiring to improve the genetics of the population, they could be forcibly sterilized. Alternatively, they could choose to have each child tested prenatally and abort any child that tests positive for cystic fibrosis. Ethical dilemmas similar to these are destined to become increasingly common as scientists develop tests for more genetic diseases.

Another dilemma arises in the case of diseases such as Huntington's disease (Huntington's chorea), which is caused by a single dominant gene and is always lethal but which does not generally cause physical symptoms until middle age or later. A parent with such a disease has a 50 percent chance of passing it on to each child. Now that people can be tested, it is possible for a child to know whether he or she has inherited the deadly gene. If a person tests positive for the disease, he or she can then choose to remain childless or opt for prenatal testing to guard against the possibility of bringing a child into the world under a death sentence.

Tests for deadly, untreatable genetic diseases in offspring have an even darker side. If the test is negative, the person may be greatly relieved; if it is positive, however, doctors can offer no hope. Is it right to let someone know that they will die sometime around middle age or shortly thereafter if there is nothing the medical community can do to help them? The psychological trauma associated with such disclosures can sometimes be severe enough to result in suicide. Additionally, who should receive information about the test, especially if it shows positive for the disease? If the information is kept confidential, a person with the disease could buy large amounts of life insurance, to the financial advantage of beneficiaries, at the same price as an unaffected person. On the other hand, if health and life insurance companies were allowed to know the results of such tests, they might use the information to refuse insurance coverage of any kind. Finally, none of the genetic tests is 100 percent accurate. There will be occasional false positives and false negatives. With so much at stake, how can doctors and genetic counselors help patients understand the uncertainties?

How Should Genetic Testing Information Be Used?

Scientists are now able to test for more than just specific, prominent genetic defects. Genetic tests are now available for determining potential risks for such things as cancer, alcoholism, Alzheimer's disease, and obesity. A positive result for the alcoholism gene does not mean that a person is doomed to be an alcoholic but rather that they have a genetic tendency toward behavior patterns that lead to alcoholism or other addictions. Knowing this, a person can then seek counseling, as needed, to prevent alcoholism and make lifestyle decisions to help prevent alcohol abuse.

Unfortunately, a positive test for genes that predispose people to diseases such as cancer may be more ominous. It is believed that people showing a predisposition can largely prevent the eventual development of cancer with aggressive early screening (for example, breast exams and colonoscopies) and lifestyle changes. Some pre-emptive strategies, however, have come under fire. For example, some women at risk for breast cancer have chosen prophylactic mastectomies. In some cases, however, cancer still develops after a mastectomy, and some studies have shown lumpectomy and other less radical treatments to be as effective as mastectomy.

Another concern centers on who should have access to the test results. Should employers be allowed to require genetic testing as a screening tool for hiring decisions? Should insurance companies have access to the records when making policy decisions? These are especially disturbing questions considering the fact that a test for one of the breast cancer genes, for example, only predicts a significantly higher probability of developing breast cancer than is typical for the general population. Making such

testing information available to employers and insurance companies would open the door to discrimination based on the probability that a prospective employee or client will become a future financial burden. A number of states have banned insurance companies from using genetic testing data for this very reason.

Impact and Applications

The long track record and accuracy of some tests, such as the tests for cystic fibrosis and Tay-Sachs disease, has led to the suggestion that they could be used to screen the general population. Although this would seem to provide positive benefits to the population at large, there is a concern about the cost of testing on such a broad scale. Would the costs of testing outweigh the benefits? What other medical needs might not receive funding if such a program were started? The medical community will have to consider the options carefully before more widespread testing takes place.

As more genetic tests become available, it will eventually be possible to develop a fairly comprehensive genetic profile for each person. Such profiles could be stored on CD-ROMs or other storage devices and be used by individuals, in consultation with their personal physicians, to make lifestyle decisions that would counteract the effects of some of the defects in their genetic profiles. The information could also be used to determine a couple's genetic compatibility before they get married. When a woman becomes pregnant, a prenatal genetic profile of the fetus could be produced; if it does not match certain minimum standards, it could be aborted. The same genetic profile could be used to shape the child's life and help determine the child's profession. Although such comprehensive testing is now prohibitively expensive, the costs should drop as the tests are perfected and made more widely available.

Access to genetic profiles by employers, insurance companies, advertisers, and law enforcement agencies could result in considerable economic savings to society, allowing many decisions to be made with greater accuracy, but at what other costs? How should the information be used? How should access be limited? How much privacy should individuals have with regard to their own genetic profiles? As genetic testing becomes more widespread, these questions will need to be answered. Ultimately, the relationship between the good of society and the rights of the individual will need to be redefined.

—*Bryan Ness*

See also: Amniocentesis and Chorionic Villus Sampling; Bioethics; Breast Cancer; Cystic Fibrosis; DNA Fingerprinting; Down Syndrome; Gene Therapy; Gene Therapy: Ethical and Economic Issues; Genetic Counseling; Genetic Screening; Genetic Testing; Hemophilia; Hereditary Diseases; Human Genetics; Huntington's Disease; In Vitro Fertilization and Embryo Transfer; Inborn Errors of Metabolism; Insurance; Linkage Maps; Paternity Tests; Phenylketonuria (PKU); Prenatal Diagnosis; RFLP Analysis; Sickle-Cell Disease; Tay-Sachs Disease.

Further Reading

Finger, Anne L. "How Would You Handle These Ethical Dilemmas?" *Medical Economics*, October 27, 1997. Presents results of a survey in which readers were asked to settle two ethical dilemmas involving genetic testing.

Marteau, Theresa, and Martin Richards, eds. *The Troubled Helix: Social and Psychological Implications of the New Human Genetics.* New York: Cambridge University Press, 1999. Offers brief personal narratives of some of the psychosocial affects of genetic testing for diseases. Illustrations, bibliography, index.

Rennie, John. "Grading the Gene Tests." *Scientific American* 273 (June, 1994). Not only focuses on the accuracy and implementation of genetic tests but also considers the problems of privacy, discrimination, and eugenics inherent in genetic testing.

Rothenberg, Karen, et al. "Genetic Information and the Workplace: Legislative Approaches and Policy Challenges." *Science* 275 (March 21, 1997). Summarizes government action designed to protect the privacy of genetic test results and outlines suggested guidelines for future legislation.

Zallen, Doris Teichler. *Does It Run in the Family? A Consumer's Guide to DNA Testing for Genetic Disorders.* New Brunswick, N.J.: Rutgers University Press, 1997. Focuses on the practical aspects of obtaining genetic information, clearly explaining how genetic disorders are passed along in families.

Web Sites of Interest

American Medical Association. http://ama-assn.org. The AMA's guidelines on the ethics of genetic testing and screening of adults and children and of multiplex testing.

National Information Resource on Ethics and Human Genetics. http://www.georgetown.edu/research/nrcbl/nirehg. Site supports links to databases, annotated bibliographies, and articles about the ethics of genetic testing and human genetics.

Genetically Modified (GM) Foods

Field of study: Genetic engineering and biotechnology

Significance: *Genetically modified foods are produced through the application of recombinant DNA technology to crop breeding, whereby genes from the same or different species are transferred and expressed in crops that do not naturally harbor those genes. While GM crops offer great potential for food production in agriculture, their release has spurred various concerns among the general public.*

Key terms

Bacillus thuringiensis (*Bt*) toxin: a toxic compound naturally synthesized by bacterium *Bacillus thuringiensis*, which kills insects

GENETIC ENGINEERING: the manipulation of genetic material for practical purposes; also referred to as recombinant DNA technology, gene splicing, or biotechnology

GENETICALLY MODIFIED ORGANISMS (GMOs): genetically modified organisms, created through the use of genetic engineering or biotechnology

HERBICIDE RESISTANCE: a trait acquired by crop plants through recombinant DNA technology that enables plants to resist chemicals designed to control weeds

In 1991, chief executive officer of CalGene Roger Salquist examines genetically modified tomatoes that are able to ripen on the vine before shipping, instead of having to be picked green. (AP/Wide World Photos)

The Technology

Genetically modified (GM) foods are food products derived from genetically modified organisms (GMOs). GMOs may have genes deleted, added, or replaced for a particular trait; they constitute one of the most important means by which crop plants will be improved in the future. The advantage of using genetic engineering is quite obvious: It allows individual genes to be inserted into organisms in a way that is both precise and simple. Using molecular tools available, DNA molecules from entirely different species can now be spliced together to form a recombinant DNA molecule.

The recombinant DNA molecule can then be introduced into a cell or tissue through genetic transformation. When a particular gene that codes for a trait is successfully introduced to an organism and expressed, that organism is defined as a transgenic or GM organism.

Most of the GM crops in production thus far have modified crop protection characteristics, mainly improving protection against insects and competition (herbicide resistance). Some have improved nutritional quality and longer shelf life. Yet others under development will lift yield caps previously not possible to overcome by conventional means. Because of the direct access to and recombination of genetic material from any source, the normal reproductive barrier among different species can now be circumvented. All these modifications offer great potential for creating transgenic animals and plants useful to humankind, but GMOs also pose the possibility of misuse and unintended outcomes.

Conceivable Benefits of GM Foods

The potential benefits of using genetic engineering to develop new cultivars are evident. Crop yields can be increased by introducing genes that increase the crop's resistance to various pathogens or herbicides and enhance its tolerance to various stresses. The increased food supply is vital to support a growing population with shrinking land. One well-known example is the introduction of *Bt* gene from the bacterium *Bacillus thuringiensis* to several crops, including corn, cotton, and soybeans. When the *Bt* gene is transferred to plants, the plant cells produce a protein toxic to some insects and hence become resistant to these insects. The grains of *Bt* maize were also found to contain low mycotoxin, thus exhibiting better food safety than non-GM corns. Another example is the successful insertion of a gene resistant to the herbicide glyphosate, reducing production costs and increasing grain purity.

Food quality can be improved in other ways. Soybeans and canola with reduced saturated fats (healthier oil) have been developed. Alterations in the starch content of potatoes and the nutritional quality of protein in maize kernels are being developed. More precise gene transfer is also being used to produce desirable products that the plant does not normally make. The potential products include pharmaceutical proteins (for example, vaccines), vitamins, and plastic compounds. "Golden rice" has been engineered to produce significantly higher vitamin A precursors. This GM rice plays an important role in alleviating vision loss and blindness caused by vitamin A deficiency among those who consume rice as their main staple food. Attempts are being made to increase nitrogen availability, a limiting factor in crop production, by transferring genes responsible for nitrogen fixation into crops such as wheat and maize. In addition, the reduction in the use of fertilizers, insecticides, and herbicides for GM crops not only saves billions of dollars in costs but also alleviates the damage to wild organisms and ecosystems.

Concerns About GM Foods

Like any other technological innovation, genetic engineering in crop breeding and production does not come without risk or controversy. Some of the common questions raised by consumers include concerns over what plant and animal organisms they are now putting into their bodies, whether these are safe, whether they have been tested, why they are not labeled as GM foods, and whether GM foods might not

Demonstrators in front of the San Diego Convention Center in 2001, dressed as "killer tomatoes," protest the annual conference of the Biotechnology Industry Organization. (AP/Wide World Photos)

contain toxins or allergens not present in their natural counterparts. Although most of these questions are understandable, the public uproar concerning the GM crops and other foods, particularly in Great Britain and Europe, are, from a scientific standpoint, an overreaction. Most of the general public does not understand much about the genetic engineering technology, and scientists need to increase their efforts to educate the public.

Second, most people are not aware of the strict regulations imposed on GM research and active safeguards by most governments. In the United States, research and chemical analyses by many scientists working with the Food and Drug Administration (FDA), the U.S. Department of Agriculture (USDA), or independently have concluded that biotechnology is a safe means of producing foods. Thousands of tests over fifteen years in the United States, along with the consumption of GM foods in the United States for four years, have revealed no evidence of harmful effects related to GM foods. Most food safety problems arise from handling (for example, microbial contamination), for GM and non-GM foods alike.

A third reason for the societal concern is rooted in negative media opinion, opposition by activists, and mistrust of the industry. Most current complaints about GM foods can be categorized into three major areas: the possible detrimental health effects, the potential environmental threats such as "superweeds," and the social, economic, and ethical implications of genetic engineering. Some activists have taken extreme measures, such as destroying field plots and even firebombing a research laboratory. Although the majority of the public do not agree with the extreme measures taken by some activists, some continue to push for mandatory labeling of all foods whose components have derived from GMOs. Activist groups and media also continue to create myths and release misinformation regarding GM foods: GMOs have no benefit to the consumer, they may harm the environment, they are unsafe to eat, the only beneficiary of GM foods is big corporations, GM crops do not benefit small farmers, or they will will drive organic farmers out of business.

Broader Issues in Biotechnology

Although some concerns are genuine—particularly ecological concerns regarding gene flow from GM plants to wild relatives—one should not ignore the fact that safety is a relative concept. Agriculture and animal husbandry have inherent dangers, as do the consumption of their products, regardless of GM or non-GM foods. In response to the demands of activist groups, the European Union (EU) and its member states adopted strict regulations over the import and release of GMOs. GM crops and foods are being subjected to more safety checks and tighter regulation than their non-GM counterparts. Through extensive studies and analyses, both the USDA and the EU have found no perceptible difference between conventional and GM foods. Of course, one cannot ensure consumers of absolute, zero risk with regard to any drug or food product, regardless of how they are produced. The demand for zero risk is more of an emotional reaction than realistically possible. Mandatory labeling on all GM foods is both impractical and technically difficult and would drive food prices to much higher levels than consumers are willing to pay. Farmers and the food industry would have to sort every GMO and store and process them separately. Realizing the complexity, federal agencies like the FDA and USDA have recommended a voluntary labeling system by which the organic and non-GM food products can be marked for consumers who are willing to pay the premium.

Where Do We Go from Here?

Development of new crops is vital for the future of the world. Since conventional breeding cannot keep up with the population explosion, biotechnology may be the best tool available to produce a greater diversity and high quality of safe food on less land, while conserving soil, water, and genetic diversity. To ensure the safety and success of GM crops, scientists and regulators will need to have open and honest communications with the public, building trust through better education and more effective regulatory oversights. In the meantime, the media will also need to convey more credible, balanced information to the public.

As Nobel laureate Norman Borlaug, father of the Green Revolution, stated, "I now say that the world has the technology that is either available or well advanced in the research pipeline to feed a population of 10 billion people. The more pertinent question today is: Will farmers and ranchers be permitted to use this new technology?"

—*Ming Y. Zheng*

See also: Biofertilizers; Biopesticides; Cell Culture: Plant Cells; Cloning; Cloning: Ethical Issues; Cloning Vectors; Genetic Engineering; Genetic Engineering: Agricultural Applications; Genetic Engineering: Historical Development; Genetic Engineering: Industrial Applications; Genetic Engineering: Risks; Genetic Engineering: Social and Ethical Issues; High-Yield Crops; Hybridization and Introgression; Lateral Gene Transfer; Transgenic Organisms.

Further Reading

Borlaug, Norman E. "Ending World Hunger: The Promise of Biotechnology and the Threat of Antiscience Zealotry." *Plant Physiology* 124 (2000): 487-490. The father of Green Revolution and Nobel Peace Prize winner speaks of his unwavering support for GMOs.

Fresco, Louise O. "Genetically Modified Organisms in Food and Agriculture: Where Are We? Where Are We Going?" Keynote Address, Conference on Crop and Forest Biotechnology for the Future, September, 2001. Falkenberg, Sweden: Royal Swedish Academy of Agriculture and Forestry, 2001. Fascinating and informative perspectives on GM foods by an EU scientist.

Potrykus, Ingo. "Golden Rice and Beyond." *Plant Physiology* 125 (2001): 1157-1161. The originator of the wonder rice presents scientific, ethical, intellectual, and social challenges of developing and using the GMOs. Illuminating and insightful.

Web Sites of Interest

AgBioWorld.org. http://www.agbioworld.org. This site advocates the use of biotechnology and GM foods.

Agriculture Network Information Center. http://www.agnic.org. Searchable by keyword or subject category, this site offers "quality" information on topics including transgenic crops.

Transgenic Crops. http://www.colostate.edu/programs/lifesciences/transgeniccrops. This richly illustrated site provides information on genetically modified (GM) foods, including news updates, the history of plant breeding, the making of transgenic plants, government regulations, and risks and concerns. This site is also available in Spanish.

Genetics, Historical Development of

Fields of study: Evolutionary biology; Genetic engineering and biotechnology; History of genetics

Significance: *Genetics is a relatively new branch of biology that explores the mechanisms of heredity. It impacts all branches of biology as well as agriculture, pharmacology, and medicine. Advances in genetics may one day eliminate a wide variety of diseases and disorders and change the way that life is defined.*

Key terms

CHROMOSOME THEORY OF HEREDITY: the theory put forth by Walter Sutton that genes are carried on cellular structures called chromosomes

MENDELIAN GENETICS: genetic theory that arose from experiments conducted by Gregor Mendel in the 1860's, from which he deduced the principles of dominant traits, recessive traits, segregation, and independent assortment

MODEL ORGANISMS: organisms, from unicellular to mammals, that are suitable for genetic research because they are small and easy to keep alive in a laboratory, reproduce a great number of offspring, and can produce many generations in a relatively short period of time

ONE GENE-ONE ENZYME HYPOTHESIS: the notion that a region of DNA that carries the information for a gene product codes for a particular enzyme, later refined to the "one gene-one protein" hypothesis and then to "one gene-one polypeptide" principle

Charles Darwin

The prevailing public attitude of the mid-nineteenth century was that all species were the result of a special creation and were immutable; that is, they remained unchanged over time. The work of Charles Darwin challenged that attitude. As a young man, Darwin served as a naturalist on the HMS *Beagle*, a British ship that mapped the coastline of South America from 1831 to 1836. Darwin's observations of life-forms and their adaptations, especially those he encountered on the Galápagos Islands, led him to postulate that living species shared common ancestors with extinct species and that the pressures of nature—the availability of food and water, the ratio of predators to prey, and competition—exerted a strong influence over which species were best able to exploit a given habitat. Those best able to take advantage of an environment would survive, reproduce, and, by reproducing, pass their traits on to the next generation. He called this response to the pressures of nature "natural selection": Nature selected which species would be capable of surviving in any given environment and, by so doing, directed the development of species over time.

When Darwin returned to England, he shared his ideas with other eminent scientists but had no intention of publishing his notebooks, since he knew that his ideas would bring him into direct conflict with the society in which he lived. However, in 1858, he received a letter from a young naturalist named Alfred Russel Wallace. Wallace had done the same type of collecting in Malaysia that Darwin had done in South America, had observed the same phenomena, and had drawn the same conclusions. Wallace's letter forced Darwin to publish his findings, and in 1858, a joint paper by both men on the topic of evolution was presented at the London meeting of the Linnean Society. In 1859, Darwin reluctantly published *On the Origin of Species by Means of Natural Selection*. The response was immediate and largely negative. While the book became a best-seller, Darwin

found himself under attack from religious leaders and other prominent scientists. In his subsequent works, he further delineated his proposals on the emergence of species, including man, but was never able to answer the pivotal question that dogged him until his death in 1882: If species are in fact mutable (capable of change over long periods of time), by what mechanism is this change possible?

Gregor Mendel

Ironically, it was only six years later that this question was answered, and nobody noticed. Today, Gregor Mendel is considered the "father" of genetics, but, in 1865, he was an Augustinian monk in a monastery in Brunn, Austria (now Brno, Czech Republic). From 1856 to 1863, he conducted a series of experiments using the sweet pea (*Pisum sativum*), in which he cultivated more than twenty-eight thousand plants and analyzed seven different physical traits. These traits included the height of the plant, the color of the seed pods and flowers, and the physical appearance of the seeds. He cross-pollinated tall plants with short plants, expecting the next generation of plants to be of medium height. Instead, all the plants produced from this cross, which he called the F_1 (first filial) generation, were tall. When he crossed plants of the F_1 generation, the next generation of plants (F_2) were both tall and short at a 3:1 ratio; that is, 75 percent of the F_2 generation of plants were tall, while 25 percent were short. This ratio held true whether he looked at one trait or multiple traits at the same time. He coined two phrases still used in genetics to describe this phenomenon: He called the trait that appeared in the F_1 generation "dominant" and the trait that vanished in the F_1 generation "recessive." While he knew absolutely nothing about chromosomes or genes, he postulated that each visible physical trait, or phenotype, was the result of two "factors" and that each parent contributed one factor for a given trait to its offspring. His research led him to formulate several statements that are now called the Mendelian principles of genetics.

Mendel's first principle is called the principle of segregation. While all body cells contain two copies of a factor (what are now called genes), gametes contain only one copy. The factors are segregated into gametes by meiosis, a specialized type of cell division that produces gametes. The principle of independent assortment states that this segregation is a random event. One factor will segregate into a gamete independently of other factors contained within the dividing cell. (It is now known that there are exceptions to this rule: Two genes carried on the same chromosome will not assort independently.)

To make sense of the data he collected from twenty-eight thousand plants, Mendel kept detailed numerical records and subjected his numbers to statistical analysis. In 1865, he presented his work before the Natural Sciences Society. He received polite but indifferent applause. Until Mendel, scientists rarely quantified their findings; as a result, the scientists either did not understand Mendel's math or were bored by it. In either case, the scientists completely overlooked the significance of his findings. Mendel published his work in 1866. Unlike Darwin's work, it was not a best-seller. Darwin himself died unaware of Mendel's work, in spite of the fact that he had an unopened copy of Mendel's paper in his possession. Mendel died in 1884, two years after Darwin, with no way of knowing the eventual impact his work was to have on the scientific community. That impact began in 1900, when three botanists, working in different countries with different plants, discovered the same principles as had Mendel. Hugo De Vries, Carl Correns, and Erich Tschermak von Seysenegg rediscovered Mendel's paper, and all three cited it in their work. Sixteen years after his death, Mendel's research was given the respect it deserved, and the science of genetics was born.

Pivotal Research in Genetics

In 1877, Walter Fleming identified structures in the nuclei of cells that he called chromosomes; he later described the material of which chromosomes are composed as "chromatin." In 1900, William Bateson introduced the term "genetics" to the scientific vocabulary. Wilhelm Johannsen expanded the terminology the following year with the introduction of the terms "gene," "genotype," and "pheno-type." In fact, 1901 was an exciting year in the history of genetics: The ABO blood group was discovered by Karl Landsteiner; the role of the X chromosome in determining gender was described by Clarence McClung; Reginald Punnett and William Bateson discovered genetic linkage; and De Vries introduced the term "mutation" to describe spontaneous changes in the genetic material. Walter Sutton suggested a relationship between genes and chromosomes in 1903. Five years later, Archibald Garrod, studying a strange clinical condition in some of his patients, determined that their disorder, called alkaptonuria, was caused by an enzyme deficiency. He introduced the concept of "inborn errors of metabolism" as a cause of certain diseases. That same year, two researchers named Godfrey Hardy and Wilhelm Weinberg published their extrapolations on the principles of population genetics.

From 1910 to 1920, Thomas Hunt Morgan, with his graduate students Alfred Sturtevant, Calvin Bridges, and Hermann Müller, conducted a series of experiments with the fruit fly *Drosophila melanogaster* that confirmed Mendel's principles of heredity and also confirmed the link between genes and chromosomes. The mapping of genes to the fruit fly chromosomes was complete by 1920. The use of research organisms such as the fruit fly became standard practice. For an organism to be suitable for this type of research, it must be small and easy to keep alive in a laboratory and must produce a great number of offspring. For this reason, bacteria (such as *Escherichia coli*), viruses (particularly those that infect bacteria, called bacteriophages), certain fungi (such as *Neurospora*), and the fruit fly have been used extensively in genetic research.

During the 1920's, Müller found that the rate at which mutations occur is increased by exposure to X-ray radiation. Frederick Griffith described "transformation," a process by which genetic alterations occur in pneumonococci bacteria. In the 1940's, Oswald Avery, Maclyn McCarty, and Colin MacLeod conducted a series of experiments that showed that the transforming agent Griffith had not been able to identify was, in fact, DNA. George Beadle and Edward Tatum proposed the concept of "one gene, one enzyme"; that is, a gene or a region of DNA that carries the information for a gene product codes for a particular enzyme. This concept was further refined to the "one gene, one protein" hypothesis and then to "one gene, one polypeptide." (A polypeptide is a string of amino acids, which is the primary structure of all proteins.)

During the 1940's, it was thought that proteins were the genetic material. Chromosomes are made of chromatin; chromatin is 65 percent protein, 30 percent DNA, and 5 percent RNA. It was a logical conclusion that if the chromosomes were the carriers of genetic material, that material would make up the bulk of the chromosome structure. By the 1950's, however, it was fairly clear that DNA was the genetic material. Alfred Hershey and Martha Chase were able to prove in 1952 that DNA is the heredi-

tary material in bacteriophages. From that point, the race was on to discover the structure of DNA.

For DNA or any other substance to be able to carry genetic information, it must be a stable molecule capable of self-replication. It was known that along with a five-carbon sugar and a phosphate group, DNA contains four different nitrogenous bases (adenine, thymine, cytosine, and guanine). Erwin Chargaff described the ratios of the four nitrogenous bases in what is now called Chargaff's rule: adenine in equal concentrations to thymine, and cytosine in equal concentrations to guanine. What was not known was the manner in which these constituents bonded to each other and the three-dimensional shape of the molecule. Groups of scientists all over the world were working on the DNA puzzle. A group in Cambridge, En-

James Watson (left) and Francis Crick pose with a model of the double-helical structure of DNA. They won the 1962 Nobel Prize in Physiology or Medicine, along with Maurice Wilkins. (Hulton Archive)

Maurice Wilkins poses with a model of a DNA molecule at a London celebration of the fiftieth anniversary of the discovery of the double helix. Wilkins, with Rosalind Franklin, was able to elucidate the molecule's physical structure using X-ray crystallography. (AP/Wide World Photos)

gland, was the first to solve it. James Watson and Francis Crick, supported by the work of Maurice Wilkins and Rosalind Franklin, described the structure of DNA in a landmark paper in *Nature* in 1953. They described the molecule as a double helix, a kind of spiral ladder in which alternating sugars and phosphate groups make up the backbone and paired nitrogenous bases make up the rungs. Arthur Kornberg created the first synthetic DNA in 1956. The structure of the molecule suggested ways in which it could self-replicate. In 1958, Matthew Meselson and Franklin Stahl proved that DNA replication is semiconservative; that is, each new DNA molecule consists of one template strand and one newly synthesized strand.

The Information Explosion

Throughout the 1950's and 1960's, genetic information grew exponentially. This period saw the description of the role of the Y chromosome in sex determination; the description of birth defects caused by chromosomal aberrations such as trisomy 21 (Down syndrome), trisomy 18 (Edwards' syndrome), and trisomy 13 (Patau syndrome); the description of operon and gene regulation by François Jacob and Jacques Monod in 1961; and the deciphering of the genetic code by Har Gobind Khorana, Marshall Nirenberg, and Severo Ochoa in 1966.

The discovery of restriction endonucleases (enzymes capable of splicing DNA at certain sites) led to an entirely new field within genetics called biotechnology. Mutations, such as the sickle-cell mutation, could be identified using restriction endonucleases. Use of these enzymes and DNA banding techniques led to the development of DNA fingerprinting. In 1979, human insulin and human growth hormone were synthesized in *Escherichia coli*. In 1981, the first cloning experiments were successful when the nucleus from one mouse cell was transplanted into an enucleated mouse cell. By 1990, cancer-causing genes called oncogenes had been identified, and the first attempts at human gene therapy had taken place. In 1997, researchers in England successfully cloned a living sheep. As the result of a series of conferences between 1985 and 1987, an international collaboration to map the entire human genome began in 1990. A comprehensive, high-density genetic map was published in 1994, and in 2003 the human genome was completed.

Impact and Applications

The impact of genetics is immeasurable. In less than one hundred years, humans went from complete ignorance about the existence of genes to the development of gene therapies for certain diseases. Genes have been manipulated in certain organisms for the production of drugs, pesticides, and fungicides. Genetic analysis has identified the causes of many hereditary disorders, and genetic counseling has aided innumerable couples in making difficult decisions about their reproductive lives. DNA analysis has led to clearer understanding of the manner in which all species are linked. Techniques such as DNA fingerprinting have had a tremendous impact on law enforcement.

Advances in genetics have also given rise to a wide range of ethical questions with which humans will be struggling for some time to come. Termination of pregnancies, in vitro fertilization, and cloning are just some of the technologies that carry with them serious philosophical and ethical problems. There are fears that biotechnology will make it possible for humans to "play God" and that the use of biotechnology to manipulate human genes may have unforeseen consequences for humankind. For all the hope that biotechnology offers, it carries with it possible societal changes that are unpredictable and potentially limitless. Humans may be able to direct their own evolution; no other species has ever had that capability. How genetic technology is used and the motives behind its use will be some of the critical issues of the future.

—*Kate Lapczynski*

See also: Central Dogma of Molecular Biology; Chromosome Theory of Heredity; Classical Transmission Genetics; DNA Structure and Function; Evolutionary Biology; Genetic Code, Cracking of; Genetic Engineering: Historical Development; Genetics in Television and Films; Genomics; Human Genome Project; Lamarckianism; Mendelian Genetics; Sociobiology; Biographical Dictionary of Important Geneticists (appendix); Nobel Prizes for Significant Discoveries in Genetics (appendix); Time Line of Major Developments in Genetics (appendix).

Further Reading

Ayala, Francisco J., and Walter M. Fitch, eds. *Genetics and the Origin of Species: From Darwin to Molecular Biology Sixty Years After Dobzhansky.* Washington, D.C.: National Academies Press, 1997. Papers presented on Theodosius Dobzhansky's theory of evolution, which argued for a genetics perspective on Darwin's theory of evolution. Illustrations, maps.

Corcos, A., and F. Monaghan. *Mendel's Experiments on Plant Hybrids: A Guided Study.* New Brunswick, N.J.: Rutgers University Press, 1993. Covers the seminal work of Gregor Mendel, along with a biography.

Darwin, Charles. *The Variation of Animals and Plants Under Domestication.* Rev. 2d ed. London: J. Murray, 1875. Anticipating discovery of the genetic basis for phenotypic variation, Darwin describes the remarkable variability of domesticated plants and animals. Bibliography, index.

Edelson, Edward. *Gregor Mendel and the Roots of Genetics.* New York: Oxford University Press, 1999. Story of Mendel's research into the inheritance of traits in the garden pea. Illustrations (including botanical drawings), bibliography, index.

Fujimura, Joan H. *Crafting Science: A Sociohistory of the Quest for the Genetics of Cancer.* Cambridge, Mass.: Harvard University Press, 1996. Provides a medical history of how cancer research shifted in the 1970's from viewing cancer as a set of heterogeneous diseases to a disease of human genes.

King, Robert C., and William D. Stansfield. *A Dictionary of Genetics.* 6th ed. New York: Oxford University Press, 2001. Organized to provide a quick understanding to students and nonspecialists, including more than sixty-five hundred definitions of terms and species names relevant to the study of genetics, as well as a chronology that spans four hundred years of genetic study. Extensive bibliography.

Sturtevant, A. H. *A History of Genetics.* 1965. Reprint. Introduction by Edward B. Lewis. Cold Spring Harbor, N.Y.: Cold Spring Harbor Laboratory Press, 2001. Details Thomas Morgan's research, which laid the foundations for modern, chromosomal genetics.

Tudge, Colin. *The Engineer in the Garden: Genes and Genetics, From the Idea of Heredity to the Creation of Life.* New York: Hill and Wang, 1995. Provides a historical overview of genetics and explores the potential ramifications of past, present, and future genetic advances. Illustrations, bibliography, index.

_____. *In Mendel's Footnotes: An Introduction to the Science and Technologies of Genes and Genetics from the Nineteenth Century to the Twenty-Second.* London: Jonathan Cape, 2000. Investigates the world of biotechnologies, including cloning, genomics, and genetic engineering. Bibliography, index.

Watson, James. *The Double Helix.* 1968. Reprint. New York: Simon and Schuster, 2001. Discusses the race to solve the structure of the DNA molecule.

Web Sites of Interest

Dolan DNA Learning Center, DNA from the Beginning. http://www.dnaftb.org. Sponsored by the Cold Spring Harbor Laboratory, an animated site aimed at those looking for a general introduction to DNA, genes, genetics, and heredity, and their scientific histories. Organized by key concepts.

Electronic Scholarly Publishing Project, Classic Genetics: Foundations. http://www.esp.org. A collection of classic papers marking the development of genetics. Includes a time line.

Genetics in Television and Films

Field of study: History of Genetics; Human genetics and social issues

Significance: *Popular culture expresses attitudes regarding genetics. Films and television programs present biotechnology in extremes of either promoting genetics as a valuable investigative and reproductive tool or demeaning it as a dangerous science which is hazardous to people and environments. Most genetic depictions in these media are more entertaining than accurate.*

Key terms

EUGENICS: the selective application of genetics to produce superior offspring

GENETIC DETERMINISM: how genes might influence behavioral characteristics

Science Fiction

In the 1950's, science-fiction films and television programming gradually incorporated references to genetics. The expansion of biotechnology research in the 1970's inspired fictional plots that focused on genetics to amuse audiences more than educate them. Box-office successes such as *Jurassic Park* (1993) and hit television series, including *The X-Files* (1993-2002), popularized genetics and how it can be applied to transform, extend, and enrich lives. With the exception of documentaries such as *The DNA Revolution* (1998), few films and programs featuring genetics are realistic and accurate.

Science-fiction films and television programs usually depict genetics as a wondrous endeavor that can abruptly go awry. Genetics is often appropriated to provoke rather than to resolve dilemmas. In film and television, genetic engineering is usually equated with power—power that genetically superior characters occasionally abuse. Plots frequently contrast extremes, such as good and evil scientists pitted against each other or combating corrupt administrators and greedy entrepreneurs. Many depictions of genetics perpetuate stereotypes such as mad scientists isolated in laboratories and unaccountable to humankind for their research and creations. A host of biotechnological monsters and mutants populate films.

DNA and Identity

CSI: Crime Scene Investigation, a television series that first aired in 2000, is representative of crime-based television shows that became popular in the late 1990's, in part because of public fascination with the O. J. Simpson murder trial and other high-profile cases in which DNA evidence was showcased in the media. Both episodic drama programming and true-crime shows such as *Cold Case Files* rely on sets that are filled with genetic tools. Scenes depict characters collecting DNA samples from crime scenes and evaluating the tissues in laboratories to identify victims, prove criminals' guilt, or exonerate the falsely accused.

Soap-opera writers often appropriate genetics as a plot device. Characters test DNA to confirm paternity, establish identity, or prove a person's presence at a crime scene. These daytime serials usually restrict access to DNA knowledge to medical and police personnel. Some characters manipulate DNA evidence by switching samples or tampering with laboratory records. In 2002, *Days of Our Lives* introduced a story line involving the genetically engineered Gemini Twins, who displayed previously undocumented DNA patterns.

Cloning Characters

Clones are often depicted as evil creatures that prey on humans. The feature film *The Boys from Brazil* (1978) reveals the potential horrific results if Nazi sympathizers successfully cloned Adolf Hitler. Clones are sometimes shown to be dutiful, almost robotic, helpers. In *Star Wars Episode II: Attack of the Clones* (2002), thousands of clonetroopers are created as soldiers during the clone wars. In these movies, cloning concepts are more futuristic than realistic.

Jurassic Park and its two sequels captured worldwide attention for cloning. Those movies are based on the concept that scientists cloned dinosaurs from DNA preserved in amber. Scientists criticize this movie's premise of cloning a dinosaur from fragments of ancient genetic material as improbable, stating that locating an egg and host animal capable of transforming the DNA into a dinosaur would be difficult if not impossible. More important, DNA of the age required for dinosaurs (more than 65 million years old), even if recovered, would almost certainly be far too degraded to make cloning possible.

The Brazilian soap opera *O clone* (2001-2002; the clone) chronicles Dr. Albieri secretly creating the clone Leo. Albieri is concerned for Leo's health, referring to cloned sheep Dolly's premature aging, and addresses ethical issues related to cloning. Leo suffers identity problems, questions his potential life span, and resents unwanted public attention.

Designer Plots

Hollywood is intrigued with the idea of human genetic modification. Films and programs explore the possibilities of manipulating genes to give characters unnatural advantages. Often these genetic changes create designer bodies in an almost eugenic effort to attain physical perfection and perceived superiority. These presentations usually simultaneously address determinism and how genes might control both positive and negative behaviors unrelated to appearance.

In *Gattaca* (1997), genetically altered characters have power in a futuristic society over normal characters who have not benefited from biotechnology and are relegated to an underclass because of their imperfections. Vincent, a frustrated janitor who aspires to become an astronaut, uses DNA to adopt the appearance of the elite. His genetic transformation enables him to achieve his professional ambition. Vincent's emotional traits are shown to be superior to the physical beauty of the genetically engineered people.

Beginning in 2002, MTV aired *Clone High*, a controversial cartoon featuring clones of significant historical leaders. These characters are presented as angst-ridden teenagers whom the scripts hint represent genetic determinism. For example, Joan of Arc is an atheist, suggesting that she might have been genetically prone to that behavior if she had not been influenced by cultural factors.

Reactions

Genetic-based movie and television programming impacts audiences by influencing

The film Jurassic Park *(1993) posited that dinosaurs could be cloned from ancient DNA—theoretically plausible, but practically impossible due to the extreme degradation of DNA the age of dinosaurs. (AP/ Wide World Photos)*

how people perceive and accept or reject biotechnology. Although these media expand awareness of genetics, they usually are not reliable educational resources and perpetuate misunderstandings. Advertisements for the fictional *O clone*, designed as news broadcasts, were so realistic that many viewers thought an actual person had been cloned.

Errors detract from programs being credible cinema. Movies and series offer simplified depictions of complex scientific processes, suggesting they require minimal time and effort and consistently produce positive results. Viewers develop unrealistic expectations that genetics can quickly solve mysteries because of the immediacy of DNA testing in brief episodes.

Dr. J. Craig Venter criticizes popular culture's concentration on genetic determinism because such emphasis and negative cinematic portrayals might cause people to reject biotechnology instead of recognizing its merits. Experts worry about cinema geneticists acting irresponsibly and unprofessionally. In an effort to improve depictions, some scientists have served as genetics advisers for film and television productions.

—*Elizabeth D. Schafer*

See also: Ancient DNA; Biological Determinism; Chromosome Theory of Heredity; Classical Transmission Genetics; Cloning; Cloning: Ethical Issues; Criminality; DNA Fingerprinting; Eugenics; Eugenics: Nazi Germany; Evolutionary Biology; Forensic Genetics; Genetic Code, Cracking of; Genetic Engineering: Historical Development; Genetic Engineering: Social and Ethical Issues; Human Genetics; Human Genome Project; Lamarckianism; Mendelian Genetics; Patents on Life-Forms; Paternity Tests; Race; Sociobiology.

Further Reading

DeSalle, Robert, and David Lindley. *The Science of "Jurassic Park" and "The Lost World."* New York: BasicBooks, 1997. Authors reveal how the cloning of dinosaurs would be impossible to achieve.

Glassy, Mark C. *The Biology of Science Fiction Cinema.* Jefferson, N.C.: McFarland, 2001. Cancer researcher critiques films for plausibility of biotechnology and explains scientific principles and whether the results could be duplicated off film.

Simon, Anne. *The Real Science Behind the X-Files: Microbes, Meteorites, and Mutants.* New York: Simon & Schuster, 1999. The official science adviser to the television series discusses the authenticity of many of the genetic plots.

Turney, Jon. *Frankenstein's Footsteps: Science, Genetics, and Popular Culture.* New Haven, Conn.: Yale University Press, 1998. Science communication expert analyses how people perceive genetics as presented in films.

Web Site of Interest

The Science Behind the *X-Files.* http://huah .net/scixf/xeve.html. Describes the genetics-related science presented in each episode of this television series, and provides relevant links to scientific experts and research institutes.

Genome Size

Field of study: Molecular genetics
Significance: *Genome size, the total amount of genetic material within a cell of an organism, varies 200,000-fold among species. Since the 1950's it has been clear that there is no obvious link between an organism's complexity and the size of its genome, although numerous hypotheses to explain this paradox exist.*

Key terms

C-VALUE: the characteristic genome size for a species

CHROMOSOME: a self-replicating structure, consisting of DNA and protein, that contains part of the nuclear genome of a eukaryote; also used to describe the DNA molecules comprising the prokaryotic genome

GENOME: the entire genetic complement of an organism

JUNK DNA: a disparaging (and now known to be inaccurate) characterization of the non-coding DNA content of a genome

REASSOCIATION KINETICS: a technique that uses hybridization of denatured DNA to reveal DNA classes differing in repetition frequency

REPETITIVE DNA: a DNA sequence that is repeated two or more times in a DNA molecule or genome

Genome Sizes in Prokaryotes vs. Eukaryotes

Wide variation in genome size exists among species, from 580,000 bases in the bacterium *Mycoplasma genitalium* to 670 billion bases in the protist *Amoeba dubia*. In general, prokaryotic genomes are smaller than the genomes of eukaryotes, although a few prokaryotes have genomes that are larger than those of some eukaryotes. The largest known prokaryotic genome (10 million bases in the cyanobacterium *Nostoc punctiforme*) is several times larger than the genomes of parasitic eukaryotic microsporidia, with genome sizes of approximately 3 million bases. Within the prokaryotes, the archaea have a relatively small range of genome sizes, with the majority of species in the 1- to 3-million-base range, while bacterial species have been found with genomes differing by twentyfold.

Contrary to expectations, there is no obvious correlation between genome size and organismal complexity in eukaryotes. For example, the genome of a human is tenfold smaller than the genome of a lily, twenty-five-fold smaller than the genome of a newt, and two-hundred-fold smaller than the genome of an amoeba. The characteristic genome size of a species is called the C-value; the lack of relationship between genome size, number of genes, and organismal complexity has been termed the "C-value paradox."

Reasons for Size Differences

The majority of DNA in most eukaryotes is noncoding. Previously known as "junk" DNA, this DNA (comprising up to 98.5 percent of some genomes) does not contain the coding sequences for proteins. The complexity of DNA can be characterized using a technique called reassociation kinetics. DNA is sheared into pieces of a few hundred bases, heated to denature into single strands, then allowed to renature during cooling. The rate of renaturation is related to the sequence complexity: DNA sequences present in numerous copies will renature more rapidly than unique DNA sequences. Unique DNA sequences usually represent protein-coding regions, whereas repetitive DNA generally does not encode traits. In many genomes, three types of DNA can be identified by reassociation kinetics: highly repetitive DNA, middle repetitive DNA, and unique DNA. Prokaryotes have little or no repetitive DNA. Among eukaryotes, the amounts of the three types of DNA varies. The share of the genome dedicated to genes is relatively constant, whereas the amount of repetitive DNA, 10-70 percent of the total, varies widely even within families of organisms. The existence of noncoding DNA appears to account for the lack of correlation between genome size and complexity because complexity may be more directly related to number of genes, a number which does appear to have more correlation to organismal complexity.

The variation in the amount of repetitive DNA, even within families, may be related to the spontaneous rate of DNA loss. Small genomes may be small because they throw away junk DNA very efficiently, whereas large genomes may be less able to weed out unnecessary DNA. Studies on several invertebrates support this hypothesis: Species within a family with large genomes have substantially lower spontaneous DNA losses.

Genome size does have a positive correlation with cell size and a negative correlation with cell division rate in a number of taxa. Because of these correlations, genome size is associated with developmental rate in numerous species. This correlation is not exact, however. For some organisms (particularly plants) with relatively simple developmental complexity, developmental rate is constrained by external factors such as seasonal changes, while for others (amphibians with time-limited morphogenesis) developmental complexity overwhelms the effects of developmental rate.

Differences in Chromosome Number

The genomes of eukaryotes are organized into sets of two or more linear DNA molecules, each contained in a chromosome. The number of chromosomes varies from 2 in females of the ant species *Myrmecia pilosula* to 46 in humans to

94 in goldfish. These numbers represent the diploid number of chromosomes. A genome that contains three or more full copies of the haploid chromosome number is polyploid. As a general rule polyploids can be tolerated in plants but are rarely found in animals. One reason is that the sex balance is important in animals and variation from the diploid number results in sterility. Chromosome number appears to be unrelated to genome size or to most other biological features of the organism.

For most of the prokaryotes studied, the prokaryotic genome is contained in a single, circular DNA molecular, with the possible addition of small, circular, extrachromosomal DNA molecules called plasmids. However, some prokaryotes have multiple chromosomes, some of which are linear; and some prokaryotes have several very large plasmids, nearly the size of the bacterial chromosome.

—*Lisa M. Sardinia*

See also: Ancient DNA; Evolutionary Biology; Gene Families; Genomics; Human Genetics; Molecular Clock Hypothesis; Noncoding RNA Molecules; Plasmids; Pseudogenes; Repetitive DNA; Transposable Elements.

Further Reading

Petrov, Dmitri A. "Evolution of Genome Size: New Approaches to an Old Problem." *Trends in Genetics* 17, no. 1 (2001): 23-28. Petrov, a longtime researcher on genome complexity, reviews current theories of genome complexity and offers new explanations for the lack of relationship between genome size and organismal complexity.

Petsko, Gregory A. "Size Doesn't Matter." *Genome Biology* 2, no. 3 (2001): comment 1003.1-1003.2. Expands the discussion of genome size to proteome, or functional, size.

Genomic Libraries

Fields of Study: Bioinformatics; Techniques and methodologies
Significance: *A genomic library is a collection of clones of DNA sequences, each containing a relatively short piece of the genome of an organism. All of the clones together contain most or all of the genome. To find a specific gene, scientists can screen the library using labeled probes of various kinds.*

Key terms

GENOME: all the genetic material carried by a cell
LAMBDA (λ) PHAGE: a virus that infects bacteria and then makes multiple copies of itself by taking over the infected bacterium's cellular machinery
LIGATION: the joining together of two pieces of DNA using the enzyme ligase

What Is a Genomic Library?

Scientists often need to search through all the genetic information present in an organism to find a specific gene. It is thus convenient to have collections of genetic sequences stored so that such information is readily available. These collections are known as genomic libraries.

The library metaphor is useful in explaining both the structure and function of these information-storage centers. If one were interested in finding a specific literary phrase, one could go to a conventional library and search through the collected works. In such a library, the information is made up of letters organized in a linear fashion to form words, sentences, and chapters. It would not be useful to store this information as individual words or letters or as words collected in a random, jumbled fashion, as the information's meaning could not then be determined. The more books a library has, the closer it can come to having the complete literary collection, although no collection can guarantee that it has every piece of written word. The same is true of a genomic library. The stored pieces of genetic information cannot be individual bits but must be ordered sequences that are long enough to define a gene. The longer the string of information, the easier it is to make sense of the gene they make up, or "encode." The more pieces of genetic information a library has, the more likely it is to contain all the information present in a cell. Even a large collection of sequences, however, cannot guarantee that it contains every piece of genetic information.

How Is a Genomic Library Created?

In order for a genomic library to be practical, some method must be developed to put an entire genome into discrete units, each of which contains sufficiently large amounts of information to be useful but which are also easily replicated and studied. The method must also generate fragments that overlap one another for short stretches. The information exists in the form of chromosomes composed of millions of units known as base pairs. If the information were fragmented in a regular fashion—for example, if it were cut every ten thousand base pairs—there would be no way to identify each fragment's immediate neighbors. It would be like owning a huge, multivolume novel without any numbering system: It would be almost impossible to determine with which book to

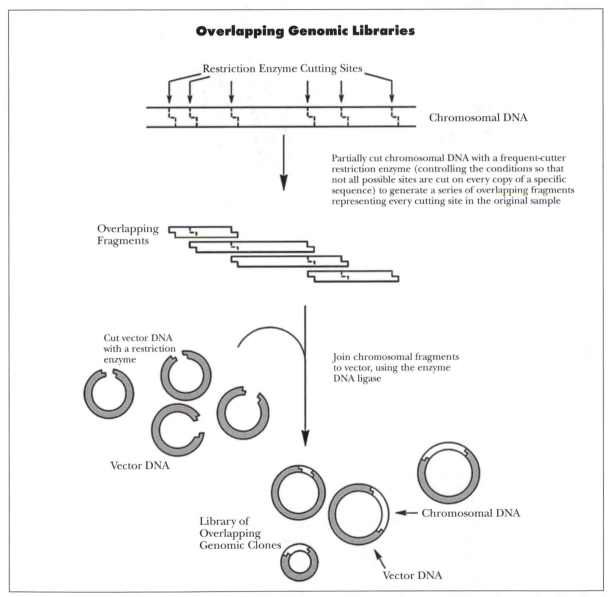

Overlapping Genomic Libraries

Restriction Enzyme Cutting Sites

Chromosomal DNA

Partially cut chromosomal DNA with a frequent-cutter restriction enzyme (controlling the conditions so that not all possible sites are cut on every copy of a specific sequence) to generate a series of overlapping fragments representing every cutting site in the original sample

Overlapping Fragments

Cut vector DNA with a restriction enzyme

Join chromosomal fragments to vector, using the enzyme DNA ligase

Vector DNA

Library of Overlapping Genomic Clones

Chromosomal DNA

Vector DNA

Genomic libraries are collections of clones of chromosomal DNA segments. These must be constructed in such a way that the order of the segments can be determined. To accomplish this, regions of each cloned segment overlap with other segments. (U.S. Department of Energy Human Genome Program, http://www.ornl.gov/hgmis)

DNA samples are stored in the "Big Bertha" freezer at the Armed Forces Institute of Pathology in Gaithersburg, Maryland. (AP/ Wide World Photos)

start and which to proceed to next. Similarly, without some way of tracking the order of the genetic information, it would be impossible to assemble the sequence of each subfragment into the big continuum of the entire chromosome. The fragments are thus cut so that their ends overlap. With even a few hundred base pairs of overlap, the shared sequences at the end of the fragments can be used to determine the relative position of the different fragments. The different pieces can then be connected into one long unit, or sequence.

There are two common ways to fragment DNA, the basic unit of genetic information, to generate a library. The first is to disrupt the long strands of DNA by forcing them rapidly through a narrow hypodermic needle, creating forces that tear, or shear, the strands into short fragments. The advantage of this method is that the fragment ends are completely random. The disadvantage is that the sheared ends must be modified for easy joining, or ligation. The other method is to use restriction endonu-

cleases, enzymes that recognize specific short stretches of DNA and cleave the DNA at specific positions. To create a library, scientists employ restriction enzymes that recognize four-base-pair sequences for cutting. Normally, the result of cleavage with such an enzyme would be fragments with an average size of 256 base pairs. If the amount of enzyme in the reaction is limited, however, only a limited number of sites will be cut, and much longer fragments can be generated. The ends created by this cleavage are usable for direct ligation into vectors, but the distribution of cleavage sites is not as random as that produced by shearing.

In a conventional library, information is imprinted on paper pages that can be easily replicated by a printing press and easily bound into a complete unit such as a book. Genetic information is stored in the form of DNA. How can the pieces of a genome be stored in such a way that they can be easily replicated and maintained in identical units? The answer is to take the DNA fragments and attach, or ligate, them into lambda phage DNA. When the phage infects bacteria, it makes copies of itself. If the genomic fragment is inserted into the phage DNA, then it will be replicated also, making multiple exact copies (or clones) of itself.

To make an actual library, DNA is isolated from an organism and fragmented as described. Each fragment is then randomly ligated into a lambda phage. The pool of lambda phage containing the inserts is then spread onto an agar plate coated with a "lawn" or confluent layer of bacteria. Wherever a phage lands, it begins to infect and kill bacteria, leaving a clear spot, or "plaque," in the lawn. Each plaque contains millions of phages with millions of identical copies of one fragment from the original genome. If enough plaques are generated on the plate, each one containing some random piece of the genome, then the entire genome may be represented in the summation of the DNA present in all the plaques. Since the fragment generation is random, however, the completeness of the genomic library can only be estimated. It takes 800,000 plaques containing an average genomic fragment of 17,000 base pairs to give a 99 percent probability that the total will contain a specific human gene. While this may sound like a large number, it takes only fifteen teacup-sized agar plates to produce this many plaques. A genetic library pool of phage can be stored in a refrigerator and plated out onto agar petri dishes whenever needed.

How Can a Specific Gene Be Pulled out of a Library?

Once the entire genome is spread out as a collection of plaques, it is necessary to isolate the one plaque containing the specific sequences desired from the large collection. To accomplish this, a dry filter paper is laid onto the agar dish covered with plaques. As the moisture from the plate wicks into the paper, it carries with it some of the phage. An ink-dipped needle is pushed through the filter at several spots on the edge, marking the same spot on the filter and the agar. These will serve as common reference points. The filter is treated with a strong base that releases the DNA from the phage and denatures it into single-stranded form. The base is neutralized, and the filter is incubated in a salt buffer containing radioactive single-stranded DNA. The radioactive DNA, or "probe," is a short stretch of sequence from the gene to be isolated. If the full gene is present on the filter, the probe will hybridize with it and become attached to the filter. The filter is washed, removing all the radioactivity except where the probe has hybridized. The filters are exposed to film, and a dark spot develops over the location of the positive plaque. The ink spots on the filter can then be used to align the spot on the filter with the positive plaque on the plate. The plaque can be purified, and the genomic DNA can then be isolated for further study.

It may turn out that the entire gene is not contained in the fragment isolated from one phage. Since the library was designed so that the ends of one fragment overlap with the adjacent fragment, the ends can be used as a probe to isolate neighboring fragments that contain the rest of the gene. This process of increasing the amount of the genome isolated is called genomic walking.

—J. Aaron Cassill

See also: Bioinformatics; cDNA Libraries; DNA Fingerprinting; DNA Sequencing Technology; Forensic Genetics; Genetic Testing: Ethical and Economic Issues; Genetics, Historical Development of; Genomics; Human Genome Project; Icelandic Genetic Database; Linkage Maps; Proteomics; Restriction Enzymes; Reverse Transcriptase.

Further Reading

Bird, R. Curtis, and Bruce F. Smith, eds. *Genetic Library Construction and Screening: Advanced Techniques and Applications*. New York: Springer, 2002. A laboratory manual.

Bishop, Martin J., ed. *Guide to Human Genome Computing*. 2d ed. San Diego: Academic Press, 1998. Guides researchers with organizing, analyzing, storing, and retrieving information about genome organization, DNA sequence information, and macromolecular function.

Cooper, Necia Grant, ed. *The Human Genome Project: Deciphering the Blueprint of Heredity*. Foreword by Paul Berg. Mill Valley, Calif.: University Science Books, 1994. Chapters include "Understanding Inheritance: An Introduction to Classical and Molecular Genetics," "Mapping the Genome: The Vision, the Science, the Implementation," "DNA Libraries: Recombinant Clones for Mapping and Sequencing," and "Computation and the Genome Project: A Shotgun Wedding."

Hoogenboom, H. R. "Designing and Optimizing Library Selection Strategies for Generating High-Affinity Antibodies." *Trends in Biotechnology* 15 (1997). Contains detailed information about laboratory techniques used to engineer monoclonal antibodies.

Sambrook, Joseph, et al. *Molecular Cloning: A Laboratory Manual*. 3d ed. Cold Spring Harbor, N.Y.: Cold Spring Harbor Laboratory Press, 1989. A standard manual for more than twenty years, provides complete descriptions of 250 laboratory protocols in DNA science, including techniques for isolating, analyzing and cloning both large and small DNA molecules; cDNA cloning and exon trapping, amplification of DNA, mutagenesis, and DNA sequencing; and methods to screen expression libraries, analyze transcripts and proteins, and detect protein-protein interactions.

Sandor, Suhai, ed. *Theoretical and Computational Methods in Genome Research*. New York: Plenum Press, 1997. Covers topics such as mathematical modeling, three-dimensional modeling of proteins, and applications, such as drug design, construction and use of databases, techniques of sequence analysis and functional domains, and approaches to linkage analysis.

Watson, James, et al. *Recombinant DNA*. New York: W. H. Freeman, 1992. Uses accessible language and exceptional diagrams to give a concise background on the methods, underlying concepts, and far-reaching applications of recombinant DNA technology.

Web Sites of Interest

Johns Hopkins University. The Genome Database. http://gdbwww.gdb.org. The official central storage center for gene mapping data compiled in the Human Genome Initiative, an international effort to decode and analyze human DNA. Intended for scientists, the site presents information in three categories: regions, maps, and variations of the human genome.

National Center for Biotechnology Information. http://www.ncbi.nlm.nih.gov. A central repository for biological information, including links to genome projects and genomic science. Maintains GenBank, a comprehensive, annotated collection of publicly available DNA sequences.

Genomics

Field of study: Molecular genetics
Significance: *Genomics involves studying the entire complement of genes that an organism possesses. A genomic approach to biology uses modern molecular and computational techniques in conjunction with large-scale experimental approaches to sequence, identify, map, and determine the function of genes. It is also concerned with the structure and evolution of the genome as a whole.*

Key terms

BACTERIAL ARTIFICIAL CHROMOSOMES (BACs): cloning vectors that hold inserts of 100-200 kilobase pairs of foreign DNA

EXPRESSED SEQUENCE TAGS (EST) LIBRARY: a survey of expressed sequence tags, which are partial sequences from messenger RNA (mRNA)

Definition

A genome comprises all of the DNA that occurs in each cell of an organism. For prokaryotes, which are always single-celled, it comprises all of the DNA within the bacterial cell that is specific to that species. Other DNA molecules may also reside in a bacterial cell, such as plasmids (small extra pieces of circular DNA) and bacteriophage DNA (bacterial virus DNA). These extra pieces are not considered a part of the genome. In eukaryotes, the genome typically includes just the DNA in the nucleus, which is composed of linear chromosomes. All eukaryotic cells also have DNA in their mitochondria, the organelle that carries out a portion of cellular respiration. It is a circular molecule and is sometimes referred to as the mitochondrial genome or simply mitochondrial DNA (mtDNA). Plants and some single-celled organisms have, in addition to mitochondria, another type of organelle called a chloroplast, which also has a circular DNA molecule. This DNA is called the chloroplast genome, or simply chloroplast DNA (cpDNA).

Sequenced Organisms

Many genomes—from vertebrate mitochondria at about 16,000 base pairs (bp) to mammals at more than 3 billion bp—have been completed, and although there still is no one repository for all these data, the National Center for Biotechnology Information maintains GenBank, which keeps track of many. Prokaryotic genomes (both *Euacteria*, or simply *Bacteria*, and *Archaea*) are now relatively minor projects on the order of 0.6-8 megabase pairs (Mbp), and the number completed is now in the hundreds, because large sequencing centers are capable of completing thirty or more per month.

Compared to the prokaryotes, eukaryotic genomes generally involve much more work. Vertebrate genomes that have either been completed or mostly sequenced and are awaiting assembly include *Homo sapiens* (humans) at about 3.3 billion bp, *Mus musculus* (the mouse) at about 3 billion bp, *Rattus norvegicus* (the rat) at about 2.8 billion bp, *Danio rerio* (the zebra fish) at about 1.7 bp, *Fugu rubripes* (the pufferfish) at about 3.6 million bp, and *Tetraodon nigroviridis* (another form of pufferfish) at 3.8 million bp. Projects for which more than 90 percent of sequencing may be complete by 2010 (if not sooner) include *Pan paniscus* (the chimpanzee) at about 3.3 billion bp, *Macaca mulatta* (the rhesus monkey), *Papio cynocephalus* (the yellow baboon), *Bos taurus* (the cow), *Sus scrofa* (the pig), *Canis familiaris* (the dog), *Felis catus* (the cat), *Equus caballus* (the horse), *Oryctolagus cuniculus* (the rabbit), *Gallus gallus* (the chicken), *Xenopus tropicalis* (a frog), and *Xenopus laevis* (another species of frog). These include most of the well-known experimental vertebrates as well as others of commercial importance. As in the Human Genome Project, annotation, closing gaps, and checking assemblies may require additional years.

Beyond the next few years, there is strong advocacy for genomic sequences of less well-known experimental animals, including *Peromyscus* (the deer mouse) and *Tupaia* (the tree shrew), as well as representatives of distinct evolutionary lineages such as elephants. Sequencing the genomes of such animals is important, since the best animals for comparative genomics are not necessarily experimentally or commercially important. For example, the small size of the *Fugu* genome or the intermediate size of the marsupial genome makes these valuable because of their uniqueness, while at the same time they possess copies of different variants of many of the same genes. Such comparisons may provide insights into gene function and interactions among genes and their products.

Nonvertebrate animal genomes have been sequenced for *Ciona intestinalis* (the sea-squirt), *Anopheles gambiae* (the malaria mosquito), *Drosophila melanogaster* (the fruit or vinegar fly), and *Caenorhabditis briggsae* and *C. elegans* (nematode worms). Projects soon to be completed include *Apis mellifera ligustica* (the honeybee), *Culex* and *Aedes* (mosquitoes), *Glossina morsitans* (the tsetese fly), and *Brugia malayi* (the nematode that causes elephantiasis). For comparative reasons a cnidarian and a mollusk would be valuable.

Fungi projects include *Aspergillus* species, *Candida albicans* (which causes thrush infections), *Cryptococcus neoformans*, *Neurospora crassa* (orange bread mold), *Phanerochaete chrysosporium* (white wood rot), *Saccharomyces cerevisiae* (baker's and brewer's yeast), *Schizosaccharomyces pombe* (fission yeast), and *Pneumocystis carinii* (which causes pneumonia). Many more are soon to start.

Plants often have very large genomes because of duplication events (tetraploidy). *Arabidopsis thaliana* (thale cress) at about 115 Mbp and *Oryza sativa* (rice) at about 430 Mbp have been completed, and large-scale EST sequencing projects are under way for wheat, potato, cotton, tomato, barley and corn, which all have much larger genomes.

A wide variety of parasites are also being sequenced: *Cryptosporidium parvum*, which causes diarrhea; *Plasmodium falciparum*, which causes malaria; *Toxoplasma gondii*, a microsporidian; *Encephalitozoon cuniculi*, kinetoplastids; *Leishmania major*, which causes leishmaniasis; *Trypanosoma brucei*, which causes sleeping sickness; *Trypanosoma cruzi*, which causes Chagas' disease; *Thalassiosira pseudonana*, a diatom; *Dictyostelium discoideum*, a slime mold; and *Entamoeba histolytica*, which causes amebic dysentery.

—Peter J. Waddell and Michael J. Mclachlan

Because the genome includes all of the genes that are expressed in an organism, knowing its nucleotide sequence is considered the first step in a complete understanding of the genetics of an organism. It must be emphasized that much more work follows this first step, because just knowing the nucleotide sequence of all the genes does not necessarily identify their function or how they interact with other genes. One important side benefit of having the complete genome sequence is that it can greatly speed the discovery of mutant genes. The human genome sequence, completed by the Human Genome Project in 2003, has already enabled medical geneticists to find a number of genes for genetic defects.

Sequencing Whole Genomes

A number of complementary strategies are involved in sequencing a genome. One approach is the shotgun sequencing of mapped clones. Large sections of DNA are cloned into vectors such as bacterial artificial chromosomes (BACs). A physical map of each BAC is made using techniques such as restriction mapping or the assignment of previously known sequence elements. The BAC maps are compared to identify overlapping clones, forming a map of long contiguous regions of the genome. BACs are selected from this map and the inserts are randomly fragmented into short pieces, 1-2 kilobase pairs (kb), and subcloned into vectors. Subclones are selected at random and sequenced. Many subclones are sequenced (often enough to provide sevenfold coverage of the clone) and then assembled to yield the contiguous sequence of the original BAC insert. The sequences from overlapping BACs are then assembled. In the finishing stage, additional bridging sequences are obtained to close gaps where there were no overlapping clones.

Whole genome shotgun sequencing involves randomly fragmenting the whole genome and sequencing clones without an initial map. Small clones (up to around 2 kb) are sequenced and assembled into contiguous regions with the help of sequences from larger (10-50 kb) clones that form a scaffold. The sequence is then linked to a physical map of the organism's chromosomes. This method works effectively on bacterial genomes because of their small size and lack of repetitive DNA. However, the amount of repetitive sequence in eukaryotes can lead to difficulties for sequence assembly and gap filling. Therefore, a mapped, clone-based approach may be needed to finish such sequences. A genomic sequencing project may use a combination of the mapped clone method and whole genome shotgun sequencing to produce a completed genomic sequence.

An important adjunct to the genomic sequence is an extensive catalog of expressed sequence tags (ESTs), or full-length mRNAs from many different tissue types. This is achieved by reverse transcribing mRNA to complementary DNA (cDNA) and then sequencing. If a genome is impractically large to sequence at present (due to large amounts of noncoding DNA), this stage alone can yield much useful information.

Annotation

The annotation process involves gathering and presenting information about the location of genes, regulatory elements, structural elements, repetitive DNA, and other factors. It is important to integrate any previously known information regarding the genome, such as location of ESTs, at this stage. A powerful approach to identifying genes is to map ESTs and mRNAs to the genome. This will identify many of the protein-coding genes and can reveal the intron-exon structure plus possible alternative splicing of the gene. It will not identify most functional RNA genes, and how to do so effectively is an open question. Indeed, how many functional RNA genes there may be in eukaryotic genomes is unclear. For example, in humans approximately thirty-five thousand protein-coding genes have been identified, but there is evidence of many more transcribed sequences, and exactly what these are is unknown.

Some genes can be identified in the genomic sequence by the comparative approach—that is, by showing significant sequence similarity (for example, via BLAST algorithm) with annotated genes from other organisms. Such an approach becomes more powerful as the genomes of more organisms are published.

Computational methods can also be used to predict regions of the sequence that may represent genes. These rely on identifying patterns in the genomic sequence that resemble known properties of protein-coding genes, such as the presence of an open reading frame or sequence elements associated with promoters, intron-exon boundaries, and the 3' tail.

Functional Genomics

The availability of information identifying the majority of genes in an organism allows new kinds of experiments to be devised, and on a larger scale than ever before. Functional genomics, for example, aims to assign a functional role to each gene and identify the tissue type and developmental stage at which it is expressed. Identifying all genes in a genome makes it possible to determine the effect of altering the expression of each gene, through the use of knockouts, gene silencing, or transgenic experiments. Technologies such as microarray analysis allow mRNA expression levels to be measured for tens of thousands of genes simultaneously, while proteomic methods such as mass spectroscopy are beginning to allow high-throughput measurements of proteins. In these areas genomics overlaps with transcriptomics, proteomics, and specialties such as glycomics.

Structural Genomics

Structural genomics touches upon proteomics in the need to consider structural changes when there are post-translational changes or binding with other molecules. Structural genomics aims to define the three-dimensional folding of all protein products that an organism produces. The structure of a protein can provide insights into its function and mode of action. Identifying all the genes in a genome allows the amino acid sequence of each protein to be inferred from the DNA, and comparisons between them allow proteins (or characteristic sections of a protein, called folds or domains) to be identified and classified into families.

Structural prediction typically proceeds via each gene being cloned and then expressed. The protein product is then purified, and its

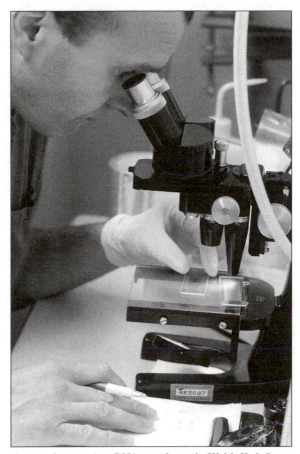

A researcher examines DNA samples at the W. M. Keck Center for Comparative and Functional Genomics at the University of Illinois, Urbana. (AP/Wide World Photos)

structures are experimentally determined using methods such as X-ray crystallography and nuclear magnetic resonance (NMR) spectroscopy. Computational methods of structural prediction, either *ab initio* (from the beginning) or alternatively by computational prediction, aided by the known structure of a related protein, are generally inferior to direct experimental approaches, but these fields are rapidly advancing and are the key to the future.

Comparative Genomics

Comparative genomics expands knowledge through the comparison of different organisms' genomes. This is essential to the annotation of genomic sequences. For example, both otherwise unknown genes and particularly regulatory elements in humans and mice were first revealed by identifying conserved regions of their genomic sequences. This can identify genes homologous to those in other species or identify a new member of a gene family. Comparing genomes can give insights into evolutionary questions about a particular gene or the organisms themselves. Important information can also be discovered about the regulation of different genes, the effects of different gene expression patterns between different species, and how the genome of each species came to be the way it is. Comparative genomics essentially rests upon phylogenetic methodology to describe the pattern and process of molecular evolution (phylogenomics).

—*Peter J. Waddell and Michael J. Mclachlan*

See also: Bioinformatics; cDNA Libraries; Chromosome Walking and Jumping; DNA Sequencing Technology; Gene Families; Genetic Engineering; Genome Size; Genomic Libraries; Human Genome Project; Molecular Clock Hypothesis; Protein Structure; Protein Synthesis; Proteomics; Reverse Transcription; RNA World.

Further Reading

International Human Genome Sequencing Consortium. "Initial Sequencing and Analysis of the Human Genome." *Nature* 409, no. 6822 (2001): 860-921. The publication of the first draft of the Human Genome Project. The whole journal issue contains many other papers considering the structure, function, and evolution of the human genome.

Venter, J. C., et al. "The Sequence of the Human Genome." *Science* 291, no. 5507 (2001): 1304-1351. Report on the Celera Genomics human genome project.

Web Sites of Interest

Department of Energy. Joint Genome Institute. http://www.jgi.doe.gov. A collaboration between the Department of Energy's Lawrence Berkeley, Lawrence Livermore, and Los Alamos National Laboratories. Includes an introduction to genomics, a research time line that starts with Darwin's work in 1859, and links.

Human Genome Sequencing Center. http://www.hgsc.bcm.tmc.edu. Baylor College of Medicine. Posts an ongoing "counter" of human genome sequencing completed worldwide.

National Center for Biotechnology Information. http://www.ncbi.nlm.nih.gov. A central repository for biological information, including links to genome projects and genomic science. Maintains GenBank, a comprehensive, annotated collection of publicly available DNA sequences.

Hardy-Weinberg Law

Field of study: Population genetics

Significance: *The Hardy-Weinberg law is the foundation for theories about evolution in local populations, often called microevolution. First formulated in 1908, it continues to be the basis of practical methods for investigations in fields from plant breeding and anthropology to law and public health.*

Key terms

ALLELE FREQUENCY: the proportion of all the genes at one chromosome location (locus) within a breeding population

GENE FLOW: movement of alleles from one population to another by the movement of individuals or gametes

GENE POOL: the total set of all the genes in all individuals in an interbreeding population

GENETIC DRIFT: random changes in allele frequencies caused by chance events

Introduction

The Hardy-Weinberg law can be phrased in many ways, but its essence is that the genetic makeup of a population, which meets certain assumptions, will not change over time. More important, it allows quantitative predictions about the distribution of genes and genotypes within and among generations. It may seem strange that theories about fundamental mechanisms of evolution are based on a definition of conditions under which evolution will not occur. It is the nature of science that scientists must make predictions about the phenomena being studied. Without something with which to compare the results of experiments or observations, science is impossible. Sir Isaac Newton's law of inertia plays a similar role in physics, stating that an object's motion will not change unless it is affected by an outside force.

After the rediscovery of Mendelian genetics in 1900, some scientists initially thought dominant alleles would become more common than recessive alleles, an error repeated in each generation of students. In 1908, Godfrey Hardy published his paper "Mendelian Proportions in a Mixed Population" in the journal *Science* to counteract that belief, pointing out that by themselves, sexual reproduction and Mendelian inheritance have no effect on an allele's commonness. Implicit in Hardy's paper was the idea that populations could be viewed as conglomerations of independent alleles, what has come to be called a "gene pool." Alleles randomly combine in pairs to make up the next generation. This simplification is similar to Newton's view of objects as simple points with mass.

Hardy, an English mathematician, wrote only one paper in biology. Several months earlier, Wilhelm Weinberg, a German physician, independently and in more detail had proposed the law that now bears both their names. In a series of papers, he made other contributions, including demonstrating Mendelian heredity in human families and developing methods for distinguishing environmental from genetic variation. Weinberg can justifiably be regarded as the father of human genetics, but his work, like Mendel's, was neglected for many years. The fact that his law was known as Hardy's law until the 1940's is an indictment of scientific parochialism.

The Hardy-Weinberg Paradigm

The Hardy-Weinberg "law" is actually a paradigm, a theoretical framework for studying nature. Hardy and Weinberg envisioned populations as collections of gametes (eggs and sperm) that each contain one copy of each gene. Most populations consist of diploid organisms that have two copies of each gene. Each generation of individuals can be regarded as a random sample of pairs of gametes from the previous generation's gamete pool. The proportion of gametes that contain a particular allele is the "frequency" of that allele.

Genome Frequencies

Genotype	Number	Genotype frequency
AA	36	36/100 = **0.36**
AB	48	48/100 = **0.48**
BB	16	16/100 = **0.16**
Total	100	1.00

Imagine a population of one hundred individuals having a gene with two alleles, *A* and *a*. There are three genotypes (combinations of alleles) in the population: *AA* and *aa* (homozygotes), and *Aa* (heterozygotes). If the population has the numbers of each genotype listed in the table "Genome Frequencies," then the genotype frequencies can be computed as shown.

The individuals of each genotype can be viewed as contributing one of each of their alleles to the gene pool, which has the composition shown in the table headed "Gene Pool Composition."

Gene Pool Composition

Genotype	*A* gametes	*B* gametes	Genotype contributions
AA	36 + 36 = 72		72
AB	48	48	96
BB		16 + 16 = 32	32
Total	120	80	200
Allele Frequency	**120/200 = 0.6**	**80/200 = 0.4**	**200/200 = 1.0**

This population can be described by the genotype ratio $AA:Aa:aa = 0.36:0.38:0.16$ and the allele frequencies $A:a = 0.6:0.4$. Note that allele frequencies must total 1.0, as must genotype frequencies.

The Hardy-Weinberg Law and Evolution

Allele and genotype frequencies would be of little use if they only described populations. By making a Punnett square of the gametes in the population and using allele frequencies, the table showing predicted genotype frequencies in the next generation will be obtained.

The predicted frequencies of homozygotes are 0.36 and 0.16; the frequency of *Aa* is 0.48 (adding the frequencies of *Aa* and *aA*). These are the same as the previous generation.

Hardy pointed out that if the frequency of $A = p$ and the frequency of $a = q$, then $p + q = 1$. Random mating can be modeled by the equation $(p + q) \times (p + q) = 1$, or more compactly $(p + q)^2 = 1$. This can be expanded to provide the genotype frequencies: $p^2 + 2pq + q^2 = 1$. In other words, the

ratio of $AA:Aa:aa = p^2:2pq:q^2$. Substituting 0.6 for p and 0.4 for q produces the figures shown in the preceding table, but more compactly and easily. The Hardy-Weinberg concept may also be extended to genes with more than two alleles. Therefore, three predictions may be made for a Hardy-Weinberg population: Frequencies of alleles p and q sum to 1.0 and will not change; the frequencies of genotypes *AA*, *Aa*, and *aa* will be $p^2:2pq:q^2$ respectively, will sum to 1.0, and will not change (that is, they are in equilibrium); and if the genotype frequencies are not initially at equilibrium ratios, they will eventually reach equilibrium.

There are within-generation and between-generation predictions. Within any one generation, the ratios of the genotypes are predictable if allele frequencies are known; if the frequency of a genotype is known, allele frequencies can be estimated. Between generations, allele and genotype frequencies will not change, as long as the following assumptions are met: (1) there are no mutations, (2) there is no gene flow with other populations, (3) mating is totally random, (4) the population is of infinite size, and (5) there is no natural selection. Violations of these assumptions define the five major evolutionary forces: mutation, gene flow, nonrandom mating, genetic drift, and natural selection, respectively.

Despite its seeming limitations, the Hardy-Weinberg law has been crucially useful in three major ways. First, its predictions of allele and genotype frequencies in the absence of evolution provide what statisticians call the "null hypothesis," which is essential for statistically rigorous hypothesis tests. If measured frequencies do not match predictions, then evolution is occurring. This redefines evolution from a vague "change in species over time" to a more useful, quantitative "change in allele or genotype frequencies." However, it is a definition that cannot be used in the domain of "macroevolution" and paleontology above the level of biological

Predicted Genotype Frequencies

Sperm	Eggs	
	A (frequency = 0.6)	B (frequency = 0.4)
A (frequency = 0.6)	AA (frequency = 0.6 × 0.6 = **0.36**)	BA (frequency = 0.6 × 0.4 = **0.24**)
B (frequency = 0.4)	AB (frequency = 0.6 × 0.4 = **0.24**)	BB (frequency = 0.4 × 0.4 = **0.16**)

species. Similarly, Newton's definition of a moving object does not apply in quantum physics. Second, Hardy-Weinberg provides a conceptual framework for investigation. If evolution is happening, a checklist of potential causes of evolution can be examined in turn. Finally, the Hardy-Weinberg paradigm provides the foundation for mathematical models of each evolutionary force. These models help biologists determine whether a specific evolutionary force could produce observed changes.

Using the Hardy-Weinberg Law

Sickle-cell disease is a severe disease of children characterized by reduced red blood cell number, bouts of pain, fever, gradual failure of major organs, and early death. In 1910, physicians noticed the disease and associated it with distortion ("sickling") of red blood cells. They realized that victims of the disease were almost entirely of African descent. Studies showed that the blood of about 8 percent of adult American blacks exhibited sickling, although few actually had the disease. By the 1940's, they knew sickling was even more common in some populations in Africa, India, Greece, and Italy.

In 1949, James Neel proved the disease was caused by a recessive gene: Children homozygous for the sickle allele developed the disease and died. Heterozygotes showed the sickle trait but did not develop the disease. Using the Hardy-Weinberg law, Neel computed the allele frequency among American blacks as follows: Letting p = the frequency of the sickle allele, $2pq$ is the frequency of heterozygotes (8 percent of adult African Americans). Since $p + q = 1$, $q = 1 - 0p$ and $2p(1 - p) = 0.08$. From this he computed $p = 0.042$ (about 4 percent). From the medical literature, Neel knew the frequency of the sickle trait in several African populations and computed the sickle allele frequency to be as high as 0.10 (since then the frequency has been found to be as high as 0.20). These are extraordinarily high frequencies for a lethal recessive allele and begged the question: Why was it so common?

The Hardy-Weinberg assumptions provided a list of possibilities, including nonrandom mating (mathematical models based on Hardy-Weinberg showed nonrandom mating distorts genotype frequencies but cannot change allele frequencies), mutation (for the loss of sickle alleles via death of homozygotes to be balanced by new mutations, scientists estimated the mutation rate from normal to sickle allele would have to be about three thousand times higher than any known human mutation rate, which seemed unlikely), and gene flow (models showed gene flow reduces differences between local populations caused by other evolutionary forces; gene flow from African populations caused by slavery explained the appearance of the sickle allele in North America but not high frequencies in Africa).

Another possibility was genetic drift. Models had shown deleterious alleles could rise to high frequencies in very small populations (smaller than one thousand). It was possible the sickle allele "drifted" to a high frequency in a human population reduced to small numbers by some catastrophe (population "bottleneck") or started by a small number of founders (the "founder effect"). If so, the population had since grown far above the size at which drift is significant. Moreover, drift was random; if there had been several small populations, some would have drifted high and some low. It was unlikely that drift would maintain high frequencies of a deleterious allele in so many large populations in different locations. Therefore, the remaining possibility, natural selection, was the most reasonable possibility: The heterozygotes must have some selective advantage over the normal homozygotes.

A few years later, A. C. Allison was doing field work in Africa and noted that the incidence of the sickle-cell trait was high in areas where malaria was prevalent. A search of the literature showed this was also true in Italy and Greece. In 1954, Allison published his hypothesis: In heterozygotes, sickle-cell alleles significantly improved resistance to malaria. It has been repeatedly confirmed. Scientists have found alleles for several other blood disorders that also provide resistance to malaria in heterozygotes.

Impact and Applications

The Hardy-Weinberg law has provided scientists with a more precise definition of evolution: change in allele or genotype frequencies. It allows them to measure evolution, provides a conceptual framework for investigation, and continues to serve as the foundation for the theory of microevolution. Beyond population genetics and evolution, the Hardy-Weinberg paradigm is used in such fields as law (analysis of DNA "fingerprints"), anthropology (human migration), plant and animal breeding (maintaining endangered species), medicine (genetic counseling), and public health (implementing screening programs). In these and other disciplines, the Hardy-Weinberg law and its derivatives continue to be useful.

The Hardy-Weinberg law also has implications for social issues. In the early twentieth century, growing knowledge of genetics fueled a eugenics movement that sought to improve society genetically. Eugenicists in the 1910's and 1920's promoted laws to restrict immigration and promote sterilization of "mental defectives," criminals, and other "bad stock." The Hardy-Weinberg law is often credited with the decline of eugenics. The ratio $2pq/q^2 = 1$ makes it clear that if a recessive trait is rare (as most deleterious alleles are), most copies of a recessive allele are hidden in apparently normal heterozygotes. Selecting against affected individuals will be inefficient at best. However, a host of respected scientists championed eugenics into the 1920's and 1930's, long after the implications of Hardy-Weinberg were understood. It was really the reaction to the horrors of Nazi leader Adolf Hitler's eugenics program that made eugenics socially unacceptable. More-over, it is premature to celebrate the end of the disturbing questions raised by eugenics. Progress in molecular biology makes it possible to detect deleterious alleles in heterozygotes, making eugenics more practical. Questions of whether genes play a major role in criminality and mental illness are still undecided. Debate about such medical and social issues may be informed by knowledge of the Hardy-Weinberg law, but decisions about what to do lie outside the domain of science.

—*Frank E. Price*

See also: Consanguinity and Genetic Disease; Eugenics; Eugenics: Nazi Germany; Evolutionary Biology; Genetic Load; Genome Size; Heredity and Environment; Inbreeding and Assortative Mating; Natural Selection; Polyploidy; Population Genetics; Punctuated Equilibrium; Quantitative Inheritance; Sickle-Cell Disease; Sociobiology; Speciation.

Further Reading

Provine, William. *The Origins of Theoretical Population Genetics.* 1971. 2d ed. Chicago: University of Chicago Press, 2001. Provides a comprehensive overview of the history of population genetics, including the Hardy-Weinberg law.

Heart Disease

Field of study: Diseases and syndromes
Significance: *Individual susceptibility to cardiovascular (heart) disease involves the interaction of complex genetic traits, as well as factors loosely defined as "environmental." Only in rare circumstances do specific mutations result in disease; more often, quantitative differences in gene products reflect a minimum threshold necessary for overt disease.*

Key terms

ANGINA PECTORIS: chest pain that can be indicative of heart disease
ANGIOTENSIN-CLEAVING ENZYME (ACE): a protein indirectly involved in regulation of blood pressure

APOLIPOPROTEIN A1 (APO A1): a molecule that binds cholesterol in the bloodstream, facilitating its removal by the liver

ATHEROSCLEROSIS: narrowing of coronary arteries that results from plaque formation in the vessel

CHOLESTEROL: a steroid derivative that is a normal constituent of cell membranes, but which plays a role in plaque formation in arteries

INTEGRIN B3 (ITGB3): a glycoprotein found in platelet membranes that plays a role in adherence to capillary surfaces

LOW-DENSITY LIPOPROTEIN RECEPTOR (LDLR): a protein on surface of liver cells that removes cholesterol and other lipids from the bloodstream

QUANTITATIVE TRAIT LOCUS (QTL) (*pl.* QUANTITATIVE TRAIT LOCI): a group of genes that interact in defining physical or biochemical characteristics such as development of disease

Forms and Genetic Basis

Coronary heart disease (CHD) is arguably the leading cause of death among older adults. While CHD may take a variety of forms, most commonly it is associated with a narrowing of the coronary arteries that supply oxygen and other nutrients to heart tissue. Eventually, the artery may be completely blocked. While cholesterol or other lipids certainly play a role in the process, other factors or systems are also involved. Most of these involve genetically encoded proteins or protein-utilizing systems.

Since heart disease often runs in families, there is clearly a genetic element in its development. Generally, development of disease in populations can be described as a continuum, with the quantitative level and rate of CHD development varying among individuals; even accounting for the role of the environment, some individuals are more susceptible than others. The genes that are involved can be mapped, in some cases to specific sites on chromosomes, and until the genes are actually identified they are referred to as quantitative trait loci (QTLs).

At least 250 genes or QTLs have now been linked in some manner with CHD development. This is not to say that environment plays only a limited role. The form or quantity of the gene product may define susceptibility to disease; often, however, the environment may play a major role in determining the significance played by that product.

Cholesterol

Cholesterol was the first component in blood in which the concentration could be correlated with risk for heart disease. While other molecules found in the bloodstream are now known to also play undefined roles in CHD, elevated cholesterol concentration remains one of the more important risk factors.

The concentration of cholesterol is responsive to a variety of processes under genetic control. These take the form of either cholesterol "packaging" or removal. Once in the bloodstream, cholesterol becomes linked with any of a variety of proteins or other molecules. Low-density lipoprotein (LDL) is often referred to as "bad" cholesterol, while high-density lipoprotein (HDL) is called the "good" cholesterol. The apo A1 protein, product of the *APOA1* gene, binds with cholesterol in forming HDLs. At least certain forms of CHD are associated with variants of the gene product apo A1, which form lower concentrations of HDL. Likewise, the *APOA1* gene can express variants that increase the concentration of LDLs, with a concordant increase in risk of CHD.

The concentration of "packaged" cholesterol is also reflected by its rate of removal from the bloodstream. In part, this is a function played by the liver. Liver cells have LDL receptors, the function of which is to bind and remove LDLs. At least 350 genetic variants have been described for the *LDLR* gene, resulting in significant quantitative differences among persons in their efficiency of LDL removal. Persons with those receptor variants that function inefficiently often are at increased risk of heart disease. For example, hypercholesterolemia, a condition in which cholesterol may be three to four times the normal level, is often the result of reduced numbers of LDL receptors. Persons with this condition are at extremely high risk for development of atherosclerosis and may suffer heart attacks even as young adults.

Defects in the Blood-Clotting System

While the buildup of plaque in coronary arteries is a major factor in heart disease, a heart attack is often triggered by clot formation at the site of narrowing. The uneven nature or structure of the plaque may itself be sufficient for clot formation. Nevertheless, the discovery that persons with variants of certain clot-associated or inflammatory factors may show increased risk of clot formation suggests genetic factors may also play a role.

Clot formation begins when blood platelets attach to the surface at the site of an injury. Among the molecules found in the cell membrane of platelets is a glycoprotein encoded by the *ITGB3* gene. The normal function of this molecule is to enable the platelet to attach to the surface of the blood capillary. However, a large proportion of middle-aged adults suffering from CHD have been shown to express an unusual variant of this gene, resulting in an increased capacity of platelets to initiate clot formation.

A second gene in the blood-clotting category encodes the protein thrombospondin (TSP), one of a family of proteins that regulate adhesion of cells to capillary surfaces. Epidemiological studies of persons with coronary artery disease have shown that certain variations of this gene are found in a large proportion of patients, while other variants seem associated with a decreased risk of disease.

Quantitative Trait Loci

Quantitative traits are those expressed at varying levels in the population. To date, most QTL studies have involved genetic crosses of rats or mice; mice in particular are simple to inbreed and share some genetic similarities with humans. Other genes have been found as a result of the Human Genome Project.

Approximately thirty QTLs have been defined in humans. Individual genes that make up QTLs are themselves heterogeneous, existing as multiple variants or alleles. Clearly, CHD is a complex disorder involving the interaction of many gene products of numerous alleles, and the specific role played by such loci in most forms of CHD remains to be explained. The association of one type of QTL linked to a risk factor for CHD, that of left ventricular hypertrophy (LVH), provides a prototype for understanding the interaction of gene products in development of disease.

LVH represents a condition in which the mass of the lower portion of the heart increases, raising the level of blood pressure and increasing the strain on the heart. Since LVH is a significant risk factor in devel-

At the Nebraska Heart Institute in August, 2002, Dr. Vish Bhoopalam discusses gene therapy used to stimulate growth of new capillaries near the heart. (AP/Wide World Photos)

opment of coronary disease, an understanding of the process is important in its prevention.

Using crosses among inbred strains of rats, researchers have found that genetic markers for LVH can be mapped to certain QTLs. The quantitative expression of a protein, atrial natriuretic factor (ANF), appeared to correlate with the extent of LVH in rats: The higher the concentration of ANF, the lower was the ventricular mass. A specific gene, natriuretic peptide precursor A (*Nppa*), was found in the region, and it encoded the ANF precursor protein.

The relationship between QTLs and CHD remains more elusive. Human populations obviously do not lend themselves to similar forms of crosses, so identification of candidate QTLs has generally been limited to studies between twins, or at least siblings. These studies have demonstrated that both genetics and the environment play roles of indeterminate importance in CHD. If one sibling develops disease, the other is at significantly higher risk for the same. At the same time, certain behavioral risk factors such as smoking or obesity also increase the incidence of cardiovascular disease.

Most candidate QTLs have been limited to an association with hypertension; mutations in certain such genes have been observed in hypertensive cases responsive to reduced levels of salt intake. The actual function of the gene products has not been determined. Since there exists a clear relationship between elevated blood pressure and development of heart disease, these would qualify as candidate loci.

Blood Pressure

The regulation of blood pressure by the body involves a number of complementary systems. One of these mechanisms involves the molecule angiotensin II. Produced within the liver, angiotensin II constricts small arteries, increasing blood pressure, and regulates salt uptake by the kidney.

Angiotensin II is first synthesized as an inactive precursor molecule, angiotensinogen; subsequent cleavage is required for activation. Regulation of angiotensin concentration is in part a function of two different gene products. The product of the *ACE* gene, an angiotensin-cleaving enzyme, converts the inactive precursor to the active form. Inactivation of angiotensin II is carried out by the product of the *ACE2* gene, angiotensin-converting enzyme 2.

It is known that high blood pressure in some individuals may be aggravated by diets high in salt. Animals that exhibit similar characteristics are often found to express decreased concentrations of the ACE2 enzyme, resulting in greater salt retention as well as higher blood pressure. Whether a similar situation exists in humans is unclear, though there is evidence that reduced levels of ACE2 production may play an analogous role in humans.

Nature vs. Nurture

For most populations, the risk of CHD represents a continuum, with some individuals at low risk and others at increasingly elevated risk. Excluding those situations in which specific genetic variation is directly the cause of CHD, as in the situation of hypercholesterolemia, in most persons the genetic pattern merely relates to the ability of environmental factors to trigger disease. For example, while elevated blood pressure is a significant factor in long-term cardiovascular injury, environmental factors such as obesity, diet, level of stress, and level of exercise may themselves determine the extent of hypertension.

The single most important environmental factor associated with CHD that can be controlled is that of tobacco use. While genetic variation may play a role in susceptibility to the effects of tobacco smoke, there is no question that a cause-and-effect relationship between the extent of smoking and increased risk for disease exists.

As the role played by specific genes in development of CHD becomes more apparent, not only is there the potential for improved treatments, but methods both for screening and prevention become possible. Persons at greater risk may be identified on the basis of possessing certain genetic variants, and methods of treatment may be determined as a result of knowing a specific cause.

—*Richard Adler*

See also: Congenital Defects; Diabetes; Genetic Testing; Hereditary Diseases; Heredity and Environment; Human Genetics; Human Genome Project; Hypercholesterolemia; Organ Transplants and HLA Genes; Prenatal Diagnosis.

Further Reading

Braunwald, Eugene, Douglas P. Zipes, Peter Libby, and Douglas D. Zipes. *Heart Disease: A Textbook of Cardiovascular Medicine.* 6th ed. Philadelphia: W. B. Saunders, 2001. Contains a large section on the role of genetics and cardiovascular disease as well as a thorough discussion of heart disease in general.

Crackower, M. A., et al. "Angiotensin-Converting Enzyme Is an Essential Regulator of Heart Function." *Nature* 417 (2002): 822-828. The authors mapped the *ACE2* gene to a particular QTL, demonstrating a role for the gene in regulation.

Danielli, Gian Antonia, and Gina Antonia Danielli. *Genetics and Genomics for the Cardiologist.* Boston: Kluwer Academic, 2002. Specialized overview of the topic addressed primarily to clinicians. However, the book does emphasize applications of recent techniques in screening for CHD-associated genes.

Gambaro, G., F. Anglani, and A. D'Angelo. "Association Studies of Genetic Polymorphisms and Complex Disease." *The Lancet* 355 (January 22, 2000): 308-311. General discussion on methods to correlate genetic markers and disease. The primary example is that of ACE polymorphisms and CHD.

Goldbourt, Uri, Kare Berg, and Ulf de Faire, eds. *Genetic Factors in Coronary Heart Disease.* Boston: Kluwer Academic, 1994. Emphasizes epidemiologic features of CHD. However, portions of the book also address genetic factors known at the time.

Marian, Ali J. *Genetics for Cardiologists: The Molecular Genetic Basis of Cardiovascular Disorders.* London: ReMEDICA, 2000. Addresses the molecular genetic basis of heart disease. Color illustrations.

Pirisi, Angela. "Researchers Find Genetic Clues to Coronary Artery Disease." *The Lancet* 358 (December 1, 2001): 1879. Review of work on possible roles of thromboplastin variants and CHD.

Web Sites of Interest

American Heart Association. http://www.americanheart.org. The AHA's Web site offers education regarding the different forms of heart disease, symptoms, and treatments. A page devoted to congenital disorders as well as a search on "gene" and similar words offer information and articles on specific hereditary defects and conditions.

National Institutes of Health, National Library of Medicine. http://www.nlm.nih.gov. This site offers comprehensive information on heart disease, including genetics and research.

Hemophilia

Field of study: Diseases and syndromes
Significance: *Hemophilia is a sex-linked inherited genetic disorder in which the blood does not clot adequately. Although incidents of hemophilia are relatively rare, the study of this disease has yielded important information about genetic transmission and the factors involved in blood clotting.*

Key terms

HEMOPHILIA A: a blood disease with a deficiency of clotting factor VIII

HEMOPHILIA B: a blood disease with a deficiency of clotting factor IX

HEMOSTASIS: the process by which blood flow is stopped at an injury site

Causes and Symptoms

When an injury occurs that involves blood loss, the body responds by a process known as hemostasis. Hemostasis involves several steps that result in the blood clotting and stopping the bleeding. With hemophilia, an essential substance is absent. For blood to clot, a series of chemical reactions must occur in a "domino effect." The reaction starts with a protein called the Hageman factor or factor XII, which cues factor XI, which in turn cues factor X and so on until factor I is activated. Each factor is expressed by a different gene. If one of the genes is defective, the blood will not clot properly. Hemophilia A is the most common type, affect-

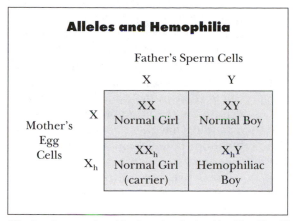

Alleles and Hemophilia

		Father's Sperm Cells	
		X	Y
Mother's Egg Cells	X	XX Normal Girl	XY Normal Boy
	X_h	XX_h Normal Girl (carrier)	X_hY Hemophiliac Boy

The daughters produced by the union depicted in this table will be physically normal, but half will be carriers of hemophilia. Half the sons produced by the union will be hemophiliacs.

ing over 80 percent of all hemophiliacs and resulting when clotting factor VIII is deficient. Hemophilia B (also known as Christmas disease) affects about 15 percent of hemophiliacs and results when clotting factor IX is deficient.

Hemophilia affects males almost exclusively because it is an X-linked (often called sex-linked) recessive trait. Although it is possible for women to have hemophilia, it is extremely rare, because women must have two copies of the defective gene to be affected. A female has two X chromosomes, and a male has an X and Y chromosome. Even though the trait is recessive, because men have a single X chromosome, recessive X-linked genes are expressed as if they were dominant. Thus, hemophilia in males is inherited, along with their X chromosome, from the mother. The daughter of a hemophiliac father will carry the disease because she inherits one X chromosome (with the abnormal gene) from the father and one from the mother. Any son born to a carrier has a 50 percent chance of having hemophilia, since she will either pass on the X chromosome with the normal gene or the one with the abnormal gene. In order for a female to have hemophilia, she would have to inherit the abnormal gene on the X chromosomes from both her mother and her father.

Hemophilia can be mild, moderate, or severe, depending on the extent of the clotting factor deficiency. Mild hemophilia may not be evident until adulthood when prolonged bleeding is observed after surgery or a major injury. The symptoms of moderate or severe hemophilia often appear early in life. These symptoms may include easy bruising, difficulty in stopping minor bleeding, bleeding into the joints, and internal bleeding without any obvious cause (spontaneous bleeding). When bleeding occurs in the joints, the person experiences severe pain, swelling, and possible deformity in the affected joint. The weight-bearing joints such as ankles and knees are usually affected. Internal bleeding requires immediate hospitalization and could result in death if severe. People who experience prolonged or abnormal bleeding are often tested for hemophilia. Testing the specific blood-clotting factors can determine the type and severity of hemophilia. Although a family history of hemophilia may help in the diagnosis, approximately 20 percent of hemophiliacs have no such history of the disease.

Impact and Applications

Hemophilia is not curable, although advances in the treatment of the disease are prolonging life and preventing crippling deformities. Symptoms of hemophilia can be reduced by replacing the deficient clotting factor. People with hemophilia A may receive antihemophilic factors to raise their blood-clotting factor above normal levels so that the blood clots appropriately. People with hemophilia B may receive clotting factor IX during bleeding episodes in order to increase the clotting factor levels. The clotting factors may be taken from plasma (the fluid part of blood), although it takes a great deal of plasma to produce a small amount of the clotting factors. Risks include infection by the hepatitis virus or human immunodeficiency virus (HIV), although advanced screening procedures have greatly reduced such risks. In 1993 the FDA approved a new recombinant form of factor VIII and in 1997 the FDA approved a new recombinant form of factor IX for treating individuals with hemophilia A and B, respectively. The advantage of recombinant factors is that they are automatically free of plasma-derived viruses, thus reducing one of the primary risks endured by previous hemophiliacs. Patients with mild hemophilia may be treated with a synthetic hormone known as desmopressin acetate (DDAVP).

Treatment with the plasma clotting factors has increased longevity and quality of life. In addition, many patients are able to treat bleeding episodes as outpatients with home infusions or self-infusions of the clotting factors. However, problems do exist with the treatment of hemophilia. Various illnesses, such as HIV, liver disease, or cardiovascular disease, have resulted from contamination of the clotting factors. Several techniques are used to reduce the risk of contamination, and most difficulties were largely eliminated by the mid-1990's. Bleeding into the joints is often controlled by the use of elastic bandages and ice. Exercise is recommended to help strengthen and protect the joint. Painkillers are used to reduce the chronic pain associated with joint swelling and inflammation, although hemophiliacs cannot use products containing aspirin or antihistamines because they prolong bleeding. Patients and their families have also benefited from genetic education, counseling, and testing. Hemophilia centers can provide information on how the disease is transmitted, potential genetic risks, and whether a person is a carrier. This knowledge provides options for family planning as well as support in coping with the disease.

—Virginia L. Salmon, updated by Bryan Ness

Recombinant Factor VIII

Prior to the development of recombinant factor VIII, patients with hemophilia were treated with coagulation factors prepared from the blood of thousands of different donors. While these coagulation factor concentrates were highly effective in treating acute bleeding episodes, they also proved to be the source of infection with hepatitis and human immunodeficiency (HIV) viruses. Many patients with hemophilia became seriously ill and died from a treatment that was designed to save their lives.

Once the risk of viral infection from these pooled donations was recognized in the early 1980's, biomedical manufacturers introduced measures to inactivate the viruses during the process of preparing the concentrates. The next, even more important, step in improving hemophilia treatment was the development of recombinant factors VIII and IX using DNA technology. Early studies demonstrated that the recombinant factors were as effective as the pooled blood concentrates and had few adverse effects.

The first recombinant factor VIII concentrate was introduced in 1987. Large-scale multinational studies of the safety and effectiveness of recombinant factor VIII began in human subjects in 1989. All of these studies are classified as "prospective" or "cohort" studies where patients are enrolled, treated, and followed through many years. Since prospective studies are considered the most methodologically sound, they yield scientific information that is highly respected.

The results are encouraging. Previously untreated patients with hemophilia who have had severe bleeding episodes have responded well to recombinant products. The majority of the bleeds (71-91 percent) in most studies resolved with a single dose. Patients rarely have side effects, and those they experience are mild. About one-third of patients developed inhibitors to recombinant factor VIII, but several of these inhibitors disappeared over time. No one has found evidence of the transmission of infectious agents in the recombinant factor concentrates. Newer studies show that treatment at home by the patients themselves, preventive treatment prior to necessary surgery, and treatment in previously treated patients are effective and safe, with minimal adverse effects.

In the United States, recombinant factor VIII was licensed for use in 1992. These products are now used in United States, Canada, Europe, Japan, and elsewhere. Recombinant factors are considered in most areas the treatment of choice for the treatment of patients with severe hemophilia. Unfortunately, these products are not readily available and are extremely costly, meaning that physicians must select which patients are most appropriate for using recombinant factor VIII. In general, patients who have not been treated before and who are not infected with hepatitis or HIV viruses are the candidates most likely to receive these products until the supplies are greater and the costs lower.

—Rebecca Lovell Scott

See also: Amniocentesis and Chorionic Villus Sampling; Bacterial Genetics and Cell Structure; Chromosome Mutation; Cloning; Gene Therapy; Gene Therapy: Ethical and Economic Issues; Genetic Counseling; Genetic Engineering; Genetic Engineering: Medical Applications; Genetic Testing; Hereditary Diseases.

Further Reading

Buzzard, Brenda, and Karen Beeton, eds. *Physiotherapy Management of Haemophilia*. Malden, Mass.: Blackwell, 2000. Examines, among other topics, principles of assessment and pain mechanisms; techniques in hydrotherapy, electrotherapy, exercise, and sport; rehabilitation in developing countries; and physiotherapy following orthopedic surgery.

Jones, Peter. *Living with Haemophilia*. 5th ed. New York: Oxford University Press, 2002. Provides an understandable discussion of hemophilia and its transmission, symptoms, and management. Illustrated.

Monroe, Dougald M., et al., eds. *Hemophilia Care in the New Millennium*. New York: Kluwer Academic/Plenum, 2001. Explores the management of hemophilia, providing background and resources. Illustrated.

Potts, D. M., and W. T. W. Potts. *Queen Victoria's Gene: Haemophilia and the Royal Family*. Stroud: Sutton, 1999. Explores the source of hemophilia in the royal families of Europe and the effect it had on history. Illustrations, plates, genealogical tables, map.

Resnik, Susan. *Blood Saga: Hemophilia, AIDS, and the Survival of a Community*. Berkeley: University of California Press, 1999. Details the social history of hemophilia in the United States, beginning in the early twentieth century, when most hemophilia patients did not live past their teens. Illustrated, extensive glossary and bibliography, and statistical data.

Rodriguez-Merchan, E. C., N. J. Goddard, and C. A. Lee, eds. *Musculoskeletal Aspects of Haemophilia*. Malden, Mass.: Blackwell, 2000. Topics include hemostasis, orthopedic surgery, rehabilitation and physiotherapy, gait corrective devices, burnout syndrome in staff, and anti-inflammatory drugs from the view of a rheumatologist.

Web Sites of Interest

Dolan DNA Learning Center, Your Genes Your Health. http://www.ygyh.org. Sponsored by the Cold Spring Harbor Laboratory, this site, a component of the DNA Interactive Web site, offers information on more than a dozen inherited diseases and syndromes, including hemophilia.

National Hemophilia Foundation. http://www.hemophilia.org. Site that includes information on research and links to related organizations.

Hereditary Diseases

Field of study: Diseases and syndromes
Significance: *Scientists are discovering the genetic bases of an ever-increasing number of diseases affecting children and adults. The Human Genome Project was begun in 1990 with the goal of determining and mapping all human genes by the year 2005, a task which was largely completed by April, 2003. As knowledge about the genetics underlying different diseases is gained, opportunities should increase for the diagnosis, prevention, and treatment of these diseases.*

Key terms

CHROMOSOMAL DEFECTS: defects involving changes in the number or structure of chromosomes

CONGENITAL DEFECTS: birth defects, which may be caused by genetic factors, environmental factors, or interactions between genes and environmental agents

HEMIZYGOUS: characterized by being present only in a single copy, as in the case of genes on the single X chromosome in males

MENDELIAN DEFECTS: also called single-gene defects; traits controlled by a single gene pair

MITOCHONDRIAL DISORDERS: disorders caused by mutations in mitochondrial genes

MODE OF INHERITANCE: the pattern by which a trait is passed from one generation to the next

MULTIFACTORIAL DISORDERS: disorders determined by one or more genes and environmental factors

Causes and Impact of Hereditary Diseases

Twentieth century medicine was hugely successful in conquering infectious diseases. Elimination, control, and treatment of diseases such as smallpox, measles, diphtheria, and plague have greatly decreased infant and adult mortality. Improved prenatal and postnatal care have also decreased childhood mortality. Shortly after the rediscovery of Mendelism in the early 1900's, reports of genetic determination of human traits began to appear in medical and biological literature. For the first half of the twentieth century, most of these reports were regarded as interesting scientific reports of isolated clinical diseases that were incidental to the practice of medicine. The field of medical genetics is considered to have begun in 1956 with the first description of the correct number of chromosomes in humans (forty-six). Between 1900 and 1956, findings were accumulating in cytogenetics, Mendelian genetics, biochemical genetics, and other fields that began to draw medicine and genetics together.

The causes of hereditary diseases fall into four major categories:

(1) single-gene defects or Mendelian disorders (such as cystic fibrosis, Huntington's disease [Huntington's chorea], color blindness, and phenylketonuria)
(2) chromosomal defects involving changes in the number or alterations in the structure of chromosomes (such as Down syndrome, Klinefelter syndrome, and Turner syndrome)
(3) multifactorial disorders, caused by a combination of genetic and environmental factors (such as congenital hip dislocation, cleft palate, and cardiovascular disease)
(4) mitochondrial disorders caused by mutations in mitochondrial genes (such as Leber hereditary optic neuropathy)

These four categories are relatively clear-cut. It is likely that genetic factors also play a less well-defined role in all human diseases, including susceptibility to many common diseases and degenerative disorders. Genetic factors may affect a person's health from the time before birth to the time of death.

Congenital defects are birth defects and may be caused by genetic factors, environmental factors (such as trauma, radiation, alcohol, infection, and drugs), or the interaction of genes and environmental agents. Alan Emery and David Rimoin noted that the proportion of childhood deaths attributed to nongenetic causes was estimated to be 83.5 percent in London in 1914 but had declined to 50 percent in Edinburgh by 1976, whereas childhood deaths attributed to genetic causes went from 16.5 percent in 1914 to 50 percent in 1976. These changes reflect society's increased ability to treat environmental causes of disease, resulting in a larger proportion of the remaining diseases being caused by genetic defects. Rimoin, J. Michael Connor, and Reed Pyeritz estimate that single-gene disorders have a lifetime frequency of 20 in 1000, chromosomal disorders have a frequency of 3.8 in 1000, and multifactorial disorders have a frequency of 646 in 1000. It is evident that hereditary diseases are and will be of major concern for some time.

Single-Gene Defects

Single-gene defects result from a change or mutation in a single gene and are referred to as Mendelian disorders or inborn errors of metabolism. In 1865, Gregor Mendel described the first examples of monohybrid inheritance. In a trait governed by a single locus with two alleles, individuals inherit one allele from each parent. If the alleles are identical, the individual is said to be homozygous. If the alleles are different, the individual is said to be heterozygous. Single-gene defects are typically recessive. A single copy of a dominant allele will be expressed the same in homozygous and heterozygous individuals. A recessive allele, on the other hand, is expressed in homozygous individuals (often called homozygotes). In heterozygotes, the dominant allele "hides" or masks the expression of the recessive allele. This helps explain why recessive single-gene defects predominate. Dominant single-gene defects are always expressed when present and never remain "hidden." As a result, natural selection quickly removes these defects from the population.

Genes can be found either on sex chromosomes or non-sex chromosomes (called auto-

Some Genetic Disorders

Disorder	Genetic Characteristics	Disorder	Genetic Characteristics
Achondroplasia	Autosomal dominant disorder	Hemochroma-tosis	Autosomal recessive disorder
Albinism	Autosomal recessive disorder	Hemophilia	X-linked recessive disorder
Alzheimer's disease, familial early onset	Mutations in *PS1*, *PS2*	Huntington's disease	Autosomal dominant disorder
Alzheimer's disease, late onset	Mutations in *APOE*	Hypercholestero-lemia	Autosomal dominant disorder
Angelman syndrome	Deletion in chromosome 15	Klinefelter syn-drome	Males that are XXY autosomal dominant disorder
Beta-thalassemia	Mutations in or impaired expression of the gene for beta-globin	Kuru	Prion disease
		Lactose intoler-ance	Autosomal recessive disorder
Breast cancer	Mutations in *BRCA1*, *BRCA2*, *p53* cause predisposition	Marfan syndrome	Autosomal dominant disorder
Burkitt's lymphoma	Reciprocal translocation involving chromosomes 8 and 14 (or occasionally 22 or 2)	Metafemale (multiple X syndrome)	Females with more than two X chromosomes
Cancer	Mutations in proto-oncogenes and tumor-suppressor genes or in the control regions of these genes cause predisposition	Neurofibromato-sis (NF)	Types 1 and 2 both autosomal dominant disorders
		Phenylketonuria (PKU)	Autosomal recessive disorder
Color blindness (common form)	Sex-linked recessive disorder	Polycystic kidney disease	Autosomal dominant disorder
Creutzfeldt-Jakob syndrome	Prion disease	Prader-Willi syndrome	Deletion in chromosome 15
Cystic fibrosis	Autosomal recessive disorder	Pseudohermaph-roditism	Autosomal recessive disorder
Diabetes, Type I	Mutations in the gene for insulin	Sickle-cell disease	Autosomal incompletely dominant disorder (sometimes considered autosomal recessive)
Diabetes, Type II	Mutations in the gene for insulin		
Down syndrome	Trisomy 21	Tay-Sachs disease	Autosomal recessive disorder
Down syndrome, familial	Translocation of part of chromosome 21	Testicular feminization syndrome	Form of pseudohermaphroditism, and also an autosomal recessive disorder
Dwarfism (achon-droplasia)	Autosomal dominant disorder	Turner syndrome	Monosomy
Fragile X syndrome	X-linked showing imprinting	XYY syndrome	Males with an extra Y chromosome

—Bryan Ness

somes). One pair of chromosomes (two chromosomes of the 46 in humans) have been designated sex chromosomes because the combination of these two chromosomes determines the sex of the individual. Human males have an unlike pair of sex chromosomes, one called the X chromosome and a smaller one called the Y chromosome. Females have two X chromosomes. Genes on the X or Y chromosomes are considered sex-linked. However, since Y chromosomes contain few genes, "sex-linked" usually refers to genes on the X chromosome; when greater precision is required, genes on the X chromosome are referred to as "X-linked." Inheritance patterns for X-linked traits are different than for autosomal traits. Because males only have one X chromosome, any allele, whether normally recessive or dominant, will be expressed. Therefore, recessive X-linked traits are typically much more common in men than in women, who must have two recessive al-

leles to express a recessive trait. Additionally, a male inherits X-linked alleles from his mother, because he only gets a Y chromosome from his father.

Chromosomal Disorders

Chromosomal disorders are a major cause of birth defects, some types of cancer, infertility, mental retardation, and other abnormalities. They are also the leading cause of spontaneous abortions. Deviations from the normal number of forty-six chromosomes, or structural changes, usually result in abnormalities. Variations in the number of chromosomes may involve just one or a few chromosomes, a condition called aneuploidy, or complete sets of chromosomes, called polyploidy. Polyploidy among live newborns is very rare, and the few polyploid babies who are born usually die within a few days of birth as a result of severe malformations. The vast majority of embryos and fetuses

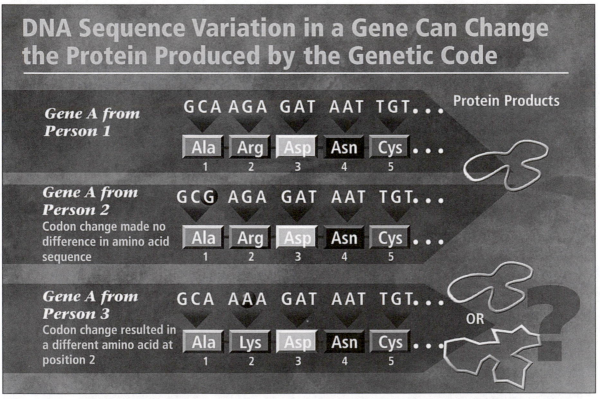

Whereas some variations in an individual's genetic code will not affect the protein produced, others will, possibly resulting in disease or sensitivity to environmental triggers for the disease. (U.S. Department of Energy Human Genome Program, http:// www.ornl.gov/hgmis)

with polyploidy are spontaneously aborted.

Aneuploidy typically involves the loss of one chromosome from a homologous pair, called monosomy, or possession of an extra chromosome, called trisomy. Monosomy involving a pair of autosomes usually leads to death during development. Individuals have survived to birth with forty-five chromosomes, but they suffered from multiple, severe defects. Most embryos and fetuses that have autosomal trisomies abort early in pregnancy. Invariably, trisomics that are born have severe physical and mental abnormalities. The most common trisomy involves chromosome 21 (Down syndrome), with much rarer cases involving chromosome 13 (Patau syndrome) or chromosome 18 (Edwards' syndrome). Infants with trisomy 13 or 18 have major deformities and invariably die at a very young age. Down syndrome is the most common (about one in seven hundred births) and is the best known of the chromosomal disorders. Individuals with Down syndrome are short and have slanting eyes, a nose with a low bridge, and stubby hands and feet; about one-third suffer severe mental retardation. The risk of giving birth to a child with Down syndrome increases dramatically for women over thirty-five years of age.

Variations in the number of sex chromosomes are not as lethal as those involving autosomes. Turner syndrome is the only monosomy that survives in any number, although 98 percent of them are spontaneously aborted. Patients have forty-five chromosomes consisting of twenty-two pairs of autosomes and only one X chromosome. They are short in stature, sterile, and have underdeveloped female characteristics but normal or near-normal intelligence. Other diseases caused by variations in the number of sex chromosomes include Klinefelter syndrome, caused by having forty-seven chromosomes, including two X and one Y chromosome (affected individuals are male with small testes and are likely to have some female secondary sex characteristics such as enlarged breasts and sparse body hair) and multiple X syndrome, or metafemale (affected individuals are females whose characteristics are variable; some are sterile or have menstrual irregularities or both).

Variations in the structure of chromosomes include added pieces (duplications), missing pieces (deletions), and transfer of a segment to a member of a different pair (translocation). Most deletions are likely to have severe effects on developing embryos, causing spontaneous abortion. Only those with small deletions are likely to survive and will have severe abnormalities. The cri du chat ("cry of the cat") syndrome produces an infant whose cry sounds like a cat's meow. There is also a form of Down syndrome, called familial Down syndrome, that is caused by a type of reciprocal translocation between two chromosomes.

Multifactorial Traits

Multifactorial traits (sometimes referred to as complex traits) result from an interaction of one or more genes with one or more environmental factors. Sometimes the term "polygenic" is used for traits that are determined by multiple genes with small effects. Multifactorial traits do not follow any simple pattern of inheritance and do not show distinct Mendelian ratios. Such diseases show an increased recurrence risk within families. "Recurrence risk" refers to the likelihood of the trait showing up multiple times in a family; in general, the more closely related someone is to an affected person, the higher the risk. Recurrence risk is often complicated by factors such as the degree of expression of the trait (penetrance), the sex of the affected individual, and the number of affected relatives. For example, pyloric stenosis, a disorder involving an overgrowth of muscle between the stomach and small intestine, is the most common cause of surgery among newborns. It has an incidence of about 0.2 percent in the general population. Males are five times more likely to be affected than females. For an affected male, there is a 5 percent chance his first child will be affected, whereas for a female, there is a 16 percent chance her first child will be affected.

It is necessary to develop separate risks of recurrence for each multifactorial disorder. Multifactorial disorders are thought to account for 50 percent of all congenital defects. In addition, they play a significant role in many adult disorders, including hypertension and other

Sequencing Targets and Associated Diseases

Chromosome

16

Hb H mental retardation syndrome
Carbohydrate Deficient Glycoprotein syndrome
T cell lymphoma

Tuberous Sclerosis
Polycystic kidney disease
Familial Mediterranean fever
Rubinstein-Taybi syndrome

Acute Myeloid leukemia
Myxoid Liposarcoma

Liddle's syndrome

Batten disease

Blau syndrome

Cylindromatosis (Turban tumour)
Acute Myeloid leukemia

Phoshorylase Kinase Beta deficiency
Towns-Brocks syndrome
Crohn's disease
CETP deficiency
Bardet-Biedl Syndrome

Aldolase A deficiency
Richrner-Hanhart syndrome

Chronic Granulomatous disease

Breast and Prostate cancer
Marner's Cataract
Spinocerebellar Ataxia
Morquio syndrome A
APRT deficiency
Breast and Prostate cancer
Fanconi anemia

5

Homocystinuria-megaloblastic anemia

Attention-deficit hyperactivity disorder
Craniometaphyseal dysplasia
Leigh syndrome

Cri du Chat syndrome

Salt-resistant hypertension

Hirschsprung disease
Severe combined immunodeficiency

Complement component 6-9 deficiency

Ketoacidosis due to SCOT deficiency
Laron dwarfism

Malignant hyperthermia
susceptibility 6

Chondrocalcinosis with early onset osteoarthritis

Endometrial carcinoma

Combined pituitary hormone deficiency

Klippel-Feil syndrome
Megaloblastic anemia
Spinal muscular atrophy
Basel cell carcinoma

Schizophrenia
Wagner syndrome

Cortisol resistance
Autosomal dominant deafness

Colorectal cancer
Recessive cutis laxa
Corneal dystrophies

Bronchial asthma
Myelodysplasia (MDS)
Susceptibility to allergy and asthma

Familial Eosinophilia
Hereditary Capillary Hemangioma
Plasmodium falciparum parasitemia intensity
Schistosoma mansoni/susceptibility/resistance

Acute myelogenous leukemia (AML)

Autosomal recessive retinitis pigmentosa
Diastrophic dysplasia

GM2-gangliosidosis, AB variant
Startle disease
Treacher Collins syndrome

Susceptibility to obesity
Susceptibility to nocturnal asthma

Limb-girdle muscular dystrophy
Factor XII deficiency
Carnitine deficiency
Myeloid malignancy
Neurogenic arthrogryposis multiplex congenita

Atrial septal defect
Craniosynostosis, type 2
Leukotriene C4 synthase deficiency

Hereditary I Lymphedema

Acute promyelocytic leukemia
Cockayne syndrome-1
Complex I deficiency

19

Peutz-Jeghers syndrome
Diabetes mellitus

Mullerian duct syndrome
Lymphoid leukemia

Mannosidosis

Atherosclerosis
Familial hypercholesterolemia

Familial Hemiplegic Migraine
CADASIL
Immunodeficiency, HLA (II)
Multiple epiphysial dysplasia
Pseudoachondroplasia

Hemolytic anemia
Congenital Nephrotic Syndrome

Central core disease
Malignant hyperthermia

Polio susceptibility

Xeroderma pigmentosum, D
Cockayne's syndrome
DNA ligase I deficiency

Maple syrup urine disease
Hyperlipoproteinemia (IIIb, II)
Myotonic dystrophy
Hypogonadism

Glutaricacidurea, IIB

This poster from the Joint Genome Institute shows the location of genes associated with diseases in three human chromosomes. (U.S. Department of Energy's Joint Genome Institute, Walnut Creek, California, http://www.jgi.doe.gov)

cardiovascular diseases, rheumatoid arthritis, psychosis, dyslexia, epilepsy, and mental retardation. In total, multifactorial disorders account for more genetic diseases than do single-gene and chromosome disorders combined.

Impact and Applications

In 2003, the Human Genome Project achieved its goal of mapping the entire human genome. The complete specifications of the genetic material on each of the twenty-two autosomes and the X and Y chromosomes will improve the understanding of the biological and molecular bases of hereditary diseases. Once the location of a gene is known, it is possible to make a better prediction of how that gene is transmitted within a family and of the probability that an individual will inherit a specific genetic disease.

For many hereditary diseases, the protein produced by the gene and its relation to the symptoms of the disease are not known. Locating a gene facilitates this knowledge. It becomes possible to develop new diagnostic tests and therapies. The number of hereditary disorders that can be tested prenatally and in newborns will increase dramatically. In the case of those single genes that do not produce clinical symptoms until later in life, many more of these disorders will be diagnosed before symptoms appear, opening the way for better treatments and even prevention. Possibilities will exist to develop the means of using gene therapy to repair or replace the disease-causing gene. The identification and mapping of single genes and those identified as having major effects on multifactorial disorders will greatly affect hereditary disease treatment and genetic counseling techniques. It is evident that knowledge of genes, both those that cause disease and those that govern normal functions, will begin to raise many questions about legal, ethical, and moral issues.

—Donald J. Nash, updated by Bryan Ness
See also: Albinism; Alcoholism; Alzheimer's Disease; Autoimmune Disorders; Breast Cancer; Burkitt's Lymphoma; Cancer; Color Blindness; Congenital Defects; Consanguinity and Genetic Disease; Cystic Fibrosis; Diabetes; Down Syndrome; Dwarfism; Emerging Diseases; Fragile X Syndrome; Gender Identity; Heart Disease; Hemophilia; Hermaphrodites; Homosexuality; Human Genetics; Human Genome Project; Huntington's Disease; Hypercholesterolemia; Inborn Errors of Metabolism; Infertility; Klinefelter Syndrome; Lactose Intolerance; Metafemales; Mitochondrial Diseases; Monohybrid Inheritance; Neural Tube Defects; Phenylketonuria (PKU); Prader-Willi and Angelman Syndromes; Prion Diseases: Kuru and Creutzfeldt-Jakob Syndrome; Pseudohermaphrodites; Sickle-Cell Disease; Smallpox; Swine Flu; Tay-Sachs Disease; Testicular Feminization Syndrome; Thalidomide and Other Teratogens; Turner Syndrome; XYY Syndrome.

Further Reading

Dykens, Elisabeth M., Robert M. Hodapp, and Brenda M. Finucane. *Genetics and Mental Retardation Syndromes: A New Look at Behavior and Interventions.* Baltimore: Paul H. Brookes, 2000. Reviews the genetic and behavioral characteristics of nine mental retardation syndromes, giving in-depth information on genetic causes, prevalence, and physical and medical features of Down, Williams, fragile X, and Prader-Willi syndromes, as well as five other less frequently diagnosed syndromes.

Faraone, Stephen V., Ming T. Tsuang, and Debby W. Tsuang. *Genetics of Mental Disorders: A Guide for Students, Clinicians, and Researchers.* New York: Guilford Press, 1999. Reviews the complex interplay of genes and environmental factors involved in the causation and expression of frequently encountered disorders including schizophrenia, bipolar disorder, depression, and Alzheimer's disease.

Gilbert, Patricia. *Dictionary of Syndromes and Inherited Disorders.* 3d ed. Chicago: Fitzroy Dearborn, 2000. Lists syndromes and inherited disorders with notes on alternative names, incidence, causes, characteristics, management implications, and future prospects for individuals with these conditions; contact information for self-help groups; and a glossary.

Goldstein, Sam, and Cecil R. Reynolds, eds. *Handbook of Neurodevelopmental and Genetic Disorders in Children.* New York: Guilford Press, 1999. Highlights the role of genetics in shaping the development and lives of many children and surveys disorders primarily affecting learning and behavior and those with broader-spectrum effects, including attention-deficit hyperactivity disorder, Tourette's syndrome, and autism.

Jorde, Lynn B., et al. *Medical Genetics.* 2d ed. St. Louis, Mo.: Mosby, 1999. Explains basic molecular genetics, chromosomal and single-gene disorders, immunogenetics, cancer genetics, multifactorial disorders, and fetal therapy.

Lewis, Ricki. *Case Workbook in Human Genetics.* 2d ed. New York: McGraw-Hill, 2000. Presents problems based on specific diseases, including acute leukemia, alcoholism, fragile X syndrome, gonadal dysgenesis, muscular dystrophy, nephrolithiasis, Tangier disease, Tay-Sachs disease, and thyroid cancer.

McKusick, Victor A., comp. *Mendelian Inheritance in Man.* Baltimore: Johns Hopkins University Press, 1994. Comprehensive catalog of Mendelian traits in humans. Filled with medical terminology, clinical descriptions, and fascinating accounts of many traits.

Pasternak, Jack J. *An Introduction to Human Molecular Genetics: Mechanisms of Inherited Diseases.* Bethesda, Md.: Fitzgerald Science Press, 1999. Discusses treatment advances, fundamental molecular mechanisms that govern human inherited diseases, the interactions of genes and their products, and the consequences of these mechanisms on disease states in major organ systems such as muscles, the nervous system, and the eyes. Also addresses cancer and mitochondrial disorders. Illustrations (some color), chapter summaries, review questions, glossary.

Pierce, Benjamin A. *The Family Genetic Sourcebook.* New York: John Wiley & Sons, 1990. An introduction to the principles of heredity and a catalog of more than one hundred human traits. Topics include heredity, inheritance patterns, chromosomes and chromosomal abnormalities, genetic risks, genetic counseling, and family history. Written for the general reader, with short descriptions, and includes suggested readings, appendices, glossary, and index.

Scriver, Charles, et al., eds. *The Metabolic and Molecular Bases of Inherited Disease.* 4 vols. 8th ed. New York: McGraw-Hill, 2001. An authority on heredity of disease and genetic inheritance, covering genetic perspectives, basic concepts, how inherited diseases occur, diagnostic approaches, and the effects of hormones.

Wynbrandt, James, and Mark D. Ludman. *The Encyclopedia of Genetic Disorders and Birth Defects.* 2d ed. New York: Facts On File, 2000. Six hundred entries cover the spectrum of clinical and research information on hereditary conditions and birth defects in a nontechnical manner. Illustrated.

Web Sites of Interest

Centers for Disease Control, Office of Genomics and Disease Prevention. http://www .cdc.gov/genomics/default.htm. Offers information on the genetic discoveries and prevention of diseases in humans. Includes links to related resources.

Dolan DNA Learning Center, Your Genes Your Health. http://www.ygyh.org. Sponsored by the Cold Spring Harbor Laboratory, this site, a component of the DNA Interactive Web site, offers information on more than a dozen inherited diseases and syndromes.

Genetic Alliance. http://www.geneticalliance .org. An international advocacy group of those with genetic conditions. This site provides links to information on the diseases, public policy, and support organizations for a broad array of hereditary diseases.

Medline Plus. http://www.nlm.nih.gov/medline plus. Medline, sponsored by the National Institutes of Health, is one of the first stops for any medical question, and it offers information and references on most genetic diseases, birth defects, and disorders.

Heredity and Environment

Field of study: Human genetics

Significance: *"Heredity and environment" is the modern incarnation of the age-old debate on the effects of nature versus nurture. Research in the field has implications ranging from the improvement of* *crop plants to the understanding of the heritability of behavioral traits in humans.*

Key terms

GENOTYPE: the genes that are responsible for physical or biochemical traits in organisms

HERITABILITY: a measure of the genetic variation for a quantitative trait in a population

PHENOTYPE: the physical and biochemical traits of a plant or animal

PHENOTYPIC PLASTICITY: the ability of a genotype to produce different phenotypes when exposed to different environments

QUANTITATIVE TRAIT LOCUS (QTL) MAPPING: a molecular biology technique used to identify genes controlling quantitative traits in natural populations

REACTION NORM: the graphic illustration of the relationship between environment and phenotype for a given genotype

Nature vs. Nurture and the Origin of Genetics

Is human behavior controlled by genes or by environmental influences? The "nature vs. nurture" controversy has raged throughout human history, eventually leading to the modern antithesis between hereditarianism and environmentalism in biological research. These two schools of thought have shaped a dispute that is at once a difficult scientific problem and a thorny ethical dilemma. Many disciplines, chiefly genetics but also the cognitive sciences, have contributed to the scientific aspect of the discussion. At the same time, racist and sexist overtones have muddled the inquiry and inextricably linked it to the implementation of social policies. Nevertheless, the relative degree of influence of genes and environments in determining the characteristics of living organisms is a legitimate and important scientific question, apart from any social or ethical consideration.

At the beginning of the twentieth century, scientists rediscovered the laws of heredity first formulated by Gregor Mendel in 1865. Mendel understood a fundamental concept that underlies all genetic analyses: Each discrete trait in a living organism, such as the color of peas, is influenced by minute particles inside the body that behave according to simple and predictable patterns. Mendel did not use the term "gene" to refer to these particles (he called them "factors"), and his pioneering work remained largely unknown to the scientific community for the remainder of the nineteenth century. Immediately following the rediscovery of Mendel's laws in 1900, the Danish biologist Wilhelm Johannsen proposed the fundamental distinction between "phenotype" and "genotype." The phenotype is the ensemble of all physical and biochemical traits of a plant or animal. The composite of all the genes of an individual is its genotype. To some extent, the genotype determines the phenotype.

Reaction Norm: Environments and Genes Come Together

It was immediately clear to Johannsen that the appearance of a trait is the combined result of both the genotype and the environment, but to understand how these two factors interact took the better part of the twentieth century and is still a preeminent field of research in ecological genetics. One of the first important discoveries was that genotypes do not always produce the same phenotype but that this varies with the particular environment to which a genotype is exposed. For instance, if genetically identical fruit flies are raised at two temperatures, there will be clear distinctions in several aspects of their appearance, such as the size and shape of their wings, even though the genes present in these animals are indistinguishable.

This phenomenon can be visualized in a graph by plotting the observed phenotype on the *y*-axis versus the environment in which that phenotype is produced on the *x*-axis. A curve describing the relationship between environment and phenotype for each genotype is called a reaction norm. If the genotype is insensitive to environmental conditions, its reaction norm will be flat (parallel to the environmental axis); most genotypes, however, respond to alterations in the environment by producing distinct phenotypes. When the latter case occurs, that genotype is said to exhibit phenotypic plasticity. One can think of plasticity as the degree of responsiveness of a given genotype to changes in its environment: The more responsive the genotype is, the more plasticity it displays.

The first biologist to fully appreciate the importance of reaction norms and phenotypic plasticity was the Russian Ivan Schmalhausen, who wrote a book on the topic in 1947. Schmalhausen understood that natural selection acts on the shape of reaction norms: By molding the genotype's response to the environment, selection can improve the ability of that genotype to survive under the range of environmental conditions it is likely to encounter in nature. For example, some butterflies are characterized by the existence of two seasonal forms. One form exists during the winter, when the animal's activity is low and the main objective is to avoid predators. Accordingly, the coloration of the body is dull to blend in with the surroundings. During the summer, however, the butterflies are very active, and camouflage would not be an effective strategy against predation. Therefore, the summer generation develops brightly colored "eyespots" on its wings. The function of these spots is to attract predators' attention away from vital organs, thereby affording the insect a better chance of survival. Developmental geneticist Paul Brakefield demonstrated, in a series of works published in the 1990's, that the genotype of these butterflies codes for proteins that sense the season by using environmental cues such as photoperiod and temperature. Depending on the perceived environment, the genotype directs the butterfly developmental system to produce or not produce the eyespots.

Quantitative Genetics of Heredity and Environment

An important aspect of modern science is the description of natural phenomena in mathematical form. This allows predictions on future occurrences of such phenomena. In the 1920's, Ronald Fisher developed the field of quantitative genetics, a major component of which is a powerful statistical technique known as analysis of variance. This allows a researcher to gather data on the reaction norms of several genotypes and then mathematically partition the observed phenotypic variation (V_p) into its three fundamental constituents:

$$V_p = V_g + V_e + V_{ge}$$

where V_g is the percentage of variation caused by genes, V_e is the percentage attributable to environmental effects, and V_{ge} is a term accounting for the fact that different genotypes may respond differently to the same set of environmental circumstances. The power of this approach is in its simplicity: The relative balance among the three factors directly yields an answer to any question related to the nature-nurture conundrum. If V_g is much higher than the other two components, genes play a primary role in determining the phenotype ("nature"). If V_e prevails, the environment is the major actor ("nurture"). However, when V_{ge} is more significant, this suggests that genes and environments interact in a complex fashion so that any attempt to separate the two is meaningless. Anthony Bradshaw pointed out in 1965 that large values of V_{ge} are indeed observable in most natural populations of plants and animals.

The quantity V_g is particularly important for the debate because when it is divided by V_p, it yields the fundamental variable known as heritability. Contrary to intuition, heritability does not measure the degree of genetic control over a given trait but only the relative amount of phenotypic "variation" in that trait that is attributable to genes. In 1974, Richard Lewontin pointed out that V_g (and therefore heritability) can change dramatically from one population to another, as well as from one environment to another, because V_g depends on the frequencies of the genes that are turned on (active) in the individuals of a population. Since different sets of individuals may have different sets of genes turned on, every population can have its own value of V_g for the same trait. Along similar lines, some genes are turned on or off in response to environmental changes; therefore, V_g for the same population can change depending on the environment in which that population is living. Accordingly, estimates of heritability cannot be compared between different populations or species and are only valid in one particular set of environmental conditions.

Molecular Genetics

The modern era of the study of nature-nurture interactions relies on the developments in molecular genetics that characterized the whole of biology throughout the second half of the twentieth century. In 1993, Carl Schlichting and Massimo Pigliucci proposed that specific genetic elements known as plasticity genes supervise the reaction of organisms to their surroundings. A plasticity gene normally encodes a protein that functions as a receptor of environmental signals; the receptor gauges the state of a relevant environmental variable such as temperature and sends a signal that initiates a cascade of effects eventually leading to the production of the appropriate phenotype. For example, many trees shed their leaves at the onset of winter in order to save energy and water that would be wasted by maintaining structures that are not used during the winter months. The plants need a reliable cue that winter is indeed coming to best time the shedding process. Deciduous trees use photoperiod as an indicator of seasonality. A special set of receptors known as phytochromes sense day length, and they initiate the shedding whenever day length becomes short enough to signal the onset of winter. Phytochromes are, by definition, plasticity genes.

Research on plasticity genes is a very active field in both evolutionary and molecular genetics. Johanna Schmitt's group has demonstrated that the functionality of photoreceptors in plants has a direct effect on the fitness of the organism, thereby implying that natural selection can alter the characteristics of plasticity genes. Harry Smith and collaborators have contributed to the elucidation of the action of photoreceptors, uncovering an array of other genes that relate the receptor's signals to different tissues and cells so that the whole organism can appropriately respond to the change in environmental conditions. Similar research is ongoing on an array of other types of receptors that respond to nutrient availability, water supply, temperature, and a host of other environmental conditions.

From an evolutionary point of view, it is important not only to uncover which genes control a given type of plasticity but also to find out if and to what extent these genes are variable in natural populations. According to neo-Darwinian evolutionary theory, natural selection is effective only if populations harbor different versions of the same genes, thereby providing an ample set of possibilities from which the most fit combinations are passed to the next generation. Thomas Mitchell-Olds pioneered a combination of statistical and molecular techniques known as quantitative trait loci (QTL) mapping, which allows researchers to pinpoint the location in the genome of those genes that are both responsible for phenotypic plasticity and variable in natural populations. These genes are the most likely targets of natural selection for the future evolution of the species.

Complex Traits: Behavior and Intelligence

The most important consequence of nature-nurture interactions is their application to the human condition. Humans are compelled to investigate questions related to the degree of genetic or environmental determination of complex traits such as behavior and intelligence. Unfortunately, such a quest is a potentially explosive mixture of science, philosophy, and politics, with the latter often perverting the practice of the first. For example, the original intention of intelligence quotient (IQ) testing in schools, introduced by Alfred Binet at the end of the nineteenth century, was simply to identify pupils in need of special attention in time for remedial curricula to help them. Soon, however, IQ tests became a widespread tool to support the supposed "scientific demonstration" of the innate inferiority of some races, social classes, or a particular gender (with the authors of such studies usually falling into the "superior" race, social class, or gender). During the 1970's, ethologist Edward Wilson freely extrapolated from behavioral studies on ant colonies to reach conclusions about human nature; he proposed that genes directly control many aspects of animal and human behavior, thereby establishing the new and controversial discipline of sociobiology.

The reaction against this trend of manipulating science to advance a political agenda has, in some cases, overshot the mark. Some well-intentioned biologists have gone so far as to imply either that there are no genetic differences among human beings or that they are at least irrelevant. This goes against everything that is known about variation in natural populations of any organism. There is no reason to think that humans are exceptions: Since humans can measure genetically based differences in behavior and problem-solving ability in other species and relate these differences to fitness, the argument that such differences are somehow unimportant in humans is based on social goodwill rather than scientific evidence.

The problem with both positions is that they do not fully account for the fact that nature-nurture is not a dichotomy but a complex interaction. In reality, genes do not control behavior; their only function is to produce a protein, whose only function is to interact with other proteins at the cellular level. Such interactions do eventually result in what is observed as a phenotype—perhaps a phenotype that has a significant impact on a particular behavior—but this occurs only in a most indirect fashion and through plenty of environmental influences. On the other hand, plants, animals, and even humans are not infinitely pliable by environmental occurrences. Some behaviors are indeed innate, and others are the complex outcome of a genotype-environment feedback that occurs throughout the life span of an organism. In short, nature-nurture is not a matter of "either/or" but a question of how the two relate and influence each other.

As for humans, it is very likely that the precise extent of the biological basis of behavior and intelligence will never be determined because of insurmountable experimental difficulties. While it is technically feasible, it certainly is morally unacceptable to clone humans and study their characteristics under controlled conditions, the only route successfully pursued to experimentally disentangle nature and nurture in plants and animals. Studies of human twins help little, since even those separated at birth are usually raised in similar societal conditions, with the result that the effects of heredity and environment are hopelessly confounded from a statistical standpoint. Regardless of the failure of science to answer these questions fully, the more compelling argument that has been made so far is that the actual answer should not matter to society, in that every human being is entitled to the same rights and privileges of any other one, regardless of real and sometimes profound differences in genetic makeup. Even the best science is simply the wrong tool to answer ethical questions.

—*Massimo Pigliucci*

See also: Aggression; Alcoholism; Altruism; Artificial Selection; Behavior; Biological Clocks; Biological Determinism; Criminality; Developmental Genetics; Eugenics; Gender Identity; Genetic Engineering: Medical Applications; Genetic Engineering: Social and Ethical Issues; Genetic Screening; Genetic Testing; Genetic Testing: Ethical and Economic Issues; Heredity and Environment; Homosexuality; Human Genetics; Inbreeding and Assortative Mating; Intelligence; Miscegenation and Antimiscegenation Laws; Natural Selection; Sociobiology; Twin Studies; XYY Syndrome.

Further Reading

Carson, Ronald A., and Mark A. Rothstein. *Behavioral Genetics: The Clash of Culture and Biology.* Baltimore: Johns Hopkins University Press, 1999. Experts from a range of disciplines—genetics, ethics, neurosciences, psychiatry, sociology, and law—address the cultural, legal, and biological underpinnings of behavioral genetics.

Cartwright, John. *Evolution and Human Behavior: Darwinian Perspectives on Human Nature.* Cambridge, Mass.: MIT Press, 2000. Offers an overview of the key theoretical principles of human sociobiology and evolutionary psychology and shows how they illuminate the ways humans think and behave. Argues that humans think, feel, and act in ways that once enhanced the reproductive success of our ancestors.

Clark, William R., and Michael Grunstein. *Are We Hardwired? The Role of Genes in Human Behavior.* New York: Oxford University Press, 2000. Explores the nexus of modern genetics and behavioral science, revealing that few elements of behavior depend upon a single gene; instead, complexes of genes, often across chromosomes, drive most of human heredity-based actions. Asserts that genes and environment are not opposing forces but work in conjunction.

DeMoss, Robert T. *Brain Waves Through Time: Twelve Principles for Understanding the Evolution of the Human Brain and Man's Behavior.* New York: Plenum Trade, 1999. Provides an accessible examination on what makes humans unique and delineates twelve principles that can explain the rise of humankind and the evolution of human behavior.

Gould, Stephen Jay. *The Mismeasure of Man.* Rev. ed. New York: W. W. Norton, 1996. A noted biologist provides a fascinating account of the misuse of biology in supporting racial policies.

Plomin, Robert, et al. *Behavioral Genetics.* 4th ed. New York: Worth, 2001. Introductory text that explores the basic rules of heredity, its DNA basis, and the methods used to find genetic influence and to identify specific genes.

Wright, William. *Born That Way: Genes, Behavior, Personality.* New York: Knopf, 1998. Uses twin and adoption studies to trace the evolution of behavioral genetics and discusses the corroborating research in molecular biology that underlines the links between genes and personality.

Dawkins, Richard. *The Selfish Gene.* New York: Oxford University Press, 1989. Argues that the world of the selfish gene revolves around competition and exploitation and yet acts of apparent altruism do exist in nature. A popular account of sociobiological theories that revitalized Darwin's natural selection theory.

Hermaphrodites

Field of study: Diseases and syndromes
Significance: *Hermaphrodites are people born with both male and female sexual parts. Early identification and thorough medical evaluation of these individuals can help them lead relatively normal lives.*

Key terms

GENOTYPE: an organism's complete set of genes

GONAD: an organ that produces reproductive cells and sex hormones; termed ovaries in females and testes in males

KARYOTYPE: a description of the chromosomes of an individual's cells, including the number of chromosomes and a physical description of them (normal female is 46,XX and normal male is 46,XY)

PHENOTYPE: the physical and biochemical characteristics of an individual based on the interaction of genotype and environment

Early Human Sexual Development

Up to the ninth week of gestation, the external genitalia (external sexual organs) are identical in appearance in both male and female human embryos. There is a phallus that will become a penis in males and a clitoris in females and labioscrotal swelling that will become a scrotum in males and labial folds in females. A person's development into a male or female is governed by his or her sex chromosome constitution (the X and Y chromosomes). An individual who has two X chromosomes normally develops into a female, and one who has one X and one Y chromosome normally develops into a male. It is the Y chromosome that determines the development of a male. The Y chromosome causes the primitive gonads (the gonads that have not developed into either an ovary or a testis) to develop into testes and to produce testosterone (the male sex hormone). It is testosterone that acts on the early external genitalia and causes the development of a penis and scrotum. If testosterone is not present, regardless of the chromosome constitution of the embryo, normal female external genitalia will develop.

Hermaphrodites

Hermaphrodites are individuals who have both male and female gonads. At birth, hermaphrodites can have various combinations of external genitalia, ranging from completely female to completely male genitalia. Most hermaphrodites have external genitalia that are ambiguous (genitalia somewhere between normal male and normal female) and often consist of what appears to be an enlarged clitoris or a small penis, hypospadias (urine coming from the base of the penis instead of the tip), and a vaginal opening. The extent to which the genitalia are masculinized depends on how much testosterone was produced by the testicular portion of the gonads during development. The gonadal structures of a hermaphrodite can range from a testis on one side and an ovary on the other side, to testes and ovaries on each side, to an ovotestis (a single gonad with both testicular and ovarian tissue) on one or both sides.

Hermaphroditism has different causes. The chromosomal or genotypic sex of a hermaphrodite can be 46,XX (58 percent have this karyotype), 46,XY (12 percent), or 46,XX/46,XY (14 percent), while the rest have different types of mosaicism, such as 46,XX/47,XXY or 45,X/46,XY. Individuals with a 46,XX/46,XY karyotype are known as chimeras. Chimerism usually occurs through the merger of two different cell lines (genotypes), such as when two separate fertilized eggs fuse together to produce one embryo. This can result in a single embryo with some cells being 46,XX and some being 46,XY. Mosaicism means having at least two different cell lines present in the same individual, but the different cell lines are caused by losing or gaining a chromosome from some cells early in development. An example would be an embryo that starts out with all cells having a 47,XXY chromosome constitution and then loses a single Y chromosome from one of its cells, which then produces a line of 46,XX-containing cells. This individual would have a karyotype written as 46,XX/47,XXY. In a chimera or mosaic individual, the proportion of developing gonadal cells with Y chromosomes determines the appearance of the external genitalia. More cells with a Y chromosome mean that more testicular cells are formed and more testosterone is produced.

The cause of hermaphroditism in the majority of affected individuals (approximately 70 percent) is unknown, although it has been postulated that those hermaphrodites with normal male or female karyotypes may have hidden chromosome mosaicism in just the gonadal tissue.

Impact and Applications

Hermaphrodites with ambiguous genitalia are normally recognized at birth. It is essential that these individuals have a thorough medical evaluation, since other causes of ambiguous genitalia besides hermaphroditism can be life-threatening if not recognized and treated promptly. Once hermaphroditism is diagnosed in a child, the decision must be made whether to raise the child as a boy or a girl. This decision is made by the child's parents working with specialists in genetics, endocrinology, psychology, and urology. Typically, the karyotype and appearance of the external genitalia of the child are the major factors in deciding the sex of rearing. Previously, most hermaphrodites with male karyotypes who had either an absent or an extremely small penis were reared as females. The marked abnormality or absence of the penis was thought to prevent these individuals from having fulfilling lives as males. This practice has been challenged by adults who are 46,XY but who were raised as females. Some of these individuals believe that their conversion to a female gender was the wrong choice, and they prefer to think of themselves as male. Hermaphrodites with a female karyotype and normal or near-normal female external genitalia are typically reared as females.

The debate over what criteria should be used to decide sex of rearing of a child is ongoing. An increasingly important part of this debate is the concept of gender identity, which describes what makes people male or female in their own minds rather than according to what sex their genitalia are. This is an especially important issue for those individuals with chimerism or mosaicism who have both a male and female karyotype. Currently, the decision to raise these individuals as boys or girls is made primarily on the basis of the degree to which their external genitalia are masculinized or feminized.

Those hermaphrodites who have normal female or male genitalia at birth are at risk for developing abnormal masculinization in the phenotypic females or abnormal feminization in the phenotypic males at puberty if both testicular and ovarian tissue remains present. Thus it is usually necessary to remove the gonad that is not specific for the desired sex of the individual. An additional reason to remove the abnormal gonad is that the cells of the gonad(s) that have a 46,XY karyotype are at an increased risk of becoming cancerous.

—*Patricia G. Wheeler*

See also: Gender Identity; Homosexuality; Metafemales; Pseudohermaphrodites; Steroid Hormones; Testicular Feminization Syndrome; X Chromosome Inactivation; XYY Syndrome.

Further Reading

Dreger, Alice Domurat. *Hermaphrodites and the Medical Invention of Sex.* Cambridge, Mass.: Harvard University Press, 1998. Traces the evolution of what makes a person male or female and shows how the answer has changed historically depending on when and where the question was asked.

Gilbert, Ruth. *Early Modern Hermaphrodites: Sex and Other Stories.* New York: Palgrave, 2002. Examines the conceptions and depictions of hermaphrodites between the sixteenth and eighteenth centuries in a range of artistic, mythological, scientific, and erotic contexts.

Moore, Keith. *The Developing Human: Clinically Oriented Embryology.* 7th ed. Amsterdam: Elsevier Science, 2003. Details embryology from a clinical perspective, providing discussions of the stages of organs and systems development, including the genital system.

Zucker, Kenneth J. "Intersexuality and Gender Identity Differentiation." *Annual Review of Sex Research* 10 (1999). An extensive overview of intersexuality, gender identity formation, psychosexual differentiation, concerns about pediatric gender reassignment, hermaphroditism and pseudohermaphroditism, gender socialization, childrearing, and more. Includes a discussion of terminology, a summary, tables, and a bibliography.

Web Sites of Interest

Androgen Insensitivity Syndrome Support Group. http://www.medhelp.org/www/ais. A comprehensive educational and informational site with links to related resources.

Intersex Society of North America. http://www.isna.org. The society is "a public awareness, education, and advocacy organization which works to create a world free of shame, secrecy, and unwanted surgery for intersex people (individuals born with anatomy or physiology which differs from cultural ideals of male and female)." Includes links to information on such conditions as clitoromegaly, micropenis, hypospadias, ambiguous genitals, early genital surgery, adrenal hyperplasia, Klinefelter syndrome, androgen insensitivity, and testicular feminization.

Johns Hopkins University, Division of Pediatric Endocrinology, Syndromes of Abnormal Sex Differentiation. http://www.hopkinsmedicine.org/pediatricendocrinology. Site provides a guide to the science and genetics of sex differentiation, including a glossary. Click on "patient resources."

National Organization for Rare Disorders (NORD). http://www.rarediseases.org. Offers information and articles about rare genetic conditions and diseases, including true hermaphrodism, in several searchable databases.

High-Yield Crops

Field of study: Genetic engineering and biotechnology

Significance: *The health and well-being of the world's large population is primarily dependent on the ability of the agricultural industry to produce high-yield food and fiber crops. Advances in the production of high-yield crops will have to continue at a rapid rate to keep pace with the needs of an ever-increasing population.*

Key terms

CULTIVAR: a subspecies or variety of plant developed through controlled breeding techniques

GREEN REVOLUTION: the introduction of scientifically bred or selected varieties of grain (such as rice, wheat, and maize), which, with high enough inputs of fertilizer and water, greatly increased crop yields

MONOCULTURE: the agricultural practice of continually growing the same cultivar on large tracts of land

The Historical Development of High-Yield Crops

No one knows for certain when the first crops were cultivated, but by six thousand years ago, humans had discovered that seeds from certain plants could be collected, planted, and later gathered for food. As human populations continued to grow, it was necessary to select and produce higher-yielding crops. The Green Revolution of the twentieth century helped to make this possible. Agricultural scientists developed new, higher-yielding varieties, particularly grains that supply most of the world's calories. In addition to greatly increased yields, the new crop varieties also led to an increased reliance on monoculture, the practice of growing only one crop over a vast number of acres. Current production of high-yield crops is extremely mechanized and highly reliant on agricultural chemicals such as fertilizers and pesticides. It also requires less human power, and encourages extensive monocropping.

Methods of Developing High-Yield Crops

The major high-yield crops are wheat, corn, soybeans, rice, potatoes, and cotton. Each of these crops originated from a low-yield native plant. The two major ways to improve yield in agricultural plants is to produce a larger number of harvestable parts (such as fruits or leaves) per plant or to produce plants with larger harvestable parts. For example, to increase yield in corn, the grower must either produce more ears of corn per plant or produce larger ears on each plant. Numerous agricultural practices are required to produce higher yields, but one

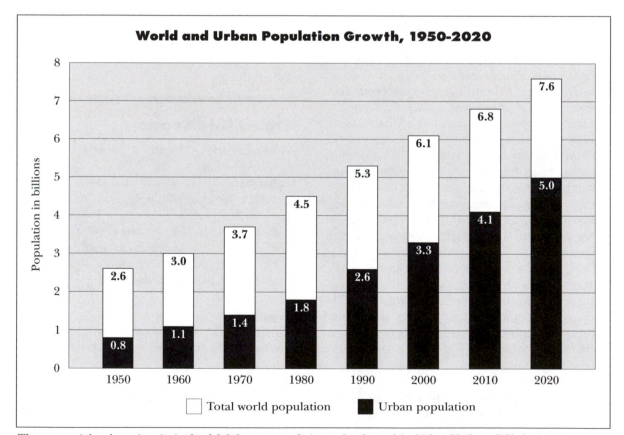

The exponential and ongoing rise in the globe's human population makes the need for high-yield, dependable food crops ever more compelling.

Source: Data are from U.S. Bureau of the Census International Data Base and John Clarke, "Population and the Environment: Complex Interrelationships," in *Population and the Environment* (Oxford, England: Oxford University Press, 1995), edited by Bryan Cartledge.

of the most important is the selection and breeding of genetically superior cultivars.

Throughout most of history, any improvement in yield was primarily based on the propagation of genetically favorable mutants. When a grower observed a plant with a potentially desirable gene mutation that produced a change that improved some yield characteristic such as more or bigger fruit, the grower would collect seeds or take cuttings (if the plant could be propagated vegetatively) and propagate them. This selection process is still one of the major means of improving yields. Sometimes a high-yield cultivar is developed which has other undesirable traits, such as poor flavor or undesirable appearance. Another closely related cultivar may have good flavor or desirable appearance, but low yield. Traditional breeding techniques can be used to form hybrids between two such cultivars, in hopes that all the desirable traits will be combined in a new hybrid cultivar.

Genetic Modification

The advent of recombinant DNA technology has brought greater precision into the process of producing high-yield cultivars and has made it possible to transfer genetic characteristics between any two plants, regardless of how closely related. The first step generally involves the insertion of a gene or genes that might increase yield into a piece of circular DNA called a plasmid. The plasmid is then inserted into a bacteria, and the bacteria is then used as a vector to transfer the gene into the DNA of another plant. This technology has resulted in genetically modified crops such as "golden rice" (fortified with vitamin A), herbicide-resistant soybeans, and new strains such as triticale, which promise to ameliorate world hunger at the same time that they threaten to reduce biodiversity and alter other plants through genetic drift.

Impact and Applications

As the human population grows, pressure on the world's food supply will increase. Consequently, researchers are continually seeking better ways to increase food production. In order to accomplish this goal, advances in the

Grain crops such as rice, wheat, and (above) corn, grown here for research by DeKalb Genetics Corporation, are among those that have been genetically modified to increase yield and nutritional value. (AP/Wide World Photos)

production of high-yield crops will have to continue at a rapid rate to keep pace. New technologies will have to be developed, and many of these new technologies will center on advances in genetic engineering. It is hoped that such advances will lead to the development of new high-yield crop varieties that require less water, fertilizer, and chemical pesticides.

—*D. R. Gossett, updated by Bryan Ness*
See also: Biofertilizers; Biopesticides; Cell Culture: Plant Cells; Cloning; Cloning: Ethical Issues; Cloning Vectors; Genetic Engineering; Genetic Engineering: Agricultural Applications; Genetic Engineering: Historical Development; Genetic Engineering: Industrial Applications; Genetic Engineering: Risks; Genetic Engineering: Social and Ethical Issues; Geneti-

cally Modified (GM) Foods; Hybridization and Introgression; Lateral Gene Transfer; Transgenic Organisms.

Further Reading

Avery, Dennis T. *Saving the Planet with Pesticides and Plastic: The Environmental Triumph of High-Yield Farming*. 2d ed. Indianapolis, Ind.: Hudson Institute, 2000. Argues that high-yield agriculture using chemical pesticides, fertilizers, and biotechnology is the solution to environmental problems, not a cause of them, as environmental activists have averred.

Bailey, L. H., ed. *The Standard Cyclopedia of Horticulture*. 2d ed. 3 vols. New York: Macmillan, 1963. Since the 1920's, a standard reference that still offers basic information; its original subtitle reads: "a discussion, for the amateur, and the professional and commercial grower, of the kinds, characteristics and methods of cultivation of the species of plants grown in the regions of the United States and Canada for ornament, for fancy, for fruit and for vegetables; with keys to the natural families and genera, descriptions of the horticultural capabilities of the states and provinces and dependent islands, and sketches of eminent horticulturists."

Chrispeels, Maarten J., and David E. Sadava. *Plants, Genes, and Agriculture*. Boston: Jones and Bartlett, 1994. A textbook on the use of biotechnology in crop production. Contains sections related to the use of biotechnology to transfer desirable traits from one plant to another.

Janick, Jules. *Horticultural Science*. 4th ed. New York: W. H. Freeman, 1986. Contains sections on horticultural biology, environment, technology, and industry and covers the fundamentals associated with the production of high-yield crops.

Lynch, J. M. *Soil Biotechnology: Microbiological Factors in Crop Productivity*. Malden, Mass.: Blackwell, 1983. Contains some excellent information on the potential for genetically engineering microorganisms to improve crop production.

Metcalfe, D. S., and D. M. Elkins. *Crop Production: Principles and Practices*. 4th ed. New York: Macmillan, 1980. A text for the introductory agriculture student, which serves as one of the most valuable sources available on the practical aspects of the production of high-yield crops.

Web Site of Interest

Food and Agriculture Organization of the United Nations. Biotechnology in Food and Agriculture. http://www.fao.org. Addresses the role of biotechnology in worldwide food production.

Homeotic Genes

Field of study: Developmental genetics

Significance: *Embryonic development and organogenesis proceed by way of a complex series of cascaded gene activities, which culminate in the activation of the homeotic genes to specify the final identities of body parts and shapes. The discovery of homeotic genes has provided the key to understanding these patterns of development in multicellular organisms. Knowledge of homeotic genes not only is helping scientists understand the variety and evolution of body shapes (morphology) but also is providing new insights into genetic diseases and cancer.*

Key terms

PROMOTER: the control switch in genes where transcription factors bind to activate or repress the conversion of DNA information into proteins

TRANSCRIPTION FACTOR: a protein with specialized structures that binds specifically to the promoters in genes and controls the gene's activity

The Discovery of Homeotic Genes

One of the most powerful tools in genetic research is the application of mutagenic agents (such as X rays) that cause base changes in the DNA of genes to create mutant organisms. These mutants display altered appearances, or phenotypes, giving the geneticist clues about how the normal genes function. Few geneticists have used this powerful research tool as well as the recipient of the 1995 Nobel Prize in

Physiology or Medicine recipient Christiane Nüsslein-Volhard (who shared the award with Edward B. Lewis and Eric Wieschaus). She and her colleagues, analyzing thousands of mutant *Drosophila melanogaster* fruit flies, discovered many of the genes that functioned early in embryogenesis.

Among the many mutant *Drosophila* flies studied by these and other investigators, two were particularly striking. One mutant had two sets of fully normal wings; the second set of wings, just behind the first set, displaced the normal halteres (flight balancers). The other mutant had a pair of legs protruding from its head in place of its antennae. These mutants were termed "homeotic" because major body parts were displaced to other regions. Using such mutants, Lewis was able to identify a clustered set of three genes responsible for the extra wings and map or locate them on the third chromosome of *Drosophila*. He called this gene cluster the bithorax complex (*BX-C*). The second mutation was called antennapedia, and its complex, with five genes, was called *ANT-C*. If all the *BX-C* genes were removed, the fly larvae had normal head structures, partially normal middle or thoracic structures (where wings and halteres are located), but very abnormal abdominal structures that appeared to be nothing more than the last thoracic structure repeated several times. From these genetic studies, it was concluded that the *BX-C* genes controlled the development of parts of the thorax and all of the abdomen and that the *ANT-C* genes controlled the rest of the thorax and most of the head.

The *BX-C* and *ANT-C* genes were called homeotic selector genes: "selector" because they acted as major switch points to select or activate whole groups of other genes for one developmental pathway or another (for example, formation of legs, antennae, or wings from small groups of larval cells in special compartments called imaginal disks). Although geneticists knew that these homeotic selector genes were arranged tandemly in two clusters on the third *Drosophila* chromosome, they did not know the molecular details of these genes or understand how these few genes functioned to cause such massive disruptions in the *Drosophila* body parts.

The Molecular Properties of Homeotic Genes

With so many mutant embryos and adult flies available, and with precise knowledge about the locations of the homeotic genes on the third chromosome, the stage was set for an intensive molecular analysis of the genes in each complex. In 1983, William Bender's laboratory used new, powerful molecular methods to isolate and thoroughly characterize the molecular details of *Drosophila* homeotic genes. He showed that the three bithorax genes constituted only 10 percent of the whole *BX-C* cluster. What was the function of the other 90 percent if it did not contain genes? Then William McGinnis' and J. Weiner's laboratories made another startling discovery: The base sequences (the order of the nucleotides in the DNA) of the homeotic genes they examined contained nearly the same sequence in the terminal 180 bases. This conserved 180-base sequence was termed the "homeobox." What was the function of this odd but commonly found DNA sequence? What kind of protein did this homeobox-containing gene make?

Soon it was discovered that homeotic genes and homeoboxes were not confined to *Drosophila*: All animals had them, both vertebrates, such as mice and humans, and invertebrates, such as worms and even sea sponges. The homeobox sequence was not only conserved within homeotic and other developmental genes, but it was also conserved throughout the entire animal kingdom. All animals seemed to possess versions of an ancestral homeobox gene that had duplicated and diverged over evolutionary time.

New discoveries about homeobox genes flowed out of laboratories all over the world in the late 1980's and early 1990's; it was discovered that the order of the homeobox genes in the gene clusters from all animals was roughly the same as the order of the eight genes found in the original *BX-C* and *ANT-C* homeotic clusters of *Drosophila*. In more complex animals such as mice and humans, the two *Drosophila*-type clusters were duplicated on four chromosomes instead of just one. Mice have thirty-two homeotic genes, plus a few extra not found in *Drosophila*. Frank Ruddle hypothesized that the more anatomically complex the animal, the more homeotic genes it will have in its chromosomes. Experimental evidence from several laboratories has supported Ruddle's hypothesis.

The questions posed earlier about the functions of extra DNA in the homeotic clusters and the role of the homeobox in gene function were finally answered. It seems that all homeotic genes code for transcription factors, or proteins that control the activity or expression of other genes. The homeobox portion codes for a section of protein that binds to base sequences in the promoters of other genes, thus stimulating those genes to express their proteins. The earlier idea of homeotic genes as selector genes makes sense; the protein products of homeotic genes bind to the promoter control regions of many other genes and activate them to make complex structures such as legs and wings. The homeotic genes themselves are under the control of other genes making transcription factors that bind to the extra DNA in the homeotic clusters. The bound transcription factors control the differential expression of homeotic genes in many different cellular environments throughout the developing embryo, all along its anterior to posterior axis. Embryonic development and organogenesis proceed by way of a complex series of cascaded gene activities, which culminate in the activation of the homeotic genes to specify the final identities of body parts and shapes.

Impact and Applications

In a 1997 episode of the television series *The X-Files*, a mad scientist transforms his brother into a monster with two heads. Federal Bureau of Investigation (FBI) agent Dana Scully patiently explains to her partner Fox Mulder that the scientist altered his brother's homeobox genes, causing the mutant phenotype. Science fiction indeed—but with the successful cloning of Dolly the sheep in 1997, the prospect of manipulating homeobox genes in embryos is no longer far-fetched.

The first concern of scientists is to elucidate more molecular details about the actual processes by which discrete genes transform an undifferentiated egg cell into a body with per-

fectly formed, bilateral limbs. Sometimes mutations in homeobox genes cause malformed limbs, extra digits on the hands or feet, or fingers fused together, conditions known as synpolydactyly; often limb and hand deformities are accompanied by genital abnormalities. Several reports in 1997 provided experimental evidence for mutated homeobox genes in certain leukemias and cancerous tumors. Beginning in 1996, the number of reports describing correlations between mutated homeobox genes and specific cancers and other developmental abnormalities increased dramatically. Although no specific gene-based therapies have been proposed for treating such diseases, the merger between the accumulated molecular knowledge of homeotic genes and the practical gene manipulation technologies spawned by animal cloning will likely lead to new treatments for limb deformities and certain cancers.

—*Chet S. Fornari*

See also: Developmental Genetics; Evolutionary Biology; Model Organism: *Drosophila melanogaster.*

Further Reading

DeRobertis, Eddy. "Homeobox Genes and the Vertebrate Body Plan." *Scientific American* 269 (July, 1990). A classic article on homeobox gene studies.

Lewin, B. *Genes VII.* New York: Oxford University Press, 2001. Provides an integrated account of the structure and function of genes and incorporates all the latest research in the field, including topics such as accessory proteins (chaperones), the role of the proteasome, reverse translocation, and the process of X chromosome inactivation. More than eight hundred full-color illustrations.

Lodish, Harvey, et al. *Molecular Cell Biology.* 4th ed. New York: W. H. Freeman, 2000. Contains a clear, detailed discussion of homeotic genes.

Raff, Rudolf. *The Shape of Life: Genes, Development, and the Evolution of Animal Form.* Chicago: University of Chicago Press, 1996. A detailed but readable account of how genes and evolution influence the shape of animal bodies.

Homosexuality

Field of study: Human genetics and social issues

Significance: *The debate over whether individuals choose to whom they are attracted or their orientation is determined primarily by genetic or social factors is ongoing. Interest persists in part because individuals' sexual orientation appears to extend beyond sexuality to influence gender and in part because individuals erroneously believe that social acceptance and treatment of homosexuals may differ depending upon whether gay and lesbian individuals are free agents or are responding to biological imperatives.*

Key terms

CONCORDANCE: the presence of a trait in both members of a pair of twins

HERITABILITY: the proportion of phenotypic variation that is due to genes rather than the environment

SEX-LINKED TRAITS: Characteristics that are encoded by genes on the X or Y chromosome

Biological vs. Environmental Factors

Sexual orientation is a fundamental aspect of human sexuality that usually results in females mating with males (heterosexuality). Sexual orientation may be closely linked to sexual experience, but many factors (social, religious, or logistic) can decrease the correlation. As a result, the frequency of homosexuality (a sexual orientation or attraction to persons of the same sex) varies from approximately 2 to 10 percent of the population, depending on how homosexuality is defined and measured. In general, there appears to be a continuum, from exclusive heterosexuality (90-92 percent) to exclusive homosexuality (1-4 percent) with many people falling somewhere between. Like most complex behaviors, homosexuality is probably influenced by both biological and environmental factors. The exact mechanism may differ for individuals who appear to exhibit similar behavioral patterns.

Genetic Influences

The genetic basis of homosexuality has been assessed using twin studies and pedigree analysis. Lesbians are approximately three times as likely as heterosexual women to have lesbian sisters and generally have more lesbian relatives as well, which suggests that genes as well as environmental factors influence homosexuality in women. Similarly, among men, concordance in sexual orientation among monozygotic (MZ) twins is greater than that for dizygotic (DZ) twins or nontwin brothers. Since MZ twins share 100 percent of their genes but are not always either both straight or both gay, sexual orientation cannot be 100 percent due to genes.

Heritability of homosexuality has been estimated at 30-75 percent for men and at 25-76 percent for women. The different rates of heritability and frequency, with lesbians typically representing a smaller proportion of the population than gay men, suggests that men's and women's sexuality may have different origins. The X-linked locus associated with homosexuality in some men (*Xq28*, according to Hamer et al., 1993) does not appear to be associated with lesbianism (Hu et al., 1995). Further, research suggests that men's orientation is bimodal in distribution relative to the Kinsey scale of sexual orientation, whereas women's orientation is distributed more continuously and is more likely than men's to change through adulthood.

Neurohormonal Influences

Adult homosexuals do not differ from their heterosexual counterparts in terms of circulating levels of sex hormones. Instead, the neuroendocrine theory predicts that prenatal exposure to high levels of androgens masculinizes brain structures and influences sexual orientation. Consistent with this, women with congenital adrenal hyperplasia (CAH) who experience atypically high levels of androgens prenatally appear to be somewhat more likely to engage in same-sex sexual fantasies and behavior compared to heterosexual women, whereas XY women with complete androgen insensitivity syndrome (cAIS) do not exhibit increased expression of lesbianism. Exposure to the synthetic estrogen DES, which is also thought to have a demasculating effect on the brain, also appears to influence women's sexuality modestly and to induce higher levels of homosexuality.

Stress hormones generally reduce the production of sex hormones. The level and timing of stress experienced by women during pregnancy may therefore also affect the amount of sex hormones experienced prenatally and hence the sexual differentiation and organizational phase of early brain development. Studies suggest that some women who experience stress during pregnancy may be more likely to have homosexual children, but the data are still preliminary.

Given that most homosexuals do not have one of the aforementioned hormonal conditions and most individuals who do have them are heterosexual, the neuroendocrine theory alone does not appear to account for the origin of homosexuality.

Neuroanatomical Influences

Although stereotypes exist, there is no overall lesbian or gay physique. There is some evidence that gay men's brains may differ from heterosexual men's in some structures where sexual dimorphism also occurs (for example, interstitial nuclei of the anterior hypothalmus 3, suprachiasmatic nucleus in the anterior hypothalamus and the anterior commissure), presumably due to the organizational effects of sex hormones. Structure size varies considerably both within and between sexes; however, all three structures appear to differ significantly in size for gay versus heterosexual men. It is not yet clear whether these differences cause homosexual activity or are caused by it.

Evolutionary Perspective

Evolutionary biologists have suggested that homosexuality may persist because there is little cost associated with the behavior. In situations in which homosexuality is not exclusive (that is, most individuals engage in heterosexual as well as homosexual liaisons) homosexuals would experience little or no decline in reproductive success. This could occur when marriage is compulsory, where there are strict gender roles and religious requirements, or when homosexual behavior is situational or opportunistic. Similarly, in situations in which in-

dividuals are exclusively homosexual and experience no direct individual fitness (that is, no offspring are produced), homosexuals can reduce the reproductive cost by increasing their inclusive fitness via contributions to relatives' offspring. Consistent with the latter hypothesis, there is some evidence that gay men exhibit increased levels of empathy, an accepted indicator of altruism.

Homosexuality is one of the three most common expressions of human sexual orientation and has been observed throughout human history and across religions and cultures. Like other complex behavioral traits, sexual orientation appears to be influenced by both biological and environmental factors. There is some evidence that situational or opportunistic homosexuality may differ from obligatory homosexuality and that the mechanisms influencing sexual orientation may be different in gay men and lesbians.

—Cathy Schaeff

See also: Behavior; Biological Clocks; Gender Identity; Heredity and Environment; Hermaphrodites; Human Genetics; Metafemales; Pseudohermaphrodites; Steroid Hormones; Testicular Feminization Syndrome; X Chromosome Inactivation; XYY Syndrome.

Further Reading

Diamant, L., and R. McAnuity, eds. *The Psychology of Sexual Orientation, Behavior, and Identity: A Handbook.* Westport, Conn.: Greenwood Press, 1995. Draws from biological and psychological research to provide a comprehensive overview of the major theories about sexual orientation; to summarize recent developments in genetic and neuroanatomic research; to consider the role of social institutions in shaping current beliefs; and to discuss the social construction of gender, sexuality, and sexual identity.

Hamer, D. H., S. Hu, et al. "A Linkage Between DNA Markers on the X Chromosome and Male Sexual Orientation." *Science* 261 (1993): 321-327. The first study to identify genetic markers for male sexual orientation.

Haynes, Felicity, and Tarquam McKenna. *Unseen Genders: Beyond the Binaries.* New York: Peter Lang, 2001. Explores the effects of binary stereotypes of sex and gender on transsexuals, homosexuals, cross-dressers, and transgender and intersex people.

Hu, S., et al. "Linkage Between Sexual Orientation and Chromosome Xq28 in Males but Not in Females." *Nature Genetics* 11 (1995): 248-256. Determined that the DNA marker on the X chromosome does not correspond to lesbianism.

McWhirter, David P., et al. *Homosexuality/Heterosexuality: Concepts of Sexual Orientation.* New York: Oxford University Press, 1990. Discusses sexual orientation and the current usefulness of the Kinsey Scale. Includes other scales proposed by contributors to this work.

Web Sites of Interest

Parents, Families, and Friends of Lesbians and Gays. http://www.pflag.org. Site includes a section on frequently asked questions and information about local chapters, news, and public advocacy.

Sexuality Information and Education Council of the United States. http://www.siecus.org. A vast resource on all aspects of sex and sexuality. Includes links for teenagers, public policy issues, school health, a searchable bibliography database, and more.

Human Genetics

Fields of study: Human genetics and social issues

Significance: *Human genetics is concerned with the study of the human genome. The study of human genetics includes identifying and mapping genes; determining their function, mode of transmission, and inheritance; and detecting mutated or nonfunctioning genes. Important aspects of modern human genetics include gene testing or genetic screening, gene therapy, and genetic counseling.*

Key terms

BIOINFORMATICS: The science of compiling and managing genetic and other biology data using computers, requisite in human genome research

DYSMORPHOLOGY: Abnormal physical development resulting from genetic disorder

FORENSIC GENETICS: the application of genetics, particularly DNA technology, to the analysis of evidence used in civil cases, criminal cases, and paternity testing

GENE THERAPY: the use of a viral or other vector to incorporate new DNA into a person's cells with the objective of alleviating or treating the symptoms of a disease or condition

GENE TRANSFER: Using a viral or other vector to incorporate new DNA into a person's cells. Gene transfer is used in gene therapy

GENETIC SCREENING: the use of the techniques of genetics research to determine a person's risk of developing, or his or her status as a carrier of, a disease or other disorder

GENETIC TESTING: the process of investigating a specific individual or population of people to detect the presence of genetic defects

GENOMICS: the branch of genetics dealing with the study of the genetic sequences of organisms, including the human being

PHARMACOGENOMICS: The branch of human medical genetics that evaluates how an individual's genetic makeup influences his or her response to drugs

PROTEOMICS: the study of how proteins are expressed in different types of cells, tissues, and organs

TOXICOGENOMICS: evaluating ways in which genomes respond to chemical and other pollutants in the environment

Human Genome Project

Human genetics is the discipline concerned with identifying and studying the genes carried by humans, the control and expression of traits caused by these genes, their transmission from generation to generation, and their expression in offspring. Modern human genetics properly begins with the elucidation of the structure of DNA in 1953 by James D. Watson and Francis H. Crick. This discovery led to very rapid advances in acquisition of genetic information and ultimately spawned the Human Genome Project (HGP), which was initiated in 1986 by the DOE (Department of Energy). In 1990 the DOE combined efforts with the National Institutes of Health (NIH) and private collaborators, including the Wellcome Trust of the United Kingdom, along with private companies based in Japan, France, Germany, and China. The ultimate goal of HGP was to determine the precise genetic makeup of humans as well as explore human genetic variation and human gene function. The first high-quality draft of the human genetic sequence was completed in April of 2003, thereby providing a suitable salute to the fiftieth anniversary of the discovery of DNA, which opened the modern era of human genetics.

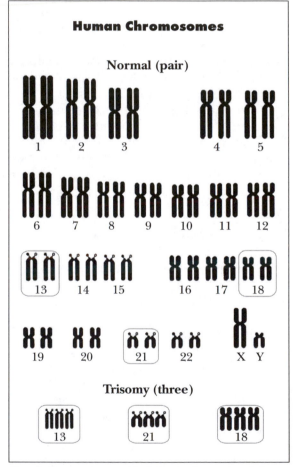

Human Chromosomes

Normal (pair)

Trisomy (three)

Genetic diseases are caused by defects in the number of chromosomes, in their structure, or in the genes on the chromosome (mutation). Shown here is the human complement of chromosomes (23 pairs) and three errors of chromosome number (trisomies) that lead to the genetic disorders Patau's syndrome (trisomy 13), Edward's syndrome (trisomy 18), and the more common Down syndrome (trisomy 21). (Hans & Cassidy, Inc.)

Almost all modern human genetics is directly related to the enormous mass of genetic data obtained and made available by the HGP. Some of the many themes now being explored include medical genetics, genetic bioinformatics, proteomics, toxicogenomics, the inheritance and prevention of gene-related cancers and other diseases, and policy and ethical issues related to genetic concerns of humans.

The human genome consists of genes located in chromosomes, along with a much smaller gene content, found in mitochondria, that is called mitochondrial DNA or mtDNA. About 99.7 percent of the human genome is located in the chromosomes, and another 0.3 percent consists of the mtDNA genome, which encodes for a number of enzymes involved in cellular respiration. The mtDNA is inherited almost entirely through the female line, so its genetic transmission and expression differ from that of classical Mendelian genetics. Studies of human mtDNA have revealed a number of medical pathologies associated with this unique mode of inheritance transmission. Studies have also proven useful in determining significant trends in the evolutionary development of *Homo sapiens* and elucidating relationships with the near-species *Homo neanderthalensis* (the now extinct Neanderthals).

The HGP effort decoded the genetic arrangement—the gene sequence of roughly 3 billion nucleotide base pairs of between 25,000 and 45,000 genes that collectively form the human genome. Many, but not all, of these have been sequenced and their locations on chromosomes mapped. Structurally, base-sequencing studies reveal that human genes showed great variations in size, ranging from several thousand base pairs to some genes comprising nearly half a million base pairs. The genetic functions have been determined for about half of the human genes that have been identified and sequenced. HGP provided so much information that a new field called bioinformatics was developed to handle the enormous amounts of genetic sequencing data for the human genome.

Bioinformatics

The purpose of bioinformatics is to help organize, store, and analyze genetic biological information in a rapid and precise manner, dictated by the need to be able to access genetic information quickly. In the United States the online database that provides access to these gene sequences is called GenBank, which is under the purview of the National Center for Biotechnology Information and has been made available on the Internet. In addition to human genome sequence records, GenBank provides genome information about plants, bacteria, and other animals.

Proteomics

Bioinformatics provides the basis for all modern studies of human genetics, including analysis of genes and gene sequences, determining gene functions, and detecting faulty genes. The study of genes and their functions is called proteomics, which involves the comparative study of protein expression. That is, exactly what is the metabolic and morphological relationship between the protein encoded within the genome and how that protein works. Geneticists are now classifying proteins into families, superfamilies, and folds according to their configuration, enzymatic activity, and sequence. Ultimately proteomics will complete the picture of the genetic structure and functioning of all human genes.

Toxicogenomics

Another newly developing field that relies on bioinformatics is the study of toxicogenomics, which is concerned with how human genes respond to toxins. Currently, this field is specifically concerned with evaluating how environmental factors negatively interact with messenger RNA (mRNA) translation, resulting in disease or dysfunction.

Medical Genetics

Almost all of modern human medical genetics rests on the identification of human gene sequences that were provided by the HGP and made accessible through bioinformatics. Human medical genetics begins with recognition of defective genes that are either nonfunctioning or malfunctioning and that cause diseases or tissue malformation. Once defective genes have been identified and cataloged, patients

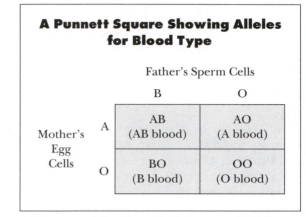

A heterozygous AO mother and a heterozygous BO father can produce children with any of the four blood types.

can be screened with gene testing procedures to determine if they carry such genes. Following detection of a defective gene, several options may be explored and implemented, including genetic counseling, gene therapy, and pharmacogenetics.

At least four thousand diseases of humans are known to have a genetic basis and can be passed from generation to generation. In addition to many kinds of human cancers, all of which have a genetic basis, human genetic disorders include diabetes, heart disease, and cystic fibrosis. Other diseases and disorders that have been directly linked to human genetic anomalies include predispositions for colon cancer, Alzheimer's disease, and breast cancer.

Gene Testing

In a gene-testing protocol, a sample of blood or body fluids is examined to detect a genetic anomaly such as the transposition of part of a chromosome or an altered sequence of the bases that comprise a specific gene, either of which can lead to a genetically based disorder or disease. Currently more than six hundred tests are available to detect malfunctioning or nonfunctioning genes. Most gene tests have focused on various types of human cancers, but other tests are being developed to detect genetic deficiencies that cause or exacerbate infectious and vascular diseases.

The emphasis on the relationship between genetics and cancer lies in the fact that all human cancers are genetically triggered by genes or have a genetic basis. Some cancers are inherited as mutations, but most result from random genetic mutations that occur in specific cells, often precipitated by viral infections or environmental factors not yet well understood.

At least four types of genetic problems have been identified in human cancers. The normal function of oncogenes, for example, is to signal the start of cell division. However, when mutations occur or oncogenes are overexpressed, the cells keep on dividing, leading to rapid growth of cell masses. The genetic inheritance of certain kinds of breast cancers and ovarian cancers results from the nonfunctioning tumor-suppressor genes that normally stop cell division. When genetically altered tumor-suppressor genes are unable to stop cell division, cancer results. Conversely, the genes that cause inheritance of colon cancer result from the failure of DNA repair genes to correct mutations properly. The accumulation of mutations in these "proofreading" genes makes them inefficient or less efficient, and cells continue to replicate, producing a tumor mass.

If a gene screening reveals a genetic problem several options may be available, including gene therapy and genetic counseling. If the detected genetic anomaly results in disease, then pharmacogenomics holds promise of patient-specific drug treatment.

Gene Therapy

The science of gene therapy uses recombinant DNA technology to cure diseases or disorders that have a genetic basis. Still in its experimental stages, gene therapy may include procedures to replace a defective gene, repair a defective gene, or introduce healthy genes to supplement, complement, or augment the function of nonfunctional or malfunctioning genes. Several hundred protocols are being used in gene therapy trials, and many more are under development. Current trials focus on two major types of gene therapy, somatic gene therapy and germ-line gene therapy.

Somatic gene therapy focuses on altering a defective gene or genes in human body cells in an attempt to prevent or lessen the debilitating impact of a disease or other genetic disorder. Some examples of somatic cell gene therapy protocols now being tested include ones for adenosine deaminase (ADA) deficiency, cystic fibrosis, lung cancer, brain tumors, ovarian cancer, and AIDS.

In somatic gene therapy a sample of the patient's cells may be removed and treated, and then reintegrated into body tissue carrying the corrected gene. An alternative somatic cell therapy is called gene replacement, which typically involves insertion of a normally functioning gene. Some experimental delivery methods for gene insertion include use of retroviral vectors and adenovirus vectors. These viral vectors are used because they are readily able to insert their genomes into host cells. Hence, adding the needed (or corrective) gene segment to the viral genome guarantees delivery into the cell's nuclear interior. Nonviral delivery vectors that are being investigated for gene replacement include liposome fat bodies, human artificial chromosomes, and naked DNA (free DNA, or DNA that is not enclosed in a viral particle or any other "package").

Another type of somatic gene therapy involves blocking gene activity, whereby potentially harmful genes such as those that cause Marfan syndrome and Huntington's disease are disabled or destroyed. Two types of gene-blocking therapies now being investigated include the use of antisense molecules that target and bind to the messenger RNA (mRNA) produced by the gene, thereby preventing its translation, and the use of specially developed ribozymes that can target and cleave gene sequences that contain the unwanted mutation.

Germ-line therapy is concerned with altering the genetics of male and female reproductive cells, the gametes, as well as other body cells. Because germ-line therapy will alter the individual's genes as well as those of his or her offspring, both concepts and protocols are still very controversial. Some aspects of germ-line therapy now being explored include human cloning and genetic enhancement.

The next steps in human genetic therapy involve determining the underlying mechanisms by which genes are transcribed, translated, and expressed, which is called proteomics.

Clinical Genetics

Clinical genetics is that branch of medical genetics involved in the direct clinical care of people afflicted with diseases caused by genetic disorders. Clinical genetics involves diagnosis, counseling, management, and support. Genetic counseling is a part of clinical genetics directly concerned with medical management, risk determination and options, and decisions regarding reproduction of afflicted individuals. Support services are an integral feature of all genetic counseling themes.

Clinical genetics begins with an accurate diagnosis that recognizes a specific, underlying genetic cause of a physical or biochemical defect following guidelines outlined by the NIH Counseling Development Conference. Clinical practice includes several hundred genetic tests that are able to detect mutations such as those associated with breast and colon cancers, muscular dystrophies, cystic fibrosis, sickle-cell disease, and Huntington's disease.

Genetic counseling follows clinical diagnosis and focuses initially on explaining the risk factors and human problems associated with the genetic disorder. Both the afflicted individual and family members are involved in all counseling procedures. Important components include a frank discussion of risks, of options such as preventive operations, and of options involved in reproduction. All reproductive options are described along with their potential consequences, but genetic counseling is a support service rather than a directive mode. That is, it does not include recommendations. Instead, its ultimate mission is to help both the afflicted individuals and their families recognize and cope with the immediate and future implications of the genetic disorder.

Pharmacogenomics

That branch of human medical genetics dealing with the correlation of specific drugs to fit specific diseases in individuals is called pharmacogenomics. This field recognizes that different individuals may metabolically respond differentially to therapeutic medicines based on their genetic makeup. It is anticipated that testing human genome data will greatly speed the development of new drugs that not only target specific diseases but also will be tailored to the specific genetics of patients.

Policy and Ethical Concerns and Issues in Human Genetics

The "new genetics" of humans has raised a number of critical concerns that are currently being addressed on a number of levels. Some of these concerns are related to the ownership of genetic information obtained by the Human Genome Project, privacy issues, and use of genetic information in risk assessment and decision making.

Privacy issues have focused on psychological impact, possible discrimination, and stigmatization associated with identifying personal genetic disorders. For example, policy guarantees must be established to protect the privacy of persons with genetic disorders to prevent overt or covert societal discrimination against the affected individual. Another question arising from this is exactly who has the right to the genetic information of persons.

Use of information obtained by the Human Genome Project has provided entrepreneurial opportunities that will undoubtedly prove economically profitable. That is, the limits of commercialization of products, patents, copyrights, trade secrets, and trade agreements have to be determined. If patents of DNA sequences are permitted, will they limit accessibility and free scientific interchange among and between peoples of the world? This question becomes critical when it is recognized that the human genome is properly the property of all humans.

Noncoding "Junk" DNA

Like that of other organisms, the human genome consists of long segments of DNA that contain noncoding sequences called introns (intervening sequences). These vary from a few hundred to several thousand base pairs in length and often consist of repetitive DNA elements with no known function; that is, they do not code for proteins. Because they appear functionless but take up valuable chromosomal space, these noncoding sequences have been considered useless and have been termed junk DNA or selfish DNA. Recent studies, however, lend strong support to the possibility that the seemingly useless repetitive DNA may actually play a number of important genetic roles, from providing a substrate on which new genes can evolve to maintaining chromosome structure and participating in some sort of genetic control. Consequently, it is now out of fashion among geneticists to refer to these parts of the genome as junk DNA, but rather as DNA of unknown function.

Forensic Genetics

Law enforcement agencies are increasingly relying on a branch of human genetics called forensic genetics. The aims of forensic genetics typically are to determine the identity or nonidentity of suspects in crimes, based on an analysis of DNA found in hair, blood, and other body substances retrieved from the scene of the crime in comparison with that of suspects. Popularly called DNA fingerprinting, forensic genetics relies on the fact that the DNA of every human carries unique tandem repeats of 20 or more kilobase pairs that can be compared and identified using radioactive probes. Thus, comparisons can establish identity or nonidentity to a very high level of probability. DNA fingerprinting is also used in recognizing genetic parentage of children, identifying victims—sometimes from fragments of bodies—and identifying relationships of missing children.

Phylogeny and Evolution

Another rapidly developing field in human genetics is the use of human gene sequences in both nuclear and mitochondrial DNA (mtDNA) to explore questions of human origins, evolution, phylogeny, bioarchaeology, and past human migration patterns.

Much of the analytical work has involved mtDNA to study relationships. Because it is inherited strictly through the egg line or female component, mtDNA is somewhat more useful, but comparisons of DNA sequences along the Y chromosome of human populations have also yielded valuable information regarding human origins and evolution.

One of the more interesting of these studies involves comparing mtDNA over a broad spectrum of global human populations. Comparisons of DNA sequencing of these populations

has revealed differences in DNA sequences o about 0.33 percent, which is considerably less than seen in other primate species. These minor differences strongly suggest that all members of the human species, *Homo sapiens*, are far more closely related to one another than are members of many other vertebrate species.

A separate study compared human gene sequences among different human populations across the globe. This study revealed that the highest variations in DNA sequences are found among the modern human populations of Africa. Since populations that exhibit the highest genome variations are thought to be the oldest populations (because chance mutations have a longer time to accumulate in older populations as opposed to younger populations), these results strongly suggest that modern humans originated in Africa and subsequently dispersed into other regions of the world. This "out of Africa" theory has received compelling support from the DNA evidence, and the theory also explains why all other human populations are so remarkably similar. Since all other global human populations show minimal DNA sequence differences, it is hypothesized that a small group of humans emigrated from Africa to spread across and eventually colonize the other continents. Tests of gene sequences along Y chromosomes show similar patterns, leading to the proposal that all humans today came from a mitochondrial Eve and a Y chromosome Adam who lived between 160,000 and 200,000 years ago.

DNA-based phylogeny studies are also shedding light on the relationship between the Neanderthals (*Homo neanderthalensis*), a species that disappeared between 30,000 and 60,000 years ago, and the modern Cro-Magnon humans (*Homo sapiens*) that replaced them. Comparisons of mtDNA between the two *Homo* species indicate that Neanderthals began diverging from modern humans half a million years ago and were significantly different in genomic content to be placed in a separate species. These findings also support the suggestion that Neanderthals were ecologically replaced by modern humans rather than genetically amalgamated into present human populations, as was once proposed. Although such arguments are not universally accepted, many more geneticists, paleoanthropologists, and forensic scientists are now using comparative analysis of DNA sequences among and between human populations to study questons of human evolutionary history.

—*Dwight G. Smith*

See also: Aggression; Aging; Bioethics; Bioinformatics; Biological Determinism; Criminality; DNA Fingerprinting; Eugenics; Eugenics: Nazi Germany; Evolutionary Biology; Forensic Genetics; Gender Identity; Gene Therapy; Gene Therapy: Ethical and Economic Issues; Genetic Counseling; Genetic Screening; Genetic Testing; Genetic Testing: Ethical and Economic Issues; Human Genome Project; Human Growth Hormone; In Vitro Fertilization and Embryo Transfer; Insurance; Intelligence; Miscegenation and Antimiscegenation Laws; Patents on Life-Forms; Paternity Tests; Prenatal Diagnosis; Race; Sterilization Laws.

Further Reading

Andrews, Lori B. *The Clone Age: Adventures in the New World of Reproductive Technology.* New York: Henry Holt, 1999. A lawyer specializing in reproductive technology, Andrews examines the legal ramifications of human cloning, from privacy to property rights.

Baudrillard, Jean. *The Vital Illusion.* Edited by Julia Witwer. New York: Columbia University Press, 2000. A sociological perspective on what human cloning means to the idea of what it means to be human.

Fridell, Ron. *DNA Fingerprinting: The Ultimate Identity.* New York: Scholastic, 2001. The history of the technique, from its discovery to early uses. Aimed at younger readers and nonspecialists.

Hartwell, L. H., L. Hood, M. L. Goldberg, A. E. Reynolds, L. M. Silber, and R. C. Veres. *Genetics: From Genes to Genomes.* 2d ed. New York: McGraw-Hill, 2004. A comprehensive textbook on genetics, including human genetics discussed in a comparative context.

Hekimi, Siegfried, ed. *The Molecular Genetics of Aging.* New York: Springer, 2000. Part of the Results and Problems in Cell Differentiation series. Illustrated.

Jorde, L. B., J. C. Carey, M. J. Bamshad, and R. L. White. *Medical Genetics.* 2d ed. St. Louis, Mo.: Mosby, 2000. Provides both an introduction to the field of human genetics and chapters on clinical aspects of human genetics such as gene therapy, genetic screening, and genetic counseling.

Pasternak, Jack J. *An Introduction to Human Molecular Genetics: Mechanisms of Inherited Diseases.* Bethesda, Md.: Fitzgerald Science Press, 1999. Discusses treatment advances, fundamental molecular mechanisms that govern human inherited diseases, the interactions of genes and their products, and the consequences of these mechanisms on disease states in major organ systems such as muscles, the nervous system, and the eyes. Also addresses cancer and mitochondrial disorders.

Rudin, Norah, and Keith Inman. *An Introduction to Forensic DNA Analysis.* Boca Raton, Fla.: CRC Press, 2002. An overview of many DNA typing techniques, along with numerous examples and a discussion of legal implications.

Shostak, Stanley. *Becoming Immortal: Combining Cloning and Stem-Cell Therapy.* Albany: State University of New York Press, 2002. Examines the question of whether human beings are equipped for potential immortality.

Wilson, Edward O. *On Human Nature.* Cambridge, Mass.: Harvard University Press, 1978. A look at the significance of biology and genetics on how we understand human behaviors, including aggression, sex, and altruism and the institution of religion.

Web Sites of Interest

American Society of Human Genetics (ASHG). http://www.faseb.org/genetics/ashg/ashg menu.htm. Founded in 1948, a group of several thousand physicians, genetic counselors, researchers, publishing the *American Journal of Human Genetics.*

Association of Professors of Human or Medical Genetics (APHMG). http://www.faseb.org/genetics/aphmg/aphmg1.htm. This site of the North American group of academicians in medical and graduate schools features information on core curricula and workshops.

Johns Hopkins University. The Genome Database. http://gdbwww.gdb.org. The official central storage center for gene mapping data compiled in the Human Genome Initiative, an international effort to decode and analyze human DNA. Intended for scientists, the site presents information in three categories: regions, maps, and variations of the human genome.

National Center for Biotechnology Information. http://www.ncbi.nlm.nih.gov. Maintains GenBank, a comprehensive, annotated collection of publicly available DNA sequences.

Sanger Centre, Wellcome Trust. http://www.sanger.ac.uk. One of the premier genome research centers, focusing on large-scale sequencing projects and analysis. Offers many data resources, software, databases, and information on career opportunities.

Human Genome Project

Fields of study: History of genetics; Human genetics; Techniques and methodologies

Significance: *The Human Genome Project will have a profound effect in the twenty-first century, providing the means to identify disease-causing mutations (including those involved in cancer), to design new drugs, to provide human gene therapy, to learn how genes control development, and to understand the origins and evolution of the human race.*

Key terms

GENOME: the entire complement of genetic material (DNA) in a cell

GENOMICS: that branch of genetics dealing with the study of genetic sequences

PROTEOMICS: that branch of genetics dealing with the expression, function, and structure of proteins

SINGLE NUCLEOTIDE POLYMORPHISM (SNP): differences at the individual nucleotide level among individuals

Perspective

April 25, 2003, was the fiftieth anniversary of the publication of the double helix model of DNA by James Watson and Francis Crick, based on the experimental data of Rosalind Franklin and others. It was fitting then, that fifty years later, in April of 2003, the complete sequence of the human genome was published, marking probably one of the greatest achievements not only in genetics but in all of science. During the years ahead, thousands of scientists will mine these data for information about the human body, how its genes shape development and behavior, and the role mutations play in diseases.

Origins of the Human Genome Project

The Human Genome Project (HGP) began as a result of the catastrophic events of World War II: the dropping of atomic bombs on the Japanese cities of Nagasaki and Hiroshima. There were many survivors who had been exposed to high levels of radiation, known to cause mutations. Such survivors were stigmatized by society and were considered poor marriage prospects, because of potential genetic damage. The U.S. Atomic Energy Commission of the U.S. Department of Energy (DOE) established the Atomic Bomb Casualty Commission in 1947 to assess mutations in such survivors. However, there were no suitable methods to measure these mutations, and it would be many years before suitable techniques would be developed. Knowing the sequence of the human genome would be the greatest tool for identifying human mutations.

Advances in Molecular Biology

As in all areas of science, progress in molecular biology was limited by available technology. Many advances in molecular biology made feasible the undertaking of the HGP. Starting in the 1970's, techniques were developed to isolate and clone individual genes. By 1977, Walter Gilbert and Frederick Sanger had independently developed methods for sequencing DNA, and in 1977 Sanger's group published the sequence of the first genome, the small bacterial virus Phi X174. In 1985, Kary Mullis and colleagues developed the method of polymerase chain reaction (PCR), in which extremely small amounts of DNA could be amplified billions of times, providing significant amounts of specific DNA for analysis. Finally, in 1986, Leroy Hood and Applied Biosystems developed

Craig Venter of Celera Genomics (at the microphone) and Francis Collins, Director of the National Institutes of Health (right), announce the initial sequencing of the human genome on June 26, 2000, with President Bill Clinton in attendance. (AP/Wide World Photos)

an automated DNA sequencer that could sequence DNA hundreds of times faster than was previously possible. Additional advances in computer technology now made it possible to sequence the human genome.

The "Holy Grail" of Molecular Biology

In 1985 a conference of leading scientists was held at the University of California, Santa Cruz, to discuss the feasibility of sequencing the entire human genome. Biologists were looking for the equivalent of a Manhattan Project for biology. The Manhattan Project was the concerted effort of physicists to develop atomic weapons during World War II and resulted in huge increases of government funding for physics research. Walter Gilbert called the HGP the Holy Grail of molecular biology. With impetus

from the DOE and the National Research Council, the Human Genome Project was launched in 1990 with James Watson as head. The goal was completion in 2005 at a cost of $1.00 per base pair. In 1992, Watson resigned over a controversy surrounding the patenting of human sequences. Francis Collins took over as head of the HGP at the National Institutes of Health (NIH). The sequencing of genetic model organisms, in addition to the human genome, was one of the goals of the HGP. This included genomes of the bacterium *Escherichia coli*, yeast, the fruit fly *Drosophila melanogaster*, the round worm *Caenorhabditis elegans*, and other organisms. Moreover, 10 percent of the funding was to be directed toward studies of the social, ethical, and legal implications of learning the human genome.

Competition Between the Public and Private Sectors

Craig Venter, a former National Institutes of Health researcher, left the NIH and formed a private company, The Institute for Genome Research (TIGR). TIGR, using a different approach (known as the shotgun method) was able to sequence the 1.8 million-base-pair genome of the first free-living organism, the bacterium *Haemophilus influenzae*, in less than a year. In 1998 Venter along with Perkin-Elmer Corporation formed the biotech company Celera Genomics to sequence the human genome privately. Celera had more than three hundred of the world's fastest automated sequencers and a supercomputer to analyze data. Meanwhile, public funds supported scientists in the United States, the United Kingdom, Japan, Canada, Sweden, and fourteen other countries working on HGP sequencing. The public sector was now in competition with Celera. To assure free access, each day new sequence data from the public projects were made available on the Internet.

The Human Genome Project Is Completed

In 2001 the first draft of the human genome was published in the February 15 issue of *Nature* and the February 16 issue of *Science*. There are many short, repeated sequences of DNA in the genome, and certain regions that were difficult to sequence needed to be sequenced again for accuracy. Thus in April, 2003, the final sequence of the human genome was achieved. It is remarkable that a government-funded project was completed two and a half years ahead of schedule and under budget. April 25, 2003, was designated National DNA Day.

Findings from the Human Genome Project

Perhaps the most surprising finding from the HGP is the relatively low number of human genes. Scientists had predicted the human genome would contain about 100,000 genes, yet the actual number of protein-coding sequences is approximately 21,000, representing only about 1 percent of the entire genome. In comparison, yeast has about 6,000 genes, the fruit fly about 13,000, and the *Caenorhabditis*

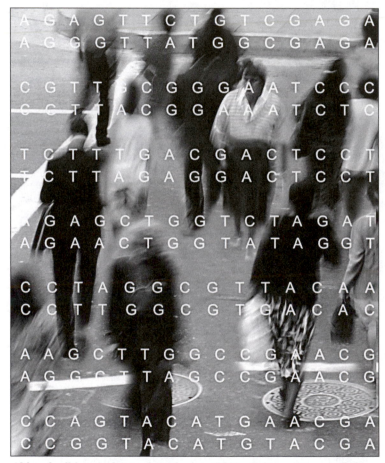

Although all human beings share the same DNA, slight variations in DNA sequences, including single nucleotide polymorphisms (SNPs), occur commonly across individuals. One individual, for example, might have the base A (adenine) where another has the base C (cytosine); several different combinations of these bases can often code for the same amino acid and hence protein, so the differences often have little or no affect. However, these SNPs can account for variations in our reactions to pathogens, drugs, and other environmental conditions. Knowing these variations may help researchers identify the genes associated with complex conditions such as cancer, diabetes, and cardiovascular diseases. (U.S. Department of Energy Human Genome Program, http://www.ornl.gov/hgmis.)

about 18,000. It was surprising that a complex human had less than twice the number of genes as the lowly roundworm. The human genome also contains 740 genes that encode stable RNAs. The genome of the mouse, another model genetic organism, will provide interesting comparisons to the human genome.

Whose Genome Is It?

Although more than 99.99 percent of the DNA sequences of all humans are identical, there are many differences. An important question is, Whose genome is it? Craig Venter has acknowledged that Celera has been sequencing mostly his DNA. However, the final sequence database is an "average" or "consensus" genome. Every human carries many and perhaps even hundreds of mutations. Even before the HGP was completed, databases listing single nucleotide polymorphisms were being established. These databases list the types of genetic variations that occur at individual nucleotides in the genome. For example, a cancer gene database lists the types of mutations that have been identified in specific cancer-causing genes and the frequency of such mutations. Mutations in genes such as *BRCA1* and *BRCA2* are responsible for breast and ovarian cancers, while mutations in the tumor-suppressor gene *p53* have been found in the majority of human tumors.

The Future: Genomics and Proteomics

The Human Genome Project has given rise to two new fields of study. Genomics is the study of genomes. To do so requires databases and search engines to seek out information from these sequences. Today there are hundreds of such databases already established. Scientists can search for complete gene sequences if they know only a short segment of a gene. They can look for related sequences within the same genome or among different species. From such information one can study the evolution of particular genes.

The next step is to define the human proteome, giving rise to the field of proteomics. Proteomics seeks to determine the expression patterns of genes, the functions of the proteins produced, and the structure of specific pro-

teins derived from their DNA sequence. If a particular protein is involved in a disease process, specific drugs to interfere with it may be designed. Humanity is just beginning to reap the benefits from the Human Genome Project.

—*Ralph R. Meyer*

See also: Behavior; Bioinformatics; Chromosome Theory of Heredity; Genetic Code, Cracking of; Genetic Engineering; Genomic Libraries; Genomics; Hereditary Diseases; Human Genetics; Icelandic Genetic Database; Polymerase Chain Reaction; Proteomics; Race.

Further Reading

Collins, Francis, and Karin G. Jegalian. "Deciphering the Code of Life." *Scientific American* 281, no. 6 (1999): 86-91. A description in lay terms of the progress and goals of the HGP.

Dennis, Carina, and Richard Gallagher. *The Human Genome.* London: Palgrave Macmillan, 2002. Written by two editors of the British journal *Nature*, the book gives a description of the HGP in lay terms and provides some of the information from the first draft of the human genome.

International Human Genome Sequencing Consortium. "Initial Sequencing and Analysis of the Human Genome." *Nature* 409, no. 6822 (2001): 860-921. The publication of the first draft of the Human Genome Project. The whole journal issue contains many other papers considering the structure, function, and evolution of the human genome.

_____. "The Human Genome." *Science* 291 (2001): 1145-1434. This special issue provides data from the first draft of the human genome sequence.

Sulston, John, and Georgina Ferry. *The Common Thread: A Story of Science, Politics, Ethics, and the Human Genome.* Washington, D.C.: Joseph Henry Press, 2002. A chronicle of the race for the HGP from the perspective of a British Nobel Laureate Sir John Sulston, head of Sanger Centre, the British research unit involved in the HGP. Describes the effort to ensure public access to the genome data.

Wolfsberg, Tyra G., Kris A. Wetterstrand, Mark S. Guyer, Francis S. Collins, and Andreas D.

Baxevanis. "A User's Guide to the Human Genome." *Nature Genetics Supplement* 32 (2002): 1-79. This supplement nicely illustrates how one can search the human genome database. It is set up as a series of questions with step-by-step color Web page illustrations of such searches. The supplement also lists major Web resources and databases.

Web Sites of Interest

Department of Energy. Office of Science. http://doegenomes.org. Along with the National Human Genome Research Institute, conducted the Human Genome Project. Site includes discussion of the ethical, legal, and social issues surrounding the project, a genome glossary, and "Genetics 101."

National Center for Biotechnology Information. http://www.ncbi.nlm.nih.gov/gene map99. Starting with a general introduction to the human genome and the process of gene mapping, this site provides charts of the known genes on each chromosome, articles about the Genome Project and gene-related medical research, and links to other genome sites and databases.

National Human Genome Research Institute. http://www.genome.gov. One of the major gateways to the Human Genome Project, with a brief but thorough introduction to the project, fact sheets, multimedia education kits for teachers and students, a glossary, and links.

New York University/Bell Atlantic/Center for Advanced Technology. The Student Genome Project. http://www.cat.nyu.edu/sgp/parent .html. Uses interactive multimedia and three-dimensional technology to present tutorials and games related to the human genome and genetics for middle and high school students.

The Institute for Genomic Research (TIGR). http://www.tigr.org. The organization founded by Craig Venter, focusing on structural, functional, and comparative analysis of genomes and gene products. Provides databases, gene indices, and educational resources.

Human Growth Hormone

Field of study: Human genetics and social issues; Molecular genetics

Significance: *Human growth hormone (HGH) determines a person's height, and abnormalities in the amount of HGH in a person's body may cause conditions such as dwarfism, giantism, and acromegaly. Genetic research has led to the means to manufacture enough HGH to correct such problems and expand the understanding of HGH action and endocrinology.*

Key terms

ENDOCRINE GLAND: a gland that secretes hormones into the circulatory system

HYPOPHYSECTOMY: surgical removal of the pituitary gland

PITUITARY GLAND: an endocrine gland located at the base of the brain; also called the hypophysis

TRANSGENIC PROTEIN: a protein produced by an organism using a gene that was derived from another organism

Growth Hormones and Disease Symptoms

The pituitary (hypophysis) is an acorn-sized gland located at the base of the brain that makes important hormones and disseminates stored hypothalamic hormones. The hypothalamus controls the activity of the pituitary gland by sending signals along a network of blood vessels and nerves that connects them. The main portion of the pituitary gland, the adenohypophysis, makes six trophic hormones that control many body processes by causing other endocrine glands to produce hormones. The neurohypophysis, the remainder of the pituitary, stores two hypothalamic hormones for dissemination.

Dwarfism is caused by the inability to produce growth hormone. When humans lack only human growth hormone (HGH), resultant dwarfs have normal to superior intelligence. However, if the pituitary gland is surgically removed (hypophysectomy), the absence of other pituitary hormones causes additional mental and gender problems. The symptoms of dwarfism are inability to grow at a normal rate or attain adult size. Many dwarfs are two to three feet tall. In contrast, some giants have reached heights of more than eight feet. The advent of gigantism often begins with babies born with pituitary tumors that cause the production of too much HGH, resulting in continued excess growth. People who begin oversecreting HGH as adults (also caused by tumors) do not grow taller. However, the bones in their feet, hands, skull, and brow ridges overgrow, causing disfigurement and pain, a condition known as acromegaly.

Dwarfism that is uncomplicated by the absence of other pituitary hormones is treated with growth hormone injections. Humans undergoing such therapy can be treated with growth hormones from humans or primates. Growth hormone from all species is a protein made of approximately two hundred amino acids strung into a chain of complex shape. However, differences in amino acids and chain arrangement in different species cause shape differences; therefore, growth hormone used for treatment must be extracted from a related species. Treatment for acromegaly and gigantism involves the removal of the tumor. In cases where it is necessary to remove the entire pituitary gland, other hormones must be given in addition to HGH. Their replacement is relatively simple. Such hormones usually come from animals. Until recently, the sole source of HGH was pituitaries donated to science. This provided the ability to treat fewer than one thousand individuals per year. Molecular genetics has solved that problem by devising the means to manufacture large amounts of transgenic HGH.

Growth Hormone Operation and Genetics

In the mid-1940's, growth hormone was isolated and used to explain why pituitary extracts increase growth. One process associated with HGH action involves cartilage cells at the ends of long bones (such as those in arms and legs). HGH injection causes these epiphysial plate cells (EPCs) to rapidly reproduce and stack up. The EPCs then die and leave a layer of protein, which becomes bone. From this it

has been concluded that growth hormone acts to cause all body bones to grow until adult size is reached. It is unclear why animals and humans from one family exhibit adult size variation. The differences are thought to be genetic and related to production and cooperation of HGH, other hormones, and growth factors.

Genetic research has produced transgenic HGH in bacteria through the use of genetic engineering technology. The gene that codes for HGH is spliced into a special circular piece of DNA called a plasmid expression vector, thus producing a recombinant expression vector. This recombinant vector is then put into bacterial cells, where the bacteria express the HGH gene. These transgenic bacteria can then be grown on an industrial scale. After bacterial growth ends, a huge number of cells are harvested and HGH is isolated. This method enables isolation of enough HGH to treat anyone who needs it.

Impact and Applications

One use of transgenic HGH is the treatment of acromegaly, dwarfism, and gigantism. The availability of large quantities of HGH has also led to other biomedical advances in growth and endocrinology. For example, growth hormone does not affect EPCs in tissue culture. Ensuing research, first with animal growth hormone and later with HGH, uncovered the EPC stimulant somatomedin. Somatomedin stimulates growth in other tissues as well and belongs to a protein group called insulin-like growth factors. Many researchers have concluded that the small size of women compared to men is caused by estrogen-diminished somatomedin action on EPCs. Estrogen, however, stimulates female reproductive system growth by interacting with other insulin-like growth factors.

Another interesting experiment involving HGH and genetic engineering is the production of rat-sized mice. This venture, accomplished by putting the HGH gene into a mouse chromosome, has important implications for understanding such mysteries as the basis for species specificity of growth hormones and maximum size control for all organisms. Hence, experiments with HGH and advancements in genetic engineering technology have led to, and should continue to lead to, valuable insights into the study of growth and other aspects of life science.

—*Sanford S. Singer*

See also: Cloning; Dwarfism; Genetic Engineering: Historical Development; Genetics, Historical Development of; Prader-Willi and Angelman Syndromes; Turner Syndrome.

Further Reading
Eiholzer, Urs. *Prader-Willi Syndrome: Effects of Human Growth Hormone Treatment.* New York: Karger, 2001. Discusses the therapeutic use of somatotropin, among other topics.

Flyvbjerg, Allan, Hans Orskov, and George Alberti, eds. *Growth Hormone and Insulin-like Growth Factor I in Human and Experimental Diabetes.* New York: John Wiley & Sons, 1993. Discusses advances regarding the effects of growth hormone and insulin-like growth factors in relation to metabolism in diabetes and the development of complications.

Shiverick, Kathleen T., and Arlan L. Rosenbloom, eds. *Human Growth Hormone Pharmacology: Basic and Clinical Aspects.* Boca Raton, Fla.: CRC Press, 1995. Describes the research on and clinical applicability of the human growth hormone. Illustrated.

Smith, Roy G., and Michael O. Thorner, eds. *Human Growth Hormone: Research and Clinical Practice.* Totowa, N.J.: Humana Press, 2000. Provides recent findings about regulation of the hormone and its action at the molecular level.

Thorner, M., and R. Smith. *Human Growth Hormone: Research and Clinical Practice.* Vol. 19. Totowa, N.J.: Humana Press, 1999. Examines the use of human growth hormone therapies in the treatment of short stature and various diseases.

Ulijaszek, J. S., M. Preece, and S. J. Ulijaszek. *Cambridge Encyclopedia of Human Growth and Development Growth Standards.* New York: Cambridge University Press, 1998. Broadly discusses genetic growth anomalies in relation to environmental, physiological, social, economic, and nutritional influences on human growth.

Web Sites of Interest
ABC News Online, Standing Tall. http://abc news.go.com/sections/living/goodmorning america/growthhormone_030619.html. A June, 2003, news report on children using human growth hormone to increase their stature.

National Institutes of Health, National Library of Medicine, Growth Disorders. http://www .nlm.nih.gov/medlineplus/growthdisorders .html. Information on all aspects of growth disorders and HGH treatment.

Huntington's Disease

Field of study: Diseases and syndromes
Significance: *Huntington's disease is an autosomal dominant neurodegenerative disorder. The symptoms of this incurable, fatal condition include uncontrollable body movements and progressive dementia. The relevant gene contains a domain of repeating triplets composed of the nucleotides cytosine (C), adenine (A), and guanine (G). Mutation of this gene causes an increase in triplet number, triggering the dysfunction and death of certain neurons in the brain.*

Key terms
CAG EXPANSION: a mutation-induced increase in the number of consecutive CAG nucleotide triplets in the coding region of a gene
GENETIC ANTICIPATION: progressively earlier onset of a hereditary disease in successive generations
POLYGLUTAMINE TRACT: in a protein, an amino acid sequence consisting exclusively of glutamine, encoded by repeating CAG triplets

Characteristics
Studying an extended New York family in 1872, Dr. George Huntington first documented the heritable malady that bears his name. Huntington's disease (HD) was originally known as Huntington's chorea because of its hallmark jerky involuntary movements (the term "chorea" comes from the Greek *choros*, meaning "dance"). Patients also experience marked cognitive and psychiatric decline. HD's gradual

onset usually begins between ages thirty and forty, although its symptoms can first appear within an age range of two to eighty years. The disease typically progresses to death within fifteen or twenty years of diagnosis. HD affects about one in ten thousand people of European descent, and fewer than one in one million people in African and Japanese populations.

In HD, degeneration of neurons in specific brain regions occurs over time. Hardest hit is a particular subset of neurons in the striatum, a brain structure critical for movement control. Also affected is the frontal cortex, which is involved in cognitive processes. As the communication link between the striatum and cortex is broken through ongoing neuronal death, uncontrollable chorea as well as intellectual and psychiatric symptoms develop and worsen.

The *HD* Gene and Its Product

HD is inherited as a dominant mutation of a gene located on the short arm of chromosome 4. The cloning of the *HD* gene in 1993 provided major impetus to understanding its function. The *HD* gene encodes a 348 kDa cytoplasmic protein called huntingtin. Normally, the *HD* gene contains a stretch of repeating nucleotide triplets consisting of C (cytosine), A (adenine), and G (guanine). Healthy *HD* alleles contain anywhere from 9 to 35 CAG repeats. The CAG triplet encodes the amino acid glutamine; therefore, normal huntingtin contains a polyglutamine tract. Huntingtin is expressed throughout the brain (and indeed, the body); however, its regular function remains unclear. In neurons, huntingtin is thought to be important in counterbalancing programmed cell

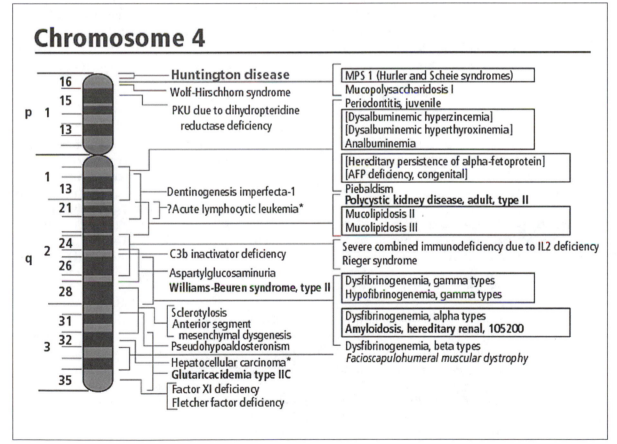

The gene for Huntington's disease is located on chromosome 4. Huntington's is one of the rare single-gene disorders, clearly detected genetically. Other genetic disease conditions have been located to chromosome 4, also shown here. (U.S. Department of Energy Human Genome Program, http://www.ornl.gov/hgmis.)

death by promoting the expression of growth factors. Huntingtin may therefore help protect striatal neurons throughout life.

The *HD* Mutation

Mutant *HD* alleles contain an expansion of the CAG repeat. The magnitude of this expansion can range from 36 to more than 60 CAG repeats (rarely, as many as 250 repeats have been observed). There is an inverse relationship between repeat number and age of disease onset: Higher repeat numbers are usually linked to younger onset. People with 36-39 CAG repeats may never show disease symptoms, whereas people with 40-60 repeats usually develop HD in mid-adulthood, and those with more than 60 repeats often experience onset at less than twenty years of age.

Although original *HD* mutations clearly must occur, they are rare and of unknown cause. However, HD's inheritance patterns shed light on the mechanisms of CAG expansion. HD exhibits genetic anticipation: Affected members of successive generations may show earlier onset, particularly when the pathogenic allele is inherited paternally. It is thought that CAG expansion occurs during the repair of DNA strand breaks, when CAG loops are retained in the nucleotide sequence during gap repair. If this happens in reproductive cells (particularly sperm), a larger CAG expansion will be present in the offspring.

Consequences of CAG Expansion

The direct result of CAG expansion within the *HD* gene is that mutant huntingtin has a polyglutamine tract of variable but abnormally long length. Misfolding and aggregation of mutant huntingtin ensues. Cleavage of the mutant protein occurs, generating a fragment that can enter the nucleus. Visible cytoplasmic and nuclear huntingtin aggregates are a key pathological feature of the striatal neurons destined to die. This aggregation represents a different (albeit toxic) function for huntingtin. The aggregates contain not only mutant huntingtin but also several other critical proteins whose functions are effectively withheld. Because some of these sequestered proteins are transcription factors, transcriptional dysregulation

may affect the expression of a host of additional proteins. In fact, the expression of huntingtin itself (from the remaining normal *HD* allele) is significantly reduced. This diminution of the availability of normal huntingtin may also contribute to neuronal demise. However, it is still unknown why only certain neurons die despite huntingtin's ubiquitous expression.

Living with HD

The cloning of the *HD* gene has enabled direct genetic testing for the *HD* mutation. With a blood test, at-risk individuals can learn not only whether they carry the CAG expansion but also its length. Knowing one's carrier status and predicted age of onset can eliminate doubt and assist in making life plans, but the prospect of developing a fatal disease can be far more stressful than the uncertainty. This may explain why a relatively low percentage of those with a family history of HD have opted to be tested. Whenever testing is performed, it is accompanied by extensive counseling both before and after the results are known.

Current treatments for HD are palliative and include antidepressants and sedatives. Strategies now under study are aimed at preventing CAG expansion, counteracting the toxic effects of mutant huntingtin, and delivering neuroprotective agents to the brain. Another tactic is to replace the dying striatal neurons with transplanted fetal neurons or stem cells. This approach has shown some promise: Following striatal grafts, a small number of HD patients have experienced improvement in motor and cognitive function.

—*Mary A. Nastuk*

See also: Behavior; Biological Clocks; Blotting: Southern, Northern, and Western; Chromatin Packaging; Chromosome Walking and Jumping; DNA Replication; Gene Therapy: Ethical and Economic Issues; Genetic Counseling; Genetic Testing; Genetic Testing: Ethical and Economic Issues; Hereditary Diseases; In Vitro Fertilization and Embryo Transfer; Inborn Errors of Metabolism; Insurance; Pedigree Analysis; Prader-Willi and Angelman Syndromes; Repetitive DNA; Stem Cells.

Further Reading

Cattaneo, Elena, et al. "The Enigma of Huntington's Disease." *Scientific American* (December, 2002): 93-97. This article provides an excellent overview of current research and hypotheses regarding the molecular biology of HD.

Huntington's Disease Collaborative Research Group. "A Novel Gene Containing a Trinucleotide Repeat That Is Expanded and Unstable on Huntington's Disease Chromosomes." *Cell* 72, no. 6 (1993): 971-983. A benchmark study in which the *HD* gene was isolated and the nature of the mutation identified.

Wexler, Alice. *Mapping Fate: A Memoir of Family, Risk, and Genetic Research.* Berkeley: University of California Press, 1996. The author's mother had HD, and her sister was part of the research group that cloned the *HD* gene. This account is striking for its immediacy, clarity, and accuracy.

Web Sites of Interest

Hereditary Disease Foundation. http://hdfoundation.org. Site devoted mainly to Huntington's disease. Includes links to research articles, organizations, and news stories.

Huntington's Disease Society of America. http://www.hdsa.org. The organization that supports research for therapies and a cure. This site offers information, support resources, publications, and ways of "getting help."

Hybridization and Introgression

Field of study: Population genetics

Significance: *Hybridization and introgression are biological processes that are essential to creating genetic variation, and hence biodiversity, in plant and animal populations. These processes occur both in natural populations and in human-directed, controlled breeding programs.*

Key terms

GENETICALLY MODIFIED ORGANISMS (GMOs): plants and animals in which techniques of recombinant DNA have been used to introduce, remove, or modify specific parts of the genome of an organism

HYBRIDIZATION: the process of mating or crossing two genetically different individuals; the resultant progeny is called a hybrid

INTROGRESSION: the transfer of genes from one species to another or the movement of genes between species (or other well-marked genetic populations) mediated by backcrossing

TRANSGENE: a gene introduced into a cell or organism by means other than sexual

Definitions and Types

Hybridization and introgression are natural biological processes. Natural hybridization is common among plant and animal species. Hybridization generally refers to the mating between genetically dissimilar individuals; parents may differ in a few or many genes. They may come from different populations or races of the same taxonomic species (interspecific hybridization) or of different species (intergeneric hybridization). In nature, hybridization can occur only if there is no barrier to crossbreeding, or when the usual barrier breaks down. Hybridization produces new genetic combinations or genetic variability. Through artificial means (controlled pollination), hybridization of both cross-pollinated and self-pollinated plants can be accomplished. Plant breeding encompasses hybridization within a species as well as hybridization between species and even genera (wide crosses). The latter are important for generating genetic variability or for incorporating a desirable gene not available within a species. There are crossing barriers, however, for accomplishing interspecific and intergeneric crosses. Josef Gottlieb Kölreuter (c. 1761) was the first to report on hybrid vigor (heterosis) in interspecific crosses of various species of *Nicotiana*, concluding that cross-fertilization was generally beneficial and self-fertilization was not.

Introgression is the introduction of genes from one species or gene pool into another species or gene pool. Introgression follows hybridization and occurs when hybrids reproduce with members of one or both of the parental species that produced the hybrids. It usually involves transfer of a small amount of DNA from one species or genus to another. Both hybridization and introgression can cause rapid evolution, that is, speciation or extinction. When introgression occurs between a common species and a rare species, the rare species is frequently exterminated.

Scientific breakthroughs relative to species-specific molecular (DNA) markers allow quantitative assessment of introgression and hybridization in natural populations. A clear distinction among species is a prerequisite to guide efforts to conserve biodiversity.

The world's second "cama," shown here with its mother, is a hybrid of a llama and a camel. (AP/Wide World Photos)

Reproductive Isolation Barriers

Isolation barriers can be divided into two types: (1) external and (2) internal. External barriers to genetic interchange between related populations prevent pollen of plants in one population from falling on stigmas of plants in another. A combination of barriers, such as geographical and ecological or ecological and seasonal (flowering time), is more common than individual barriers.

Internal barriers to genetic interchange between related populations operate through incompatibilities between physiological or cytological systems of plants from different populations. They may (1) prevent the production of F_1 (first-generation) zygotes, even if the pollen from flowers in one population falls on stigmas of flowers in the other; (2) produce F_1 hybrids that are nonviable, weak, or sterile; or (3) cause hybrid breakdown in F_2 or later generations.

The promotion of natural hybridization and introgression has, across time, increased the genetic diversity available to farmers. Traditional farmers experiment with new varieties and breed plants purposely to create new strains. They generally plant experimental plots first and integrate new varieties into their main crops only when a variety has proven itself to be of value. This constant experimentation and breeding have created the diversity of crops on which we now depend.

Transgenic Crops and Controversy

Termed "gene flow," the movement of genes between closely related plant species is quite natural and has been occurring ever since flowering plants evolved. Hybrids that are the offspring resulting from the mating of related species may then mate

One of the rarest of hybrids is the zebrass, a cross between a zebra and a donkey. (AP/Wide World Photos)

through pollen exchange with the wild-type (original) plants. Backcrossing, which is also called introgression, increases hybrids' biological fitness.

The term "transgenic" or "genetically modified organism" (GMO) has been applied to plants and animals in which techniques of recombinant DNA have been used to modify specific parts of the genome of an organism. When the procedure is successful, the resulting organism may stably express a novel protein, express a protein with novel properties, or carry a change in the regulation of some of its genes. Usually, such a change is designed to improve the ability of the organism to grow (for instance, by resisting pests or using nutrients more efficiently) or to improve the usefulness of the organism (by improving its nutritive value, using it to manufacture pharmaceutically important molecules, or employing it to carry out environmentally important processes such as digesting environmental toxins).

Hybridization and introgression may introduce novel adaptive traits. The subjects have raised controversy, because transgenes introduced into crops have the potential for spreading into related weeds or wild plants. Scientists have hypothesized that transgenes might move from the genetically modified crop plants to weeds. The possibility of spreading transgenes via introgression and bridging, from genetically modified crops to related weed species, is a concern; introduction of herbicide-resistant cultivars into commercial agriculture could lead to the creation of superweeds.

Some researchers believe that if herbicide-resistant genes were to become more common in weeds as a result of widespread use of herbicide-resistant crops, farmers who rely on herbicides to manage weeds would be forced to use greater amounts and a larger number of herbicides.

To "solve" problem of horizontal gene transfer, the producers of transgenic crops naturally turn to gene technology. They propose to reduce the risk of creating transgenic uncontrollable weeds and volunteer cultivars by linking herbicide-resistance genes to other genes that are harmless to the crop but damaging to a weed, such as genes that affect seed dormancy or prevent flowering in the next generation. Thus, if a weed did acquire an herbicide-resistance gene from a transgenic crop, its offspring would not survive to spread the herbicide resistance through the weed population. Several of the newly patented techniques sterilize seeds so that farmers cannot replant them. In addition, patent protection and intellectual property rights keep farmers from sharing and storing seeds. Thus, genetic seed sterility could increase seed industry profits; farmers would need to buy seed every season.

Maternal Inheritance

Most crops are genetically modified via insertion of genes into the nucleus. The genes can, therefore, spread to other crops or wild relatives by movement of pollen. By engineering tolerance to the herbicide glyphosate into the tobacco chloroplast genome, however, researchers not only have obtained high levels of transgene expression but also, because chloroplasts are inherited maternally in many species, have prevented transmission of the gene by pollen—closing a potential escape route for transgenes into the environment. Glyphosate is the most widely used herbicide in the world. It interferes with 5-enol-pyruvyl shikimate-3-phosphate synthase (EPSPS), an enzyme that is encoded by a nuclear gene and catalyzes a step in the biosynthesis of certain (aromatic) amino acids in the chloroplasts. Conventional strategies for producing glyphosate-tolerant plants are to insert, into the nucleus, an *EPSPS* gene from a plant or a glyphosate-tolerant bacterium (the bacterial gene is modified so that the enzyme is correctly targeted to the chloroplasts), or a gene that inactivates the herbicide.

Putting GMOs in Perspective

The prestigious Genetics Society of America has weighed in on the issue of GMOs. Part of its statement reads:

> Every year, thousands of Americans become ill and die from food contamination. This is not a consequence of using GMOs, but instead reflects contamination from food-borne bacteria. "Natural" food supplements are widely used but are generally not well-defined, purified, or studied. Although recent reports of contamination of corn meal by GMOs not approved for human consumption led to several claims of allergic response, to date, none of those individuals has been shown to contain antibodies to the GM protein.

—Manjit S. Kang

See also: Artificial Selection; Biodiversity; Chromosome Theory of Heredity; Classical Transmission Genetics; Dihybrid Inheritance; Epistasis; Extrachromosomal Inheritance; Genetic Engineering: Agricultural Applications; Genetic Engineering: Risks; Genetically Modified (GM) Foods; Hardy-Weinberg Law; High-Yield Crops; Inbreeding and Assortative Mating; Incomplete Dominance; Lateral Gene Transfer; Polyploidy; Population Genetics; Quantitative Inheritance; Repetitive DNA; Transgenic Organisms.

Further Reading

Galun, Esra, and Adina Breiman. *Transgenic Plants.* London: Imperial College Press, 1997. This is an excellent book on issues relative to transgenic crop plants.

Kang, Manjit S., ed. *Quantitative Genetics, Genomics, and Plant Breeding.* Wallingford, Oxon, United Kingdom: CABI, 2002. This is a most comprehensive book on the latest issues in crop improvement. Introgression of alien germ plasm into rice is discussed.